International Conference of Young Scientists and Specialists (Optics-2015)

Journal of Physics: Conference Series Volume 735

St. Petersburg, Russia
12 - 16 October 2015

ISBN: 978-1-5108-2914-5
ISSN: 1742-6588

Printed from e-media with permission by:

Curran Associates, Inc.
57 Morehouse Lane
Red Hook, NY 12571

Some format issues inherent in the e-media version may also appear in this print version.

Printed by Curran Associates, Inc. (2016)

Institute of Physics
Dirac House, Temple Back
Bristol BS1 6BE UK

Phone: 44 1 17 929 7481
Fax: 44 1 17 920 0979

techtracking@iop.org

Additional copies of this publication are available from:

Curran Associates, Inc.
57 Morehouse Lane
Red Hook, NY 12571 USA
Phone: 845-758-0400
Fax: 845-758-2633
Email: curran@proceedings.com
Web: www.proceedings.com

Table of contents

Volume 735

International Conference of Young Scientists and Specialists "Optics-2015"
12–16 October 2015, St Petersburg, Russia

Accepted papers received: 21 July 2016
Published online: 30 August 2016

Preface

Papers

Information Photonics Technology

Photonics materials and technologies

Nonlinear and coherent optics

Optics and photonics in biology and medicine

Photonics devices and systems

Physics of lasers and laser technologies

Physical Optics and Spectroscopy

School of Photonics and Optoinformatics

Journal of Physics: Conference Series **735** (2016) 011001 doi:10.1088/1742-6596/735/1/011001

The ninth International Conference of Young Scientists and Specialists "Optics-2015" continues the tradition of holding regular meetings of students, bachelors, masters and PhD students enrolled in the optical and related fields, and researchers and designers of new optical techniques and technologies, organized in Leningrad - St. Petersburg since 1956.

The Optics-2015 conference is organized within the framework of the International Year of Light and Light-based Technologies 2015 (IYL 2015). The United Nations (UN) General Assembly 68th Session proclaimed 2015 as the International Year of Light and Light-based Technologies. This International Year has been the initiative of a large consortium of scientific bodies together with UNESCO, and will bring together many different stakeholders including scientific societies and unions, educational institutions, technology platforms, non-profit organizations and private sector partners. In proclaiming an International Year focusing on the topic of light science and its applications, the United Nations has recognized the importance of raising global awareness about how light-based technologies promote sustainable development and provide solutions to global challenges in energy, education, agriculture and health.

INTERNATIONAL
YEAR OF LIGHT
2015

Journal of Physics: Conference Series **735** (2016) 011002 doi:10.1088/1742-6596/735/1/011002

Peer review statement

All papers published in this volume of *Journal of Physics: Conference Series* have been peer reviewed through processes administered by the proceedings Editors. Reviews were conducted by expert referees to the professional and scientific standards expected of a proceedings journal published by IOP Publishing.

Formation and decoding of the ultra-wideband data signal

V A Glukhov[1], I R Kulikov, Yu A Tolmachev
Physics faculty, Saint-Petersburg State University, 7-9 Universitetskaya nab., St.Petersburg, 199034, Russia

E-mail: vglukhov@ymail.com

Abstract. The ultra-wideband signal was encoded by the set of thin films and decoded by the correlator based on the Fresnel biprism. A white light of a thermal source was used as the carrier of information. The spectrum of the encoded signal resembles the initial spectrum of carrier but contains the code sequence based on time modulation.

1. Introduction

When a signal is transferred via the light wave, the amplitude and the phase of wave change in time. Formation, registration and measurement of these signals relate to the information optics [1]. This research shows the possibility of formation, transmitting and receiving information encoded within the ultra-wideband stochastic process, such as radiation of the thermal source. Experimental setup scheme is shown in figure 1, its main element is the interferometer based on the Fresnel biprism.

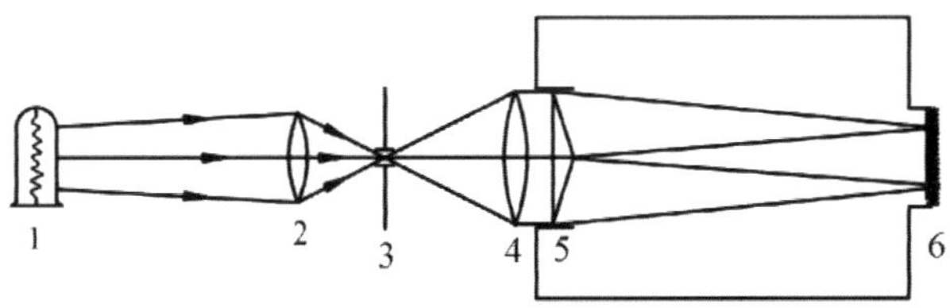

Figure 1. The setup scheme. 1 – source of light, 2, 4 - positive lenses, 3 - slit 5 – Fresnel biprism, 6 - CCD.

In its general operation mode, this is an optical autocorrelator. A divergent light wave from halogen lamp 1 is collimated by the telescopic system 2-4 comprising an adjustable slit 3. The plane wave illuminates the biprism 5 made of quartz glass to reduce the effect of chromatic aberrations. The biprism forms two beams that propagate at small angle 2α to each other. After the biprism, they form a 3D interference pattern that is analysed by the spatially sensitive 2D or 1D photodetector 6. The spatial coherence of the source is provided by appropriate width of the slit placed in the intermediate focal plane of the telescope. The entire system 5-6 was placed into a light-tight enclosure.

2. Method of the light flow operation

The basic principles of the system based on the evolution of the interference field of two inclined plane waves are described in [2], the spatial distribution of the exposure over the plane of photoreceiver is shown to represent the cross correlation function of two beams propagating in the interferometer. In particular, when the biprism is illuminated with one common beam it shows autocorrelation function. Using the Fourier transform, it permits to determine the power spectrum of signal. The designed device allows for recording not only the autocorrelation function of the signal, it is also possible to register the cross-correlation function of signals from different sources if they propagate through the different branches of interferometer.

At the initial stage of experiment, to introduce the information into the stochastic light wave we prepared two overlapping copies of the light flow separated in time. The resulting wave process saved all properties of a random signal but got a specific spectral feature: on the background of the continuous spectrum there were dark bands, the distance between them depended on the time delay between the copies of light flow, this was the so-called channelled spectrum [3, 4]. For more complicated combinations of the delay time of waves, the spectral features are expressed only slightly so that the power spectrum of the signal does not differ from the original while the function of auto-correlation permits to detect the presence of the signal in a pseudo-random stream.

For the simplest encoder we used the transparent thin film in the reflection or transmission mode. When reflected, the wave came from the front and back surface of the film. The two copies of the beam separated in time but overlapping in space went to the correlator. The autocorrelation function for that wide-band signal that contained the copy of itself was calculated in [5], it consisted of three interference pikes. We have also calculated it using a computer program (figure 2).

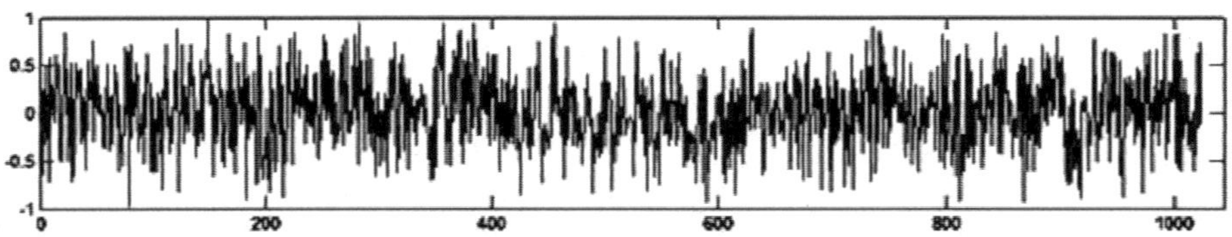

a)

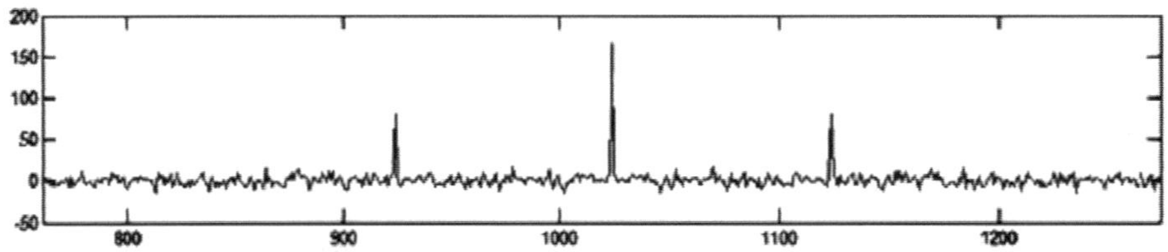

b)

Figure 2. The computer simulation of a random signal (a) consisted of the sum of two temporally separated copies, and the autocorrelation function (b) of this signal.

The calculated auto-correlation function consists of three copies of the autocorrelation function of the initial light beam. The distance between them is determined by the delay time of the wave inside the film. The experiment result is shown in figure 3, it fits the theoretical data quite well.

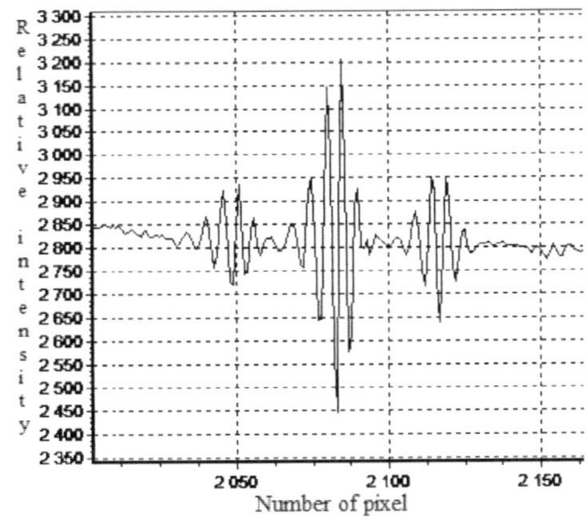

Figure 3. Part of the interferogram of white light reflected from the thin film, measured with a 1D photoreceiver.

The results of the autocorrelation function measurement demonstrate that whether the film thickness is greater than the time coherence length of white light, the light flow modulation in time can be easily reveled. By measuring the distance between the extrema, it is possible to determine the optical thickness of the film. The estimated thickness coincided with that obtained with the Rayleigh interferometer within the statistical errors.

3.Decoding

The function of autocorrelation is even, so it is impossible to determine the order in time of two copies of waves. To avoid this ambiguity, a cross-correlation was observed. The corresponded interference pattern was born by two beams, one of them was the original flow that formed the reference wave, and the other was its part that passed through the film. The beams illuminated different halves of the biprism.

Different sets of films were inserted into the object beam. The results of measurement with the different films combinations are shown in figure 4. As the obtained data show, the increasing thickness of the film package produce the linear rise of the delay time of the flux in one interferometer's arm, that is reflected in the linear increase of the shift of interference bands.

Figure 4. The results of recording of the interference pattern using the 2D receiver. 0 - position of the autocorrelation function of the original beam, 1 to 5 show the shift of pattern with the step-wise increase of the thickness of films. The package of 1 to 5 identical films was used.

A series of experiments fulfilled shows that the digital information can be encoded in a white light by the addition of time–delayed copies of a stochastic wide-band light flow. For the time shift realization, the number of thin plane films can be used. The information transferred by the light beam can be decoded by measuring the delay times even with the simple and high stable optical correlator based of a Fresnel biprism.

References

[1] Andrew M Weiner 2011 *Elsevier, Optics Communications* **284** 3669-3692
[2] Tolmachev Yu A, Wang Jue 2006 Peculiarities of the interference of ultrashort pulses *Vestnik of St.Petersurg State University* S4 vol 1 pp 39-46
[3] Born M, Wolf E 1980 *Principles of Optics* (London, Pergamon) 6th p 854
[4] Lukin K, Machekhin Yu, Tatyanko D 2011 Creation of optical frequencies grids on the basis of the spectral interferometry method *Lighting and power engineering* vol 3 pp 27-30
[5] Glukhov V A, Kulikov I R, Tolmachev Yu A 2014 Autocorrelator of single light pulses based on Frenel biprism *Conf. Laser, measurement and information – 2014 (St. Petersburg,* and *Russia, 9-11 June 2014)* vol 1 pp 32-40

Calculation of eigenfunctions of bounded waveguide with quadratic refractive index

M S Kirilenko[1,2], **R O Zubtsov**[2] **and S N Khonina**[1,2]
[1] Image Processing Systems Institute of the Russian Academy of Sciences, 151 Molodogvardeyskaya, Samara 443001, Russia
[2] Samara State Aerospace University, 34 Moskovskoye Shosse, Samara 443086, Russia

E-mail: areatangent@gmail.com

Abstract. In this work, the one-dimensional finite fractional Fourier transform has been considered in application to gradient optical waveguides. The eigenfunctions of it has been discussed. The relation between obtained functions, Hermite-Gaussian modes and spheroidal functions has been shown. The dependence of input domain width and number of nonzero eigenfunctions of transformation has been demonstrated. The functions which recover its form at given distance from the front plane have been calculated.

1. Introduction

Fractional Fourier transform (FrFT) is a set of linear transformations, which generalizes Fourier transform. The typical interpretation of Fourier transform is conversion of time domain of signal to its frequency domain.

The canonical FrFT was considered [1] as Fourier transform of α-order, where α is real value. We can define the FrFT as operation of frequency-time distribution rotation by some angle [2].

FrFT is used in differential equation solving, in quantum mechanics and quantum optics, in optical theory of diffraction, for description of optical systems and optical signal processing, including the applying of frequency filters, time filtration and multiplexing, as well as in pattern recognition, in wavelet-transformations, in operations with chirp-functions, in encryption, for neural networks creating and other applications. More detailed review of FrFT can be found in work [3] of T. Alieva et al.

One of application of FrFT is the description of laser beam propagation in gradient index media [4, 5]. In this work, we use one-dimensional FrFT for modeling of optical signal propagation in optical waveguide with parabolic dependence of the refractive index. In case of unlimited medium, an analytical solution for eigenfunctions problem of transformation is known. They are Hermite-Gaussian modes [1]. Taking bounding of gradient waveguide into account, we need to solve numerically the eigenvalues and eigenfunctions problem. In this work, we calculate eigenfunctions for various parameters of waveguide with quadratic refractive index.

2. General theory

The passing of light beam through gradient index media with finite input domain (Figure 1) is a particular case of paraxial propagation through ABCD-system [6]. It can be described by finite fractional Fourier transform [1]:

$$F_{\alpha,l}(u)=\sqrt{\frac{k}{2\pi i a\sin\alpha}}\int_{-l}^{l} f(x)\exp\left[\frac{ik}{2a\sin\alpha}\left(\cos\alpha\, x^2-2xu+\cos\alpha\, u^2\right)\right]dx\,, \tag{1}$$

where $f(x)$ – input plane optical distribution, $F_{\alpha,l}(u)$ – distribution at a given distance z from if, k – wave number, a – the parameter of refractive index $n(x)=n_0\left(1-\left(x^2/2a^2\right)\right)$, $\alpha=z/a$, $2l$ – width of input domain.

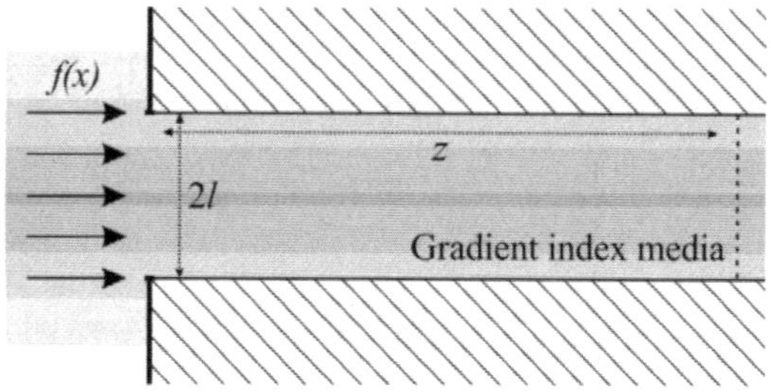

Figure 1. Scheme of optical system with bounded gradient waveguide.

The light beam propagation through the gradient index medium has property of periodicity. It is valid for both unlimited and limited (in input plane) FrFTs. The period of self-recovering is $z_T=2\pi a$. Moreover, the result of FrFT at a distance $z_{T/4}=\pi a/2$ equals to result distribution of typical Fourier transform, which is valid for lens system. The source signal "inverts" at a point $z_{T/2}=\pi a$. The applying of inverse Fourier transform to the source distribution equals to using FrFT at a distance $z_{3T/4}=3\pi a/2$.

If we consider the case of unlimited input plane domain, its eigenfunctions recovery at any distance (accurate to phase shift). Such functions are well-known as Hermite-Gaussian modes [7]:

$$h_n=\exp\left[-\frac{x^2}{2\sigma_0^2}\right]H_n\left(\frac{x}{\sigma_0}\right). \tag{2}$$

3. Simulation

The bounding of input domain of the system affects the eigenfunctions. If width l is sufficiently large that eigenfunction turns to zero on its borders, then its propagation will be without distortions and its form will be similar to Hermite-Gaussian mode. Otherwise, it will be eigenfunction only for current parameter α, wherein it was calculated. The examples of such functions propagation is shown in Figures 2 and 3. The parameters of computations are: $\lambda=1\mu\text{m}$, $a=25\lambda$, $\alpha=\pi/5$.

Besides eigenfunctions, we can consider the pair of sets of orthogonal "shifted" eigenfunctions $\{\varphi_n(x;\alpha,l)\}$ and $\{\chi_n(u;\alpha,l)\}$, which transfers to each other at distances $z_1=\alpha a$ and $z_2=(2\pi-\alpha)a$, correspondingly. The definitions of the functions are:

Journal of Physics: Conference Series **735** (2016) 012002 doi:10.1088/1742-6596/735/1/012002

$$\varphi_n(x;\alpha,l)=\psi_n(x;\alpha,l)\exp\{-ik\,x^2\operatorname{ctg}\alpha/2a\}, \tag{3}$$

$$\chi_n(u;\alpha,l)=\psi_n(u;\alpha,l)\exp\{ik\,u^2\operatorname{ctg}\alpha/2a\}, \tag{4}$$

where $\psi_n(u;\alpha,l)$ – spheroidal functions [8]. The examples of its propagation through the system are shown in Figures 4 and 5.

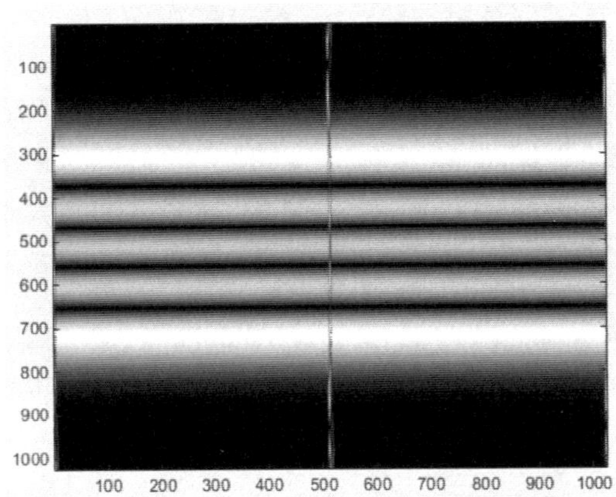

Figure 2. The propagation of eigenfunction: function turns to zero on the boundaries ($l = 12\lambda$).

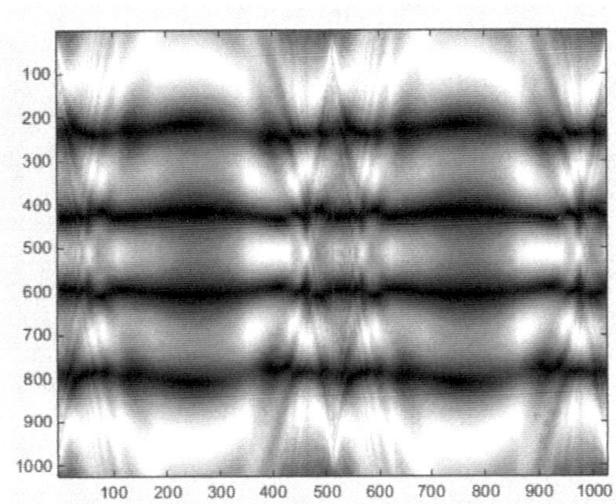

Figure 3. The propagation of eigenfunction: function doesn't turn to zero on the boundaries ($l = 6\lambda$).

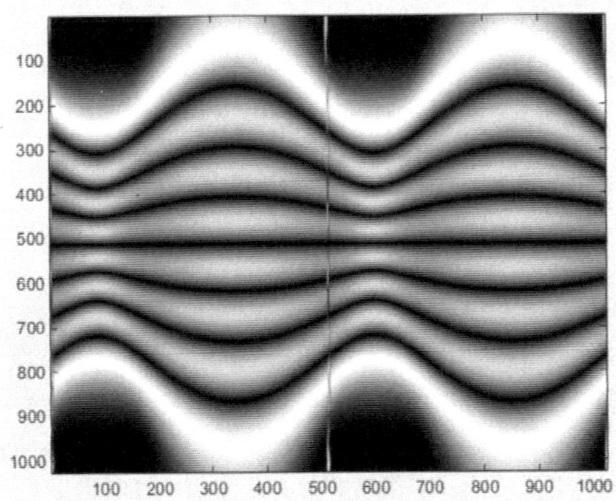

Figure 4. The propagation of shifted eigenfunction $\varphi_n(x;\alpha,l)$.

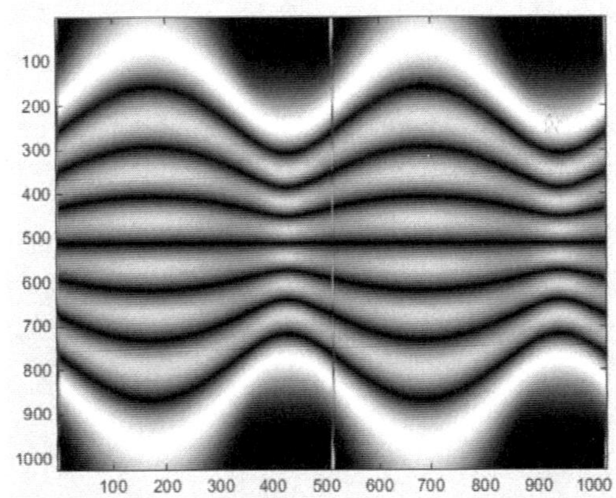

Figure 5. The propagation of shifted eigenfunction $\chi_n(u;\alpha,l)$.

The eigenvalues corresponding to transformations with different width of input domain is shown in Figures 6 and 7. The lesser input domain width is, the lesser is the number of significant eigenvalues. Therefore, the number of eigenmodes transferring through waveguide without energy lost is decreases.

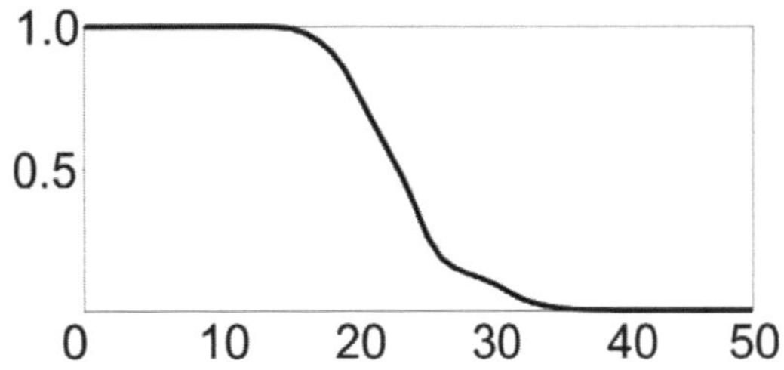

Figure 6. Moduli of eigenvalues, $l = 12\lambda$.

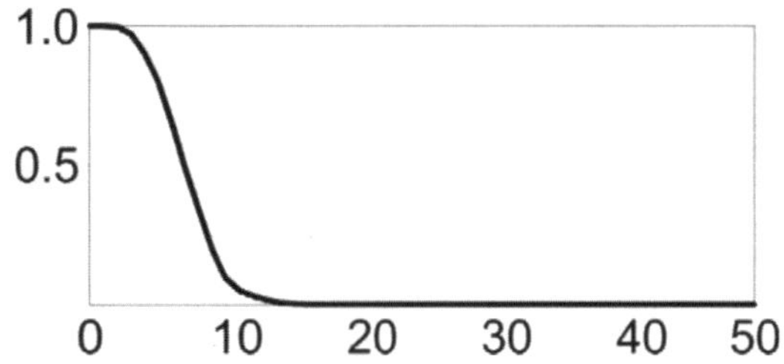

Figure 7. Moduli of eigenvalues, $l = 6\lambda$.

4. Conclusion

The one-dimensional FrFT, corresponding to gradient optical medium, is considered in the paper. We examined space-limited operator of propagation in gradient index medium, based on FrFT. The calculation of eigenfunctions at various parameters of optical waveguide with quadratic refractive index was performed.

We also considered "shifted" eigenfunctions corresponding to "input" and "output" of propagation operator.

The results of numerical simulation shown, that eigenfunctions with eigenvalues, which approximately equals to one, almost completely include in finite size of waveguide and are similar to Hermite-Gaussian modes and spheroidal functions.

Acknowledgments

The work was financially supported by the Russian Foundation for Basic Research (grant 14-01-31401 mol a).

References

[1] Namias V 1980 *J. Inst. Math. Appl.* **25** 241–265
[2] Abe S and Sheridan J. 1994 *Journal of Physics A: Mathematical and General* **27** 4179
[3] Alieva T, Bastiaans M and Calvo M 2005 *EURASIP Journal on Applied Signal Processing* **2005** 1498–1519
[4] Ozaktas H and Mendlovic D 1993 *Optics Communications* **101** 163–169
[5] Mendlovic D and Ozaktas H 1993 *J. Opt. Soc. Am.* A **10** 1875–81
[6] Goodman J 1996 *Introduction to Fourier optics* (New York: McGraw-hill)
[7] Yariv A 1983 Introduction to optical electronics (Moscow: High School)
[8] Slepian D and Sonnenblick E 1965 *Bell System Technical Journal* **44** 1745–59

Journal of Physics: Conference Series **735** (2016) 012003 doi:10.1088/1742-6596/735/1/012003

The tight focusing of ultrashort pulse with dielectric cylinder

E S Kozlova[1,2] and V V Kotlyar [1,2]
[1]Image Processing Systems Institute, Samara, Russia
[2]Samara State Aerospace University, Samara, Russia

E-mail: kozlova.elena.s@gmail.com

Abstract. The tight focusing of light is observed during the propagation of a femtosecond laser pulses duration of 17 fs and 8 fs trough the dielectric cylinder with a refractive index n=1.59 and the radius R = 0.7152λ_0. It allows to reach the focal spot width at half intensity equal 0.235 λ_0.

1. Introduction

Recently, much attention is paid to the study of subwavelength focusing laser light on microparticles, including microspheres and microcylinders radii comparable with the wavelength of light [1]. The focusщтп by elliptical [1] and multiple [2] microcylinders is discussed. The minimum size of the focus at full width at half maximum (FWHM) equal 0.46 λ_0 has been obtained [2]. Resonance focusing of TE-polarized laser light by polyester microcylinder (refractive index n=1.59) is analytically considered in [3]. The series of Bessel functions is used for this. The formation of resonant mode (whispering gallery mode) is considered in [4]. The focus for the 18th mode (whispering gallery mode) FWHM=0.22λ_0 outside of microcylinder is received by analytical method [3] and finite difference time domain (FDTD) method [4].

2. Simulation

We simulate the resonance focusing of the femtosecond pulses durations of t_s 17.08 fs and 8.54 fs by dielectric microcylinders. It is possible to calculate the resonance radius of the cylinders (Table 1) using the methods in [5].

Table 1. Radius of the cylinder.

Type of material	Polyester, *(n=1.59)*	Silica glass, *(n=1.46)*	Silica glass, *(n(ω))*
R, λ_0	0.7152	0.7588	0.7588
R, μm	0.3802	0.4037	0.4037

The materials such as polyester and silica glass are considered in this paper. The refractive index are equal 1.59 and 1.46, respectively. The simulation are performed in the package FullWAVE. The approximation steps for the spatial variables are 0.002 μm. The approximation step for the time variable is 0.0001 μm. The simulation results are shown in Figure 1-2. Tables 2-3 show the parameters of the focal spot in the considered cases.

Table 2. The parameters of the focal spot in the case for the cylinder from polyester (n=1.59).

t_s, fs	$FWHM_x$, λ_0	DOF_z, λ_0	I_{max}, a.u.
17.08	0.235	0.271	9.447
8.54	0.243	0.271	8.091

Table 3. The parameters of the focal spot in the case for the cylinder from silica glass (n=1.46).

t_s, fs	$FWHM_x$, λ_0	DOF_z, λ_0	I_{max}, a.u.
17.08	0.255	0.316	8.204
8.54	0.262	0.320	7.634

It is shown in [6] that in case of simulation of femtosecond pulses it is necessary to consider the frequency dependence of the permittivity. We make a simulation using frequency depended FDTD ($(FD)^2TD$) method implemented in the package FullWAVE. To take into account a frequency dependence of the permittivity we use the Sellmeyer 's model of silica glass [7]. Simulation parameters is the same as before. The simulation results are shown in Figure 1. Tables 4 shows the parameters of the focal spot in the considered cases.

Table 4. The parameters of the focal spot in the case for the cylinder from silica glass (n(ω)).

t_s, fs	$FWHM_x$, λ_0	DOF_z, λ_0	I_{max}, a.u.
17.08	0.255	0.316	8.241
8.54	0.262	0.316	7.605

The comparison of Table 3 and Table 4 shows that in this case, the frequency dispersion of the medium can be neglected. The deviation of the results is less than 2%.

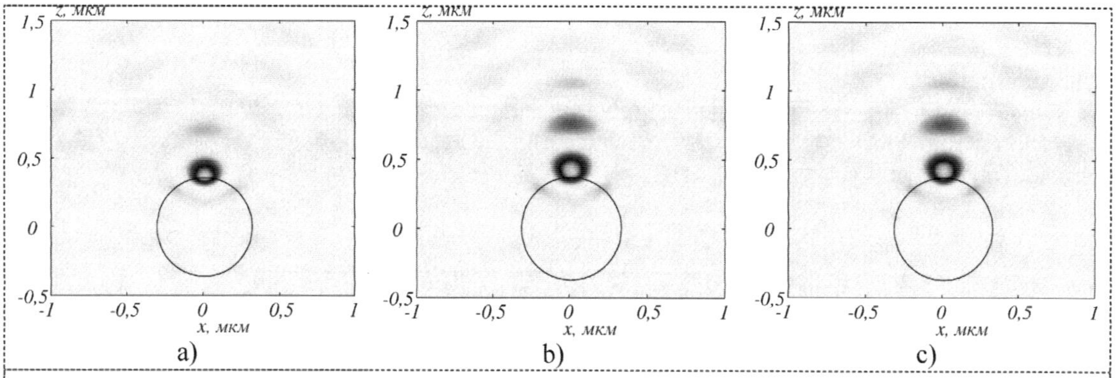

Figure 1. The diffraction pattern of the resonance focusing by the cylinder with (a) n=1.59, (b) n(ω); and (c) n=1.46 for the pulse duration of 17.08 fs

Figure 1 shows that the whispering gallery mode is formed in the waveguide. Unfortunately it quickly leaves the waveguide because of the short duration of the pulse. Tables 2-4 and Figure 2 show that during the resonance focusing of femtosecond pulse it is possible to beat the diffraction limit. The diffraction limit is equal $0.44\lambda_0/n = 0.28\lambda_0$ and is equal $0.44\lambda_0/n=0.30\lambda_0$ for polyester and silica glass, respectively.

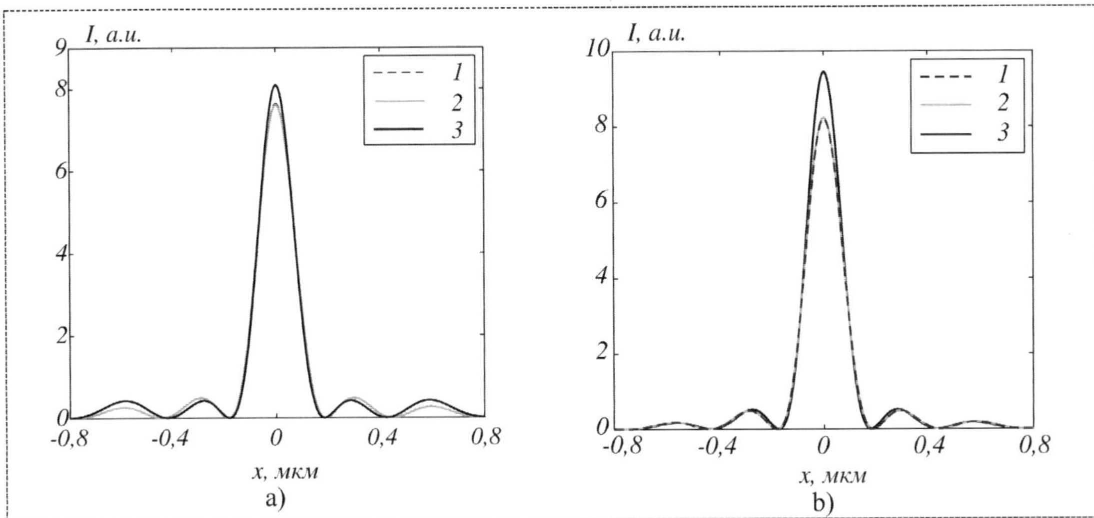

Figure 2. The intensity distribution in the focus for the different cases of cylinder medium n(ω) - line 1; n=1.46 - line 2; n=1.59 - line 3 and pulses durations of (a) 8.54 fs and (b) 17.08 fs.

For comparison, we simulate the propagation of the femtosecond pulse durations of 17.08 fs through a cylinder from the silica glass (n = 1.46) with the radius $R=0.8\lambda_0$ different from the resonance radius. The width of the focal spot at half intensity was $FWHM=0.269\lambda_0$. Thus the change in the radius of the cylinder 5% entail changing the width of the focal spot at 5%. It demonstrates that the short duration of the pulse does not give the resonance effects. Figure 3 shows the dynamics of the intensity in the focus.

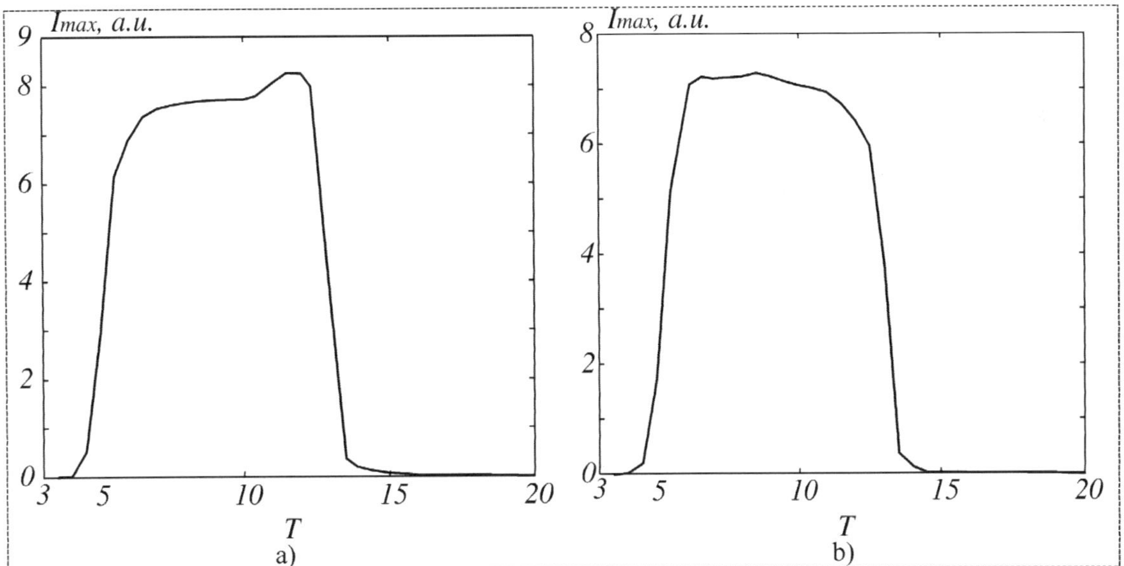

Figure 3. Dynamics of the intensity in the focus during the propagation of pulse durations of 17.08 fs through the cylinder from the silica glass with the radius (a) $R = 0.8\lambda_0$ and (b) $R = 0.7588\lambda_0$

Figure 3a shows that the intensity of focus increases when the pulse pass through the cylinder with resonance radius. The intensity reaches 8.204 a.u. during the time at which the pulse is passing. Figure

3b shows that the intensity of focus decreases when the pulse pass through the cylinder with nonresonance radius. The intensity reaches 7.3 a.u. during the time at which the pulse is passing.

3. Conclusion

In this work e shows tightly focusing of femtosecond pulse using FDTD method. We show that during the resonance focusing of femtosecond pulse it is possible to beat the diffraction limit (Table 2-4). The narrowest radius of the spot (FWHM=$0.235\lambda_0$) is obtained for the pulse duration of 17.08 fs which propagates through the cylinder from polyester (n=1.59) with resonant radius R=$0.7152\lambda_0$. Also we demonstrate that the short duration of the pulse does not give the resonance effects. But the intensity of focus increases when the pulse pass through the cylinder with resonance radius and the intensity of focus decreases when the pulse pass through the cylinder with nonresonance radius. This fact suggests that it is possible to choose the pulse width at which it will be possible all resonance effects.

Acknowledgements

This work was partially funded by the Ministry of Education and Science of the Russian Federation and Russian Foundation for Basic Research grants ## 14-07-31218, 14-29-07133, 15-07-01174

References

[1] Liu L, Chang L-J *Optik* 2014 **125** 4043.
[2] Xu B B, Jiang W X, Yu G X, Cui T J 2014 *J. Opt. Soc. Am. A.* **31** 1135
[3] Kozlov D A, Kotlyar V V 2014 *Computer Optics.* **38** 393
[4] Kotlyar V V, Kozlov D A, Kovalev A A 2015 *Computer Optics.* **39** 163
[5] Kozlova E S, Kotlyar V V, Degtyarev S A 2015 *Computer Optics.* **39** 45.
[6] Kozlova E S, Kotlyar V V 2013 *Computer Optics.* **37** 31
[7] Couairon A, Sudrie L, Franco M, Prade B, Mysyrowicz A 2005 *Phys. Rev. B.* **71** 125435

Image retrieval with the use of Bag of Words and structural analysis

R Malashin[1]

[1]National Research University of Information Technology, Mechanics and Optics, 197101, 49 Kronverkskiy av., St. Petersburg, Russia

Email: malashinroman@mail.ru

Abstract. The paper considers task of image retrieval with the use of Bag of Words (BoW). The significant improvement in robustness is achieved by using fast and robust algorithm of structural verification based on Hough transform of parameters of keypoint matches.

1. Introduction

One of important tasks in computer vision is image retrieval. There exist many factors that make the problem challenging: illumination change, view point change, content variety. Target independent analysis based on coarse structural elements [1-3] can provide robust results but these algorithms consumes too much computational resources for most of practical applications. Methods based on local features such as SIFT [4] or SURF [5] are much more effective especially when combined with Bag of Words (BoW) [6], because histogram of visual words is very compact representation of an image. The major drawback is that original BoW concept doesn't take size, orientation and location of features into account. In this paper spatial information is utilized by very fast and robust procedure of matches clustering.

2. Proposed methods

The basic idea of the structural (or spatial) verification is similar to the idea of Philbin et al. [6]. Let us consider that BoW histograms are precomputed for every image in database and there is query image. Then the retrieval algorithm consists of the following steps:

- Compute BoW histogram for the query image.
- Find N BoW histograms closest to the query histogram according to tf.idf [6].
- For every i-th of N best image candidates: a) generate keypoint-to-keypoint matches of i-th and query image; b) eliminate outliers using scene geometry constraint.
- Re-rank search results according to number of geometrically consistent matches.

Below there is a brief description of the proposed algorithms.

2.1. Algorithm to generate keypoint matches

Local matches are generated by interpreting corresponding histogram bins as a set of keypoint matches. Matches that arise from bins with several features are less reliable than those derived from bins with single feature. This information is considered by assigning weights to generated matches. Total number of generated matches is $m_1 \cdot m_2$, where m_1, m_2 – number of features assigned to specified visual word that are detected in two images (sizes of corresponding bins of two BoW histograms). The weight of every i-th match is computed as follows:

$$w_i = \frac{\min(m_1, m_2)}{m_1 \cdot m_2}. \quad (1)$$

Despite the fact that corresponding features may not be detected or can be assigned to different visual words, w_i equals probability of match i to be correct. The details on usage of these weights are discussed in the next subsection.

2.2. Scene geometry constraint

Due to different causes most of generated keypoint matches are incorrect. The general spatial constraint on image pair of static 3D scene implies usage of epipolar geometry. Unfortunately RANSAC algorithms are not able to recover fundamental matrix when number of outliers is more than 70-75%. In this work spatial information is utilized by Hough transform that is used to find clusters of matches that agree on the same parameters of similarity transform [4]. That allows successful matching with less than 5% of inliers.

Hough transform is affected by boundary effect. To avoid the problem in [4] each keypoint match votes for the 2 closest bins in each dimension, giving total of 16 entries for each hypothesis [4]. The alternative solution proposed in this paper is to organize several Hough tables that are displaced against each other in each dimension by half of bin size, every match increments appropriate bin in every table. Figure 1 illustrates this approach for 1- and 2- dimensional parameter space (similarity transform parameter space is 4D). The two approaches (several votes in single table and single vote to several tables) were compared through experiments on matching image pairs (similar to experiments described in section 3). The proposed approach with only two displaced Hough tables provide almost the same matching quality as voting for 16 hypotheses in one table (0.3% decrease in number of correctly matched images). Usage of only two tables means that boundary effect is partially ignored (figure 1b) but total number of entries to the tables decreases almost by a magnitude. That leads to a magnitude decrease of clusters found by Hough transform (see figure 2) and allows significant speedup of the program.

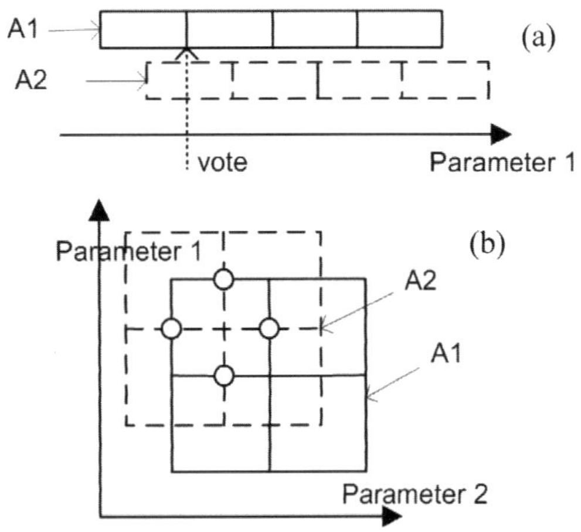

Figure 1. Usage of two Hough tables A1 and A2 to avoid boundary effect. (a) – in case of 1-dimensional parameters space, (b) – in case of 2-dimensional parameter space. In (b) circles mark points in parameter space where clusters can be split.

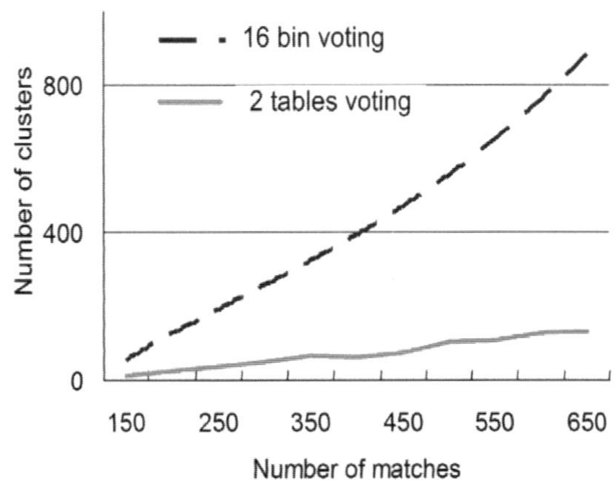

Figure 2. Number of bins with more than two votes relative to total number of votes. The number is significantly reduced by partially ignoring boundary effect using two Hough tables. The graph was generated by matching hundreds of images having non-overlapping content.

According to equation (1) different matches have different probability to be correct. Only clusters that contain at least 1% of total match weights were considered, others are eliminated. The verification procedure that is able to remove rest of 'false' clusters and incorrect matches from 'correct' clusters was developed. The procedure is based on RANSAC in similarity transform parameter space. Two matches are "randomly" picked from cluster and form a sample subset that is used to recover unknown parameters. The probability of picking a match equals its weight (equation (1)) divided by sum of weights of all matches in cluster. The latter was efficiently implemented by "cloning" indexes of more probable matches. Hypothesis that do not agree with parameters are rejected without verification of the cluster. RANSAC is followed by least square solution for the best parameters and then outlier elimination takes into account all parameters of similarity transform.

Note that the proposed algorithm described in section 2.1 generates 1-to-n matches that have higher probability to form incorrect (false) clusters of matches with relevant mutual parameters [7]. In this work we had to bear upon this fact but matches with $w_i < 0.1$ are eliminated before the Hough transform.

3. Experimental results

Experiments were carried out on the database of manually labelled indoor images with resolution 640×480. The very compact vocabulary was used (with only 2000 visual words). The experiment on image matching has shown that the proposed methods significantly outperforms baseline method (SURF [7] + epipolar geometry constraint). The results is summarized in table 1.

Table 1. Results of matching of 352 image pairs

Detector	Descriptor	Geometrical constrains	Correct keypoint matches, %	Number of pairs with more than 50% of inliers
SURF	SURF	Epipolar geometry	53,76	205
SURF	SURF + BoW	Cluster analysis	88,37	273

In the second experiment the developed algorithms were applied directly to the task of image retrieval (for database with several hundred images). It appeared that the proposed structural verification of 10 image candidates decreases false positives by more than a magnitude and false negatives by 36%. At the same time structural verification of one image pair consumes only 3 ms in average in PC environment.

Acknowledgements

This work was supported by the Ministry of Education and Science of the Russian Federation, and by the Government of the Russian Federation, Grant 074-U01.

References

[1] Lutsiv V, Malyshev I, Pepelka V and Potapov A 2002 *Proc. of SPIE* **4741** 351–62
[2] Lutsiv V, Malyshev I and Potapov A 2003 *Proc. of SPIE* **5238** 164–75
[3] Lutsiv V, Malyshev I and Pepelka V 2001. *Proc. of SPIE* **4380** 174–83
[4] Lowe D 2004 *Int. J. of Computer Vision* **60** 91-110
[5] Bay H, Tuytelaars T and Van Gool L 2006 SURF: Speeded Up Robust Features *Proc. 9th European Conf. on Computer Vision (Graz, Austria, May 7-13)* pp 404–417
[6] Philbin J, Chum O, Isard M, Sivic J, Zisserman A 2007 Object retrieval with large vocabularies and fast spatial matching Conf. on CVPR *(Minneapolis, USA, June 18–23)* pp 1–8
[7] Malashin R 2014 *J. Phys.: Conf.Ser* **536** 012018

Design of optical system with three-dimensional image visualization using an array of microlenses

I Yu Santalina, A P Toropova
Saint-Petersburg State University of Informational Technologies, Mechanics and Optics, Saint-Petersburg, Russia

E-mail: irina.santalina@yandex.ru

Abstract. The algorithm of calculation of the optical imaging system of a three-dimensional image based on the method of integral photography was given in this article. The algorithm is easy to use and allows the calculate schemes with different microlens arrays, a CCD array of the camera and the projector.

1. Introduction

A person perceives the world as a three-dimensional space filled with three-dimensional objects, and has long sought to reproduce the most plausible reality, for this purpose he is constantly inventing new ways of visualizing three-dimensional objects. One of them is the creation of three-dimensional image in the space on the basis the method of the integral photography [1].

The main problem of the reconstruction is the quality of the obtained three-dimensional image. In this paper we present the reasoning and the algorithm for calculating the basic parameters of the three-dimensional optical imaging system (integrated) images using a microlens array. The calculations take into account not only the resolution of the CCD matrix of the camera and the projector, and physiological characteristics of the human eye.

2. Schematic diagram of the recording and reconstruction of the integral image

The optical recording scheme presented in figure 1, consists of three components: the array of microlenses 1, the focusing lens 3 and the CCD camera (while recording) / projector (while reconstructing) 4.

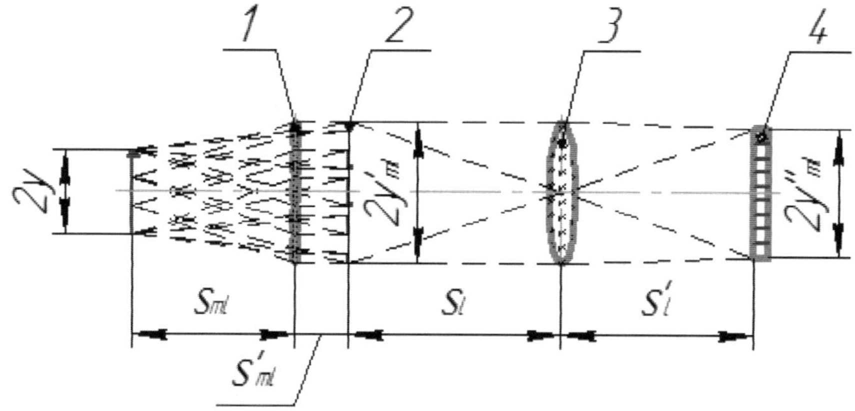

Figure 1. The optical scheme of the integral image recording: 1 microlens array; 2 - plane of elementary images; 3 - focusing lens; 4 - CCD sensor camera.

The microlenses array is a matrix consisting of a lens of small size. Each microlens captures a certain angle of the object and generates a two-dimensional image in a plane elementary image. The array of elemental images 2 with the help of focusing lens is recorded on the CCD camera.

The projector is used instead of the camera to obtain an integral image. At the same time the distances between the components of the scheme remain unchanged. In this case, the light beams passing through the microlens array cross at one point, forming an image of the object. After crossing the beams follow the observer, who perceives these diverging beams as the beams propagating from the real object.

3. The calculation of the basic parameters of the optical scheme

2.1 The choice of microlens array (figure 2)

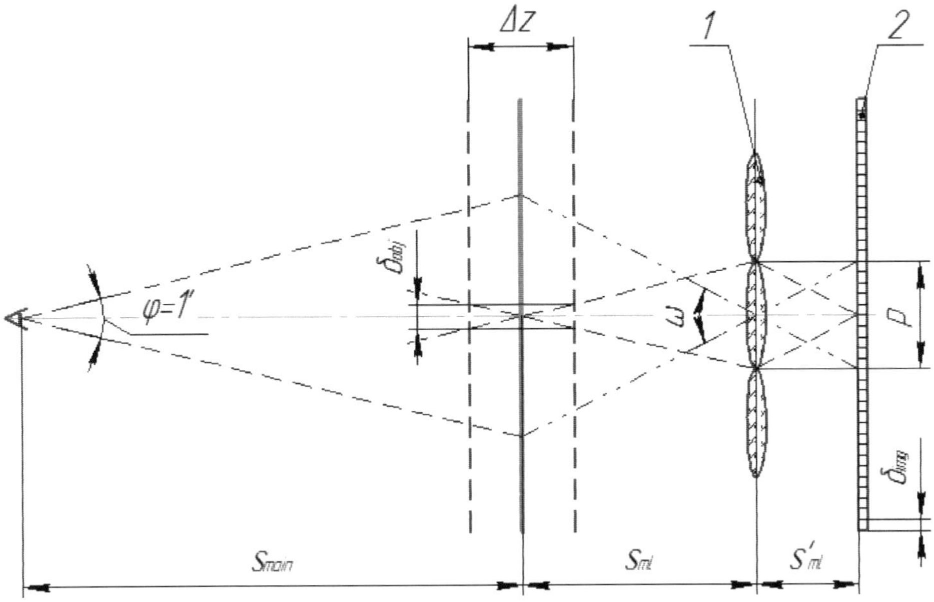

Figure 2. The choice of microlens array: 1 - microlens array;
2 - plane of elementary images

The main parameters of the microlens array are the number of microlenses in the array, the focal length f'_{ml} and the pitch p - the distance between the centers of two microlenses. The pitch size corresponds to the size of the diameter of one microlens. It is necessary to take into account the number of pixels corresponding to one microlens. This parameter defines the minimum distance between two neighbouring image points δ_{img}, which can be calculated using the following formula:

$$\delta_{img} = \frac{p}{N}, \tag{1}$$

where N - the number of vertical pixels per a microlens.

According to figure 2 the minimum distance between two points of the object will be determined by the following expression:

$$\delta_{obj} = \frac{s_{ml}}{s'_{ml}} \delta_{img} , \tag{2}$$

where s_{ml} and s'_{ml} - the front (the distance from the front surface of the microlens to the subject) and the back (the distance from the back surface of the microlens to the image of the object) line segments respectively.

The lateral resolution of the object R_{obj} and its image R_{img} can be defined as the reciprocal value of the minimum distance between two points, i.e.

$$R_{obj} = \frac{1}{\delta_{obj}}, \tag{3}$$

$$R_{img} = \frac{1}{\delta_{img}}. \tag{4}$$

Angular field microlens is defined as

$$tg\left(\frac{\omega}{2}\right) = \frac{p}{2s'_{ml}}. \tag{5}$$

Then we get

$$\omega = 2 \cdot arctg\left(\frac{p}{2s'_{ml}}\right). \tag{6}$$

The depth of the obtained image during reconstruction can be found from the following expression:

$$\Delta z = 2\frac{s_{ml}}{p}\delta_{obj}. \tag{7}$$

2.2 The choice of focusing lens

The task of the focusing lens is to project the plane of elementary images constructed by microlens array on the CCD matrix of the camera. While choosing it is necessary to pay attention to the linear increase of lens which is defined as:

$$V_l = \frac{2y''_{ml}}{2y'_{ml}}, \tag{8}$$

where $2y'_{ml}$ - the linear dimension of the plane of elementary images of the object $2y$ (see figure 1); $2y''_{ml}$ - linear dimension of the plane of elementary images on the CCD array camera.

2.3 The choice of the camera and projector

During the selection of the camera and projector, it is necessary to pay attention to their main parameters: the resolution and the size of one pixel. These parameters must be identical, otherwise the resulting integral image will be of poor quality.

To determine the value of the front and back line segments of the optical components of the system it is necessary to use well-known formulae of geometrical optics.

The distance from the back surface of the microlenses to the object image

$$s'_{ml} = (1 - V_{ml}) f'_{ml}, \tag{9}$$

where $V_{ml} = \dfrac{2y'}{2y}$ is the linear increase of the microlens; $2y$ and $2y' = p$ are the linear dimension of the object and the elementary image respectively.
The distance from the front surface of the microlens to an object is defined as

$$s_{ml} = \frac{s'_{ml}}{V_{ml}}, \tag{10}$$

The distance from the front surface of the focusing lens to the plane of elementary images

$$s'_l = (1 - V_l) f'_l, \tag{11}$$

where f'_l is the focal length of the lens.
The distance from the back surface of the focusing lens to the CCD array:

$$s_l = \frac{s'_l}{V_l}. \tag{12}$$

The distance between the microlens array and the focusing lens will be determined from the following equation

$$d = s'_{ml} + s_l. \tag{13}$$

For the high-quality reconstructed image it is necessary to take into account the physiological characteristics of the eye. The resolution of the eye is one minute. From this we can determine the maximum distance from which a person can observe the obtained three-dimensional image. In accordance with figure 2, it will match the value of the expression (14):

$$s_{main} = \frac{\delta_{obj}}{0{,}0175} + \frac{\Delta z}{2}. \tag{14}$$

4. Calculation example of the optical scheme for recording of the integral image.
This section presents the calculation of the optimal parameters of the optical system. CCD array of the camera and the projector have the following basic parameters: resolution 640 × 480, and the size of one pixel, which is $5{,}6\mu m$. Microlens array has a pitch $p = 0{,}25mm$ and focal length $f'_{ml} = 0{,}57mm$ (the array was taken as an example offered by the company Flexible Optical BV [2]). The number of pixels of the CCD per one microlens is important [3]. We assume that one microlens has 40×40 pixels. Then, the minimum distance between two neighboring points of the object and the elementary image according to the formulae (1) and (2) will be $\delta_{img} = 0{,}02mm$ and $\delta_{obj} = 0{,}5mm$. For example we take the object, that size is $20mm$. Then, linear increase of the microlens is $V_{ml} = 0{,}0125^{\times}$. The plane of the elementary images will be located at a distance equal

$s'_{ml} = 0{,}56 mm$ relatively the array of microlenses. From this, the distance from the array to the object is $s_{ml} = 45{,}03 mm$

Next, we define linear dimensions of plane of the elementary images formed by the microlenses and the CCD array camera. Let us calculate the microlens in the array. Considering the size of the CCD matrix in pixels and the fact, that it is necessary 40x40 pixels per one microlens, we receive an array of microlenses 16x12. From this, we find linear dimension of the plane of elementary images. Multiplying the number of microlens on the pitch p we get the size of the microlens array equal 4 × 3 mm. We similarly define the linear dimension of the CCD array camera. Multiply the number of pixels on the size of one pixel. We find, that the linear dimension of the CCD array is 3,6 × 2,7 mm. Thus, for the further calculations we got the parameters $2y'_{ml} = 4mm$ and $2y''_{ml} = 3{,}6mm$.

Now we define the parameters of the focusing lens with a focal length equal to, for example, $f'_l = 100 mm$. According to the formula (8) the increase of the lens is $V_l = 0{,}896^{\times}$. The back and the front line segments are equal $s'_l = 10{,}4mm$; $s_l = 11{,}6mm$ according to the formulae (11) and (12).

The distance between microlens arrays and the focusing lens in accordance with formula (13) is $d = 12{,}2mm$. The CCD array of the camera is located at a distance $s'_l = 10{,}4mm$. The angular field is 25 degrees, and the depth of the image in accordance with (7) is $\Delta z = 180{,}12mm$.

The maximum distance that a person can observe the obtained three-dimensional image in accordance with formula (14) is $s_{main} = 1724 mm$.

Thus in this paper we have described a method of creating a three-dimensional image of the object in real space, based on the method of integral photography, developed by G. Lipman in the early XX century, but has not lost its relevance nowadays.

The algorithm for calculating of the optical scheme is easy to use and allows the calculate schemes with different microlens arrays, a CCD array of the camera and the projector.

5. References.

[1] P P Sokolov, Journal of the Society of Naturalists 1911, p. 23-29
[2] Xiao X, Javidi B, Martinez-Corral M, Stern A, Journal of Applied Optics 2013, 52, №4, p. 546–560

The tight focusing of laser radiation using 4-sector polarization converter

S S Stafeev[1,2], A G Nalimov[1,2], L O'Faolain[3] and V V Kotlyar [1,2]
[1]Image Processing Systems Institute, Samara, Russia
[2]Samara State Aerospace University, Samara, Russia
[3]School of Physics and Astronomy of the University of St. Andrews, Scotland

E-mail: sergey.stafeev@gmail.com

Abstract. A laser beam with mixed linear-radial polarization formed by reflection of a linearly polarized light from a 4-sector micropolarizer was focused using a binary zone plate in a focal spot with diameters 0.35±0.02 and 0.38±0.02 of a wavelength.

1. Introduction

Using of subwavelength gratings for the manipulation of polarization state of laser light was proposed for the first time in [1]. When a linearly polarized light is incident on this grating the polarization is rotated by an angle that depends from an angle between the direction of polarization of the incident light and the direction of a grating relief. In [2-4] transmitted subwavelength gratings were used to transform polarization of infrared light. However the manufacturing of these gratings for the visible region is difficult due to the high aspect ratio. Therefore previously [5] we proposed a reflected 4-sector grating polarizer for a conversion of a linearly-polarized laser beam to a radially polarized one.
In this paper we have focused laser light with mixed linear-radial polarization formed by reflection of a linearly polarized light with wavelength 633 nm from a 4-sector polarization converter in a focal spot with diameters 0.35±0.02 and 0.38±0.02 of a wavelength using a binary Fresnel zone plate with focal length 532 nm.

2. Manufacturing

In the first step of the manufacturing a gold layer of thickness 160-180-nm was deposited on a glass substrate by electron beam evaporation. Then, the gold layer was coated with a layer of electron beam resist (ZEON ZEP520A), and the pattern of the four-sector grating polarizer projected into the resist using a 30-KV electron beam. The sample was then developed in xylene to dissolve resist fragments that had been exposed to the electron beam. The pattern of the grating polarizer was then transferred into the gold layer by sputtering with an Argon plasma. In the final stage, the remaining resist was removed with an oxygen plasma, resulting in a grating polarizer pattern engraved in gold. The duration of the reactive ion etching process was optimized so as to achieve the etch depth of the gold of about 110 nm. Figure 1 shows a SEM image of the manufactured polarizer.

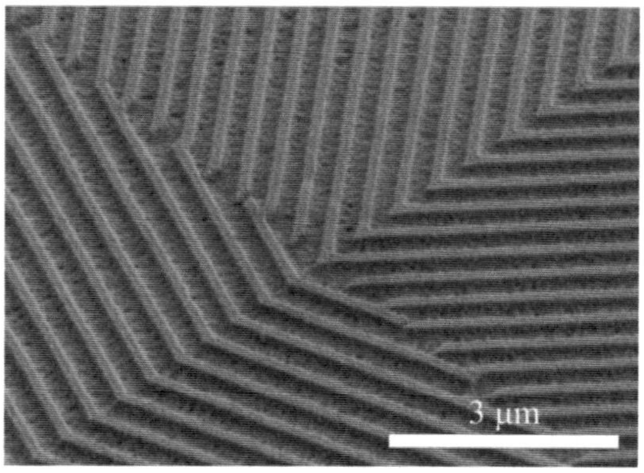

Figure 1. SEM image of the 4-sector polarization converter.

If a linearly polarized light reflects from the polarizer (Figure 1), the polarization is rotated by angle +45 or -45 relatively to direction of the incident polarization. The light reflected from the polarizer is not completely radially polarized, but is a mix of linearly polarized light and radially polarized light (in a similar way to [6]). Therefore in far field the reflected beam had a shape of a square (Figure 2a). Figures 2b and 2c show the intensity distribution in far field when an output polarizer is placed on a trajectory of the beam. Figure 2b shows the intensity of the beam when the polarizer axis is rotated by 45° and figure 2c shows the intensity when the polarizer axis is rotated by -45°.

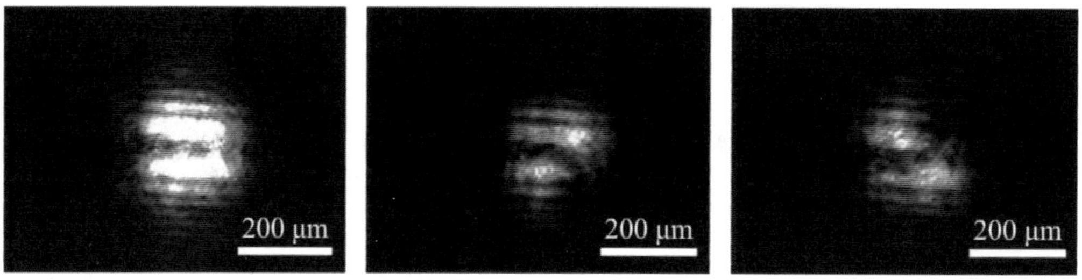

Figure 2. Far-field intensity pattern of the laser light reflected from 4-sector polarization converter: (a) without an output polarizer, with a polarizer rotated by (b) 45° and (c) -45°

In addition previously [5] we shown that the intensity of light reflected from different sectors is not equal to each other. Figure 3 shows the calculated intensity of the light reflected from the polarizer (Figure1) at the distance 200 μm; the arrows depict the direction of polarization.

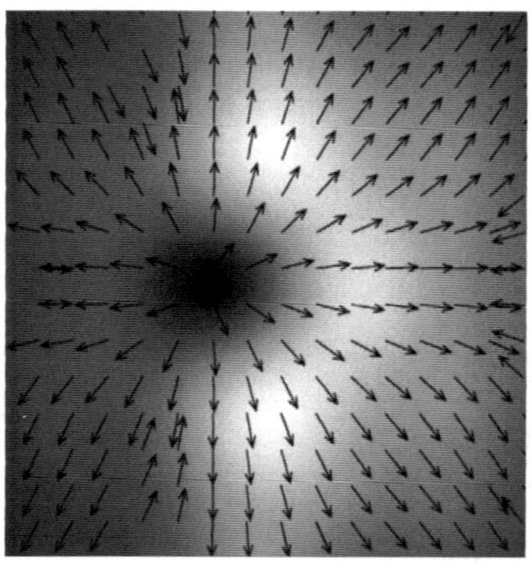

Figure 3. Intensity pattern of the light reflected from 4-sector polarizer at the distance 200 μm from the surface. The frame size is 5×5 μm. Arrows depict the direction of the polarization.

3. Simulation

Fresnel zone plate was used as a focusing object. The zone plate had a focal length 532 nm and was manufactured from ZEP520A resist (refractive index is equal to $n = 1.52$) using ion etching. The zone plate had the following parameters: focal length 532 nm, relief depth 510 nm, diameter 14 μm, the width of the last zone 266 nm. It should be noted that the radiuses of the zone plate were calculated for the wavelength equal to 532 nm, but in our research we illuminated the zone plate by the laser light with wavelength 633 nm. Previously [7] we shown that the zone plate illuminated by a linearly polarized Gaussian beam with wavelength 633 nm and radius 7 μm had a focal length less than the expected focal length (230 nm). Also the experimentally obtained diameters of the focal point were equal to $FWHM_{min} = (0,40 \pm 0,02)\lambda$, $FWHM_{max} = (0,60 \pm 0,02)\lambda$ [7]. Figure 4 shows a SEM image of the zone plate.

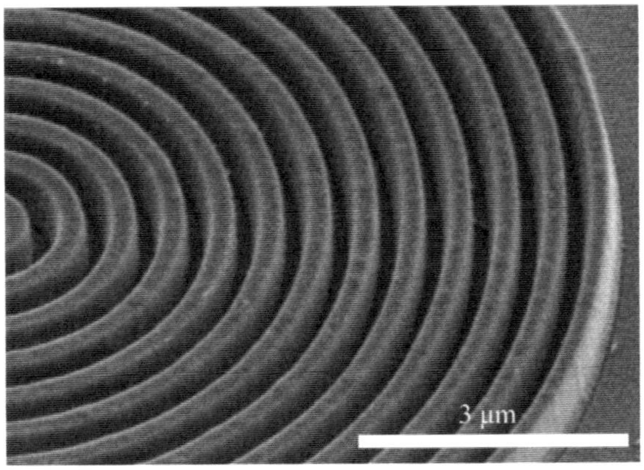

Figure 4. SEM image of the Fresnel zone plate.

Figure 5 shows the intensity profiles in the focal spot of the zone plate in Figure 4, which is illuminated by a mixed linearly and radially polarized beam (Figure 3) from the micropolarizer in Figure 1.

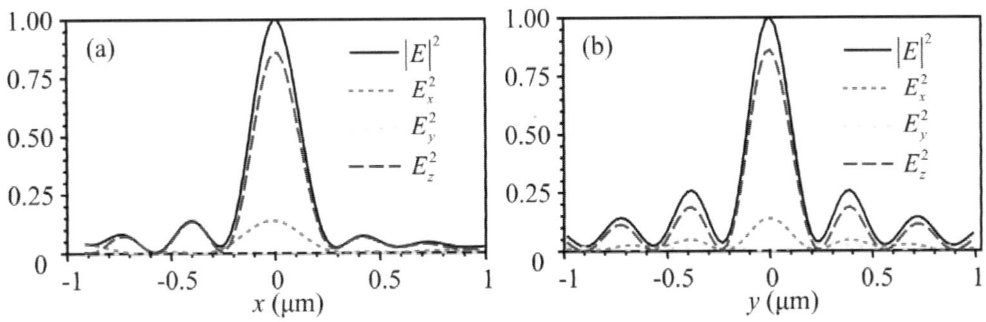

Figure 5. FDTD calculated intensity pattern in the focal plane of the zone plate. Sections along x axis (a) and y axis (b)

Figure 5 shows that distribution of intensity components differs from intensity distribution in a focal spot formed by linearly or radially polarized light. From Figure 5 the distribution of transverse intensity components is similar to distribution of intensity components in a focal point formed by a linearly polarized light. However distribution of longitudinal components is similar to distribution of intensity components of a radially polarized light. This has prompted us to define the beam under study as having a mixed, linear-radial polarization. In this case, instead of being a perfect circle (as in the case of the perfect radial polarization), the focal spot is elliptical with the axial dimensions $FWHM_{max} = (0.40\pm0.02)\lambda$ (Fig.5a) and $FWHM_{min} = (0.35\pm0.02)\lambda$ (Fig.5b). Previously [7] we shown that a metal hollow pyramid cantilever used in SNOM measurements is 3-times more sensitive to the transverse than to the longitudinal electric field component. In this case, the focal spot is seen to be somewhat larger in size: $FWHM_{min} = (0.36\pm0.02)\lambda$.

4. Experiment

Figure 6 shows the experimental setup. The laser light with wavelength 633 nm was focused with an objective O_1 on a 4-sector polarizer. Polarizers P_1 and P_2 were used to verify that the incident light is linearly polarized and to attenuate power of the incident light. Reflected from 4-sector polarizer laser light was focused on the bottom side of a substrate with the zone plate (Figure 4), propagates through the zone plate and forms a focal spot. Intensity distribution in the focal spot was measured using scanning near-field optical microscope (SNOM) Ntegra Spectra (NT-MDT). In Figure 6 SNOM is depicted as dashed rectangle.

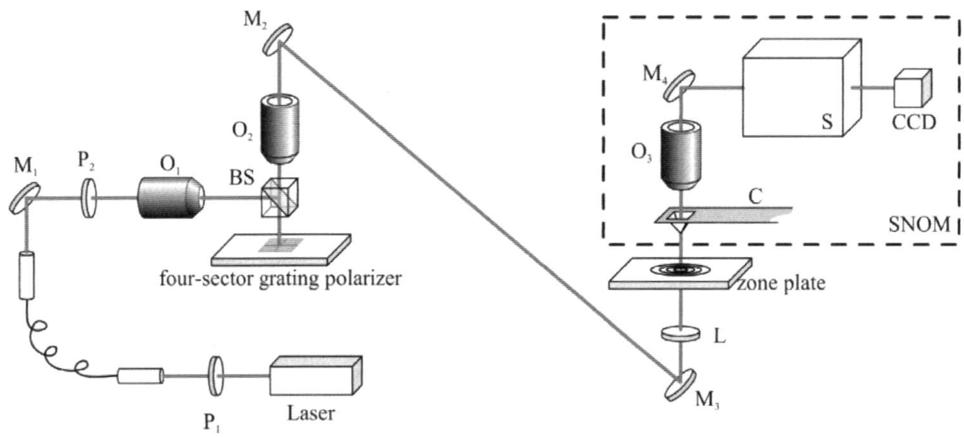

Figure 6. Experimental setup. Laser, He–Ne ($\lambda = 633$ nm); P_1, P_2: polarizers; M_1, M_2, M_3, M_4: mirrors; BS, light-splitting cube; O_1, 3.7× objective; O_2, 20× objective; O_3, 100× objective; C, cantilever; L, lens; S, spectrometer; CCD, charge-coupled device camera.

Figure 7a illustrates a SNOM-aided intensity pattern, with the minimal and maximal size of the focal spot measuring $FWHM_{max} = (0.38\pm0.02)\lambda$ (Fig.7b) and $FWHM_{min} = (0.35\pm0.02)\lambda$ (Fig.7c). To evaluate the effect of the micropolarizer, the substrate containing the micropolarizer was shifted in the transverse direction to allow the light to be reflected at the relief-free golden surface. This resulted in a microlens-aided focal spot of dimensions $FWHM_{max} = (0.50\pm0.02)\lambda$ and $FWHM_{min} = (0.40\pm0.02)\lambda$.

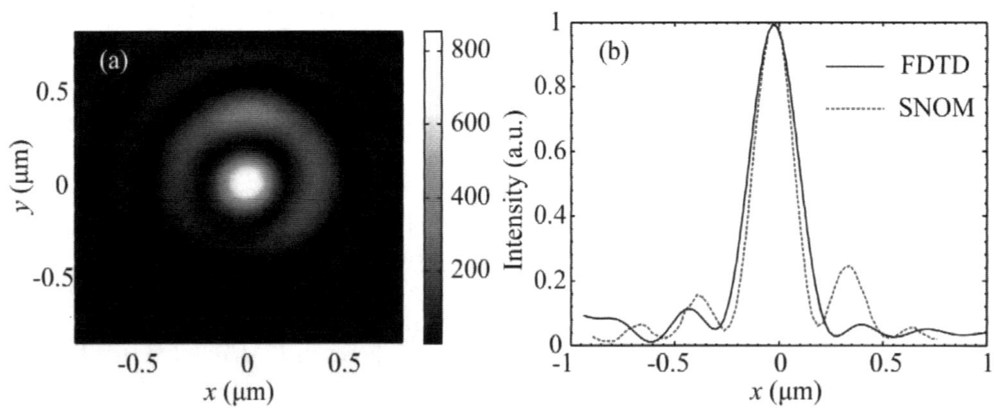

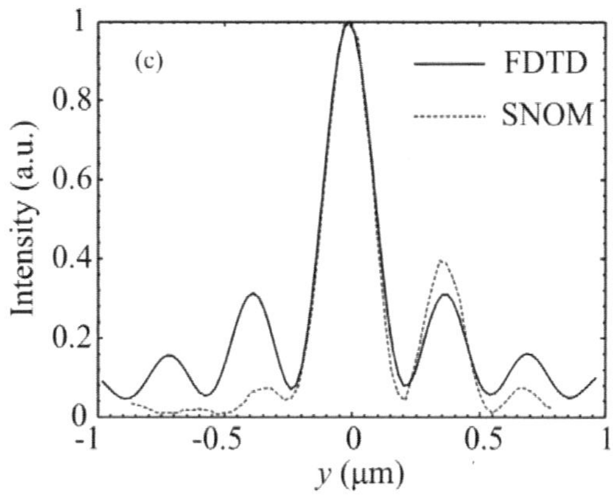

Figure 7. 2D Intensity pattern detected by SNOM (a) and its sections along *x* (b) and *y* axis (c).

It should be notes that recently [8] a focal spot with diameter 0.25λ was obtained by focusing an azimuthally polarized first-order vortex beam using a conventional immersion microlens with NA = 1.4. When reduced to the common NA, the focal spot in [8] and the one reported in this work, equal to 0.35λ, have near-same size, namely,

FWHM=0.25λ=0.25λ*1.4/NA=0.35λ/NA,

FWHM=0.35λ=0.35λ*0.997/NA=0.35λ/NA.

From the formulae above it follows that the microlens (Fig. 4) gives the same result as in [8].

5. Conclusion

Using a binary microlens with diameter 14 μm and focal length 532 nm we have focused a beam with wavelength 633 nm and mixed linear and radial polarization into a focal spot with full-width of half-maximum diameters (0.35±0.02)λ and (0.38±0.02)λ. The incident light was a mixture of linearly and radially polarized beams generated by reflecting a linearly polarized Gaussian beam at a gold 100 μm * 100 μm four-sector subwavelength diffractive optical micropolarizer, with the sector gratings of period 400 nm and microrelief depth 110 nm. The focusing of a linearly polarized laser beam (other conditions being the same) has been found to produce a larger focal ellipse measuring (0.40±0.02)λ and (0.50±0.02)λ.

Acknowledgements

This work was partially funded by the Ministry of Education and Science of the Russian Federation and Russian Federation Presidential grants for support of Young Candidate of Science (MK-4816.2014.2), and Russian Foundation for Basic Research grants ## 14-07-31218, 14-07-97039, 14-29-07133, 15-07-01174 and European Research Council Starting grant (no. 337508).

References

[1] Kotlyar V V and Zalyalov O K 1996 Optik **103**, 125–30.
[2] Niv A, Biener G, Kleiner V and Hasman E 2003 Optics Letters **28**, 510-2.
[3] Lerman G M and Levy U 2008 Optics Letters **33**, 2782-4.
[4] Mehta A, Brown J D, Srinivasan P, Rumpf R C, Johnson E G 2007 Optics Letters **32**, 1935-7.
[5] Nalimov A G, O'Faolain L, Stafeev S S, Shanina M I and Kotlyar V V 2014 Computer Optics **38**, 229–36.
[6] Dorn R, Quabis S and Leuchs G 2003 Phys. Rev. Lett. **91** 233901.
[7] Stafeev S, Kotlyar V and O'Faolain L 2013 J. Mod. Opt. **60**, 1050–9.
[8] Li X, Venugopalan P, Ren H, Hong M and Gu M 2014 Opt. Lett. **39** 5961-4

Journal of Physics: Conference Series **735** (2016) 012007 doi:10.1088/1742-6596/735/1/012007

Integral estimation of number of resolvable signal levels of digital cameras

P A Cheremkhin, N N Evtikhiev, V V Krasnov, E A Kurbatova, R S Starikov, S N Starikov

National Research Nuclear University MEPhI (Moscow Engineering Physics Institute), Faculty of Experimental and Theoretical Physics, Department of Laser Physics, Kashirskoe shosse 31, Moscow 115409, Russia

E-mail: PACheremkhin@mephi.ru, holo_mephi@mail.ru

Abstract. Number of signal levels of modern photo- and videocameras equals thousands and tens of thousands. However because of temporal and spatial camera pixels noises and linear dynamic range limitation, number of resolvable signal levels is significantly lower. Earlier iterative method of estimation of number of resolvable signal levels of cameras was proposed. In this paper integral method of estimation of number of resolvable signal levels of cameras is proposed and applied to consumer camera.

1. Introduction

Nowadays digital cameras are essential tools in different areas of science, industry and consumer applications. They are used in photography [1-2], digital holography [3-5], information processing [6-7] and etc. Nowadays cameras have analog-to-digital converter with bit depth of 12 to 16 bits. Thus number of signal levels of cameras is equal to thousands and tens of thousands (4096 levels in case of 12 bits, 65536 levels in case 16 bits). However, with decrease of pixel size and increase of bit depth of analog-to-digital converter, normalized noise value to one signal value is also increasing in spite of technology improvement. As a result, number of resolvable signal levels is significantly lower than maximum number of signal values determined by analog-to-digital converter bit depth. Earlier in [8] iterative method of estimation of the number of resolvable signal levels of cameras was proposed and applied to scientific camera. In this paper method of integral (non-iterative) estimation of resolvable signal levels of cameras is proposed and applied to consumer camera.

2. Initial characteristics for estimation of camera's resolvable signal levels

Number of resolvable signal levels shows the amount of information achievable from each pixel of the camera. It is determined by noise magnitude and value of linear dynamic range of the camera. Higher noise magnitude means wider signal intervals for each resolvable level. To obtain number of resolvable signal levels, six noise and radiometric characteristics of camera's photosensor must be known. These characteristics are:

- noise characteristics of camera's photosensor:
  - light temporal noise,
  - dark temporal noise,

  - light spatial noise,
  - dark spatial noise;
- radiometric characteristic of camera's photosensor:
  - maximum linear signal value,
  - black level offset (mean value of dark frame).

For iterative estimation of number of camera's resolvable signal levels approximated values of full noise value $\sigma_{\Sigma}(S)$ for each camera's signal value should be determined as follows (see [8]):

$$\sigma_{\Sigma}(S)=\sqrt{(\sigma_{dt})^2+(\sigma_{lt})^2+(\sigma_{ds})^2+(\sigma_{ls})^2}=\sqrt{(\sigma_{dt})^2+k\times S+(DSNU)^2+(PRNU\cdot S)^2} \quad (1)$$

where

σ_{dt} - dark temporal noise,

σ_{lt} - light temporal noise,

σ_{ds} - dark spatial noise,

σ_{ls} - light spatial noise,

S - signal value with black level offset subtracted,
k - the scaling constant for conversion of number of electrons to digital signal value,
DSNU - dark signal non-uniformity,
PRNU - photo response non-uniformity [9-10].

Light spatial noise is characterized by the *PRNU* value. Dark spatial noise is characterized by the *DSNU* value. Signal values *S* ranges from zero (black level offset is subtracted) to maximum linear signal value.

3. Description of the method of estimation of camera's resolvable signal levels

Number *G* of resolvable signal levels ΔG provided by camera should be determined as follows. Let consider the registration of a signal in a small interval from *S* to $S+\Delta S$. Quantity of ΔS digital signals will be equaled to number *G* if interval of uncertainty of value of signal is equal to 1 digital signal. The larger interval of uncertainty of value of signal is, the less quantity *G* of resolvable signal levels ΔG. Thus, number *G* for ΔS interval is equal to ratio of interval size to signal value uncertainty size. The doubled value of noise (i.e. standard deviation of noise $\sigma_{\Sigma}(S)$) at image registration is accepted as interval of uncertainty of signal value. Larger signal values lead to larger noise values. Thus in the case of little signal changes we can consider double mean noise value as signal uncertainty. But in the case of significant changes (significant part of camera dynamic range) we cannot use mean noise. Then estimation of full number of resolvable levels for signals in the range from S_{min} to S_{max} is given by the following expression:

$$G=\int_{S_{min}}^{S_{max}}\frac{dS}{2\times\sigma_{\Sigma}(S)}=\frac{1}{2}\int_{S_{min}}^{S_{max}}\frac{dS}{\sigma_{\Sigma}(S)} \quad (2)$$

Full noise $\sigma_{\Sigma}(S)$ for a single image consists of dark and light temporal and spatial noises of camera and defined as shown in the Eq. (1). Such type of dependence of full noise (see Eq. (1)) on signal value *S* allows analytical solution of the integral in Eq. (2). Obtained expression for full number of resolvable signal levels for the $PRNU\neq 0$ is:

$$G(S_{max})=\frac{1}{2\cdot PRNU}\cdot\ln\frac{S_{max}+S_0+\sqrt{(S_{max}+S_0)^2-a^2}}{S_{min}+S_0+\sqrt{(S_{min}+S_0)^2-a^2}} \quad (3)$$

where $S_0 = \dfrac{k}{2 \cdot PRNU^2}$ and $a^2 = S_0{}^2 - \dfrac{\sigma_{dt}^2 + \sigma_{ds}^2}{PRNU^2}$.

In case of absence of light spatial noise (*PRNU*=0), expression for full number of resolvable signal levels might be expressed as:

$$G(S_{\max}) = \sqrt{k} \cdot \left(\sqrt{S_{\max} + S_k} - \sqrt{S_{\min} + S_k}\right) / \sqrt{k}\,, \tag{4}$$

where $S_k = \dfrac{\sigma_{dt}^2 + \sigma_{ds}^2}{k}$.

If temporal noise might be neglected in comparison with spatial noise (for example, in case of averaging of large number of images), Eq. (2) might be rewrited into:

$$G(S_{\max}) = \frac{1}{2 \cdot PRNU} \cdot \ln \frac{\sqrt{S_{\max} + \left(S_{\max}{}^2 + S_p{}^2\right)}}{\sqrt{S_{\min} + \left(S_{\min}{}^2 + S_p{}^2\right)}}, \tag{5}$$

where $S_p = \dfrac{\sigma_{ds}}{\sqrt{1 - PRNU^2}}$.

Value of the minimum registered signal $S_{\min}$ corresponds to noise level so that signal-to-noise ratio is equal to 1. For $G(S_{\max})$ calculation, $S_{\min}$ is defined as follows:

$$S_{\min} = \sigma_\Sigma (S_{\min}). \tag{6}$$

Value of maximum signal $S_{\max}$ can't exceed value of maximum linear signal of $S_{\max Lin}$ which is the upper bound of linear dynamic range. Lower bound of dynamic range is equal to minimum registered signal $S_{\min}$, at which signal-to-noise ratio is equal to 1.

Thus, maximum number of resolvable levels of signal *G* registered by camera depends both on the value of noise and on the maximum value of linear range.

4. Experimental estimation of camera's resolvable signal levels

For experimental estimation of number of camera's resolvable signal levels, consumer camera Canon EOS 400D was used. Its main technical characteristics are given in Table 1 [11].

Noise and radiometric characteristics of this camera measured in [12,13] were used. As the camera is colour one, only its green colour channels were used in parameters measurements. Frames from the camera Canon EOS 400D had been stored in unprocessed RAW format. Using DCRAW converter [14] linear unprocessed images were obtained. Characteristics are shown in Table 2, where DN is digital numbers of signal, e^-/DN is the ratio of quantity of emitted electrons in one pixel to one signal digital number.

Obtained dependence of resolvable signal levels quantity on used part of radiometric function linear range is shown in Fig. 1 (rel.un. is relative units). Also sizes of resolvable signal levels can be estimated from the Fig. 1. As it can be seen, widths of resolvable signal levels increase with the signal magnitude. This is due to the fact that noise value increases with signal level.

Table 1. Technical characteristics of the consumer camera Canon EOS 400D.

Characteristic	Value
Sensor type	CMOS (complementary metal–oxide–semiconductor)
Quantity of pixels	3888 × 2592 (10.1 megapixels)
Pixel size	5.7 μm × 5.7 μm
Output bit depth of analog-to-digital converter	Up to 12 bits
Spectral	Colour
Minimum integration time	250 μs
Maximum integration time	30 s

Table 2. Noise and radiometric characteristics of camera Canon EOS 400D.

Characteristic	Value
Dark temporal noise	1.6 ± 0.2 DN
Camera gain (1/k)	9.2 ± 0.2 e⁻/DN
Dark spatial non-uniformity (*DSNU*)	0.377 ± 0.007 DN
Photo response non-uniformity (*PRNU*)	0.0051±0.0002
Black level offset (BLO)	256.0 ± 0.4 DN
Maximum linear signal with BLO subtracted	3070 ± 30 DN

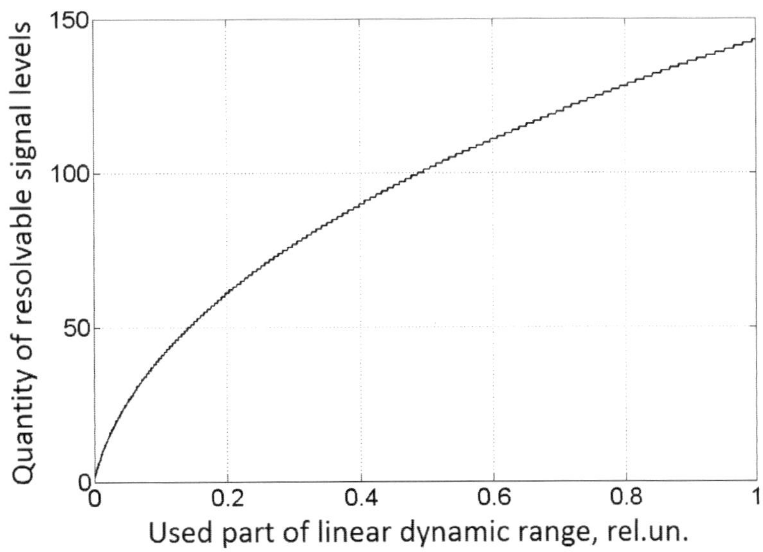

Figure 1. Obtained dependence of resolvable signal levels quantity on used part of linear dynamic range for the camera Canon EOS 400D.

Estimated maximum number of camera's resolvable signal levels is equal to 143. This value corresponds to the whole linear dynamic range of the camera. Obtained number of camera's resolvable signal levels is 21 times lower than total number of digital signal levels (3070 levels, see Table 2) defined by the bit depth of the camera analog-to-digital converter for linear part of radiometric function.

It should be noted that obtained value of number of camera's resolvable signal levels is correct for a single shots. The use of averaging over several shots [9], spatial averaging over pixels groups, compensation of light spatial noise portrait of camera's photosensor [15] or digital methods [16-21] leads to noise decreasing and, consequently, to increasing of number of resolvable signal levels.

5. Conclusion
Integral method of estimation of resolvable signal levels of photo- and videocameras is proposed. It utilizes dependence of noise values on signal values. The method was applied to consumer camera Canon EOS 400D. Estimated maximum number of camera's resolvable signal levels is equal to 143. Dependence of resolvable signal levels quantity on used part of radiometric function linear range was acquired. The proposed method can be used for cameras of different types.

Acknowledgments
This work was partially supported by grant 14-19-01751 from the Russian Science Foundation (RSF).

References
[1] Filippi A M and Guneralp I 2013 *Opt. Lett.* **38** 1676-8
[2] Marks D L, Kittle D S, Son H S, Youn S H, Feller S D, Kim J, Brady D J, Golish D R, Vera E M, Gehm M E, Stack R A, Tremblay E J and Ford J 2012 *Opt. Photonics News* **23** 31
[3] Juptner W and Schnars U, 2005 *Digital Holography: Digital Hologram Recording, Numerical Reconstruction, and Related Techniques* (Berlin-Heidelberg: Springer-Verlag)
[4] Mills G A and Yamaguchi I 2005 *Appl. Opt.* **44** 1216-25
[5] Cheremkhin P A, Evtikhiev N N, , Krasnov V V, Porshneva L A, Starikov R S and Starikov S N 2014 *J. Phys.: Conf. Ser.* **536** 012008
[6] Chen W, Chen X, Anand A and Javidi B 2013 *JOSA A* **30** 806-12
[7] Gao Y, Tang J, Hong R, Yan S, Dai Q, Zhang N and Chua T-S 2012 *IEEE Trans. Image Process.* **21** 2269-81
[8] Cheremkhin P A, Krasnov V V, Kurbatova E A, Rodin V G and Starikov S N 2014 *J. Phys.: Conf. Ser.* **536** 012023
[9] Holst G C 1998 *CCD Arrays Cameras and Displays* (Bellingham-Washington: SPIE Press)
[10] Janesick J 2007 *Photon Transfer* (Bellingham-Washington: SPIE Press)
[11] http://www.canon-europe.com/for_home/product_finder/cameras/digital_slr/eos_400d/, (accessed 26 May 2015).
[12] Evtikhiev N N, Starikov S N, Cheryomkhin P A and Krasnov V V 2012 *Proc. SPIE, Intelligent Robots and Computer Vision XXIX: Algorithms and Techniques (Burlingame, USA)* vol 8301 (SPIE) p 830113
[13] Cheremkhin P A, Evtikhiev N N, Krasnov V V., Rodin V G. and Starikov S N 2014 *Opt. Eng.* **53** 102107.
[14] Coffin D http://www.cybercom.net/~dcoffin/dcraw (accessed 26 May 2015).
[15] Evtikhiev N N, Starikov S N, Cheryomkhin P A, Krasnov V V and Rodin V G 2014 *Proc. SPIE, Intelligent Robots and Computer Vision XXXI: Algorithms and Techniques (San Francisco)* vol 9025 (SPIE) p 90250X
[16] Donoho D 1995 *IEEE Trans. Inf. Theory* **41** 613–27
[17] Buades A, Coll B, and Morel J M 2005 *Multiscale Model. Sim.* **4** 490-530
[18] Dabov K, Foi A, Katkovnik V and Egiazarian K 2007 *IEEE Trans. Image Process.* **16** 2080-95
[19] Mayer M A, Borsdorf A, Wagner M, Hornegger J, Mardin C Y and Tornow R P 2012 *Biomed. Opt. Express* **3** 572-89
[20] Rajwade A, Rangarajan A and Banerjee A 2013 *IEEE Trans. Pattern Anal. Mach. Intell.* **35** 849-62
[21] Lam F, Babacan S D, Haldar J P, Weiner M W, Schuff N and Liang Z-P 2014 *Magn. Reson. Med. Discipline* **71** 1272-84

Synthesis and study of transparent multicomponent metal oxide for use in multisensor system

E.V.Abrashova, V.A. Moshnikov, E.V.Maraeva, I.E. Kononova, D.M. Vorob'ev
Saint-Petersburg State Electrotechnical University «LETI», Saint-Petersburg, Russia

E-mail: katabr@mail.ru

Abstract. The thin films on the basis of metal oxides of SnO2-ZnO-SiO2 were prepared by sol-gel techniques, and CVD. During the experiment optical and morphological characteristics of materials were investigated. According to the study we can say that this system is promising for use in transparent multisensor devices.

1. Introduction

Functional multi-metal oxides are widely used in various branches of micro- and nanoelectronics. Today the so-called transparent electronic is one of the most promising areas of manufacturing equipment, including the topical issue of the creation of transparent conductive materials. Metal micrometals, conducting polymers, graphite, carbon nanotubes, more often metal oxides are used as the transparent conductors [1-5]. For many years the most common transparent conductor is indium oxide doped with tin (ITO) due to their successful physical and processing characteristics. But due to the shortage of indium replacement ITO more affordable and inexpensive material is a topical issue. The tin dioxide, zinc oxide and titanium dioxide are a recognized competitors of indium oxide today.

2. Experiment.

Thin films based systems oxides of tin, zinc and silicon in different ratios, obtained by means of sol-gel and CVD for possible use as the transparent conductors, as well as a gas-sensitive multisensor systems compatible with transparent electronics were investigated [6-8]. Recent studies show that films of mixed oxides and oxide films, doped metal have the highest gas sensing properties.

Therefore morphological, optical, gas sensing characteristics of system SnO_2-ZnO-SiO_2 were investigated. As precursors were chosen chlorides and nitrides of the respective metals. Tetraethoxysilane (TEOS) was chosen to improve the adhesion to the substrate in a sol-gel method, annealing was carried out at a temperature 600°C. The substrate temperature during the pyrolysis was maintained at 500 ° C.

Composition control of the obtained materials was conducted by X-ray phase analysis and diffraction of fast electrons (DFE). As an X-ray analytical equipment used desktop "DRN Farad" and small-angle electron diffraction registering EMR-102 (SELMI). Research film nanostructures morphology were conducted by top-mode vibrational technique of atomic force microscopy (AFM) by nanolaboratories NTegra Therma. Evaluation of the specific surface area was carried out by a known method with BET model. Optical properties were obtained using SF-56 spectrophotometer.

3. Results and Discussion.

Studies of the samples obtained by CVD demonstrated that formed films have a microcrystalline structure (Figure 1-6).

Figure 1. The diffraction pattern of a thin film of tin oxide produced by the CVD method.

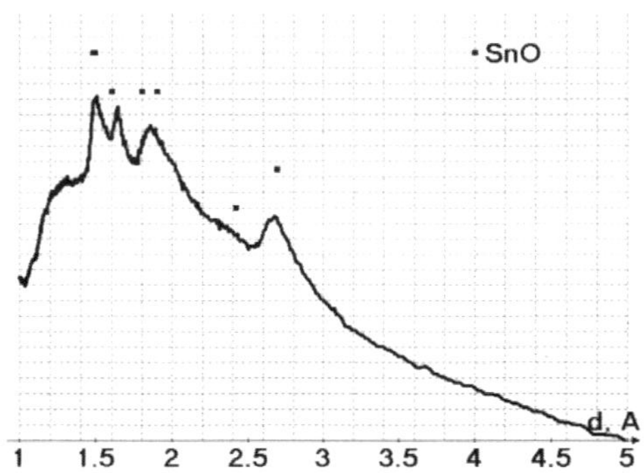

Figure 2. Peaks position of a thin film of tin oxide produced by the CVD method.

Figure 3 The diffraction pattern of a thin film of zinc oxide produced by the CVD method..

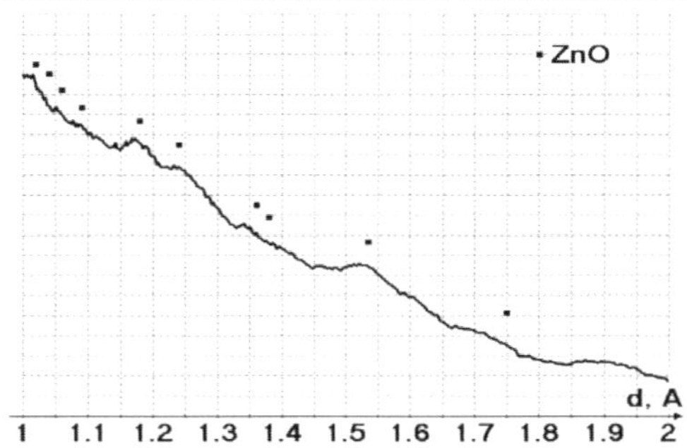

Figure 4. Peaks position of a thin film of zinc oxide produced by the CVD method.

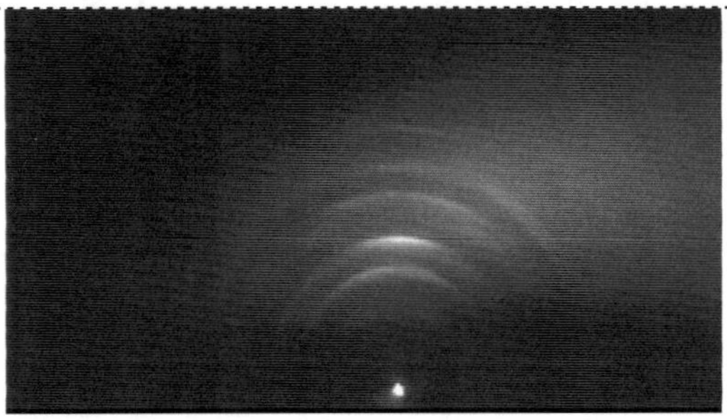

Figure 5. The diffraction pattern of a thin film of zinc stannat produced by the CVD method.

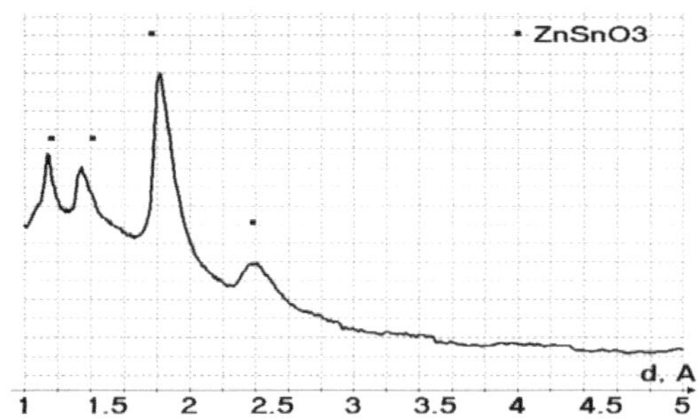

Figure 6. Peaks position of a thin film of zinc stannat produced by the CVD method.

Analysis of the morphology of the surface showed that the samples obtained using as carrier gas moistened air had a rougher surface compared with the samples obtained by using dried air. (Figure 7, 8).

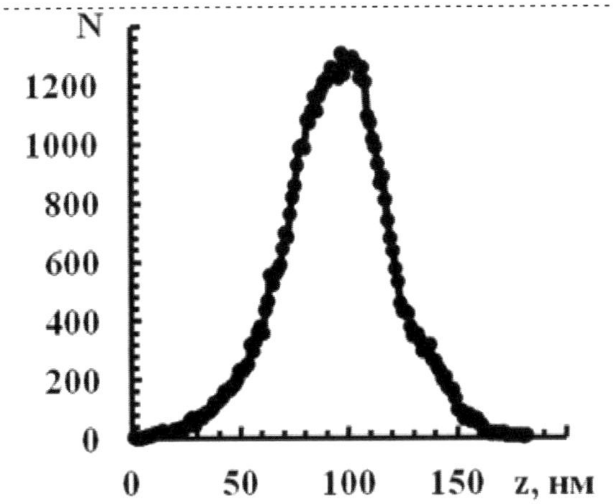

Figure 7. Histogram distribution density values of Z for the sample obtained using as carrier gas humid air.

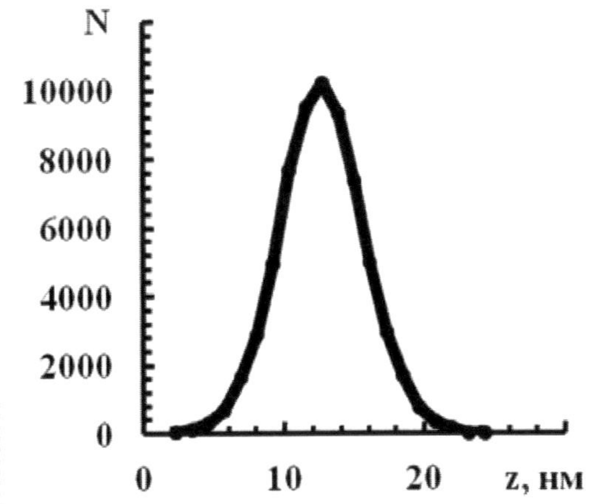

Figure 8. Histogram distribution density values of Z for the sample obtained using as carrier gas dried air.

The surface roughness of the samples was evaluated by calculating the fractal dimension of the surface by four methods: variational, by triangulation, counting cubes, by the power spectrum []. (Table 1, Figure 9)

Table 1. The values of the fractal dimension of the surface of the samples obtained by the CVD method with humidified and dry air.

The synthesis conditions	The values of fractal dimension			
	The variational method	Method of of counting of cubes	Triangulation method	Method of the power spectrum
With the participation of the humidified air	2.45	2.28	2.36	1.95
With the participation of the dried air	2.66	2.41	2.51	2.30

As a result of the calculation we can say that despite the fact that the technology involving humidified air produces large crystallites with a greater surface roughness of thin films produced in the atmosphere of dry air have a developed surface, as evidenced by the large fractal dimension, calculated 4 ways.

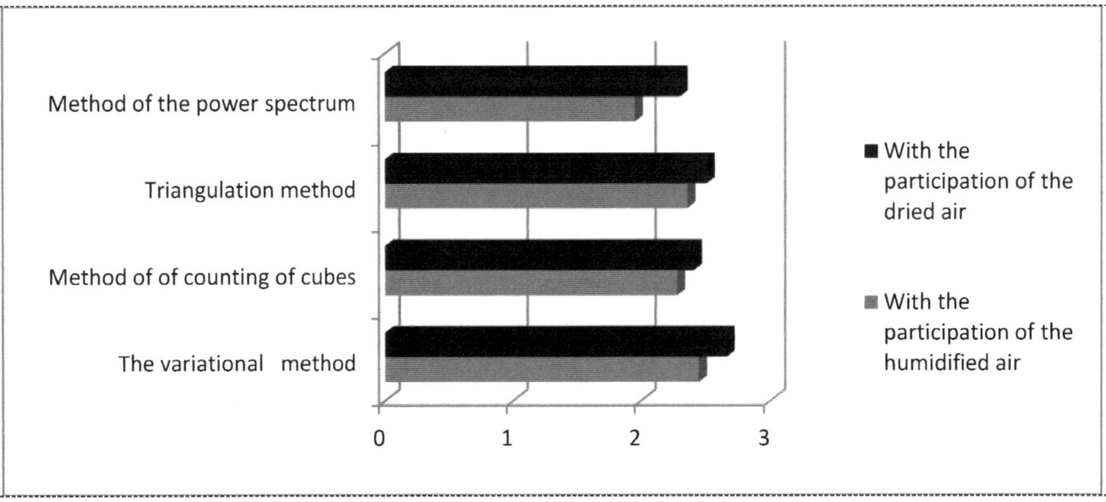

Figure 9. Histogram of the distribution of values of the fractal dimension for samples prepared with the participation of as carrier gases humidification and dried of air.

For the purpose of complex characterizing of material the layers formed by sol-gel were investigated. The technology is described in detail in [9]
The single-layer and multi-layer samples with different relative amounts of oxides of tin and zinc were obtained. AFM studies have shown that increasing the relative proportion of zinc leads to a loosening of the porous structure of the layer. (Table 2, Figure 10).

Table 2. The value of the fractal dimension of porous hierarchical structures for a different time of maturing of the sol solution, calculated in several ways.

The relative content of tin dioxide and zinc oxide	Method	The values of fractal dimension		
		150 h	360 h	1500 h
2:1	decomposition	2,79	2,77	2,98
	counting cubes	2,51	2,38	2,60
	triangulation	2,60	2,39	2,69
	power spectrum	2,59	2,95	2,99
1:1	decomposition	2,82	2,72	2,70
	counting cubes	2,37	2,39	2,50
	triangulation	2,46	2,50	2,60
	power spectrum	2,54	2,48	2,16
1:2	decomposition	2,40	2,42	2,79
	counting cubes	2,18	2,17	2,60
	triangulation	2,15	2,16	2,69
	power spectrum	2,06	2,60	2,53

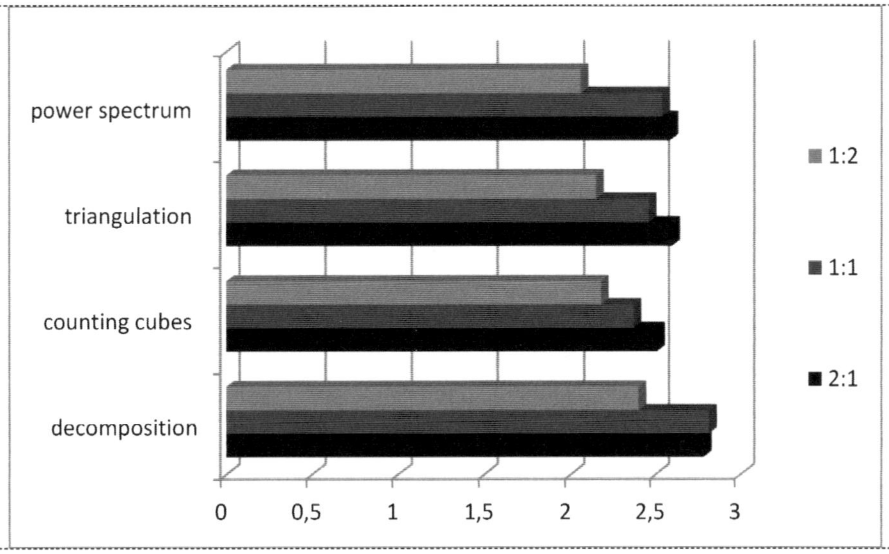

Figure 10. Gistagramma distribution fractal dimension samples depending on the relative content of tin and zinc, respectively, calculated in various ways..

Analysis of the AFM images shows that increasing the proportion of the content of zinc oxide in the solution the number of pores decreases, and the pores become larger. Fractal dimension was calculated with a relative increase in the proportion of zinc oxide is reduced Figure 10, indicating that the downward trend in the developed surface of the samples with respect to the film with a predominant content of tin dioxide. As it can be noted, that dependence of the sol is not changed with increasing exposure time.

These data correlate completely with the calculation of the specific surface obtained by the BET method. (Figure 11.)

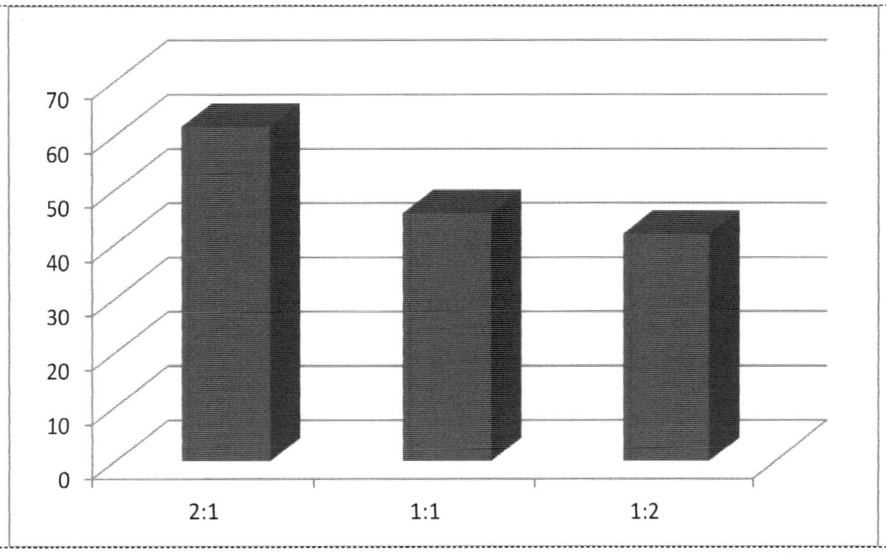

Figure 11. The specific surface area of the material of SnO2-ZnO depending on the relative content of oxides of tin and zinc, respectively

Optical characteristics of thin films of zinc stannate investigated. From the measurements it was found that the transmittance of the material is close to 1 over the entire visible range.
Thus, the results of the study we can say that the investigated thin films have a hierarchical surface morphology. The development of the surface can be controlled as a selection of process parameters and by varying the composition of the material. Investigated thin films have high transmittance, so promising for use in transparent electronics devices.

References

[1] Youngbin Lee, Graphene-based transparent conductive films, NANO: Brief Reports and Reviews , Vol. 8, No. 3 (2013) 1330001 , DOI: 10.1142/S1793292013300016
[2] Granqvist C.G., Hultaker A., Transparent and conducting ITO films: new developments and applications//Thing Solid Films, 2002,V. 411, P.1
[3] Gordon R.G., Criteria for choosing transparent conductors//Mrs Bulletin, August 2000, p. 52-57
[4] Xuan Wang, Linjie Zhi, Klaus Mullen, Transparent, Conductive Graphene Electrodes for Dye-Sensitized Solar , Nano Letters, 2008, Vol. 8, No. 1, 323-327
[5] Kin Man Yu, Marie A. Mayer, Derrick T. Speaks, Hongcai He, Ruying Zhao et al., Ideal transparent conductors for full spectrum photovoltaics, J. Appl. Phys. 111, 123505 (2012); doi: 10.1063/1.4729563
[6] I. E. Gracheva, V. A. Moshnikov, S. S. Karpova, E. V. Maraeva // Net-like structured materials for gas sensors / Journal of Physics: Conference Series. 2011. V. 291, № 1. P. 012017.Abrashova E.V., Gracheva I.E., Moshnikov V.A, Functional nanomaterials based on metal oxides with hierarchical structure, Journal of Physics: Conference Series. 2013. T. 461. C. 012019.
[7] Abrashova E.V., Kononova I.E., Moshnikov V.A., Metal oxide SnO_2 - ZnO - SiO_2 films prepared by sol-gel, Smart Nanocomposites. 2013. T. 4. № 2. C. 1-7
[8] Abrashova E.V.,Baranovskiy M.V // Preparation and Analysis of spectral characteristics of nanocomposites based on wide band gap semiconductors metal oxides of ZnO-SnO_2-SiO_2/ News of the St. Petersburg Electrotechnical University, 2013. V. 5. p. 16-21.
[9] Abrashova E.V., Changes of morphological and structural properties of thin films based on the system SnO_2- SiO_2, obtained by the sol-gel method for the modification of zinc oxide News of the St. Petersburg Electrotechnical University , 2014, V4, p. 3-7

Journal of Physics: Conference Series **735** (2016) 012009 doi:10.1088/1742-6596/735/1/012009

Theoretical study of the morphology of self-assembled amphoteric oxide colloid nanocrystals in weak electrolyte solutions

A V Alfimov, E M Aryslanova, S A Chivilikhin

ITMO University, Kronverkskiy, 49, St. Petersburg, 197101, Russian Federation

E-mail: alfimov.anton@gmail.com

Abstract. The present paper is devoted to the theoretical study of the morphology of nanocrystals formed through self-assembly of amphoteric oxide colloidal nanoparticles in weak electrolytes. A mathematical model of multi-particle colloidal interaction was developed within the framework of the Derjaguin, Landau, Verwey, Overbeak (DLVO) theory. This model accounts for the surface charge regulation during the multi-particle interaction and reveals the presence of orientation effects during nanoparticle aggregation. These effects are presumed to affect the morphology of the self-assembled nanocrystals and may present the means of controlling the structure of synthesised nanomaterials.

1. Introduction

Nanocrystalline materials are widely used in photonics and optoelectronics due to their unique properties. These properties include optical and electronic size effects, surface plasmon resonances, formation of photonic bandgaps in ordered structures, etc [1-5]. The specific parameters of synthesized nanocrystalline materials are strongly dependent on the morphology of their structural units – single nanocrystals or their assemblages [3-5]. Thus the creation of nanocrystalline materials with desired properties relies on the ability to produce the structural units of prescribed morphology.
For example, manufacturing of highly efficient nanocomposite hybrid organic-inorganic solar cells requires the synthesis of three-dimensional hyperbranched colloidal semiconductor nanocrystals of specified dimensions [4].

Self-assembly in colloid systems is among the most cost-efficient and promising methods for obtaining nanocrystalline materials with specific structure [5]. One way of implementing this technique is through the controlled coagulation of colloid nanoparticles. This allows for using the colloid self-assembly to create nanoparticle assemblages with required morphology. The obtained nanoagglomerates can be then employed for the synthesis of a variety of nanocomposite materials and photonics glasses [5].

The present paper is devoted to the theoretical study of amphoteric oxide nanoparticles' self-assembly in aqueous solutions of weak electrolytes. An original mathematical model is used to explore the effect of the colloidal system parameters on the morphology of the self-assembled agglomerates.

2. Methods

Theoretical study of the colloidal self-assembly of nanoparticles is carried out using an original colloidal multi-particle interaction model introduced by the authors in [6, 7]. The model was developed in the framework of the Derjaguin, Landau, Verwey, Overbeak (DLVO) theory. This theory assumes that the interaction between colloid particles can be represented using the attractive van der

Waals term and the repulsive electroosmotic term. The latter arises due to the overlap of particles' electric double layers (EDLs).

The developed colloidal interaction model uses an original nanoparticle surface charge regulation model along with the representation of the electric potential distribution in nanoparticle's EDL through the first few terms of its multipole expansion to solve the Debye-Huckel equation for the system of multiple interacting particles, as described by the authors in [7].

In the present work we limit ourselves to the simple case of using the monopole approximation for the potential distribution in the EDL of a colloidal particle. In this case, the developed charge regulation model yields the following expression for the nanoparticle's surface charge density σ :

$$\sigma=\left\{\sigma_0+e\left[\beta_+ n_{\mathrm{H}}\left(1-\frac{e\varphi_\Sigma}{kT}\right)-\beta_- n_{\mathrm{OH}}\left(1+\frac{e\varphi_\Sigma}{kT}\right)\right]\right\}\cdot\left\{1+\frac{e^2\left[\beta_+ n_{\mathrm{H}}+\beta_- n_{\mathrm{OH}}\right]}{4\pi\varepsilon\varepsilon_0\cdot kT}\left[\frac{1}{b(1+\kappa b)}+\frac{b-a}{ab}\right]\right\}^{-1}, \quad (1)$$

where σ_0 - constant portion of charge; $n_{\mathrm{H}}, n_{\mathrm{OH}}$ - bulk H^+ and OH^- concentrations; a,b - physical and hydrodynamic particle radii; β_+,β_- - coefficients proportional to the sums of the rate constants for the adsorption processes involved; φ_Σ - potential of the external field; e - elementary charge. The model parameters σ_0 and β_+,β_- were calculated using the experimental zeta-potential dependency on pH taken from [8].

As it has been shown in [7] by the authors, the potential energy U_α of a colloidal nanoparticle interacting with multiple other particles in the considered system can be found as:

$$U_\alpha=\frac{Q_\alpha}{4\pi\varepsilon\varepsilon_0}\cdot\sum_{\beta\neq\alpha}\left[Q_\beta\cdot\frac{\exp\left(-\kappa R_{\alpha,\beta}\right)}{R_{\alpha,\beta}}\right]+\sum_{\beta\neq\alpha}\left[U_{vdW}\left(R_{\alpha,\beta}\right)\right], \quad (2)$$

where α,β - particle numbers; ; $R_{\alpha,\beta}$ - centre-to-centre separation between particles α and β ; ε - effective dielectric permittivity in the diffuse ion layer; ε_0 - dielectric constant; κ - reciprocal of the Debye screening length; U_{vdW} - potential energy of van der Waals attraction; Q_α - equivalent screened point-charge of the particles α that can be than found by solving the following linear system of algebraic equations:

$$Q_\alpha+\sum_{\beta\neq\alpha}\theta_{\alpha,\beta}\cdot Q_\beta=Q_0 \quad (3)$$

with the coefficient given by:

$$\theta_{\alpha,\beta}=\frac{1}{4\pi\varepsilon\varepsilon_0}\cdot\frac{e^2}{kT}\cdot\frac{1}{R_{\alpha,\beta}(1+\kappa b)}\cdot\frac{\left(\beta_+ n_{\mathrm{H}}+\beta_- n_{\mathrm{OH}}\right)\cdot\exp\left[-\kappa\left(R_{\alpha,\beta}-b\right)\right]}{1+\frac{e^2\left[\beta_+ n_{\mathrm{H}}+\beta_- n_{\mathrm{OH}}\right]}{4\pi\varepsilon\varepsilon_0\cdot kT}\left[\frac{1}{b(1+\kappa b)}+\frac{b-a}{ab}\right]} \quad (4)$$

Q_0 - equivalent charge of an isolated particle.

The model validation was performed by studying its predictions for the dependency of the probability of a nanoparticle attaching to an agglomerate on the agglomerate's size. The developed model provides good qualitative agreement with experimental data for this dependency, as it has been shown in [7].

3. Results and discussion

The developed mathematical model of colloidal multi-particle interaction was used to study the morphology of self-assembled nanoparticle agglomerates. The case of a single nanoparticle attaching to an agglomerate of varied size was modelled. This allowed to explore the effects of the colloidal solution properties on the geometry of the agglomerates being formed.

Firstly, we have studied the dependency of the interaction potential energy on the location of the single nanoparticle in relation to the surface of the agglomerate. This allowed us to analyze the possible resulting agglomerate morphologies and identify the most energetically favourable of them.

Journal of Physics: Conference Series **735** (2016) 012009 doi:10.1088/1742-6596/735/1/012009

The cases of a single nanoparticle approaching the agglomerate formed by 2, 3, 4 or 5 particles were considered.

The corresponding potential energy contour plots for the electroosmotic interaction potential energy are depicted in figure 1. The figure shows that the potential barrier due to electroosmotic repulsion is anisotropic with respect to the direction of particle's approach: attachment to a site with more particles in contact results in a higher potential barrier. It should be noted that figure 1 demonstrates two options for the geometry of 3-, 4- and 5-particle agglomerates. One option is the linear structure found to be most energetically favourable with regards to electroosmotic repulsion. The other option is the close-packed structure that is most energetically favourable with regards to van der Waals attractions.

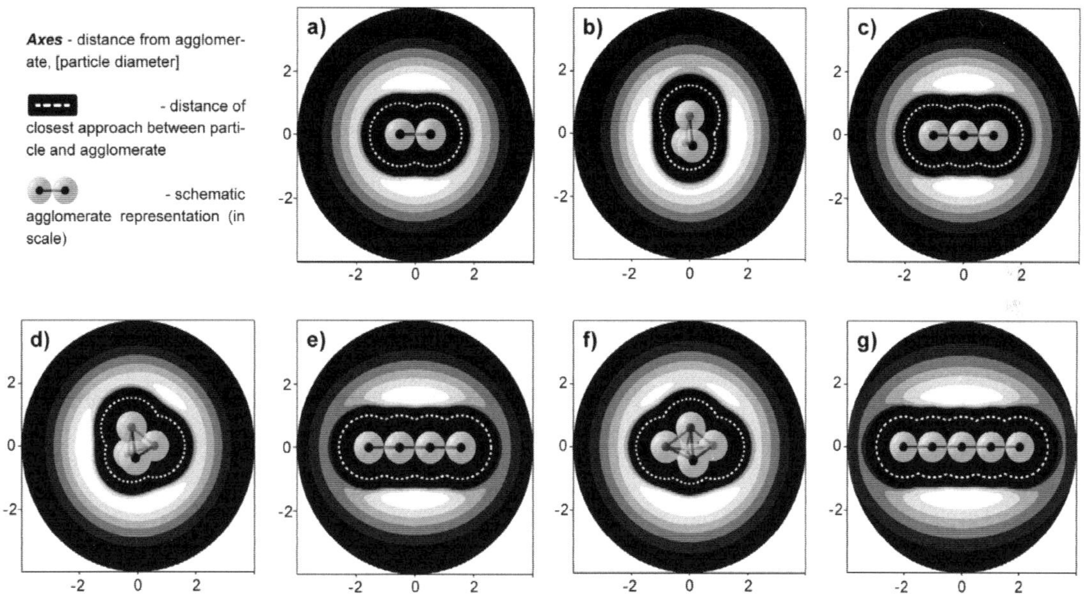

Figure 1. Contour plots of the interaction potential energy between a single mobile nanoparticle and a fixed agglomerate configuration normalized to its maximum value; brighter areas correspond to higher values of the potential energy. The considered agglomerate configurations are: a) 2-particle agglomerate; b) compact 3-particle agglomerate; c) linear 3-particle agglomerate; d) compact 4-particle agglomerate; e) linear 4-particle agglomerate; f) compact 5-particle agglomerate; g) linear 5-particle agglomerate.

The observed electroosmotic orientation effects during agglomeration are further illustrated in figure 2. This figure demonstrates the dependency of the attachment probability on the location of the attachment site on the surface of the agglomerate for different solution pH values. The attachment site position on the surface is given by its angular coordinate. According to figure 2 pH values that are further away from the neutral pH result in higher particle surface charge and, therefore, the increasing role of electroosmotic repulsion in particle interactions. This explains the corresponding increase of the attachment probability anisotropy for the approach directions found in figure 2.

It is thus shown that with the increase of the nanoparticle surface charge it becomes more and more energetically favourable to initially form linear structures. Figure 2 also shows that for linear agglomerates the attachment of each next particle in a continuation of the linear structure becomes more and more plausible.

The fate of the agglomerate after its initial formation is studied in figure 3. This figure depicts the dependency of the potential well depth on the attachment site location on the surface of the agglomerate. The attachment site position on the surface is given by its angular coordinate.

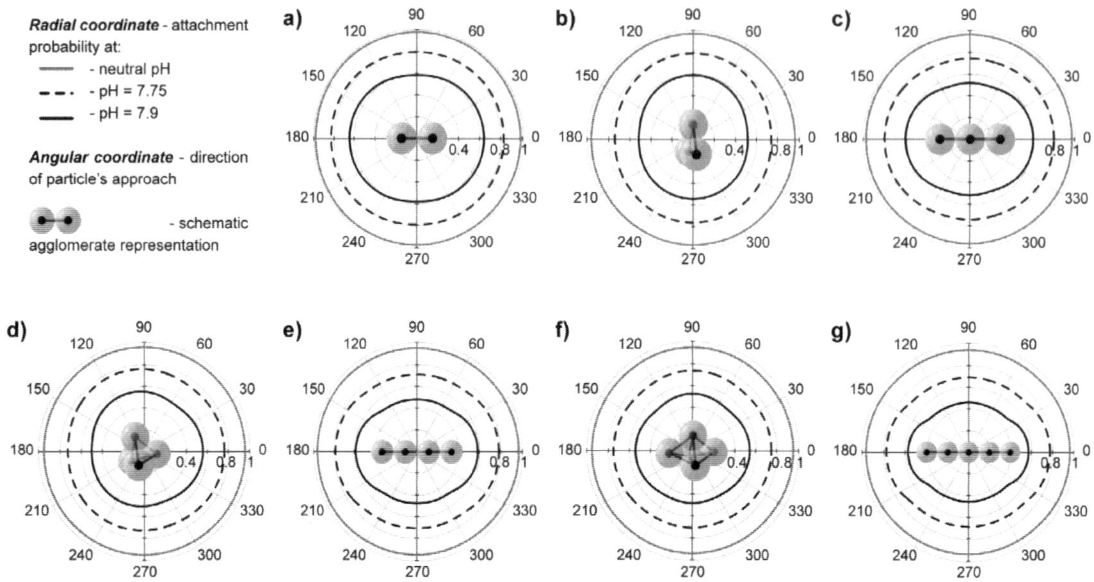

Figure 2. Probability of a nanoparticle attaching to an agglomerate of a specific morphology as a function of the particle's approach direction. The nanoparticle is moving in a plane with a selected orientation relative to the agglomerate. The considered agglomerate configurations are: a) 2-particle agglomerate; b) compact 3-particle agglomerate; c) linear 3-particle agglomerate; d) compact 4-particle agglomerate; e) linear 4-particle agglomerate; f) compact 5-particle agglomerate; g) linear 5-particle agglomerate.

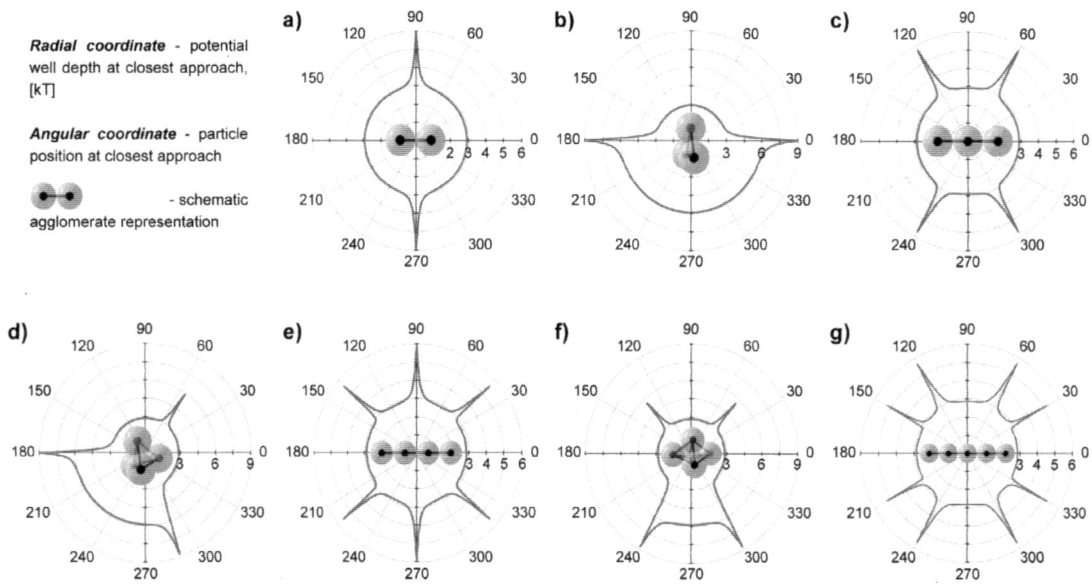

Figure 3. Potential well depth for the nanoparticle attached to the agglomerate of a specific morphology as a function of the attachment site position. The attachment sites lying in a plane with a selected orientation relative to the agglomerate are considered. The agglomerate configurations are: a) 2-particle agglomerate; b) compact 3-particle agglomerate; c) linear 3-particle agglomerate; d) compact 4-particle agglomerate; e) linear 4-particle agglomerate; f) compact 5-particle agglomerate; g) linear 5-particle agglomerate.

According to figure 3 the depth of the potential well is maximum for the attachment sites that have as many as possible particles in close contact. It is important to note that the van der Waals potential wells corresponding to the most favourable attachment sites are highly localized. Away from these sites the attractive potential energy holds constant. This means that the van der Waals attraction will not give rise to the compacting force acting on the attached particle unless it is in a close proximity with the favourable site.

Figure 3 also shows important results that arise from the absolute values of the attraction potential wells' depths. Namely, the sites in direct contact with only one particle of the agglomerate are associated with the potential well energy of $3kT$. Therefore, a particle can escape such a well due to solvent molecules collisions. Potential wells for the sites in contact with two and more particles are deep enough to be considered a permanent attachment.

Thus, attachment sites with a single particle in contact, which were shown as the most energetically favourable with regards to the electroosmotic repulsion, are unstable due to the surface-bound Brownian motion of the attached nanoparticle. The said chaotic motion can either cause the detachment of the particle, or position it near one of the more favourable van der Waals sites and cause the agglomerate compaction.

As a result, according to the proposed model the predominant type of agglomerates in the system with high nanoparticle surface charge will be determined by the balance between the initial linear structure formation and their further gradual deagglomeration or compaction due to the surface-bound Brownian motion of the constituent particles and the close-range van der Waals repulsion. The model also predicts the decrease of linear agglomerate fraction in the systems with smaller particle surface charges.

Acknowledgments
This work was financially supported by the Government of Russian Federation, Grant 074-U01.

References

[1] Shipway A N *et al* 2000 *ChemPhysChem* **1** (1) 18
[2] Talapin D V *et al* 2010 *Chem. Rev.* **110** 389
[3] Li W *et al* 2013 *J. Am. Chem. Soc.* **135** 4664
[4] Gur I *et al* 2007 *Nano Lett.* **7** (2) 409
[5] Kim S-H *et al* 2011 *NPG Asia Mater.* **3** (1) 25
[6] Alfimov A V *et al* 2014 *J. Phys.: Conf. Ser.* **541** 012063
[7] Alfimov A V *et al* 2015 *Proceedings of SPIE* **9519** 951911
[8] Berg J M *et al* 2009 *Nanotoxicology* **3** (4) 276
[9] Ohshima H 1995 *Journal of Colloid and Interface Science* **174** (1) 45

Modelling the growth of porous alumina matrix for creating hyperbolic media

E M Aryslanova[1], A V Alfimov[1], S A Chivilikhin[1]

[1]ITMO University, Kronverkskiy Ave, 49, St. Petersburg, 197101, Russia

E-mail: elizabeth.aryslanova@gmail.com

Abstract. Porous aluminum oxide is a regular self-assembled structure. During anodization it is possible to control nano-parameters of the structure using macroscopic parameters of anodization. Porous alumina films can be used as a template for the creation of hyperbolic media. In this work we consider the anodization process, our model takes into account the influence of layers of aluminum and electrolyte on the rate of growth of aluminum oxide, as well as the effect of surface diffusion. As a result of our model we obtain the minimum distance between centers of alumina pores in the beginning of anodizing process. We also present the results obtained by numerical modelling of hyperbolic media based on porous alumina film.

1. Introduction

Currently, the development of nanotechnology and metamaterials requires the ability to obtain regular self-assembled structures with different parameters [1-2]. One such structure is porous alumina film, which is a self-organizing structure that consists of hexagonally packed cylindrical pores (Figure 1). Artificially on the surface of the aluminum may be built a thick layer of porous aluminum oxide film. During anodization it is possible to control nano-parameters of the structure (such as diameter of pores and distance between them) using macroscopic parameters of anodization (pH of electrolyte, anodizing voltage, temperature, anodization time). Highly ordered pores may be obtained using two-stage anodization process, proposed in 1995 by Masuda et al. [3]. Pore size can vary from 2 to 350 nm and the distance between pores from 5 to 50 nm.

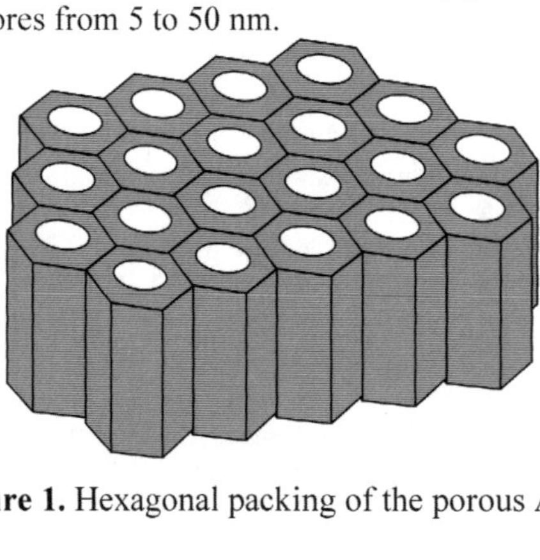

Figure 1. Hexagonal packing of the porous AAO

Anodizing is carried out in a vessel with electrolyte (aqueous solution of acid: oxalic, phosphoric acid; chromium, etc.), where are placed anode (aluminum) and cathode (inert conductive material), which are respectively connected to the positive and negative power supply output (Figure 2).

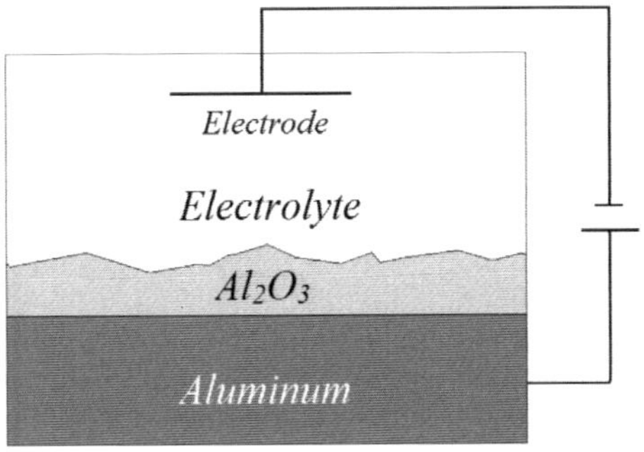

Figure 2. Anodizing process

On the surface of the aluminum grows thin dense alumina film – barrier layer. Barrier layer grows due to the migration of aluminum ions towards oxygen ions. The thickness of this barrier layer (0.01 – 0.1 nm) does not change throughout the process as it dissolves at the outer side exposed to the electrolyte [4].

The nucleation of pores occures in the centers of inhomogenities of aluminum surface, the oxide dissolves more intense the higher the inhomogeneity of the field. The thicker the structure, the more uneven grows the pore walls.

2. Model

In this paper we describe an analytical model of the growth of anodic alumina. We consider the motion of the interfaces between the electrolyte-Al_2O_3 (dissolution) and between the Al_2O_3-aluminum (oxidation) (Figure 2).

In each of the areas under consideration Laplace equation for the electric potential is solved. The process of growth of porous alumina is described by the theory of small perturbations. In zero approximation boundaries are considered flat and the speed of their movement proportional to the current density at these boundaries. The first approximation takes into account the small perturbations of the interface, which lead to small changes in the potentials and currents at these boundaries. The evolution of small perturbations of the interface is defined as a perturbation of the current density at the borders, and the process of surface diffusion.

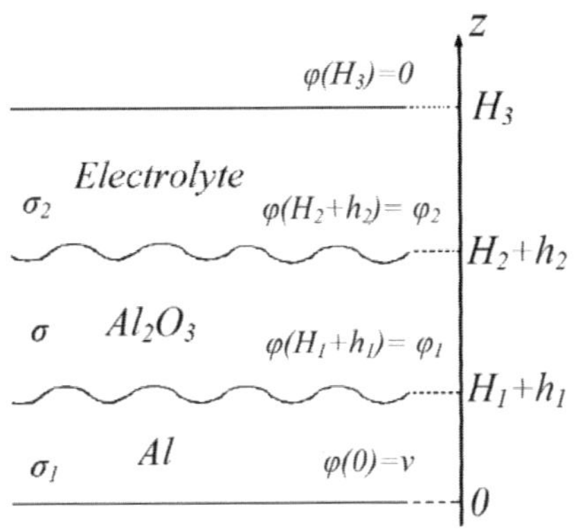

Figure 3. Inhomogeneous film with rough boundaries.

Using conditions of continuity of the current density at both interfaces we obtain the system of equations relating the potential disturbance on the interfaces:

$$\begin{cases} \left(-\sigma_2 sh(kH) - \sigma ch(kH)\right)\hat{\varphi}_{2\bar{k}} + \sigma\hat{\varphi}_{1\bar{k}} = \beta e^{kH} - \alpha \\ -\sigma\hat{\varphi}_{2\bar{k}} + \left(\sigma_1 sh(kH) + \sigma ch(kH)\right)\hat{\varphi}_{1\bar{k}} = \beta - \alpha e^{kH} \end{cases} \quad (1)$$

where $\alpha = (v/\vartheta)h_{1\bar{k}}$, $\beta = (v/\vartheta)h_{2\bar{k}}$. $\hat{\varphi}_{\bar{k}}^{r}$ - the Fourier transform of the potential disturbance to the coordinates x, y, $\varphi = \varphi^0_{(z)} + \hat{\varphi}$, $\varphi^0_{(z)}$ - stationary potential for the homogeneous problem, $\hat{\varphi}$ - potential oscillations, ϕ_1 – electrostatic potential of the aluminum layer on Al-Al_2O_3 interface, ϕ_2 – electrostatic potential of the aluminum layer on electrolyte-Al_2O_3 interface, σ_1 – conductivity of the aluminum, σ – conductivity of aluminum oxide, σ_2 – conductivity of the electrolyte, $\vartheta = \left(H/\sigma + H_1/\sigma_1 + (H_3 - H_2)/\sigma_2\right)$, v – anodization voltage, H – thickness of aluminum oxide, H_1+h_1 – thickness of the aluminum layer, $H_3-(H_2+h_2)$ - thickness of electrolyte, $h_{1\bar{k}} = \int_{-\infty}^{\infty} \exp[-ik_y y] \int_{-\infty}^{\infty} \exp[-ik_x x] h_1 dxdy$, $h_{2\bar{k}} = \int_{-\infty}^{\infty} \exp[-ik_y y] \int_{-\infty}^{\infty} \exp[-ik_x x] h_2 dxdy$.

We considered the solutions of the system (1) for different values of kH.

The rate of change of small perturbations of the boundaries Al-Al_2O_3 and Al_2O_3-electrolyte without the influence of surface diffusion is proportional to the perturbation of the current density at these interfaces $dh/dt = a\sigma\, d\varphi/dz$. Since the conductivity of the electrolyte and alumina is small compared with aluminum we see that the perturbation on the interface of Al_2O_3-electrolyte increases indefinitely with time.

The rate of change of small perturbations of the interfaces Al-Al_2O_3 and Al_2O_3-electrolyte with the influence of surface diffusion is described by the relation $\frac{\partial h}{\partial t} = V + D\Delta^2 h$, where D – is the surface diffusion coefficient.

For Al-Al_2O_3 interface we get the wavelength corresponding to the limit of stability:

$$\lambda_1 = 2\pi\sqrt[3]{a_1 v/\vartheta D_1}. \quad (2)$$

All perturbations of this interface with wavelength $\lambda > \lambda_1$ are unstable.

For Al_2O_3-electrolyte interface we get the wavelength corresponding to the limit of stability:

$$\lambda_2 = 2\pi\sqrt[3]{a_1 v/\vartheta D_2}. \quad (3)$$

All perturbations of this interface with wavelength $\lambda > \lambda_2$ are unstable.

We assume that surface diffusion coefficient D_2 on the Al_2O_3-electrolyte interface is larger than the same coefficient on D_1 the Al-Al_2O_3 interface. According to (2), (3), the correspondent critical wavelength satisfies of inequality $\lambda_2 < \lambda_1$. Therefore, for perturbations with wavelength λ in interval $\lambda_2 < \lambda < \lambda_1$ the upper boundary of alumina layer is unstable and bottom boundary is stable. The perturbations of upper boundary in this interval of wavelength are the source of porous formation. For wavelength $\lambda > \lambda_1$ the bottom boundary of alumina is unstable. The evolution of perturbations in this wavelength region determines the irregularities of porous structure.

3. Numerical investigation of hyperbolic material based on porous alumina template

We conducted numerical experiments using CST Microwave Studio. A model of an ordered system of silver nanorods with 10 nm in diameter was created.

Numerical experiments were conducted at a frequency of radiation 750 THz corresponding to the wavelength of 400 nm. The value of the dielectric constant of alumina 1.3365 for the selected frequency was specified in tabular form in accordance with the database of the company SOPRA SA [5].

In the first numerical experiment the dipole radiator was placed in the center of the medium, and the distribution of the magnetic field emitted by electromagnetic waves was detected; the result of this series is shown in Figure 4a. In the second series the medium was radiated by the dipole source and the refraction of light was observed, results of the second experiment are illustrated in Figure 4b.

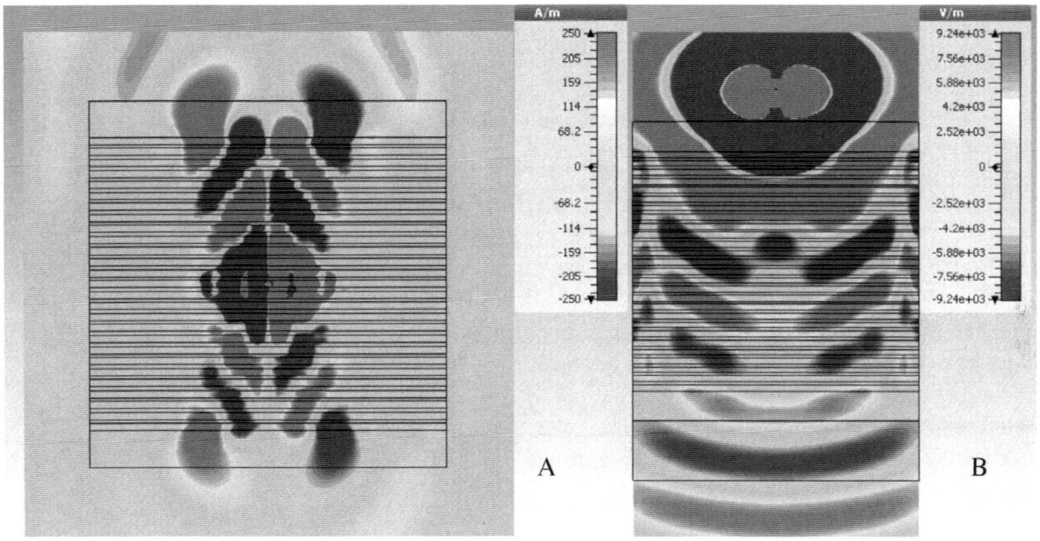

Figure 4. Spatial distribution of the intensity of magnetic field (a) and the electric field (b) of the radiation of a dipole placed in the composite material with silver rods with the 10 nm radiuses placed in porous alumina.

From the Figure 4a it is seen that in the present nanocomposite with the 10 nm rods the spatial distribution of the dipole radiation is characteristic to a hyperbolic media [6].

Figure 4b shows that the silver rods with a radius of 10 nm grown in porous alumina plate begin to act as a collecting lens, indicating a negative refraction and hence the occurrence of hyperbolic environment.

4. Conclusion

As a result of the developed model we obtained the minimum distance $2\pi\sqrt[3]{a_1 kv/\vartheta D_2}$ between centres of aluminum oxide pores in the beginning of anodizing process. The irregularities of porous structure have wavelength $\lambda > 2\pi\sqrt[3]{a_1 kv/\vartheta D_1}$.

The dependence of distance between centers of inhomogeneties on anodizing voltage is shown in Figure 5. It follows that the bigger the anodizing voltage, the less smooth the surface should be at the beginning of anodization process.

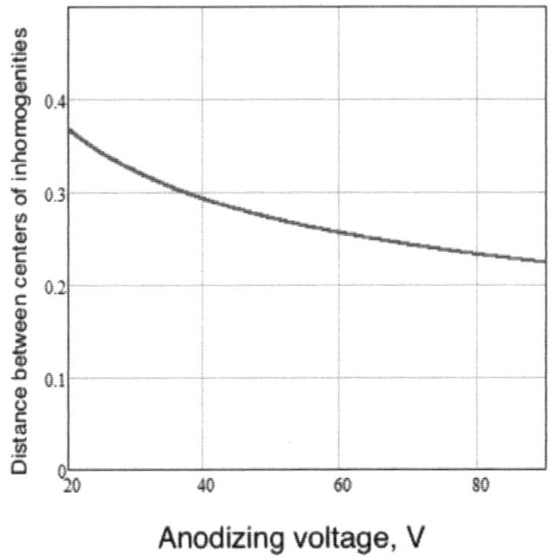

Figure 5. Dependence of distance between centers of inhomogeneties on anodizing voltage

A numerical model of a hyperbolic media grown in a porous alumina template with the pore diameter of 10 nm was created. This model shows that silver rods grown in porous alumina matrix with pore diameter of 10 nm begin to act as a collecting lens at the radiation wavelength of 400 nm, indicating a negative refraction and hence the occurrence of hyperbolic environment.

Acknowledgments
This work was financially supported by the Government of Russian Federation, Grant 074-U01.

References

[1] Limonov A G 2010 *Matem. Mod.* **22** 97–108
[2] Petukhov D I, Napolskii K S and Eliseev A A 2012 *Nanotechnology* **23** 335601
[3] Masuda H, Fukuda K 1995 *Science*, **268(9)** 1466-8
[4] Patermarakis G 2009 *Journal of Electroanalytical Chemistry* **635** 39–50
[5] Sopra S A dispersion database: http://www.sspectra.com/sopra.html
[6] Simovski C R, Belov P A, Atrashchenko A V, Kivshar Yu S 2012 *Advanced Materials* **24** 4229–48

Dynamics recording of holographic gratings in a photochromic crystal of calcium fluoride

Vladimir N Borisov, Ekaterina V Barausova, Andrey V Veniaminov, Alexandr E Andervaks, Alexandr S Shcheulin and Alexandr I Ryskin
ITMO University, 107101 St.Petersburg, Russia

E-mail: veniaminov@phoi.ifmo.ru

Abstract. Dynamics of diffraction efficiency was monitored during recording a holographic grating in additively coloured CaF_2 photochromic crystal at 180-200°C. Reciprocity failure revealed in the study was attributed to diffusion playing the crucial role in grating formation: recording at larger laser power goes faster but requires more energy. The efficiency of a recorded hologram is found to depend on the temperature; maximum diffraction is measured at the temperature far below that of recording, supposedly because of dramatic distortions suffered by the crystal along with exposure.

1. Introduction

Photochromic crystals of additively coloured calcium fluoride (fluorite) demonstrate extreme stability and spectral/spatial selectivity of holograms, but they are hard to manufacture and require extreme temperature and exposure time/energy for recording. That is why fluorite-based light-sensitive materials have poor chances to compete with photopolymers and silver halide media in mass production of holograms but, especially keeping in mind fluorite as an excellent optical material for visible and IR range, they provide a reliable basis for high-precision applications of diffractive optical elements such as metrology (e.g. holographic prisms [1]), spectroscopy (supernotch spectral selectors), etc.

The mechanism of hologram formation in fluorite crystals is based on photoionization, displacement and transformation of colour centres, in general similar to alkali halide crystals [2]. The present work is aimed at understanding the processes involved in holograms evolution in the course of exposure, and revealing the changes of their properties.

2. Experimental

Transmission-type holographic gratings with spatial periods 5 and 15 μm were recorded at different temperatures from 180 to 200°C using the second harmonic emission of a CW DPSS Nd:YAG laser (532 nm) in the bulk of a 1.2×10×20 mm plate cut from a CaF_2 crystal additively coloured in sodium vapour. During the experiments, the plate remained in a quartz-windowed metal frame whose temperature was stabilized within ±0.1°C. To monitor the diffraction efficiency in the course of exposure, one of the two recording beams was blocked with an electro-mechanical shutter, and the intensity of the second beam diffracted in the direction of the first one was measured.

The experiments could last for several hours and even days, therefore perfect stability of the interference pattern was checked using video recording of its microscopic image that did not reveal even a tiny shift.

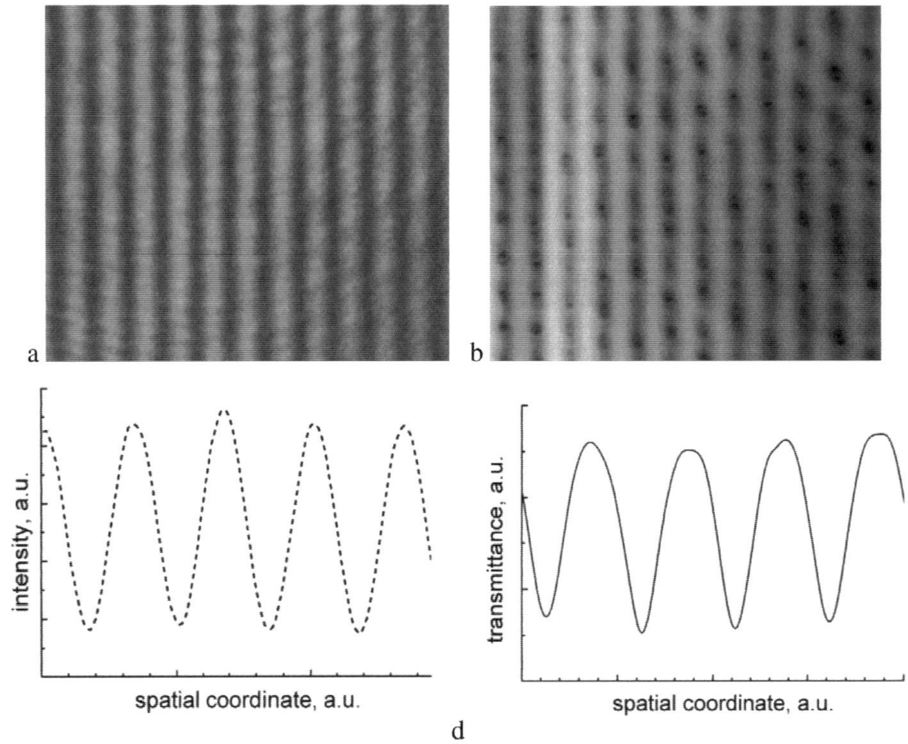

Figure 1. Microscopic images of the interference pattern (a) and the grating it produced when applied to the crystal (b); their spatial profiles (c and d, respectively).

3. Results and discussion

Although the shape of interference pattern was found expectedly almost sinusoidal, once imprinted into the crystal, it turned into a distorted distribution of colour centres with dark strips (enriched with the centres, supposedly corresponding to bright stripes of the fringe) notably narrower than those with lower concentration of the centres (figure 1). Such an intrinsically non-sinusoidal grating shape is attributed to the specific diffusion-drift mechanism of gratings formation, though nonlinear recording itself resulting from concentration saturation can be found in different materials. Another difference between the light pattern and its material record in the crystal is a sort of grainy structure of dark stripes that was recently discovered and discussed in terms of self-organization[3].

An exemplary plot of diffraction efficiency as a function of time / exposure is shown in figure 2. The diffraction efficiency may exceed 25%, however at the cost of very long-lasting and energy-inefficient recording.

In many experiments more complex nonmonotonous recording kinetics were observed, partially being a payment for the simple scheme of the experiment with the same wavelength used for recording and measurements. In this case, holograms are inevitably neither purely phase nor amplitude, providing most complex dynamics and selectivity properties.

3.1. Reciprocity failure

Brighter laser light makes holographic gratings develop faster, however the rate does not scale with laser power; in other words, reciprocity (Bunsen-Roscoe) law is not obeyed. In figure 3, the recording dynamics is presented as diffracted intensity vs. time (a) and exposure energy (b). The latter plot shows relatively more efficient, though requiring more time, growth of diffraction efficiency at lower power, in agreement with the idea of crucial role of diffusion – with finite speed – in grating formation. The rate of diffusive process involved could be evaluated by comparison of recording dynamics at different spatial periods; in this work, the diffusion coefficient of colour centres at 190°C was estimated as 200±50 nm^2s^{-1} that is one or two decades below that reported[4] for KBr at 200°C.

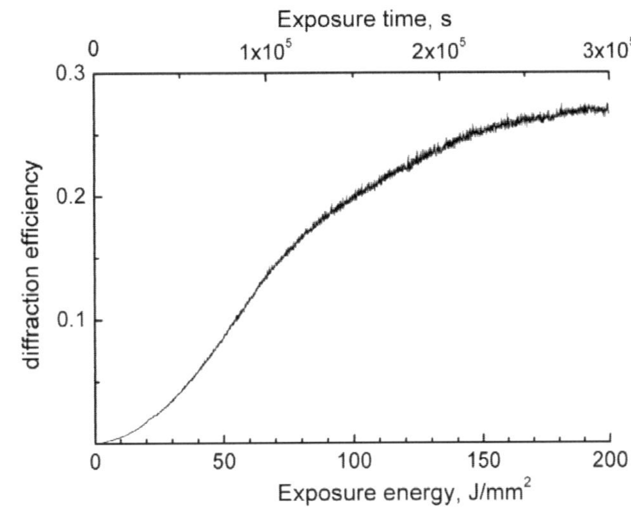

Figure 2. Illustrative long-lasting recording dynamics: diffraction efficiency at the recording wavelength 532 nm as a function of time / energy.

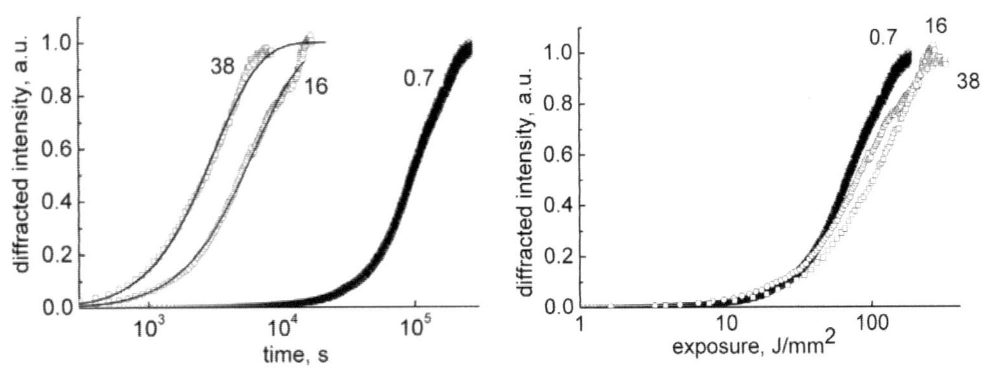

a b

Figure 3. Check of photochemical reciprocity: recording dynamics at different power density values (0.7, 16, and 38 mW/mm^2) represented as intensity of diffracted light (proportional to diffraction efficiency) vs. time (a) and total energy of laser light passed through the crystal (b). Though at large power recording goes faster, it requires more energy.

3.2. Impact of temperature: crystal swelling?

Once recorded in a fluorite crystal, holograms survive long-lasting exposure to severe heat and intense light, though they do not necessarily remain unchanged, due to various photothermal transformations of colour centres constituting them[5]. The diffraction efficiency changes with temperature in a reversible manner because of thermal expansion/shrinkage and corresponding variations of refractive index leading to Bragg detuning. One could expect the recovery of most efficient diffraction at the temperature equal to that of recording; however this is not the case. Thus, in the example presented in figure 4, the grating recorded at 180°C exhibits maximum diffraction at about 140°C. The modulation of intensity especially apparent at the temperatures above 130-140°C results from the interference of laser beams reflected from the front and back surfaces of the crystal; its period is compatible with known thermal coefficients.

The phenomenon that looks native rather for polymeric than for crystalline media, can be explained through "swelling" of the crystal caused by spatial redistribution of colour centres along with formation of a hologram – so dramatic swelling (about 10^{-3}) that cooling down by as much as 40°C is needed to compensate it. In practice, this means both desirable thermostatting of volume holographic optical elements and a convenient approach to fine tuning them.

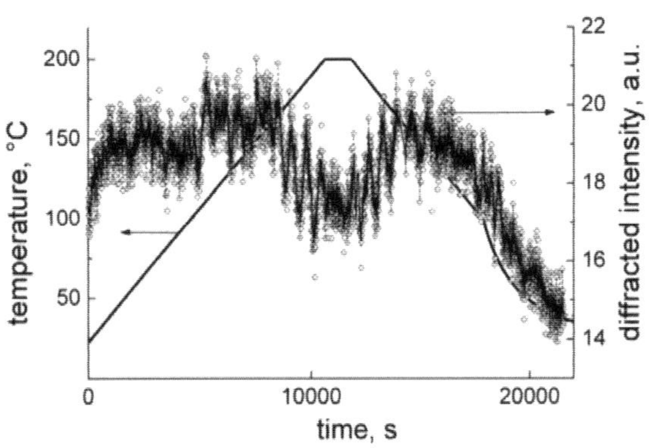

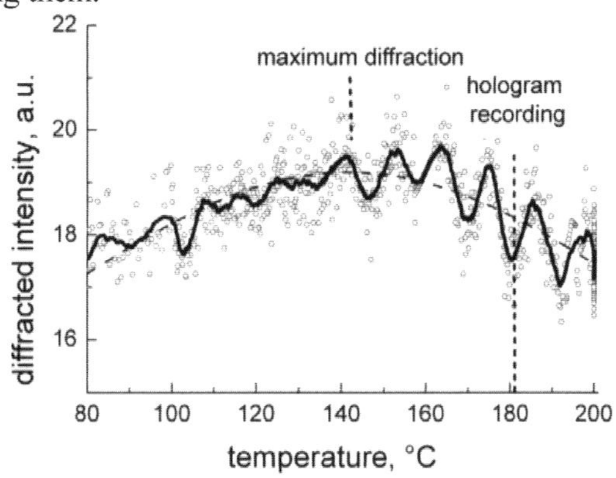

Figure 4. Original data: the temperature and the corresponding diffracted intensity as functions of time (a); the intensity of diffracted light plotted as a function of temperature (b). The solid line in (b) results from averaging the data (dots) over 20 adjacent points. The modulation is because of interference; the temperature of maximum diffraction (140°C) is found by Lorentz approximation of the data (dashed line).

References

[1] Granovskii VA, Kudryavtsev MD, Ryskin AI and Shcheulin AS 2009 Holographic prism as a new optical element: I. Principle of operation and experimental implementation *Opt. Spectrosc.* **106** 774

[2] Popov AY, Belous VM, Mandel VE, Shugailo YB and Tyurin AV 1999 Drift model of photoinduced processes in alkali-halide crystals during volume hologram recording *Proc. SPIE.* **3904** 195

[3] Shcheulin AS, Angervaks AE, Veniaminov AV, Zakharov VV, Fedorov PP, Kuznetsov SV and Ryskin AI 2015 Self-Organization of Color Centers in Holograms Recorded in Additively Colored CaF_2 Crystals *Opt. Mater.* **47** 190

[4] Salminen O, Riihola P, Ozols A and Viitala T. 1996 Spatially resolved optical studies of F-center diffusion in KBr crystals *Phys. Rev. B.* **53** 6129

[5] Shcheulin AS, Angervaks AE, Veniaminov AV, Zakharov VV and Ryskin AI 2014 Holograms convertible by an incoherent photo-thermal treatment in CaF_2 crystals with color centers *JOSA B* **31** 248

Electro-physical characteristics of MIS structures with HgTe-based single quantum wells for optoelectronics devices

S Dzyadukh, S Nesmelov, A Voitsekhovskii, D Gorn
Radiophysics faculty, Nanoelectronics and Nanophotonics Laboratory, Siberian physical-technical institute, Tomsk state university, 36 Lenina Av., 634021 Tomsk

E-mail: gorn_dim@sibmail.com

Abstract. The paper presents brief research results of the admittance of metal-insulator-semiconductor (MIS) structures based on $Hg_{1-x}Cd_xTe$ grown by molecular-beam epitaxy (MBE) method including single HgCdTe/HgTe/HgCdTe quantum wells (QW) in the surface layer. The thickness of a quantum well was 5.6 nm, and the composition of barrier layers with the thickness of 35 nm was close to 0.65. Measurements were conducted in the range of temperatures from 8 to 200 K. It is shown that for structure with quantum well based on HgTe capacitance and conductance oscillations in the strong inversion are observed. Also it is assumed these oscillations are related with the recharging of quantum levels in HgTe.

1. Introduction

HgCdTe MBE method provides new opportunities to optimize the characteristics of highly sensitive infrared detectors by growing heteroepitaxial structures based on HgCdTe with a pre-defined distribution of stochiometric composition in epitaxial films. Today the most common is creation of varyband layers with increased stochiometric composition of CdTe near the borders of the epitaxial layer. It allows reducing the influence of surface recombination on the charge carrier lifetime in the bulk of the epitaxial film, as well as to reduce the series resistance and improve the threshold characteristics of photodiodes based on p-n-junctions. Investigations of characteristics of MIS structures based on HgCdTe with inhomogeneous distribution of stochiometric composition are actual because of the need in passivation of infrared optoelectronic devices based on HgCdTe (laser diodes based on heterostructures or surface-emitting lasers [1], far-infrared range photodetectors [2]).
The aim of this scientific work is the investigation of electrophysical characteristics of MIS structures based on heteroepitaxial HgCdTe with a single HgCdTe/HgTe/HgCdTe quantum well by admittance methods in wide frequency (1 kHz till 2 MHz) and temperature (8 K to 300 K) ranges.

2. Experimental methods and samples

Investigated MIS structures were fabricated by molecular-beam epitaxy method on the basis of mercury cadmium telluride on the substrates of GaAs (013) in Institute of Semiconductor physics of Siberian branch of Russian Academy of Sciences (Novosibirsk, Russia). Some of the investigated structures did not include a quantum well in the surface region. A structure with a single quantum well had a CdTe composition in the $Hg_{1-x}Cd_xTe$ barrier layer of $x = 0.65$, and the quantum well thickness was 5.6 nm. Barrier layers had a thickness of 35 nm. A low-temperature double-layer SiO_2/Si_3N_4As dielectric applied over the CdTe 40 nm layer was used as the dielectric coating. Figure 1 shows the basic layout of the layers in the structure of the quantum well, and Figure 2 shows the distribution of CdTe composition in the $Hg_{1-x}Cd_xTe$ epitaxial film for this structure.

Measurements were carried out with the use of an automated admittance spectroscopy installation based on non-optical Janis cryostat and Agilent E4980A immittance meter. The measurements were conducted with the change in voltage from negative to positive values (forward sweep scan direction measurements).

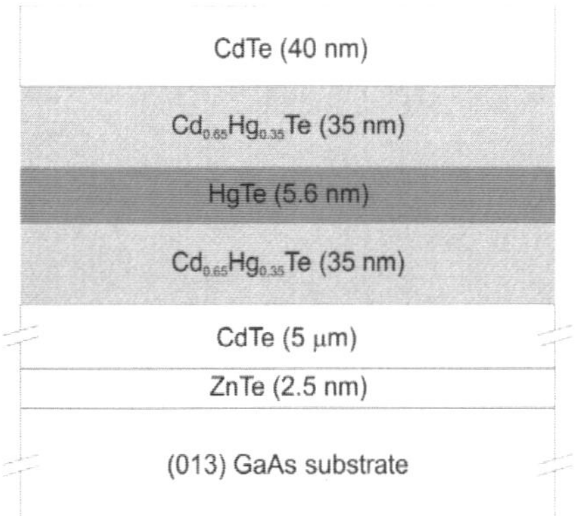

Figure 1. Scheme of the layers position in structure with single quantum well with the thickness of 5.6 nm.

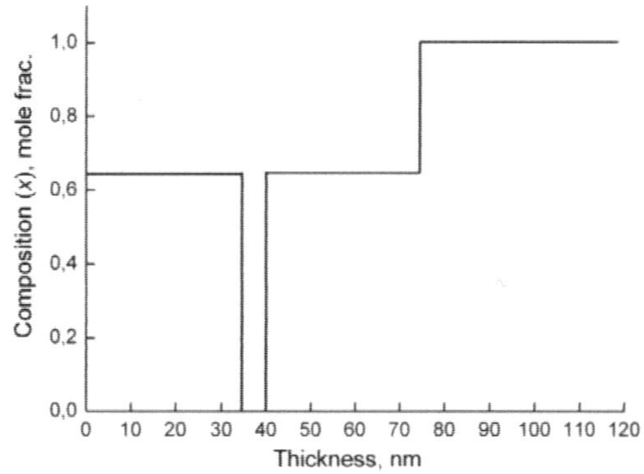

Figure 2. Composition distribution in operating layer of $Hg_{1-x}Cd_xTe$ in the structure with single quantum well with the thickness of 5.6 nm.

3. Results and discussion

Figures 3 and 4 show dependencies of capacitance and differential conductance on bias voltage for MIS structure with 5.6 nm single quantum well measured at a temperature of 7.7 K for various frequencies of alternating test signal. From Fig. 3 and 4 it may be seen that HgCdTe has a p-type conductivity, and in the area of strong inversion at the range of frequencies from 500 Hz to 100 kHz non-monotonic change in capacitance and differential conductance is observed.

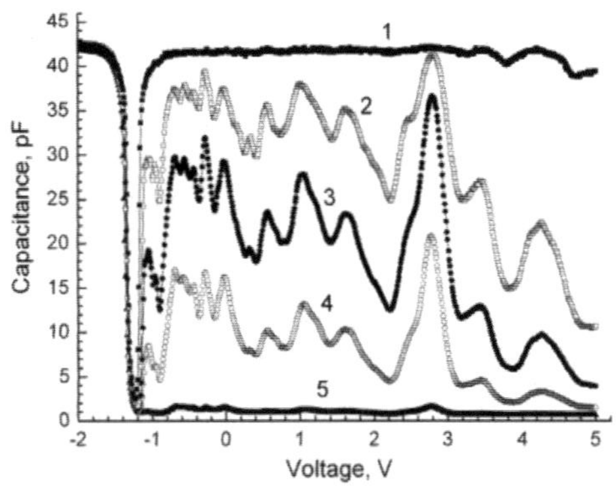

Figure 3. Capacitance-voltage characteristics of MIS structures based on HgCdTe with quantum well measured at 7.7 K under the forward direction voltage sweep at different frequencies: 1 – 500 Hz, 2 – 5 kHz, 3- 10 kHz, 4 – 20 kHz, 5 – 100 kHz.

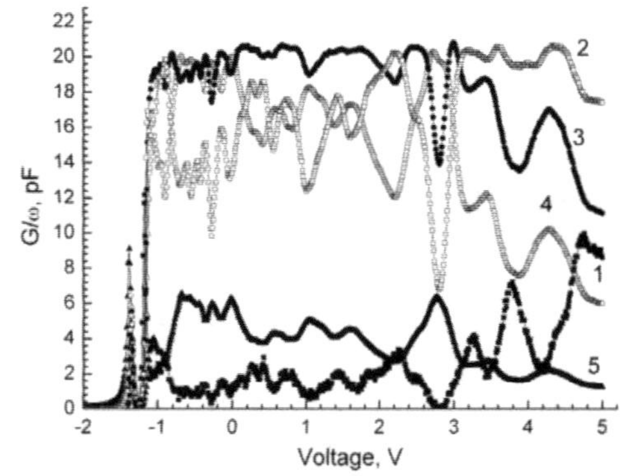

Figure 4. Dependence of differential conductance on voltage for MIS structures based on HgCdTe with quantum well measured at 7.7 K under the forward direction voltage sweep at different frequencies: 1 – 500 Hz, 2 – 5 kHz, 3- 10 kHz, 4 – 20 kHz, 5 – 100 kHz.

Journal of Physics: Conference Series **735** (2016) 012012 doi:10.1088/1742-6596/735/1/012012

Capacitance maxima in this range of frequencies are observed at the same bias voltages, but for differential conductance with the change of frequency maxima can be replaced by minima for some bias voltages (Fig. 4). At 500 Hz capacitance-voltage characteristic is of low-frequency type and maxima become less pronounced than at frequencies of 5-20 kHz. At 100 kHz capacitance-voltage characteristic is close to high-frequency type and capacitance maxima also appear less clearly, than at frequencies of 5-20 kHz. The capacitance value at 100 kHz in strong inversion is close to 0.7 pF.

Figure 5 shows capacitance-voltage characteristic for a MIS structure with a quantum well measured at 10 kHz under the forward direction voltage sweep at different temperatures in the range of temperatures from 7.8 to 77 K. From fig. 5 it is seen that capacitance maxima in strong inversion are pronounced at low temperatures, and their amplitudes and position on the voltage axis change when increasing the temperature from 7.8 to 10 K. In the range of temperatures from 30 to 77 K maxima become diffused and are pronounced weakly.

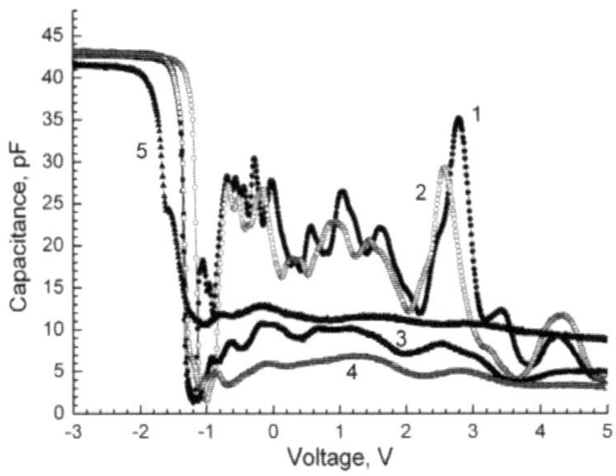

Figure 5. Capacitance-voltage characteristics of MIS structures based on HgCdTe with quantum well measured at 10 kHz under the forward direction voltage sweep at different temperatures: 1 – 7.8 K, 2 – 10 K, 3 – 30 K, 4 – 50 K, 5 – 77 K.

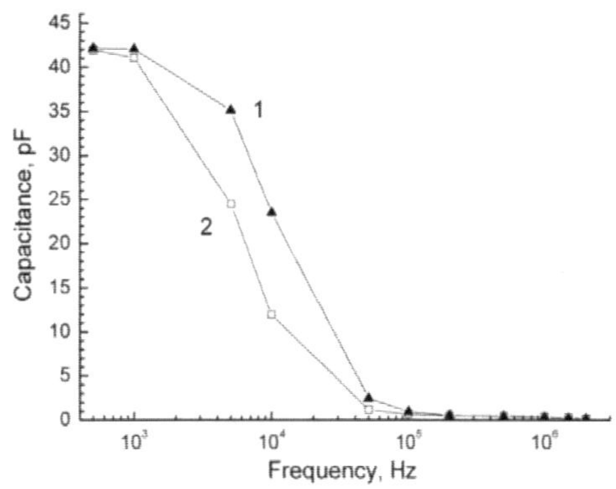

Figure 6. Frequency dependencies of capacitance of MIS structure with quantum well, measured at 8 K and different voltages: 1 – 1 V (maximum), 2 – 2 V (minimum).

Figure 6 shows capacitance dependencies on frequency for a MIS structure with a quantum well at 8 K and voltages corresponding to a maximum (1 B) and a minimum (2 B) on the capacitance-voltage characteristic. It is seen from figure 6 that a transition of capacitance-voltage characteristic from low-frequency to high-frequency type takes place with the change in frequency from 500 Hz to 100 kHz. Capacitance in the maximum at intermediate frequencies is significantly higher than in the minimum.

It must be noticed that other investigated MIS structures based on HgCdTe without quantum wells have capacitance-voltage characteristics that do not have non-monotonous changes of capacitance in strong inversion.

Thus, it is shown experimentally that capacitance maxima in strong inversion are related with the presence of a single HgTe quantum well in the structure. From the slope of capacitance-voltage characteristic in the depletion mode it is possible to define charge carrier concentration in HgCdTe, which is equal to 6.5×10^{17} m^{-3} at 7.7 K for the sample with quantum well, that is too small for the material with $x = 0.22$, but is suitable for wide band gap material with $x = 0.65$. Having known the majority charge carrier concentration it is possible to calculate capacitance value of a MIS structure in strong inversion, which was found to be 0.8 pF (experimental value is approximately 0.7 pF). The calculation of surface potential dependence on voltage for MIS structure based on HgCdTe containing quantum well is carried out.

We will briefly describe the most likely mechanism for the CVC maxima appearance in the strong inversion for the MIS structure comprising a single quantum well. Surface potential (as well as the potential in the field of quantum well) for p-HgCdTe depends on the bias voltage in the strong inversion because of the effects of degeneracy and conduction band nonparabolicity. When the Fermi level in the field of quantum well approaches to the level of quantization, quantization level recharges following a change of the AC test bias voltage. Capacity of dimensional quantization level in the quantum well contributes to the full capacity of the structure and appears at intermediate frequencies in the strong inversion. When a capacitance-voltage characteristic has a low-frequency view capacitance of the MIS structure in the strong inversion tends to the dielectric capacitance and maxima associated with dimensional quantization levels do not appear. When a capacitance-voltage characteristic has a high-frequency view the carrier concentration in the inversion layer does not have enough time to follow the test signal and contribution of the quantization levels capacitance at full capacity decreases because the quantum well is located at a distance of 35 nm from the interface. Because the charge carriers thermal energy increases with the temperature increasing dimensional quantization effects are mild.

Energies of dimensional quantization levels in a quantum well can be approximately defined by analogy with [3]. To do this, it is necessary to define the flat-band voltage in MIS structure comprising QW in terms of flat-band capacitance value and to obtain surface potentials corresponding to maxima of capacity in the strong inversion. Then it is necessary to build the potential spatial distribution in the HgCdTe surface layer and to find the potential under the given bias voltage in the quantum well. Knowing the position of the Fermi level in the flat-band mode, position of dimensional quantization levels can be find relative to the allowed energy bands in HgCdTe barriers. Crucial issue when using this method is to use the correct mathematical expressions for the main parameters of HgCdTe at low temperatures (i.e. densities of states, intrinsic concentration etc.), the proper construction of the ideal CV characteristics at low temperatures, as well as the correct numerical calculation of energy-band diagrams in the surface HgCdTe layer with taking into account the effects of degeneracy and nonparabolicity of the conduction band.

Thus the flat-band capacitance value is 33.56 pF in our case. Hence the flat-band voltage is -1.36 V. Band-gap energy of barrier layer is 0.838 eV. Fermi level location is ($E_v + 0.010$ eV) or ($E_c - 0.828$ eV). Then two first maxima of capacitance for quantum well MIS structure under the direct sweep bias voltage correspond to energies ($E_c - 0.520$ eV) and ($E_c - 0.210$ eV). Here E_c is the energy of conduction band edge of the barrier layer near the border with well.

Another way for estimation of dimensional quantization energies levels is the using of theoretical model described by us earlier in [4]. This model based on self-consistent solution of Poisson and Shrödinger equations for heterostructure comprising QW with taking into account real parameters of MBE-grown CdHgTe solid solution such as electron affinity and concentrations of ionized acceptors and donors. Calculation of dimensional quantization levels for structure under consideration at 7.7 K gives values ($E_c - 0.597$ eV) for first level of electrons E_{c1}, ($E_c - 0.347$ eV) for second level of electrons E_{c2}, ($E_c - 0.486$ eV) for first level of heavy holes E_{h1}, ($E_c - 0.517$ eV) for second level of heavy holes E_{h2}. Then two first maxima of capacitance observed on the CVC in the experiment may be associated with the presence of dimensional quantization levels. It was confirmed by estimation. But the accuracy of used calculation methods does not allow give more precise results and does not allow interpret each observed peak.

4. Conclusions

Thus the admittance of MIS structures based on MBE HgCdTe with a single 5.6 nm quantum well located in the surface layer of semiconductor was experimentally investigated. It is shown that the presence of a single quantum well thickness of 5.6 nm can lead to the appearance of capacitance-voltage and conductance-voltage characteristics peaks in the strong-inversion mode.

Capacitance maxima in the strong inversion mode are observed in an intermediate case between the low frequency and high frequency capacitance-voltage characteristics (5-50 kHz for a sample with a single quantum well). The capacitance maxima in strong inversion mode are particularly pronounced at low temperatures (8-10 K), although they are weakly apparent up to 77 K. Maxima in capacitance and conductance in strong inversion mode are not observed for samples without quantum wells. The dependences of the surface potential on voltage at various approximations are built. It is shown that considering the effects of degeneracy and non-parabolicity of the conduction band, the surface potential at strong inversion depends on the bias voltage. It is assumed that the dimensional quantization levels recharge capacitance in the quantum well contributes to the overall capacitance of the structure and appears in the capacitance-voltage characteristics in the strong inversion. An approximate method for determining the energy levels of quantum wells with the use of data obtained from the capacitance-voltage characteristics measurements is proposed.

5. Acknowledgements

This study (research grant No 8.2.10.2015) was supported by The Tomsk State University Academic D.I. Mendeleev Fund Program in 2015.

6. References

[1] Pautrat J.L., Hadji E., Bleuse J., Magnea N. 1997 *Journal of Electronics Materials* **26** (6) 667.
[2] Grein C.H., Jung H., Singh R., Flatte M.E. 2005 *Journal of Electronics Materials* **34** (6) 905.
[3] Voitsekhovskii A.V., Davydov V.N. Photoelectric MIS Structures Based on Narrow-Band Semiconductors, *Radio I Svyaz'*, 1990, Tomsk, 327 p.
[4] A.V. Voitsekhovskii, D.I. Gorn, I.I. Izhnin, A.I. Izhnin, V.D. Goldin, N.N. Mikhailov, S.A. Dvoretskii, Yu.G. Sidorov, M.V. Yakushev, V.S. Varavin. 2013 *Russian Physics Journal* **55** (8) 910.

The concentration mechanisms of cubic nonlinearity in dispersive media

V I Ivanov, G D Ivanova, S I Kirjushina and A V Mjagotin
Department of Physics and Theoretical mechanics, Far Eastern State Transport University, Khabarovsk, Serysheva street, 47, Russia

E-mail: naivi@rambler.ru

Abstract. The comparative analysis of the dynamic holograms recording efficiency in media with non-resonance mechanisms of optical nonlinearity is carried out. It is showed that the greatest values of this parameter are provided by thermodiffusion and electrostrictive mechanisms of concentration nonlinearity of liquid dispersed media.

1. Introduction

There are known many nonlinear mechanisms which are used to record dynamic holograms [1-5]. The highest values of the nonlinearity provide the resonance mechanisms - the saturation of absorption in gases, the saturation of interband transitions in semiconductors. However these nonlinearity mechanisms are limited by narrow spectral range. A non-resonant nonlinearity is free from this drawback and more promising in wide spectral range applications. The mechanisms of this type provide a sufficiently high efficiency of nonlinear interaction of light in both the visible and IR spectral ranges [1, 2].

In this paper we analyze a number of such mechanisms (and sensitive materials) in order to compare the limit characteristics of the dynamic holograms recording and identify the most promising materials for holography with the low-power laser sources.

2. The energy holographic sensitivity

It is customary to compare the different media and mechanisms of nonlinearity using a third-order nonlinear susceptibility, which characterizes the cubic nonlinear response. In the case of non-resonant non-linearity (for weakly absorbing media) is used often the other parameter - $n_2^{eff}[m^2/W]$, which characterizes the change of the refractive index of the medium under the influence of incident radiation:

$$n = n_0 + n_2^{eff} I, \tag{1}$$

where n_0 - the refraction index of the medium in the absence of radiation, I - radiation intensity, $n_2^{eff} = (dn/dI)$ - the coefficient of effective cubic nonlinearity. However this option does not allow a valid comparison of different media and mechanisms of nonlinearity. In most cases (and especially for thermal non-linearity) this coefficient is proportional to the magnitude of quasi-stationary relaxation time of the nonlinear response τ , i.e. it depends on the geometry of the system and experimental conditions. The diffraction efficiency of the hologram also depends on its thickness, which is limited by absorption or scattering losses, and the wavelength of the radiation.

Therefore we suggest the more informative complex parameter – the energy holographic sensitivity to assess the effectiveness of the recording of dynamic holograms [2]:

$$N_{2E} = 2\pi n_2^{eff} (\tau\alpha\lambda)^{-1} [m^2/J], \tag{2}$$

where α - the absorption coefficient of the medium, λ - light wavelength. For phase holograms the value of this parameter corresponds to the minimum energy of the recording radiation needed to change the optical thickness of the medium on the radiation wavelength and completely characterizes the diffraction efficiency of a thin hologram. Moreover this parameter dimension allows to compare the sensitivity of both volume and surface nonlinearity mechanisms [1]. For example, for reflection holograms the energy holographic sensitivity can be defined as:

$$N_{2E} = (d\rho / dI)\tau^{-1}, \tag{3}$$

where ρ - Fresnel amplitude reflection coefficient of the boundary between two media. Similarly, the sensitivity is determined for the amplitude thin holograms – in this case the amplitude reflection coefficient is replaced by the corresponding transmittance coefficient.

The most universal non-resonant nonlinearity that exists in any absorbing medium is the heat mechanism. Let us find the value of such sensitivity for a single component material. A simplest mechanism of thermal non-linearity is due to the thermal expansion of the medium. The expression for the coefficient of effective cubic nonlinearity can be obtained by solving the one-dimensional heat equation:

$$c_p \rho \partial T / \partial t = -\operatorname{div} J_1 + \alpha I_0 (1 + \sin Kx), \tag{4}$$

where $J_1 = -D_{11} \operatorname{grad} T$ - heat flow, T - temperature of the material, D_{11} - the thermal conductivity of the material, c_p and ρ - the specific heat capacity and density of the medium, $I = I_0(1 + \sin Kx)$ - the intensity spatial distribution along the layer of the medium (x-axis), K - the space wave vector of the elementary hologram. Using the stationary solution (3) for the amplitude of thermal lattice, we obtain [2]:

$$n_2^{eff} = \alpha \tau_T (dn/dT)(c_p \rho)^{-1}, \tag{5}$$

where (dn/dT) - the temperature coefficient of the refractive index, $\tau_T = D_{11}\mathrm{K}^{-2}$ - the thermal relaxation time. The holographic sensitivity parameter is obtained from (1):

$$N_{2E} = 2\pi (dn/dT)(c_p \rho \lambda)^{-1}. \tag{6}$$

Thus, the value of sensitivity for thermal nonlinearity of single-component fluids is defined by parameters (dn/dT), c_p and ρ. Substituting the characteristic parameters for semiconductor media $dn/dT = 2 \cdot 10^{-4} K^{-1}$, $(c_p \rho) \approx 1\ J/(cm^3 K)$, we obtain $N_{2E} \approx 1\ cm^2/J$ (for radiation with a wavelength of $\lambda = 10\ \mu m$).

However a light with an intensity of several mW/cm^2 is required the greater value of the coefficient of nonlinearity for effective dynamic hologram recording. Since the relaxation time (e.g., time diffusion concentration) increases with particle size increasing, it is natural to expect high nonlinearity parameters in nanodispersive media (colloids, suspensions, micro-emulsions) the particle size of which is limited by the wavelength of the radiation due to light scattering. The "nanodispersive" trend can be traced quite easily in the searching of large non-linearities. If in the first papers was used the nonlinear medium with a typical value $n_2^{eff} \approx 10^{-13}\ cm^2 / W$ (molecular Kerr nonlinearity), than was experimentally obtained values $n_2^{eff} \approx 10^{-8}\ cm^2 / W$ (artificial Kerr-like nonlinearity) [3].

3. The thermodiffusion and electrostrictive mechanisms of concentration nonlinearity

Multicomponent nanoparticulate medium (liquid phase mixture, suspension, emulsion) are characterized by a series of specific mechanisms nonlinearities that are not in single-component media. In particular, these include the concentration of non-linearity caused by the redistribution component in a two-phase medium in a laser field. The concentration flows in the medium can be caused by different mechanisms of interaction of radiation with matter.

A thermally induced mechanism of particle drift in a nonuniform temperature field is known as thermal diffusion (thermophoresis) in gases or Soret effect in liquid-phase binary mixtures [4, 5]. The microheterogeneous medium with different refractive index of components in the electromagnetic field is forced by electrostriction forces, which can also be the cause of concentration streams [2]. Depending on the sign of the polarizability the microparticles may be drawn (if the refractive index of the dispersed phase material is greater than the dispersion medium) or pushed out (otherwise) from areas with greater intensity of the electromagnetic wave.

Consider setting holographic sensitivity of the concentration non-linearity. In nanodispersive medium particle radius is much smaller than the radiation wavelength λ, the refractive index of the medium is proportional to the concentration of particles (for dilute systems):

$$n = n_1(1+\phi\delta), \tag{7}$$

where and $\delta = (n_2 - n_1)/n_1$; n_l *u* n_2 - the refractive indices of the substance and the dispersion medium of the dispersed phase, respectively; $\phi = (4/3)\pi r^3 C$ - the volume fraction of the dispersed phase, r - the radius of the microparticle, C - the concentration of nanoparticles. Then we have for each concentration mechanism of nonlinearity:

$$n_2^{eff} = (\mathrm{d}n/\mathrm{d}C)(\mathrm{d}C/\mathrm{d}I). \tag{8}$$

Let's consider an expression for the holographic sensitivity of the dispersion medium, taking into account both (thermodiffusion and electrostriction) concentration mechanisms. The system of balance equations for the concentration of particles and heat flow for this case is written as follows [2]:

$$c_p\rho dT / dt = -div\, J_1 + \alpha I_0(1+\sin Kx), \tag{9}$$

$$dC / dt = div(J_2 + J_3). \tag{10}$$

Here J_1 and $(J_2 + J_3)$ - thermal and concentration flows, respectively:

$$J_1 = -D_{11}\nabla T - D_{12}\nabla C, \tag{11}$$

$$J_2 = -D_{21}C\nabla T - D_{22}\nabla C, \tag{12}$$

$$J_3 = \gamma C\nabla I, \tag{13}$$

where D_{22} - diffusion coefficient of particles, D_{21} and D_{12} - coefficients describing cross-flows (the effect of thermal diffusion and Dufour-effect, respectively), J_3 - electrostrictive flow, $\gamma = (2\beta b / cn_l)$, ($\beta$ and b - polarizability and mobility of microparticle, respectively), c - the light velocity[2]. Assuming a one-dimensional problem, we seek the solution of equations (9-13) in the form of:

$$C(x,t) = C_0 + C_1(t)\sin Kx, \tag{14}$$

$$T(x,t) = T_0(t) + T_1(t)\sin Kx. \tag{15}$$

Here C_0 and T_0- the average particle concentration and temperature of the medium. Let's the amplitudes of the thermal T_1 and concentration C_1 profiles are small -$(C_1/C_0) << 1$, $(T_1/T_0) << 1$. The linearized system is easily solved and we have by taking into account (2) and (9-15):

$$N_{2E} = 2\pi n_1 \phi_0 \delta \left[D_{11}^{-1} D_{21} (c_p \rho)^{-1} - \gamma \alpha^{-1} K^2 \right] \lambda^{-1}. \tag{16}$$

The sensitivities of the concentration nonlinearity mechanisms of liquid-phase two-component medium are determined by the coefficient of thermal diffusion D_{21} (1-st term) and by microparticle polarizability in the case of the electrostrictive mechanism (2-nd term). For a non-absorbing nanodispersive medium the absorption coefficient should be replaced by the extinction coefficient. As can be seen from (16), both mechanisms may either strengthen or weaken each other depending on the sign of the thermal diffusion coefficient and polarizability of the dispersed particles. The evaluations from (6) and results of experimental studies [4-6] shows the value $N_{2E} \approx 10^2\ cm^2/J$ for both concentration nonlinear holographic sensitivity in the vicinity of the critical point of the micro-emulsion.

4. Conclusions

Thus, our analysis shows that the nanoparticulate liquid phases are characterized by high values of the holographic sensitivity. More over the dispersion media have another advantage. It is known that the main parameters of the nonlinear medium are usually interrelated: for example, the value of non-linearity usually corresponds to poor temporal resolution, spatial resolution and sensitivity of the holographic media are inversely related. Because the parameters of known materials form a discrete set, there is a problem of selection medium with optimized characteristics for particular applications. The most promising materials in this sense are also the nano-materials, because one can change the volume fraction of the various components and their composition, i.e. control (including in real time) of parameters.

Since there are known same mechanisms of nonlinearity in the nano-dispersive systems with the largest holographic sensitivity we can distinguish them as a promising new class of media for dynamic holography [2].

5. References

[1] Zeldovich B Y, Pilipeckij N F, Shkunov V V 1985 *Obrashhenie Volnovogo Fronta* (Moskow: Nauka).

[2] Ivanov V I 2006 *Thermo-induced Mechanisms of Dynamic Holograms Recording* (Khabarovsk: FESTU).

[3] Gower M G *Progr. in Quant. Electr.* 1984 **9** 100-147.

Ivanov V I, Livashvily A I, Lobov A I, Simakov S R 2006 *J. of Opt. Techn.*, **71(9)** 586-587.

[4] Leppla C, Wiegand S 2003 *Phil. Mag.* **83** 1989-1999.

[5] Ivanov V I, Okishev K N *Techn. Phys. Lett.* **32(11)** 967-968.

[6] Freysz E, Afifi M, Ducasse A, Pouligny B, Lalanne J R J *Phys. Lett.* 1985 **46** 181-187.

Journal of Physics: Conference Series **735** (2016) 012014 doi:10.1088/1742-6596/735/1/012014

Study of influence of the fiber optic coatings parameters on optical acoustic sensitivity

V S Lavrov, A V Kulikov, M U Plotnikov, M E Efimov, S V Varzhel
Saint-Petersburg National Research University of Informational Technologies, Mechanics and Optics, Saint-Petersburg, Russia

E-mail: vladimir8812@mail.ru

Abstract. The paper presents the optical fiber acoustic sensitivity dependence on the coating parameters and the thickness of coating layer. A comparison of data obtained from the theoretical research and experimental estimates of real samples sensitivity in air and water.

1. Introduction

Currently, fiber-optic technology is widM y implemented in hydroacoustics and geophysical measuring sistems[1,2]. The sensing element of phase interferometric acoustic sensors is optical fiber, when exposed to an acoustic field which changes the optical path of light propagating in its core.

One of the known methods of increasing the optical fiber acoustic sensitivity is applying coating to the fiber with certain parameters that affects acoustic sensitivity. This is caused by the fact that a coated fiber as in any composite structure more rigid components, namely the quartz core with Young's modulus is 20-30 times greater than that of the polymer coating takes a greater part of the radial load. In turn, the total axial load on the coated fiber is increased as compared with the uncoated fiber as many times as the cross-section is increased, but with a ratio of diameters of the coating material and fiber 20: 1 a saturation tendency of increasing sensitivity occurs[3].

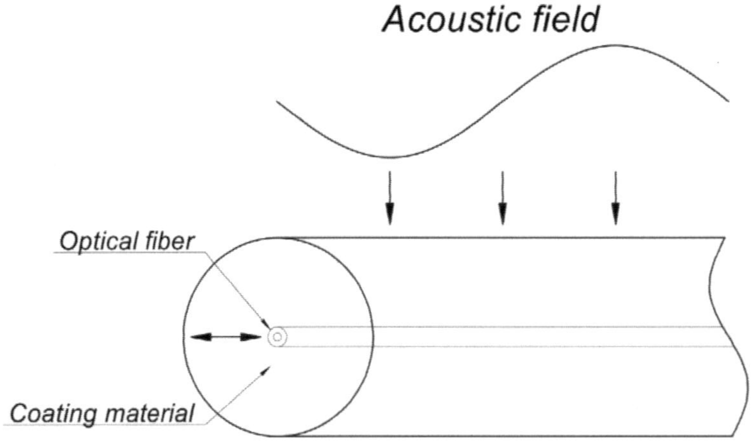

Figure.1 Coated optical fiber exposed to an acoustic field

The expression for the phase shift in response to the unit of applied pressure changes will have the form[2]:

$$\frac{\delta\Phi_s}{\delta p} \cong -\frac{\beta(1-2\mu)}{E}[1-\frac{n^2}{2}(2p_{12}+p_{11})]$$

Where $\delta\Phi_s$ phase shift, Δp the unit of applied pressure change, β the propagation constant, μ Poisson's ratio, E Young's modulus, n refractive index, p_{11},p_{12} elastooptical rates.

2. Methods

The purpose of this work is to study and identify the dependence of acoustic sensitivity of extended phase interferometric acoustic pressure sensor on the characteristics of the material and the thickness of the coating layer of the optical fibers in the air and water.

To achieve this purpose a theoretical study of the impact of the acoustic field on the optical path changing in the single-mode fiber.

Single phase interferometric acoustic pressure sensors were created and studied, which are single-mode fiber c two fiber Bragg gratings (FBG)[4] coated with polymeric material of different thickness, increasing the acoustic sensitivity of the optical fiber.

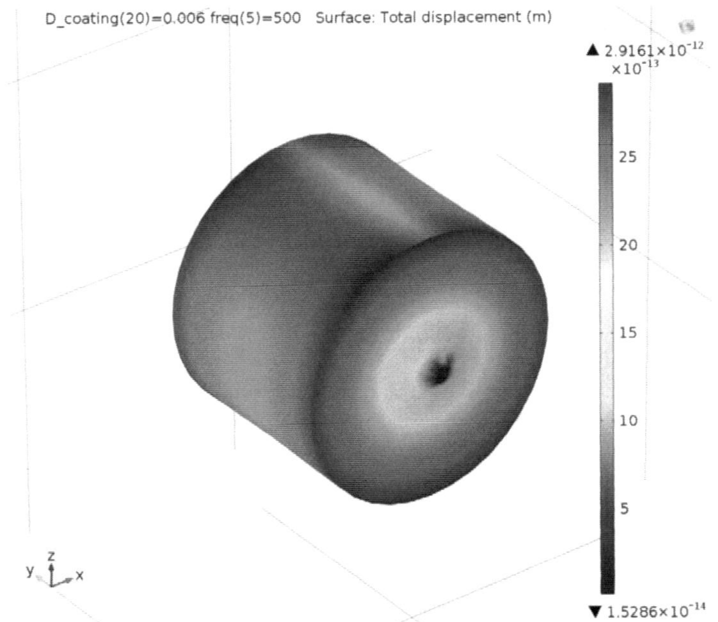

Figure 2. Modeling of the coated optical fiber exposed to an acoustic field

Measurements were carried out as follows: the light was inputted into optical fiber with two FBG that is partially reflected and partially transmitting by the first FBG.

Further, light beam reflected from the second FBG passed through the sensing element in the opposite direction to the reference arm of interferometer compensating optical path difference between light beams from the FBGs. The interference signal at the output of the interferometer incident on a specialized signal processing device, where demodulation algorithm of the interference signal has been implemented, which allows to restore the amplitude of the measured phase signal and estimate the parameters of the acoustic signal.

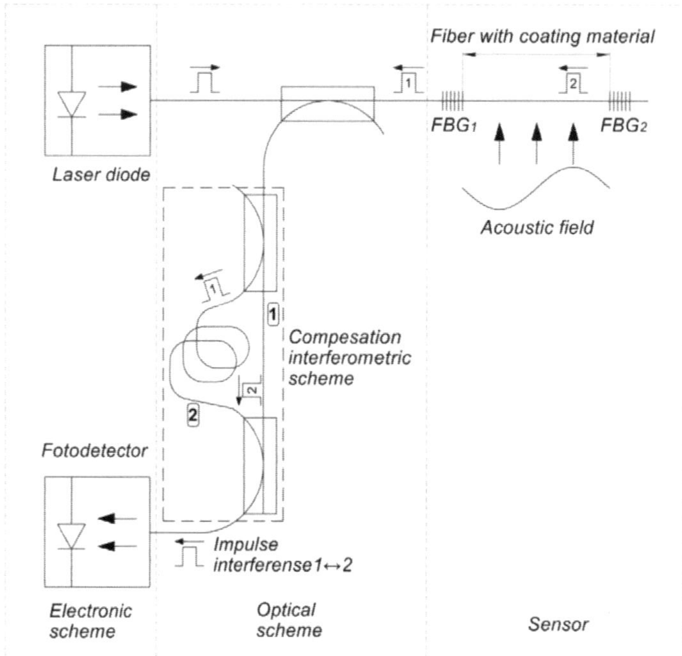

Figure 3. The research facility

3. Summary

In the course of theoretical and experimental studies the dependencies of acoustic sensitivity of optical fiber on the characteristics of the coating material and the thickness of the coating layer were obtained in the air and water.

References

[1] A.V. Kulikov, *Fiber optic acoustic sensors based on Bragg gratings*, dissertation for the degree of Candidate of Sciences, 2012, p. 144

[2] M.M. Butusov, S.L. Galkin, S.p. Orobinskiy, *Fiber optics and instrumentation,* Mechanical Engineering, 1987, p. 328

[3] G.B. Hocker, *Fiber optic sensing of pressure and temperature,* Applied Optics, **Vol.18**, 1979 p. 1445-1446

[4] V.A. Arteev, S.V. Varzhel, A.V. Kulikov, *Extended fiber optic acoustic pressure sensor based on Bragg gratings,* Proceedings of an international conference with workshops «Optics-2011» 2011 p. 509-510

[5] M.U. Plotnikov, *Fiber optic hydrophone,* dissertation for the degree of Candidate of Sciences, 2014, p. 155

Journal of Physics: Conference Series **735** (2016) 012015 doi:10.1088/1742-6596/735/1/012015

The study of the thermal annealing of the Bragg gratings induced in the hydrogenated birefringent optical fiber with an elliptical stress cladding

Munko A.S., Varzhel' S.V., Arkhipov S.V., Gribaev A.I., Konnov K.A., Belikin M.N.

Saint-Petersburg National Research University of Informational Technologies, Mechanics and Optics, Saint-Petersburg, Russia

E-mail: qveenanna-93@mail.ru

Abstract. In this work the comparative results on the dynamics of fiber Bragg gratings inscription in both the conventional and the subjected to hydrogenation birefringent optical fiber with elliptical stress cladding as well as in the same type of lightguide with the increased GeO_2 concentration are presented. Also the research on the thermal impact on the fiber Bragg gratings written in the birefringent fiber with elliptical stress cladding has been carried out. The dependences of the fiber Bragg reflectance coefficient on the time of the thermal impact, obtained by annealing of the refractive index gratings, induced in the optical fibers with increased photorefractivity, are shown.

1. Introduction

The first fiber Bragg grating (FBG) was obtained in 1978 [1]. In about 10 years FBG was inscribed through a lateral surface of optical fiber (OF) for the first time [2]. Currently the intensive implementation of fiber-optic sensors (FOS) takes place in many high-tech areas [3, 4] due to its resistance to electromagnetic disturbance, small weight, size and relatively low cost. Phase interferometric sensors (PIS) are considered to be ones of the most precise FOS [5]. Thus, acoustic monitoring systems based on the fiber optic hydrophone arrays multiplexed with FBGs, can completely replace the existing analogues based on piezoceramic elements.

During the fiber Bragg grating inscription in the conventional OF one faces the difficulties related to the low photosensitivity of the material which this fiber was made of. Several methods to increase the photorefractivity of germanium-silica fibers were proposed in literature [6]. The conventional way is to increase the GeO_2 concentration in core on the stage of OF preform formation. But this method has several disadvantages such as the growth of OF linear losses and additional losses occurring during the coupling of such lightguides to the conventional ones (that have the standard GeO_2 concentration). The method of low temperature hydrogenation is of big interest too. The hydrogenation of the fabricated OF at low temperatures and high pressure significantly raise its photorefractivity that allows to obtain Bragg grating inscription with the sufficient refractive index (RI) modulation.

The hydrogen treatment of the OF used in this work had been carried out under the following conditions: the optical fiber wound around the plastic tube with the diameter of 20 mm was placed in the gas cylinder, filled with the grade "A" hydrogen (mol. percentage of hydrogen was 99,99%) under the pressure of 10,4 MPa and was kept for 32 days at the temperature of 18 °C.

2. FBG writing scheme

The Bragg gratings studied in this work were written with KrF excimer laser Coherent COMPex Pro 102 F with the phase mask method [7]. The FBG inscription scheme is presented in the Figure 1.

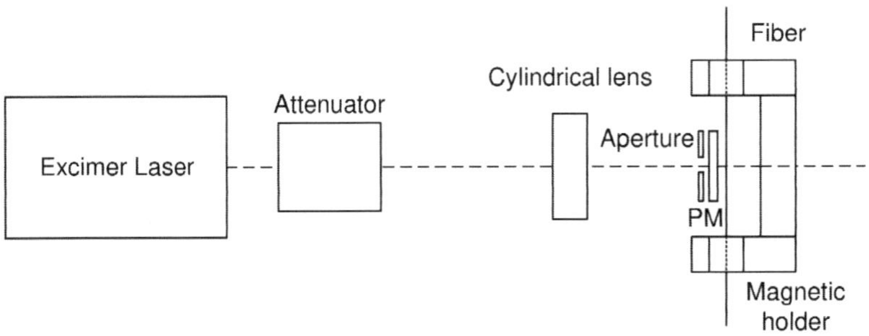

Figure 1. Phase mask method FBG inscription scheme

The laser with the pulse energy of about 250 mJ generates pulses with 20 ns duration at the 248 nm wavelength. The attenuator with the built-in shutter allows to select a single pulse out of its sequences when the laser has already been derived to a stationary mode. The slot filters out the laser beam area that is not used in the inscription process. The cylindrical lense focuses the laser beam along one of the axes to achieve the energy density needed.

In the present work the birefringent fiber with elliptical stress cladding obtained with the technology [8] was used for FBG writing. The design of the birefringent fiber with the elliptical stress cladding is presented in the Figure 2.

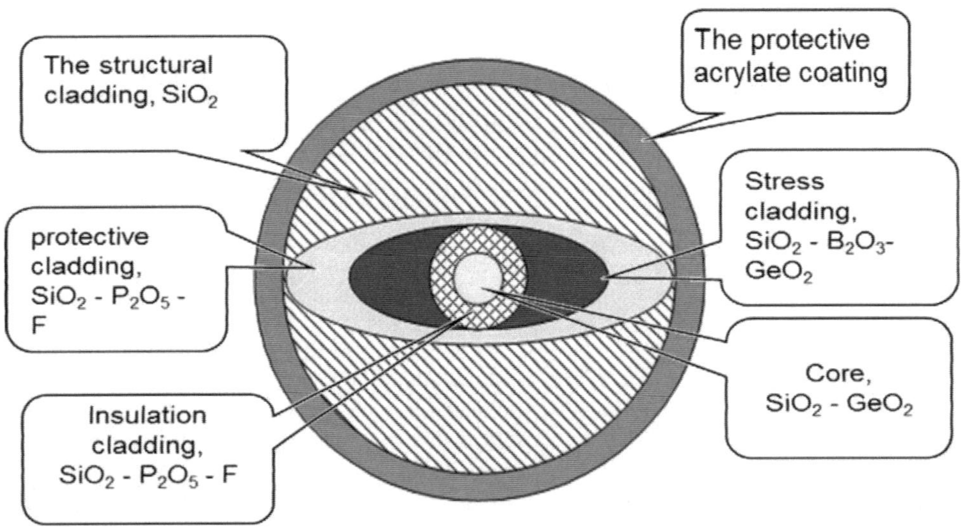

Figure 2. The birefringent OF with elliptical stress cladding design

3. FBG writing results

It is known that hydrogen, dissolved in glass, changes it's RI, which leads to some grating resonance wavelength shift. For example, in the work [9] is shown that the spectrum of the FBG, inscribed in the OF, subjected to hydrogen treatment, is shifted to longer wavelengths by about 1.2 nm. The amplitude of this wavelength shift depends on the initial hydrogen concentration and can reach a few nanometers. Also this method increases the photosensitivity of optical fibers and provides the effective FBG inscription in them.

One of the most important FBG parameters is the value of the induced RI modulation Δn. The following formula can be used to estimate the Bragg grating RI modulation amplitude (1):

$$\Delta n = \frac{\lambda_B}{\pi \cdot l} \cdot \tanh^{-1}(\sqrt{r_{max}}) \quad (1)$$

where λ_B –the central wavelength of the Bragg resonance; r_{max} -reflectance coefficient at the central grating reflectance wavelength; l - FBG length.

In Figure 3 the dynamics of the reflectance coefficient growth of the Bragg gratings induced in the birefringent fiber with elliptical stress cladding that had been subjected (ESC-4 mol.% GeO_2 (H_2)) and hadn't been subjected (ESC-4 mol.% GeO_2) to hydrogenation, as well as in the fiber with increased germanium dioxide concentration (ESC-12 mol.% GeO_2) are shown. The laser pulse energy density on the OF was approximately about 500 mJ/cm^2, the exposition time was 5 minutes at the laser pulses rate of 1 GHz.

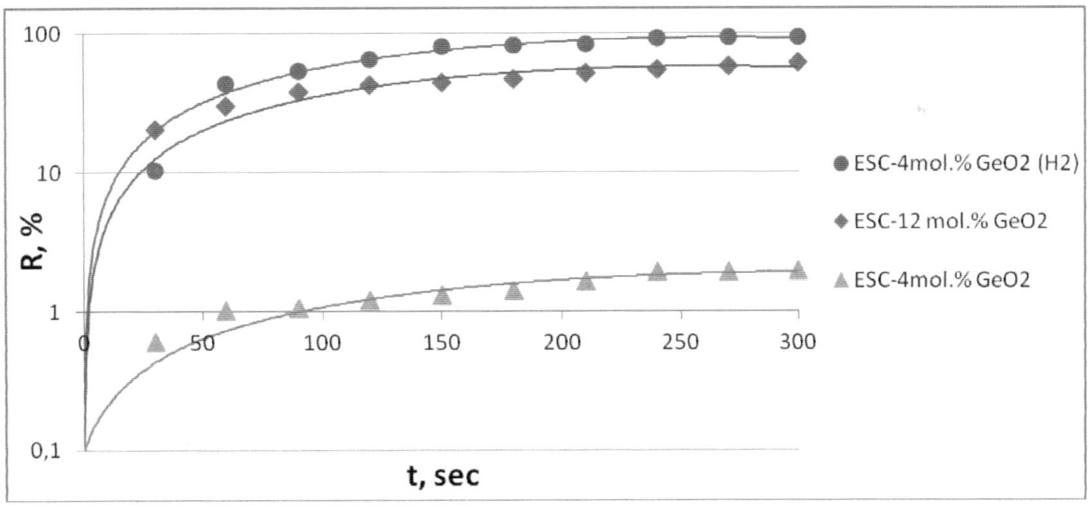

Figure 3. The dynamics of the reflectance coefficient growth of the Bragg gratings versus time

Considering the fact that the FBGs' length l in all cases is 8 mm and the central wavelength of reflection λ_B is 1550 nm, the induced RI modulation value Δn for OF ESC-4 mol.% GeO_2 is $8{,}6\times10^{-6}$, for OF ESC-12 mol.% GeO_2 is about $6{,}4\times10^{-5}$, and for OF ESC-4 mol.% GeO_2 (H_2) is about $1{,}3\times10^{-4}$.

The figure shows that both methods of the photorefratcivity amplification are effective for the FBG inscription in the birefringent OF with elliptical stress cladding. In addition in both cases the induced RI change value is sufficiently high. Nevertheless, the method of increasing of germanium dioxide concentration in the fiber core has some disadvantages related to the optical linear losses growth, therefore it is inferior to the hydrogen treatment method.

4. The research on the FBG thermal stability

After that the research on thermal stability of the FBGs induced in this type of birefringent OF subjected and not subjected to hydrogenation had been conducted.

To conduct this investigation the laboratory stand for FBG annealing and spectral properties controlling had been assembled. The experimental method of FBG thermal stability investigation is based on the subjection of the several similar fiber diffractive structures to certain heat treatment during which the spectral properties changes are registered.

During the FBG heating the Bragg resonance wavelength is shifting towards the area of longer wavelengths due to the thermal expansion of silica OF and thermooptical effect. After the reaching of the annealing temperature the smooth decrease of the reflection coefficient begins [10].

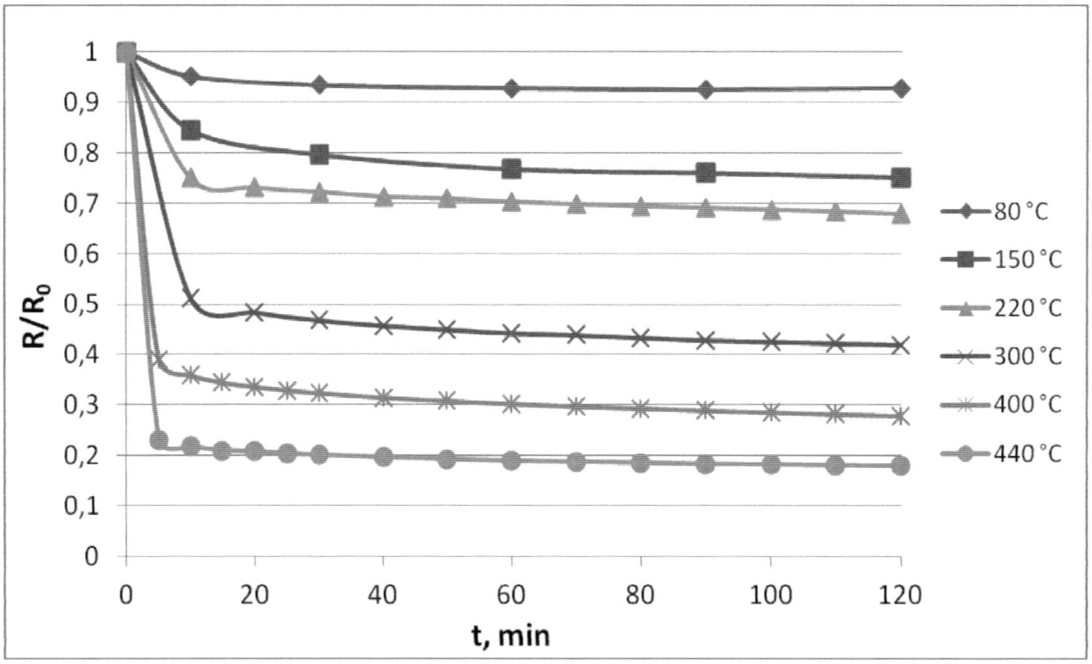

Figure 4. The dependence of the FBG normalized reflection coefficient on the time of annealing at different temperatures for OF ESC-4 mol.% GeO_2 (H_2)

In the figure 4 the change of the reflection coefficient of the Bragg gratings inscribed in the fiber ESC-4 mol.% GeO_2 (H_2) during the annealing process at different temperatures lasting for 2 hours, is presented. It is seen that that the major decrease of the reflection coefficient occurs during the first 10 minutes and after that the process stabilizes. At the end of the 2 hour annealing at the temperature of 440°C the reflection coefficient decreases by about 80%.

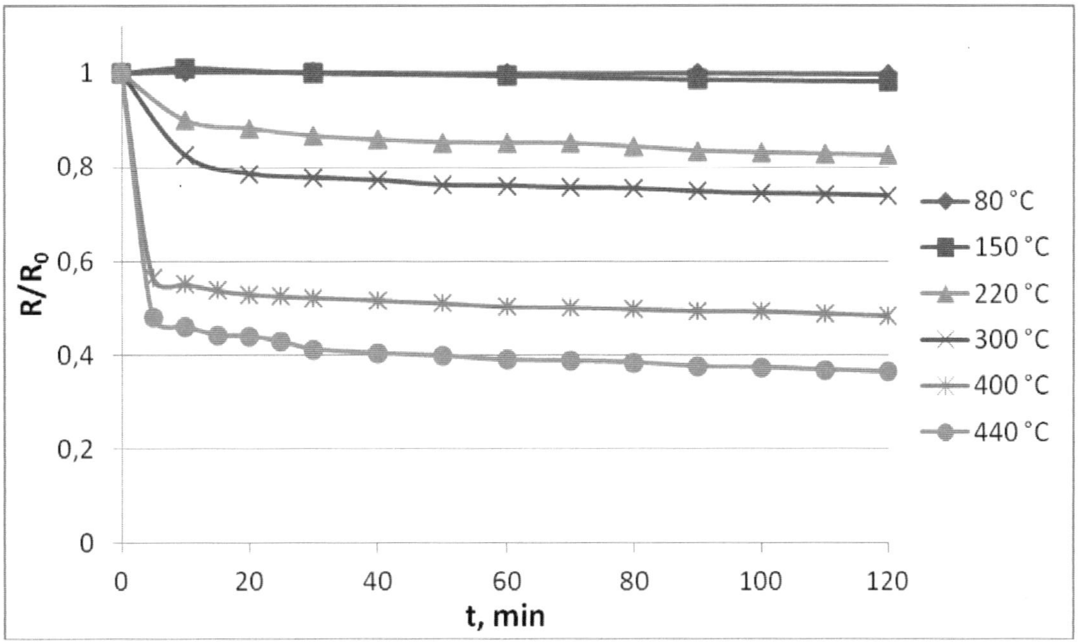

Figure 5. The dependence of the FBG normalized reflection coefficient on annealing time at different temperatures for optical fiber ESC-12 mol.% GeO_2

In figure 5 the change of the reflection coefficient of the Bragg gratings inscribed in the fiber ESC-12 mol.% GeO_2 with increased germanium dioxide concentration in the core during the annealing process at different temperatures lasting for 2 hours, is presented. It is seen that that the major decrease of the reflection coefficient occurs during the first 10 minutes and after that the process stabilizes. At the end of the 2 hour annealing at the temperature of 440°C the reflection coefficient decreases by about 60%.

5. Conclusion

The experiments that had been carried out demonstrate the effectiveness of the low-temperature hydrogen treatment for the increasing of photorefractivity of the birefringent optical fibers with elliptical stress cladding and also show that FBGs induced in the investigated fiber that was subjected to the hydrogen treatment have less thermal stability in comparison to the samples with increased GeO2 concentration in the fiber core. But at the same time the thermal stability is sufficient for many applications, i.e. as PIS reflective elements, that are operated at the temperature range from -40°C up to +70°C.

Acknowledgements

This work was done in ITMO University and was supported by the Ministry of Education and Science of the Russian Federation under the project 02.G25.31.0044.

References

[1] Hill K.O., Fujii Y., Johnson D.C., Kawasaki B.S. 1978 *J. Appl. Phys. Lett*. V. **32**. N 10. P. 647–649
[2] Meltz G., Morey W.W., Glenn W.H. 1989 *Opt. Lett*, V. **14**. N 15. P. 823–825
[3] Okosi T., Okamoto K., Otsu M., Nisihara H., Kauma K., Hatate K., Fiber Optic Sensors, Leningrad: Energosamizdat, 1991. p. 256
[4] Lefevre H. The Fiber-Optic Gyroscope. London: Artech House, 1992. p. 313

[5] S. Varzhel and V. Strigalev, Method for eliminating the noise signal influence on the sensitivity of receiving hydroacoustic antenna based on fiber bragg gratings. *Scientific and Technical Journal of Information Technologies, Mechanics and Optics*, 2010, N 5(**69**), pp. 5–9

[6] S. A. Vasil'ev, O. I. Medvedkov, I. G. Korolev, A. S. Bozhkov, A. S. Kurkov, and E. M. Dianov, Fibre gratings and their applications. *QuantumElectronics*, 2005, vol. **35**, no. 12, pp. 1085-1103

[7] I.K. Meshkovskiy, V.E. Strigalev, A.V. Kulikov, S.V. Varzhel'. *Journal of Photonics*, 2013, Article ID **936036**. –p. 4

[8] M. A. Eron'yan, Method of fabricating fiber lightguides that maintain the polarization of the radiation. *Russian Patent* no.**2155359**, 2000

[9] Munko A.S., Varzhel S.V., Arkhipov S.V., Zabiyakin A.N. Protective coatings of fiber Bragg grating for minimizing of mechanical impact on its wavelength characteristics. *Scientific and Technical Journal of Information Technologies, Mechanics and Optics*, 2015, vol.**15**, no. 2, pp. 241–245

[10] S. V. Varzhel', A. V. Kulikov, K. A. Konnov, A. I. Gribaev, *Materials of the V international research and practice conference*, 2013, P. 305-310

The concentration mechanisms of cubic nonlinearity in dispersive media

V I Ivanov, G D Ivanova, S I Kirjushina and A V Mjagotin
Department of Physics and Theoretical mechanics, Far Eastern State Transport University, Khabarovsk, Serysheva street, 47, Russia

E-mail: naivi@rambler.ru
Abstract. The comparative analysis of the dynamic holograms recording efficiency in media with non-resonance mechanisms of optical nonlinearity is carried out. It is showed that the greatest values of this parameter are provided by thermodiffusion and electrostrictive mechanisms of concentration nonlinearity of liquid dispersed media.

1. Introduction

There are known many nonlinear mechanisms which are used to record dynamic holograms [1-5]. The highest values of the nonlinearity provide the resonance mechanisms - the saturation of absorption in gases, the saturation of interband transitions in semiconductors. However these nonlinearity mechanisms are limited by narrow spectral range. A non-resonant nonlinearity is free from this drawback and more promising in wide spectral range applications. The mechanisms of this type provide a sufficiently high efficiency of nonlinear interaction of light in both the visible and IR spectral ranges [1, 2].

In this paper we analyze a number of such mechanisms (and sensitive materials) in order to compare the limit characteristics of the dynamic holograms recording and identify the most promising materials for holography with the low-power laser sources.

2. The energy holographic sensitivity

It is customary to compare the different media and mechanisms of nonlinearity using a third-order nonlinear susceptibility, which characterizes the cubic nonlinear response. In the case of non-resonant non-linearity (for weakly absorbing media) is used often the other parameter - $n_2^{eff}[m^2/W]$, which characterizes the change of the refractive index of the medium under the influence of incident radiation:

$$n = n_0 + n_2^{eff} I, \tag{1}$$

where n_0 - the refraction index of the medium in the absence of radiation, I - radiation intensity, $n_2^{eff} = (dn/dI)$ - the coefficient of effective cubic nonlinearity. However this option does not allow a valid comparison of different media and mechanisms of nonlinearity. In most cases (and especially for thermal non-linearity) this coefficient is proportional to the magnitude of quasi-stationary relaxation time of the nonlinear response τ, i.e. it depends on the geometry of the system and experimental conditions. The diffraction efficiency of the hologram also depends on its thickness, which is limited by absorption or scattering losses, and the wavelength of the radiation.

Therefore we suggest the more informative complex parameter – the energy holographic sensitivity to assess the effectiveness of the recording of dynamic holograms [2]:

$$N_{2E} = 2\pi n_2^{eff} (\tau\alpha\lambda)^{-1} [m^2/J], \tag{2}$$

where α - the absorption coefficient of the medium, λ - light wavelength. For phase holograms the value of this parameter corresponds to the minimum energy of the recording radiation needed to change the optical thickness of the medium on the radiation wavelength and completely characterizes the diffraction efficiency of a thin hologram. Moreover this parameter dimension allows to compare the sensitivity of both volume and surface nonlinearity mechanisms [1]. For example, for reflection holograms the energy holographic sensitivity can be defined as:

$$N_{2E} = (d\rho / dI)\tau^{-1}, \tag{3}$$

where ρ - Fresnel amplitude reflection coefficient of the boundary between two media. Similarly, the sensitivity is determined for the amplitude thin holograms – in this case the amplitude reflection coefficient is replaced by the corresponding transmittance coefficient.

The most universal non-resonant nonlinearity that exists in any absorbing medium is the heat mechanism. Let us find the value of such sensitivity for a single component material. A simplest mechanism of thermal non-linearity is due to the thermal expansion of the medium. The expression for the coefficient of effective cubic nonlinearity can be obtained by solving the one-dimensional heat equation:

$$c_p \rho \partial T / \partial t = -\text{div}\, J_1 + \alpha I_0 (1 + \sin Kx), \tag{4}$$

where $J_1 = -D_{11}\text{grad}T$ - heat flow, T - temperature of the material, D_{11} - the thermal conductivity of the material, c_p and ρ - the specific heat capacity and density of the medium, $I = I_0(1 + \sin Kx)$ - the intensity spatial distribution along the layer of the medium (x-axis), K - the space wave vector of the elementary hologram. Using the stationary solution (3) for the amplitude of thermal lattice, we obtain [2]:

$$n_2^{eff} = \alpha \tau_T (dn/dT)(c_p \rho)^{-1}, \tag{5}$$

where (dn/dT) - the temperature coefficient of the refractive index, $\tau_T = D_{11}\text{K}^{-2}$ - the thermal relaxation time. The holographic sensitivity parameter is obtained from (1):

$$N_{2E} = 2\pi (dn/dT)(c_p \rho \lambda)^{-1}. \tag{6}$$

Thus, the value of sensitivity for thermal nonlinearity of single-component fluids is defined by parameters (dn/dT), c_p and ρ. Substituting the characteristic parameters for semiconductor media $dn/dT = 2\cdot 10^{-4} K^{1}$, $(c_p\rho) \approx 1\ J/(cm^3 K)$, we obtain $N_{2E} \approx 1\ cm^2/J$ (for radiation with a wavelength of $\lambda = 10\ \mu m$).

However a light with an intensity of several mW/cm^2 is required the greater value of the coefficient of nonlinearity for effective dynamic hologram recording. Since the relaxation time (e.g., time diffusion concentration) increases with particle size increasing, it is natural to expect high nonlinearity parameters in nanodispersive media (colloids, suspensions, micro-emulsions) the particle size of which is limited by the wavelength of the radiation due to light scattering. The "nanodispersive" trend can be traced quite easily in the searching of large non-linearities. If in the first papers was used the nonlinear medium with a typical value $n_2^{eff} \approx 10^{-13}\ cm^2 / W$ (molecular Kerr nonlinearity), than was experimentally obtained values $n_2^{eff} \approx 10^{-8}\ cm^2 / W$ (artificial Kerr-like nonlinearity) [3].

3. The thermodiffusion and electrostrictive mechanisms of concentration nonlinearity

Multicomponent nanoparticulate medium (liquid phase mixture, suspension, emulsion) are characterized by a series of specific mechanisms nonlinearities that are not in single-component media. In particular, these include the concentration of non-linearity caused by the redistribution component in a two-phase medium in a laser field. The concentration flows in the medium can be caused by different mechanisms of interaction of radiation with matter.

A thermally induced mechanism of particle drift in a nonuniform temperature field is known as thermal diffusion (thermophoresis) in gases or Soret effect in liquid-phase binary mixtures [4, 5]. The microheterogeneous medium with different refractive index of components in the electromagnetic field is forced by electrostriction forces, which can also be the cause of concentration streams [2]. Depending on the sign of the polarizability the microparticles may be drawn (if the refractive index of the dispersed phase material is greater than the dispersion medium) or pushed out (otherwise) from areas with greater intensity of the electromagnetic wave.

Consider setting holographic sensitivity of the concentration non-linearity. In nanodispersive medium particle radius is much smaller than the radiation wavelength λ, the refractive index of the medium is proportional to the concentration of particles (for dilute systems):

$$n = n_1(1+\phi\delta), \tag{7}$$

where and $\delta = (n_2 - n_1)/n_1$; n_1 и n_2 - the refractive indices of the substance and the dispersion medium of the dispersed phase, respectively; $\phi = (4/3)\pi r^3 C$ - the volume fraction of the dispersed phase, r - the radius of the microparticle, C - the concentration of nanoparticles. Then we have for each concentration mechanism of nonlinearity:

$$n_2^{eff} = (\mathrm{d}n/\mathrm{d}C)(\mathrm{d}C/\mathrm{d}I). \tag{8}$$

Let's consider an expression for the holographic sensitivity of the dispersion medium, taking into account both (thermodiffusion and electrostriction) concentration mechanisms. The system of balance equations for the concentration of particles and heat flow for this case is written as follows [2]:

$$c_p \rho dT/dt = -div\, J_1 + \alpha I_0(1+\sin Kx), \tag{9}$$

$$dC/dt = div(J_2 + J_3). \tag{10}$$

Here J_1 and $(J_2 + J_3)$ - thermal and concentration flows, respectively:

$$J_1 = -D_{11}\nabla T - D_{12}\nabla C, \tag{11}$$

$$J_2 = -D_{21}C\nabla T - D_{22}\nabla C, \tag{12}$$

$$J_3 = \gamma C\nabla I, \tag{13}$$

where D_{22} - diffusion coefficient of particles, D_{21} and D_{12} - coefficients describing cross-flows (the effect of thermal diffusion and Dufour-effect, respectively), J_3 - electrostrictive flow, $\gamma = (2\beta b/cn_1)$, (β and b - polarizability and mobility of microparticle, respectively), c - the light velocity[2]. Assuming a one-dimensional problem, we seek the solution of equations (9-13) in the form of:

$$C(x,t) = C_0 + C_1(t)\sin Kx, \tag{14}$$

$$T(x,t) = T_0(t) + T_1(t)\sin Kx\,. \tag{15}$$

Here C_0 and T_0 - the average particle concentration and temperature of the medium. Let's the amplitudes of the thermal T_1 and concentration C_1 profiles are small $-(C_1/C_0) << 1$, $(T_1/T_0) << 1$. The linearized system is easily solved and we have by taking into account (2) and (9-15):

$$N_{2E} = 2\pi n_1 \phi_0 \delta \left[D_{11}^{-1} D_{21} (c_p \rho)^{-1} - \gamma \alpha^{-1} K^2 \right] \lambda^{-1}\,. \tag{16}$$

The sensitivities of the concentration nonlinearity mechanisms of liquid-phase two-component medium are determined by the coefficient of thermal diffusion D_{21} (1-st term) and by microparticle polarizability in the case of the electrostrictive mechanism (2-nd term). For a non-absorbing nanodispersive medium the absorption coefficient should be replaced by the extinction coefficient. As can be seen from (16), both mechanisms may either strengthen or weaken each other depending on the sign of the thermal diffusion coefficient and polarizability of the dispersed particles. The evaluations from (6) and results of experimental studies [4-6] shows the value $N_{2E} \approx 10^2\ cm^2/J$ for both concentration nonlinear holographic sensitivity in the vicinity of the critical point of the micro-emulsion.

4. Conclusions

Thus, our analysis shows that the nanoparticulate liquid phases are characterized by high values of the holographic sensitivity. More over the dispersion media have another advantage. It is known that the main parameters of the nonlinear medium are usually interrelated: for example, the value of non-linearity usually corresponds to poor temporal resolution, spatial resolution and sensitivity of the holographic media are inversely related. Because the parameters of known materials form a discrete set, there is a problem of selection medium with optimized characteristics for particular applications. The most promising materials in this sense are also the nano-materials, because one can change the volume fraction of the various components and their composition, i.e. control (including in real time) of parameters.

Since there are known same mechanisms of nonlinearity in the nano-dispersive systems with the largest holographic sensitivity we can distinguish them as a promising new class of media for dynamic holography [2].

5. References

[1] Zeldovich B Y, Pilipeckij N F, Shkunov V V 1985 *Obrashhenie Volnovogo Fronta* (Moskow: Nauka).

[2] Ivanov V I 2006 *Thermo-induced Mechanisms of Dynamic Holograms Recording* (Khabarovsk: FESTU).

[3] Gower M G *Progr. in Quant. Electr.* 1984 **9** 100-147.

Ivanov V I, Livashvily A I, Lobov A I, Simakov S R 2006 *J. of Opt. Techn.*, **71(9)** 586-587.

[4] Leppla C, Wiegand S 2003 *Phil. Mag.* **83** 1989-1999.

[5] Ivanov V I, Okishev K N *Techn. Phys. Lett.* **32(11)** 967-968.

[6] Freysz E, Afifi M, Ducasse A, Pouligny B, Lalanne J R J *Phys. Lett.* 1985 **46** 181-187.

Study of spectroscopic and thermal characteristics of nonlinear optical molecular crystals based on 4-nitrophenol

Pavlovetc I.M.[1], **Fokina M.I.**[1]
[1]ITMO University, St. Petersburg 197101, Russia

E-mail: ipavlovetc@gmail.com

Abstract. The paper presents the results of study of spectroscopic and thermal characteristics of molecular co-crystals: 2-aminopyridine-4-nitrophenol-4-nitrophenolate (2AP4N) and 2,6-diaminopyridine-4-nitrophenol-4nitrophenolate (26DAP4N). Crystals were successfully grown by slow evaporation technique. Optical transparency in the region of 190–1100 was found to be suitable for applications with cut off wavelengths 420 and 430 nm respectively. Thermogravimetric and differential thermal analysis show good quality and thermal stability for studied crystals. Kurtz and Perry powder technique proves that the crystals are acentric and have significant nonlinear optical response.

1. Introduction

Organic nonlinear optical (NLO) crystals plays important role and provides major benefits to some very important applications, such as laser frequency conversation, electro-optic modulation, frequency mixing and many others [1,2]. Three classes of NLO materials are used today: inorganic crystals (e.g. $LiNbO_3$), polymers or dendrimers doped with NLO dye molecules (e.g. DR-1, CLD-1), and organic crystals with large NLO coefficients [3,4]. Organic NLO materials have attracted some attention because of the advantages offered by organic systems including high NLO susceptibilities (compare to inorganic compounds), fast response time, relative ease of device processing, etc. [4]. Usually, organic systems with significant NLO properties are the pi-electron conjugated chromophores with strong electron donor group on the one end and acceptor on the other [4].

Creation of new efficient second-order NLO materials require optimization of donor-acceptor chromophores in noncentrosymmetric crystalline lattice or in a polymer matrix [5]. To fulfil this requirement we crystallized two organic chromophores: one acidic (4-nitrophenol) and one basic (2-aminopyridine or 2,6-diaminopyridine). 4-nitrophenol is a classic dipolar NLO chromophore and a donor-acceptor π-system, and presence of the possibility of proton transfer of the phenolic OH group of 4-nitrophenol to various organic bases can result in increasing of its molecular hyperpolarizability [6]. Aminopyridine complexes are well-known series of compounds with good optical properties, thermal stability and suitability for optoelectronics devices [7–10].

We continue our studies in the field of NLO crystals including DAST [11] and different molecular crystals [12–14]. The present work is focused on the spectroscopic and thermal properties of two herein presented organic co-crystals for nonlinear optical applications: 2-aminopyridine-4-nitrophenol-4nitrophenolate (2AP4N) and 2,6-diaminopyridine-4-nitrophenol-4-nitrophenolate (26DAP4N).

2. Experimental

2.1. Synthesis and crystal growth

To produce and study two herein presented organic compounds the reported synthetic procedure [15] was used as a basis for preparation of 2AP4N and 26DAP4N. Ethanol was used as a solvent.

For the synthesis, an ethanol solution of 4-nitrophenol was added to and ethanol solution of 2-aminopyridine or 2,6-diaminopyridine. Initial reagents has been taken in molar ration 2:1.

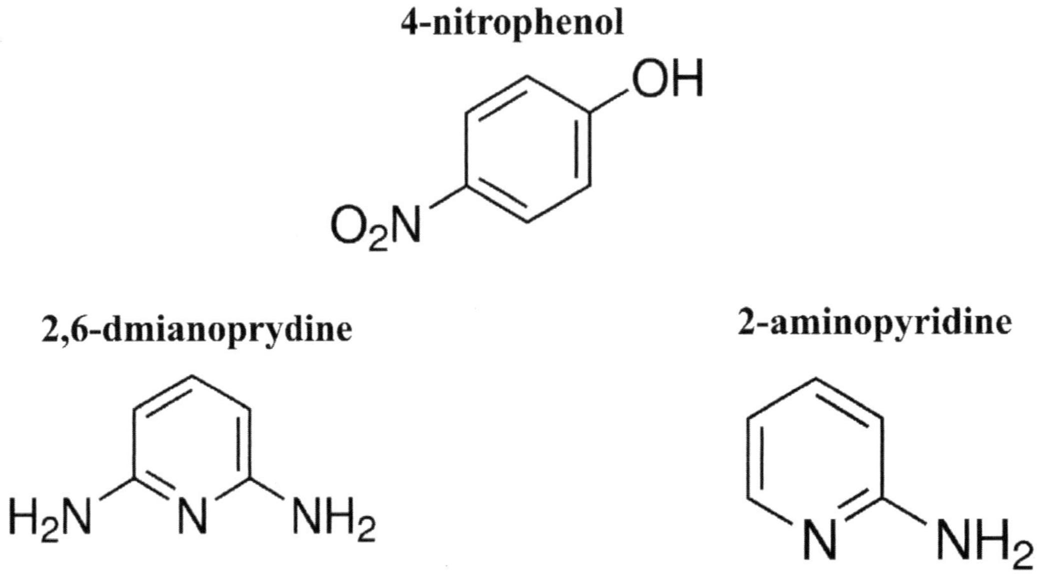

Figure 1. Structural formula of initial reagents

Preparation of 2AP4N: 4-nitrophenol (0.01 M) was dissolved in ethanol (150 ml) and mixed with ethanol solution (50 ml) of 2-aminopyridine (0.005 M).

Preparation of 26DAP4N: 4-nitrophenol (0.01 M) was dissolved in ethanol (150 ml) and mixed with ethanol solution (50 ml) of 2,6-diaminopyridine (0.005 M).

Crystals were grown by slow evaporation technique at a fixed temperature (25 ^{o}C). The mixtures of solutions were stirred, filtered, covered with pores filter paper, and placed into the thermostat for slow evaporation of the solvent. After two weeks the solutions were evaporated and the named crystals were successfully grown.

2.2 Instruments for characterization

Thermal characteristics studies were carried out using Thermo Gravimetry/Differential Thermal Analyzer (TG/DTA) HITACHI STA7200. Samples under study were weighted in and aluminium pans with a microprocessor driven temperature controller. TGA and DTA curves of the studied crystals were recorded in nitrogen atmosphere in the range 25–350 ^{o}C and are shown in Figure 2. Transmittance spectra of these crystals were recorded in the range 190–1100 nm using Shimadzu-1800 spectrophotometer and the recorded spectra are shown in Figure 3. To confirm the presence of the second-order NLO response, second harmonic generation (SHG) conversion efficiency of the grown crystals were measured by the Kurtz and Perry powder technique [16]. A Q-switched Nd:YAG laser operating on wavelength 1064 nm with 10 ns pulse width and 10 Hz pulse rate was used as a light source. Studied crystals were grounded and sieved in the particle size range between 100 and 160 μm.

SHG was confirmed by the emission of green radiation (532 nm) from the samples and collected by a photodiode with sensitivity zone 420–675 nm.

3. Results and dissucsion

TGA and DTA curves of the grown crystals are shown in Figure 2. According to DTA data, all grown crystal samples have no phase transitions between ambient temperature and melting point. DTA curves of the studied crystal samples show endothermic peaks with maxima at 99 and 149 °C for 2AP4N and 26DAP4N crystals respectively, which recognized as the melting points of these crystals. Second endothermic peaks of both DTA curves is the full evaporation temperatures of the materials, which correlates with the TGA curves.

According to TGA curves (Figure 2, blue lines) there is no weight loss between ambient temperature and the melting points for both of the crystals, which show that the obtained crystals have no inclusion of solvent (which was used for co-crystallization) of other impurities in the crystal lattices. Weight losses start from the melting point and continue up to full evaporation of the materials.

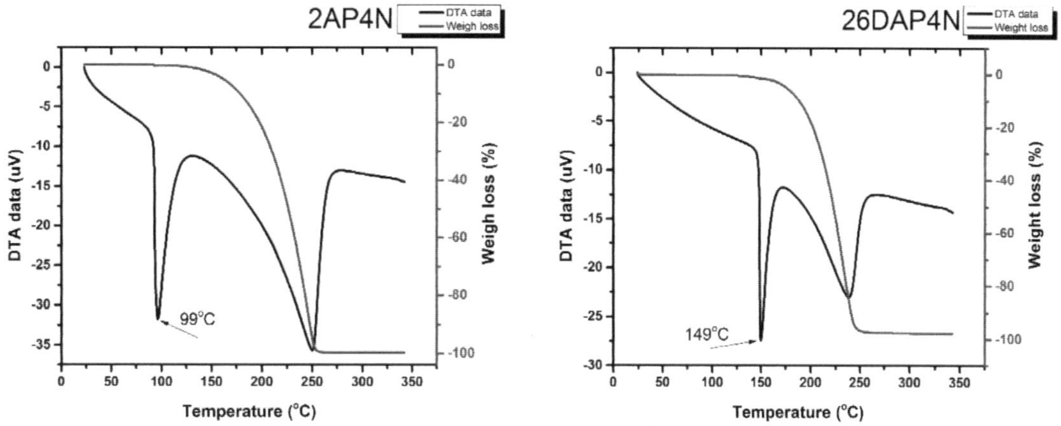

Figure 2. TGA and DTA curves for 2AP4N and 26DAP4N crystals

Transmittance in the visible and near infrared regions of the spectrum is an important parameter which determining the application areas of the material. If the crystals are used for SHG of Nd:YAG laser, the transmittance at the fundamental wavelength (1064 nm) and at the SH wavelength (532 nm) is a necessary parameter.

The obtained transmittance spectra (Figure 3) show that the 2AP4N crystal has wide range of optical transmission of above 80% up to 1100 nm from cut off wavelength (420 nm), and the 26DAP4N crystal has transmission of about 55% up to 1100 nm from cut off wavelength (430 nm). According to these spectra, all the studied crystals have almost the same level of absorption at the fundamental and SH wavelengths of Nd:YAG laser.

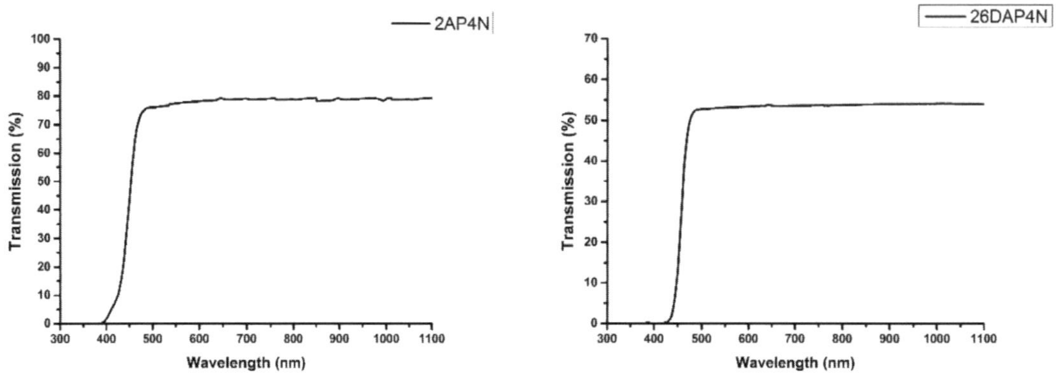

Figure 3. Transmittance spectra of studied crystals

In order to confirm the presence of the second-order NLO response, powder samples of studied crystals were subjected for the Kurtz and Perry powder test [16], which is one of the most effective ways to evaluate the effectiveness of materials for the second harmonic generation. Both the studied samples has shown the presence of a good nonlinear optical response. The relative SHG intensity of 2AP4N crystal was found to be 1.9 times more than that of 26DAP4N crystal with the same particle size range.

4. Conclusion

Good quality and transparent single crystals of 2-aminopyridine-4-nitrophenol-4-nitrophenolate and 2,6-diaminopyridine-4-nitrophenol-4-nitrophenolate were grown successfully by slow evaporation technique with ethanol as a solvent. Obtained results demonstrated that the studied crystals can be used applications below 99 oC. Second harmonic generation studies has shown the presence of significant second-order NLO response. All these properties show that the studied crystals may be promising candidates for the fabrication of NLO devices.

References

[1] Garmine E 2013 Opt. Express **21** 30532
[2] Prabhu S G and Mohar Rao P 2000 J. Cryst. Growth **210** 824
[3] Marder S R, Sohn J E and Stucky G D 1991 *Materials for Nonlinear Optics – Chemical Represpectives* (Washington: Am. Chem. Soc.)
[4] Nalwa S S and Miyata S 1997 *Nonlinear optics of organic molecules and polymers* (Boca Raton: CRC Press)
[5] Boyd R W 2008 *Nonlinear optics. Third edition* (USA: Academic Press)
[6] Evans C C, Bagieu-Beucher M, Massse R and Nicoud J-F 1998 Chem. Mater. **10** 847
[7] Al-Ahmary K M, Habeeb M M and Al-Solmy E A 2011 J. Mol. Liq. **162** 129
[8] Shkir M, Riscob B and Bhagavannarayana G 2012 Solid State Sci. **14** 773
[9] Pandi P, Peramaiyan G, Kumar M K, Kumar R M and Jayavel R, Spectrochim. Acta Mol. Biomol. Spectrosc. **88** 77
[10] Draguta S, Fonari M S, Masunov A E, Zazueta J, Sullivan S, Antipin M Yu and Timofeeva T V 2013 Cryst Eng Comm **15** 4700
[11] Burunkova J A, Denisyuk I Yu and Fokina M I 2014 Mol. Cryst. Liq. Cryst. **589** 178
[12] Draguta S, Leonova E, Fokina M, Denisyuk I and Timofeeva T V 2013 Acta Cryst. E 1280 1
[13] Denisyuk I, Fokina M, Pavlovetc I, Timofeeva T V, Fonari M and Draguta S 2014 Proc. of IEEE conference LO 2014
[14] Pavlovetc I M, Draguta S, Fokina M I, Timofeeva T V and Denisyuk I Yu 2015 Opt. Commun. doi:10.1016/j.optcom.2015.05.034

[15] Prakash M J and Radhakrishnan T P 2005 Cryst. Growth Des. 5 721
[16] Kurtz S K and Perry T T 1968 J. Appl. Phys. **39** 3798

Journal of Physics: Conference Series **735** (2016) 012018 doi:10.1088/1742-6596/735/1/012018

Plasmon-enhanced stimulated emission of chromene dye

E V Seliverstova[1] and N Kh Ibrayev

Institute of Molecular Nanophotonics, Y.A. Buketov Karaganda State University, 100028 Universitetskaya str. 28, Karaganda, Kazakhstan

Abstract. Effect of silver nanoparticles on spontaneous and stimulated emission of chromene-3 in ethanol solution is studied. It is established, that upon laser photoexcitation of solution the spontaneous fluorescence is observed, which with increasing of pumping energy transforms in stimulated emission. The increasing of fluorescence, stimulated emission and lasing threshold was observed upon addition of silver nanoparticles. The dependence of intensity of lasing of silver concentration correlates with data of fluorescence.

Introduction

During the last years the great progress was made in the field of solid-state active tunable lasers based on organic compounds. The yellow-green spectral range (550-590 nm) is most mastered1. Blue and red spectrum is less filled with active media, but research for their development are carried out actively2,3. Development of solid active media, emitting in the red region of the spectrum (over of 600 nm) is important in connection with the decision of a number of biological problems, since this radiation can penetrate deeply into biological tissue. The emission spectrum of such a range is used in photodynamic therapy of cancer.

Recently the studies of excitation of localized plasmon resonance of metal nanoparticles are performs actively4,5. Among the optical manifestations of localized plasmon resonance of nanoparticles (NPs) metals most famous is enhanced Raman scattering6. Molecules of luminophores placed near the surface of metal NPs undergo to effect of local electromagnetic fields. Thus, depending on the distance between the NPs and the molecule, the emission intensity of fluorophores is enhanced or quenched7. At short distances and direct contact of dyes and NPs the luminescence is quenched due to the prevalence of nonradiative energy transfer from the fluorescent molecules to NPs.

From a practical point of view, the interest in plasmonic effect is associated with the ability to create highly fluorescent sensors8, optoelectronic devices9, nanolasers10 and others. One of the promising areas of modern laser physics is development and the study of composite media based on the laser-active molecules and metal nanoclusters. There is evidence that the addition of metal NPs in the active medium of dye lasers leads to a decrease in the lasing threshold11-14.

Here the results of plasmon effect on photonics of Chromene-3 dye are presented. These dye is the derivatives of the lasing dye Coumarine-7. The structure of dye is shown on figure 1.

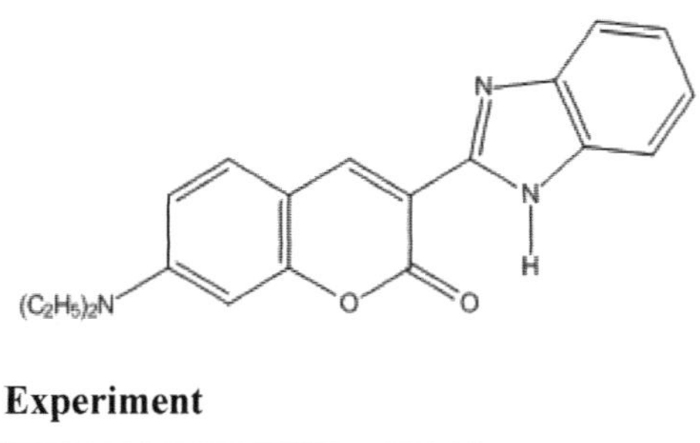

Figure 1. Structure formulae of Chromene-3, R= NH.

2. Experiment

1 E-mail: genia_sv@mail.ru

Spectral and luminescent features of dye were studied in ethanol solutions. Concentration of dye in solution was equal to 10^{-4} mol/l for Chromene-3. Ag NPs were prepared by ablation of metallic silver in eyhanol. The size of Ag nanoparticles in colloidal solutions was determined by dynamic light scattering analyzer of the sizes of submicron particles Zetasizer Nano ZS (Malvern). The average size of the silver particles was equal to 90 nm. The spread of the values of dispersibility of the silver particles was minimal and ranges from 10 to 20 nm.

Fluorescence spectra was measured with spectrofluorimeter Cary Eclipse (Agilent). The fluorescence decay kinetics of dye was measured using a pulsed spectrofluorometer with picosecond resolution and registration of the time-correlated photon counting mode (Becker & Hickl). Fluorescence excitation of samples was performed using a pulsed laser with λgen = 488 nm and a pulse width at half-maximum τ = 40 ps.

Absolute quantum yield of dye was measured in reflectometric sphere AvaSphere-50-REFL (Avantes BV) and was equal to 0.99 for Chromene-3. Measurements of the spectral and energy characteristics of induced emission of solutions of chromene-3 were performed upon excitation of samples by second harmonic of Nd:YAG laser (SOLAR LQ 215, λ_{gen} = 532 nm, E_{imp} = 90 mJ , τ = 10 ns) in the cross-version[14] (figure 2).

Pump power density was varied with the help of neutral density filter and was 0.005-0.2 MW/cm^2. Stimulated emission with the help of converging lens focused on the input of optical fiber spectrometer AvaSpec-2048.

An optical resonator, which was formed by reflection mirror (6) and the front edge of the cell with a dye solution (7) were used at the measuring of the lasing properties. Pump radiation passing diaphragm (3) by the cylindrical lenses (4,5) was focused on the lateral face of the cell in the strip with area of 0.07 cm^2. Generation of radiation arises in a narrow region near the wall of the front wall of the cell.

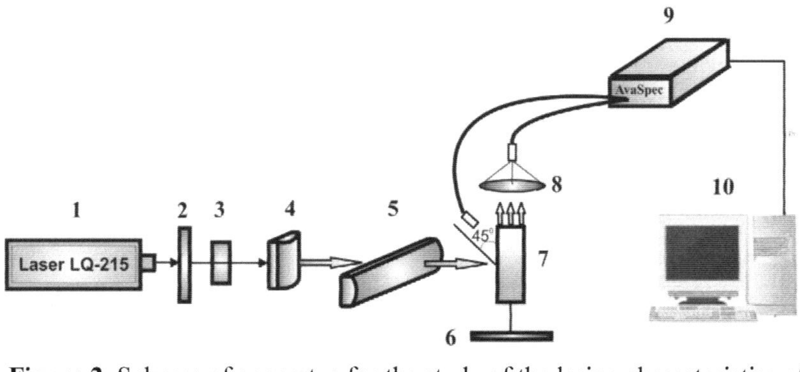

Figure 2. Scheme of apparatus for the study of the lasing characteristics of molecular systems: 1 – laser LQ 215; 2 – neutral density filters; 3 – aperture; 4,5 – cylindrical lens; 6 – nontransmitting mirror; 7 – the sample; 8 – converging lens; 9 – AvaSpec-2048 spectrometer; 10 – computer.

Pump power density was varied with the help of neutral density filter (2) and was 0.005-0.2 MW/cm^2. Stimulated emission with the help of converging lens (7) focused on the input of optical fiber spectrometer AvaSpec-2048. The lasing spectra were measured on spectrometer AvaSpec-2048 (8) connected to the computer (9). Relative error in the determination of the spectral characteristics was 3%.

3. Results and discussion

Absorption spectra of silver NPs and Chromene-3 are shown on the figure 3. It is seen that absorption spectrum of silver NPs exhibits the maximum at 452 nm. Absorption bands of Ag NPs and dye is overlapped, It is evidence that conditions of plasmonic resonance are stratified. Long-wavelength absorption spectra of ethanol solution of dye is observed as wide band in the region of 450-600 nm with two evident maxima at 517 and 550 nm. Fluorescence spectra of Chromene-3 exhibits the maximum at 580 nm and weak pronounced shoulder in the range of 615-620 nm. The position and shape of fluorescence spectra is independent of excitation wavelength. Spectral parameters of dye solution are shown in the Table 1.

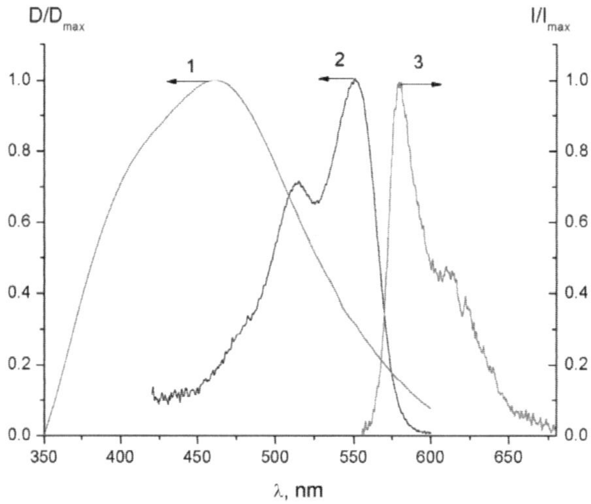

Figure 3. Absorption (1,2) and fluorescence (3) spectra of Ag NPs (1) and Chromene-3 in ethanol (2,3), $C_{dye}=10^{-4}$ mol/l.

Table 1. Spectral parameters of Chromene-3 at various concentrations of Ag.

C_{Ag}, mol/l	λ_{abs}, nm	$\Delta\lambda^{ab}_{1/2}$, nm	D	λ_{fl}, nm	$\Delta\lambda^{fl}_{1/2}$, nm	I_{fl}, a.u.	τ_{fl}, ns
0	550	68	0.45	580	27	7.7	4.1
$6\cdot10^{-6}$	550	68	0.48	580	27	7.6	4.1
$8\cdot10^{-6}$	550	68	0.52	580	27	8.0	4.1
10^{-5}	550	68	0.54	580	27	8.6	4.1
$2\cdot10^{-5}$	550	68	0.53	580	27	8.55	4.1
$4\cdot10^{-5}$	550	68	0.52	580	27	8.5	4.1
$6\cdot10^{-5}$	550	68	0.50	580	27	8.0	4.1

It is evidence that absorbance of dye molecules increases approximately on 20% with the growth of concentration of Ag NPs (figure 4). Maxima of absorption coefficient is observed for $C_{Ag}=10^{-5}$ mol/l. Optical density decreases with further increase in the concentration of NPs. Fluorescence intensity behaves similarly. The fluorescence lifetime is not changed.

Spectra of stimulated emission of ethanol solution of Chromene-3 at different densities of the pump power are shown on the figure 4.

It is seen, that upon excitation of dye with P=0.08 MW/cm^2 Chromene-3 starts to generate the stimulated emission. Maximum of spectra of stimulated emission coincides with the maximum of fluorescence spectra. The intercity of emission at mentioned pump power density increases in 6.2 times, and the half-width of emission increases of 27 to 5 nm.

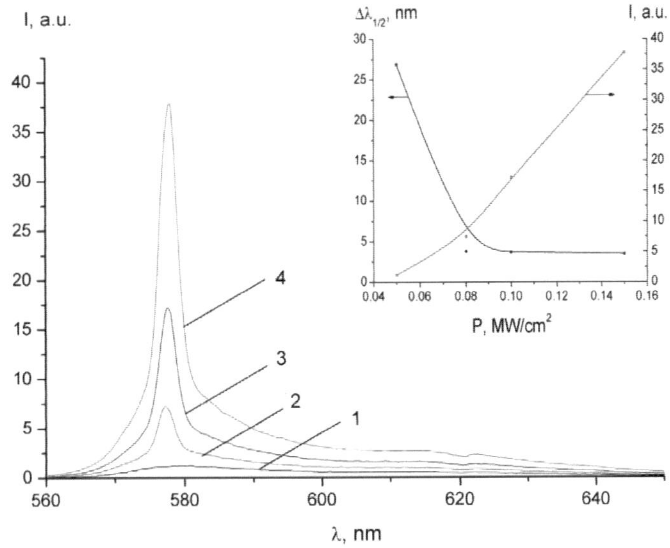

Figure 4. Emission spectra of Chromene-3 solution at various pump power densities, MW/cm^2: 1 – 0,05; 2 – 0,08; 3 – 0,1; 4 – 0,15. In the inset – dependence of the intensity and half-width of spectrum of lasing of Chromene-3 via pump power densities.

Properties of stimulated emission of dye are also dependent on the concentration of silver NPs (Table 2). Obtained dependence concerning that the lasing threshold at $C_{Ag}=10^{-5}$ mol/l decreases.

Table 2. Influence of An NPs on the lasing properties of Chromene-3

C_{Ag}, mol/l	λ_{las}, nm	$\Delta\lambda^{las}_{1/2}$, nm (P=0.05 MW/cm^2)	I_{las}, a.u.
0	580	27	12
$6\cdot10^{-6}$	580	26	13.5
$8\cdot10^{-6}$	58	22	17.5
10^{-5}	580	13	24
$2\cdot10^{-5}$	580	16	20.1
$4\cdot10^{-5}$	580	20	15.9
$6\cdot10^{-5}$	580	24	12

In the figure 5 it is shown that addition of silver NPs in dye solution leads to the increasing of intensity of emission of Chromene-3 by 2 times, and the half-width also changed in 2 times. The furthering growth in concentration leads to increasing both of intensity and half-width of band of stimulated emission of dye.

4. Conclusion

The spectral-luminescent and lasing properties of chromene dye were studied. Compounds emit light in the red range of the spectrum and could be withdrawn in biology and medicine.

Studies have shown that the addition of silver NPs in dyes ethanol solutions leads to enhancing of fluorescence intensity. The maximum of intensity of emission of Chromene-3 observed at a concentration of Ag NPs equal to 10^{-5} mol/l. A further increase in the number of NPs leads to quenching of luminescence. The life times of fluorescence for both dyes does not change with the addition of silver NP in solution.

Upon excitation of ethanol solutions of dyes by second harmonic Nd:YAG laser, the lasing of dyes was observed. For Chromene-3 lasing threshold observed when the pump source power density equal to P = 0.08 MW/cm^2. When the Ag NPs were added to the dye solutions increase in intensity and narrowing of the band of stimulated emission was observed. The lasing threshold was also reduced.

5. Acknowledgments

The authors thanks Zeynidenov A.K. and Aymukhanov A.K. for the help in the study of the properties of stimulated emission

References

[1] Garcia-Moreno I. and Costela A. 2009 *Adv. Funct. Mat.* **19(16)** 2547
[2] Xiao Y and Zhang D 2011 *Commun.* **47** 11513
[3] Shimizu S and Lino T 2013 *Commun.* **49** 1621
[4] Klimov V V 2009 *Nanoplasmonics* (Moscow: Fizmatlit) p 480
[5] Maier S A 2007 *Plasmonics*: *Fundamentals and Applications* (Berlin/Heidelberg: Springer) p 248
[6] Tian Z.Q. 2005 *J. Ram. Spect.* **36(6–7)** 466
[7] Lakowicz J.R. 2003 *Analyt. Biochem.* **320(1)**
[8] Homola J, Yee S S and Gauglitz G 1999 *Sens. Actuators B Chem.* **54(1-2)** 3
[9] Novotny L 2006 *Principles of nano-optics* (Cambridge: Cambridge University Press) p 539
[10] Oulton R F and Sorger V J 2009 *Nature* **461** 629
[11] Vasa P et al 2008 *Phys Rev Lett.* **101** 116801
[12] Kim W, Safonov V P, Shalaev V M and Armstrong R L 2009 *Phys. Rev. Lett.* **82(24)** 4811
[13] Ibrayev N Kh and Zeinidenov A K 2011 *Laser Phys. Lett.* **11(11)** 115805
[14] Ibrayev N Kh, Zeinidenov A K and Aimukhanov A K 2014 Optics and Spectr. **117(4)** 540

RESEARCH OF POLARIZATION PROPERTIES OF THE BIREFRINGENT WAVEGUIDES COUPLING POINT AT DIFFERENT POLISH ANGLES

V A Shulepov, S M Aksarin.
Saint-Petersburg National Research University of Informational Technologies, Mechanics and Optics, Saint-Petersburg, Russia

E-mail: shulepov_vladimir@mail.ru

Abstract. Integrated optical birefringent waveguides are widely used in the field of phase interferometric fiber optic sensors. In sensors integrated optical waveguides are coupled to birefringent fiber waveguides. Due to different optical properties polarization conversion occurs at waveguides coupling point which may cause parasite interference to the signal of fiber optic sensors.

1. Introduction

Fiber optic sensors (FOS) are widely used in many high-tech sectors due to their resistance to electromagnetic disturbances, possession of small weight and size parameters and relatively low cost [1]. Phase interferometric detectors (PID) are considered to belong to the most accurate ones. Among FOS a fiber optic gyro (FOG), which is based on the Sagnac interferometer [2], is recognizable. FOG is an essential instrument for naval and aircraft navigation, as well as for spacecraft orientation and stabilization. Multifunctional integrated optical scheme (MIOS) is one of the FOG's elements. It is an integrated-optical scheme comprised of channel waveguides with the X-shaped coupler topology fabricated on lithium niobate crystal, fabricated by diffusion of titanium. The advantage of the presented lithium niobate crystal is that is enables the modulation of birefringence with high efficiency, and low optical loss (about 3.5 dB), as well as coupling to a single mode fiber.

At the FOG design stage the fiber with high birefringence must be used to achieve the maximum visibility of the signal interference pattern, to improve the device accuracy and reduce the phase interferometric noise.Also, the complexity of the optical interconnection of an integrated optical waveguide to an optical birefringent fiber, since it is necessary to connect all the birefringent elements of the optical scheme with high precision matching axes [2,3]. Since the refractive indices of the integral optical waveguide and fiber lightguide are different, the back reflection occurs crystal boundary under the Fresnel's law [4]. It is partially compensated by tilting ends of the crystal by 10° [3]. Due to that the back reflection occurring on the boundary of the crystal, are partially output to the substrate but not completely.

Therefore, to avoid the parasite interference between the reflected wave trains, the waveguides are deliberately fabricated with different lengths greater than the coherence

Journal of Physics: Conference Series **735** (2016) 012019 doi:10.1088/1742-6596/735/1/012019

length of the source. The MIOS features described above are achieved due to the fact that the straight sections of the channel waveguides are positioned at the 3° angle to the optical axis of the crystal, which is displayed in Fig. 1.

Due to the refractive index differences of the fiber and crystal and its facet tilt, from the Snell's law[5], the differences between the optimal tilt angles of the fiber and crystal appear. Which are respectively shown below:

10,5° – 7 °;

19,5° – 13 °.

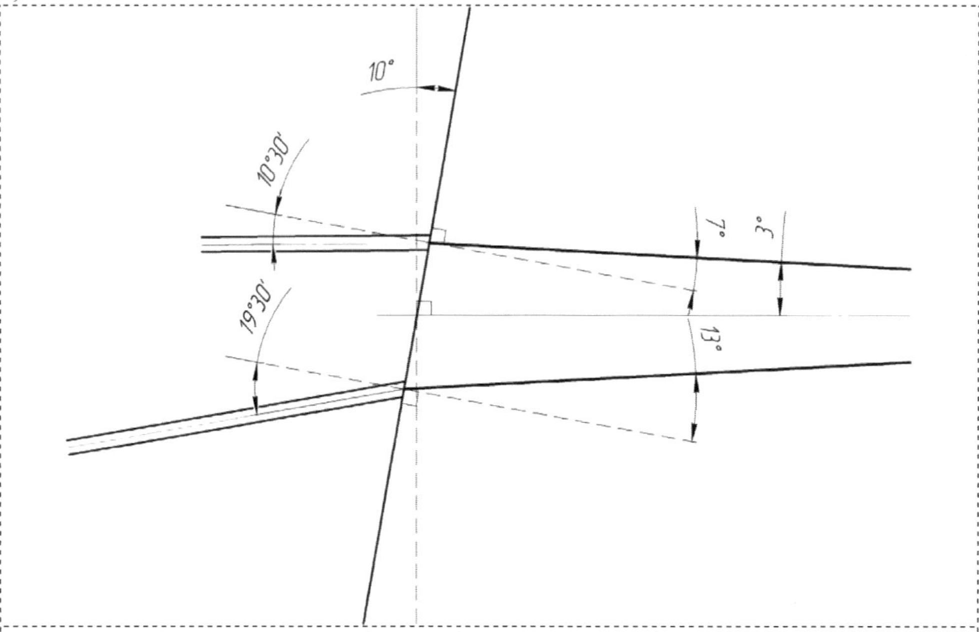

Fig. 1. The schematic figure of the MIOS to fiber coupling angles.

2. The optical scheme

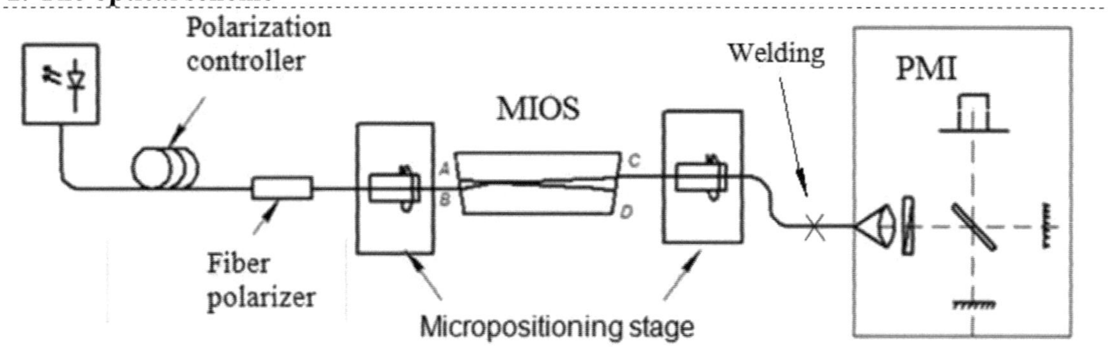

Fig. 2. The optical scheme for measuring of the conversion of the polarization state of the optical radiation.

To carry out this investigation the setup had been assembled. It consists of a light source, polarization controller, MIOS, single-mode fiber for MIOS optical radiation input/output and the polarization Michelson interferometer. The optical radiation from the source was coupled through the polarization controller through one of the MIOS ports; on the MIOS input there is the polarizer which is transmissive to one of the orthogonal components. Further, in one of the opposite ports the radiation was received with the fiber, polished at an angle 19,5°, 10,5° and 0°. After that, the optical radiation was coupled into the polarization Michelson interferometer. With the help of which, the visibility analysis had been carried out at a different displacement of one of interferometer arms, based on which the value of the coupling point polarization conversion was found.

3. The research

To avoid the of parasite interference peaks influence on the measurement results the graph of the interference pattern visibility distribution dependence on the interferometer arm displacement was obtained which is shown in Fig.1.

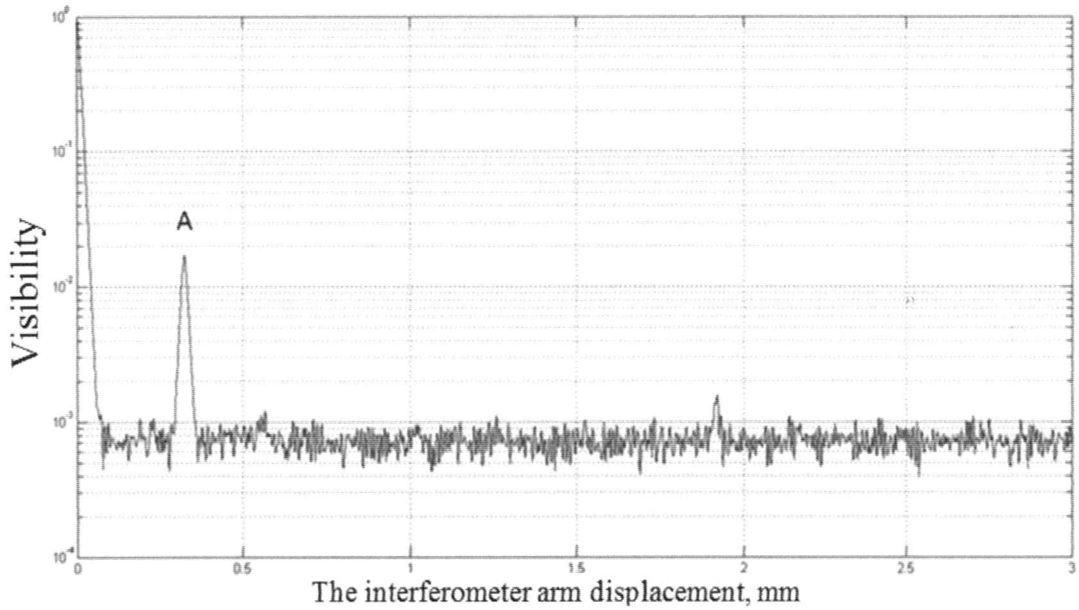

Fig. 3. The interference pattern visibility distribution dependence on the interferometer arm displacement after the light from the source passes though the polarization controller.

In fig. 3 point A - parasite interferometer peak from the radiation source.

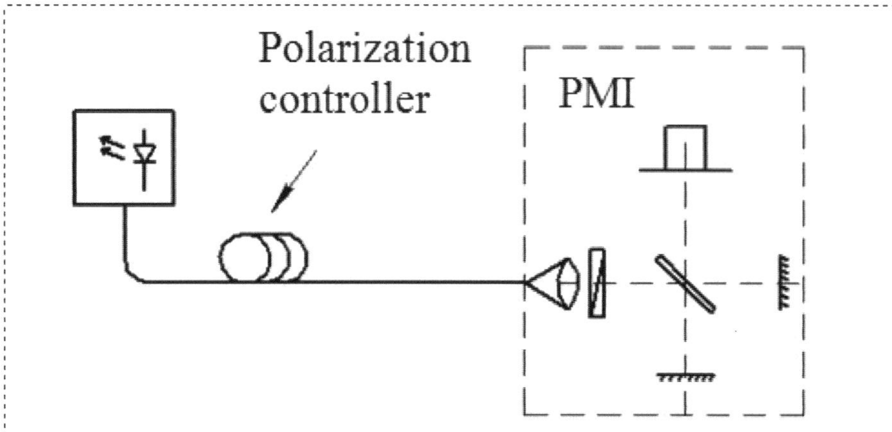

Fig. 4. Scheme, designed to eliminate the influence of parasitic interferometer peaks. PMI – polarization Michelson interferometer.

Further, the graphs of the interference pattern visibility distribution dependence on the interferometer arm displacement after the light from the source passes though the polarization controller were obtained to take measurements (Fig. 2). For the fibers with different tilt angles the extinction ratio value was measured at the coupling points with the formula (1).

$$ER = \frac{4 \cdot V_c^2}{(\sqrt{1 - 4 \cdot V_c^2} + 1)} \quad (1)$$

Where V_c - the value of the interference pattern visibility at the coupling point.
The example of scanning is shown below.

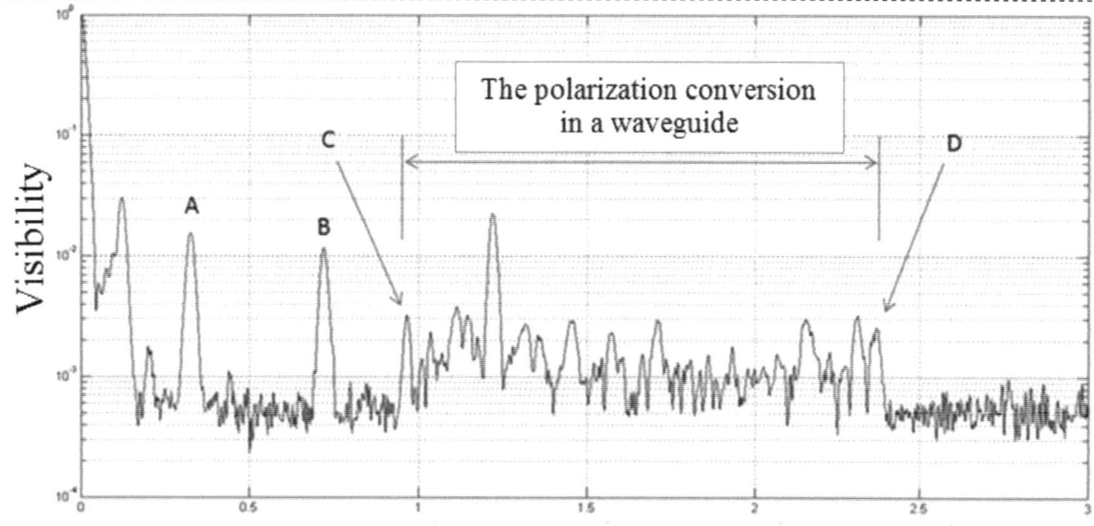

Fig. 5. The interference pattern visibility distribution dependence on the interferometer arm displacement after the light passes through the scheme used in the research.

A - parasite interferometer peak from the radiation source.
B - the conversion of the polarization state in the welding point of optical fibers.
C – the investigated coupling point of waveguide with fiber.
Segment CD – the polarization conversion in a waveguide.

The research results are presented in the table below:
Table 1. The extinction ratio at the coupling point of the fiber with the waveguide at various tilt angles of end polishing.

	Tilt angle of the waveguide	Tilt angle of end polishing	The extinction ratio at the coupling point of the fiber and the waveguide, dB
With immersion	13°	0°	-45,8
		19,5°	-57,1
Without immersion	13°	0°	-45,2
		19,5°	-50,5
With immersion	7°	0°	-46,2
		10,5°	-45,8
Without immersion	7°	0°	-39,8
		10,5°	-48,4

4. Conclusion

From the results presented in the table we can make a conclusion that the extinction ratio value degrades at the deviation from the optimum tilt angle at which the maximum radiation coupling efficiency is observed. Also, the coupling point extinction ratio with immersion it better than without it.

Acknowledgements
This work was done in ITMO University and was supported by the Ministry of Education and Science of the Russian Federation under the project 02.G25.31.0044.

References

[1] Okosi T, Okamoto K, Otsu M, Nisihara H, Kauma K and Hatate K, Fiber Optic Sensors, Leningrad: Energosamizdat, 1991. p. 256
[2] Lefevre H. The Fiber-Optic Gyroscope. London: Artech House, 1992. p. 313
[3] Аксарин С.М. *Исследование поляризационных методов и технологии согласования волоконно-оптических и интегрально-оптических волноводов.* Диссертация на соискание ученой степени к.ф.-м.н., СПб., (2014).
[4] Сивухин Д.В. *Общий курс физики. Том IV. Оптика*. Москва: "ФИЗМАТЛИТ", с. 792, (2005).
[5] Детлаф А.А., Яворский Б.М. *Курс физики. Том 3. Волновые процессы. Оптика. Атомная и ядерная физика*. Москва: "Высшая школа", с. 511, (1979).

Raman soliton generation in microstructured tellurite fiber pumped by hybrid Erbium/Thulium fiber laser system

E.A. Anashkina[1,2], M.Y. Koptev[1], S.V. Muravyev[1], V.V. Dorofeev[3], A.V. Andrianov[1], A.V. Kim[1,2]
[1]Institute of Applied Physics of the Russian Academy of Sciences, 603950 Nizhny Novgorod, Russia
[2]University of Nizhny Novgorod, 603950 Nizhny Novgorod, Russia
[3]G.G. Devyatykh Institute of Chemistry of High-Purity Substances of the Russian Academy of Sciences, 603950, Nizhny Novgorod, Russia

E-mail: elena.anashkina@gmail.com

Abstract. We demonstrate a fibre laser source generating ultrashort pulses tunable in the range 2-2.5 μm. The source is based on a hybrid Er/Tm fiber laser system and microstructured suspended-core tellurite fiber where Raman soliton shifting occurs. Nonlinear soliton dynamics is studied and possibility of tuning beyond 3 μm is shown.

1. Introduction

Ultrashort fiber laser sources widely tunable in the mid infrared (IR), including sources in the "molecular fingerprint" spectral range of 2-3 μm, are interesting for many applications: molecular spectroscopy, instrumentation, chemical sensing, material processing, ecology monitoring, biomedicine, defense and security. Soliton self-frequency shift (SSFS) in soft-glass fibers is a promising way to develop such sources via nonlinear wavelength conversion [1-3]. Tellurite glasses and fibers offer the advantages of chemical stability, wide transparency range of 0.4 – 5.5 μm, large linear ($n > 2$) and nonlinear (approximately 20 times higher than for silica glass) refractive indices, moderate melting temperature (280-440°C). A possibility of producing high quality, low loss conventional solid and microstructured tellurite glass fibers has been demonstrated successfully [2-5]. Although zero dispersion wavelength (ZDW) for the majority of tellurite glass compositions is typically beyond 2 μm, microstructured fiber geometry for dispersion management and shifting of ZDW to the range shorter than 1.5 μm can be used. This enables operation in the anomalous dispersion range with conventional Er: fiber or Tm: fiber sources, which is required for SSFS.

In this paper we experimentally and numerically study SSFS in the "molecular fingerprint" spectral range in a self-made microstructured suspended-core tellurite fiber pumped by 150-fs optical pulses at 2 μm from the hybrid Er/Tm:fiber laser source output. Wavelength tunability of the Raman solitons in the 2-2.5 μm range is achieved by pump energy varying. The verified numerical model is used to show Raman soliton shift in the range beyond 3 μm for increased pump energy.

2. Experimental and theoretical results

Figure 1 shows the optical layout of the experimental setup. It is based on a hybrid Er/Tm fiber laser source and the self-made highly nonlinear microstructured suspended-core tellurite fiber. The hybrid

Er/Tm fiber laser source consists of a passive mode-locked Er-doped fiber oscillator at 1.57 μm, Er: fiber amplifier, a silica-based fiber wavelength convertor for pulses tuning up to 2 μm due to SSFS, and Tm: fiber amplifier [6]. The hybrid Er/Tm fiber laser source can generate optical pulses at 2 μm having energy of 4 nJ and duration of 150 fs with a repetition rate of 49 MHz [6]. Suspended-core microstructured fiber with six holes has been produced from TeO_2-WO_3-La_2O_3 (TWL) glass with reduced content of 3d-transition metals and hydroxyl groups impurities by drilled preforms stretching [7]. Optical loss level in the fiber is lower than 1.5 dB/m at wavelengths of 1.1-1.6 μm with minimum of 0.7 dB/m at 1.4 μm and no more than 3.5 dB/m in the range of 1.5-2 μm. The effective core diameter of the fiber is about 3 μm.

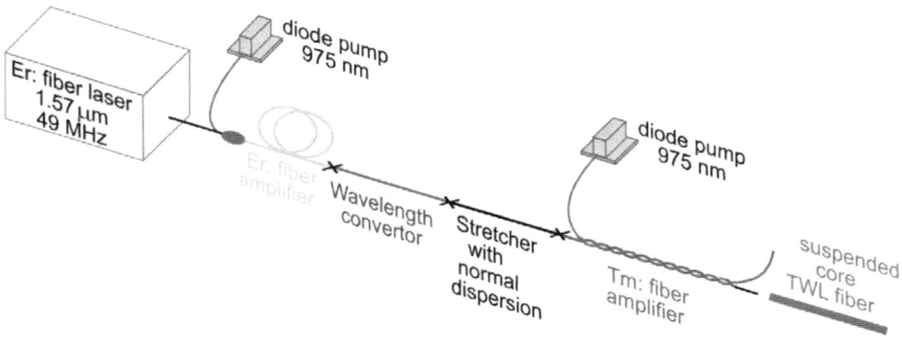

Figure 1. The optical layout of the experimental setup

To evaluate dispersion and fundamental mode field distribution of the TWL fiber, we approximate it by axially symmetric tellurite wire waveguides located in air. We applied the standard procedure for finding electric and magnetic field components and wavenumber β of the mode HE_{11} as the exact solution of Maxwell's equation for step index profiles [8]. So, the calculated, group velocity dispersion ($\beta_2 = d^2\beta/d\omega^2$, where ω is the angular frequency), changes from -260 ps^2/km for a signal at 2 μm down to -680 ps^2/km for a signal at 2.5 μm. The estimated nonlinear coefficient is 430 $(W\ km)^{-1}$ and 330 $(W\ km)^{-1}$ for these wavelengths. ZDW is about 1.55 μm.

The pulses from the hybrid Er/Tm fiber source are coupled into a 50-cm piece of TWL fiber located back-to-back. We use relatively long piece of the fiber due to low launched energy (in order of several tens of pJ). The input pulses are located in the anomalous dispersion range, the estimated soliton numbers [9] are $N = 1.5$ for 30 pJ and $N = 2.4$ for 80 pJ. So, the nonlinear dynamics associated with up to two solitons formation and their SSFS [9]. Initially we have tuned the coordinates of laser source output and TWL fiber input to maximize SSFS. After that, we have slowly changed their relative positions to waste part energy. This way of energy decreasing in the fundamental mode is very simple and allows us to save temporal and spectral structure of input signal. We obtained SSFS up to 2.5 μm. Figure 2 demonstrates experimentally measured spectra together with sech-approximations of the Raman solitons. Their spectral widths correspond to Fourier transform-limited duration in order of 100 fs.

To model pulse propagation in the TWL fiber, where ultrabroad wavelength conversion is observed, we have employed the numerical model based on the one-way wave equation dealing with the full electric field of light as in our previous paper [10]. We take into account the calculated profiles of dispersion and nonlinear coefficient, an instantaneous Kerr nonlinearity and retired Raman one (as in paper [11]), and the actual optical loss. To integrate the one-way wave equation, we use the split-step Fourier method [9].

The provided numerical simulation confirms that the red-shifted tunable pulses are high-quality sech-shape solitons. A good agreement between experimental and theoretical Raman soliton spectra for varying input energies is achieved. Indeed, the calculated pulses are Fourier transform-limited ones with duration in order of 100 fs (see Fig. 2). The experimentally measured spectral components up to

2.2 μm may correspond to signal which propagates not only in the fundamental mode but also in the cladding or in the higher modes.

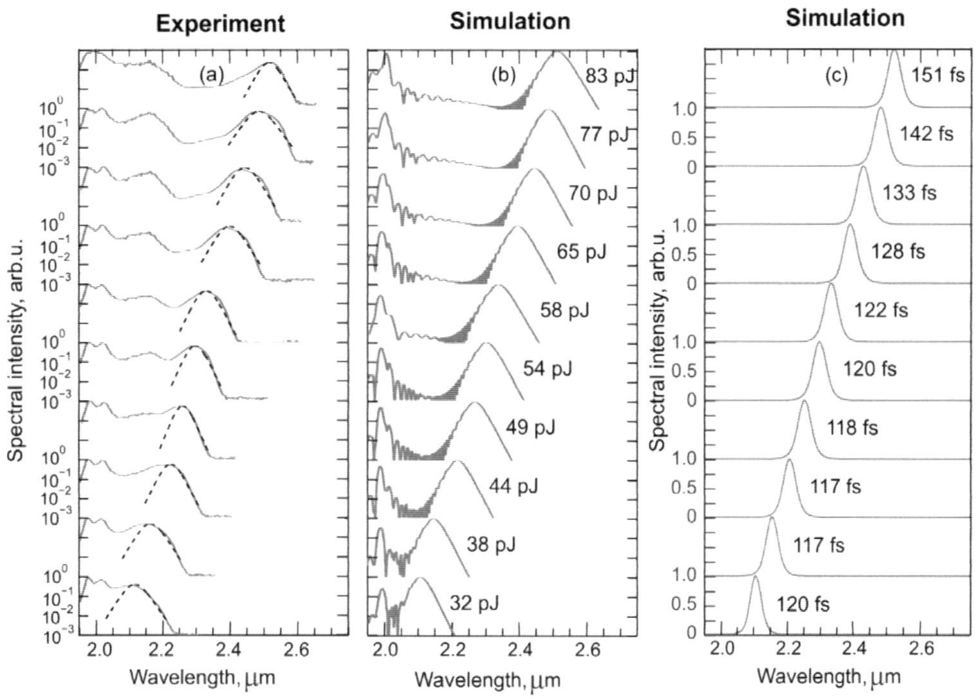

Figure 2. Wavelength tunability of the Raman solitons which is achieved by pump energy varying at the output of 50-cm piece of TWL fiber: (a) experimentally measured spectra together with sech-approximations of the Raman solitons, (b) simulated spectra with represented input energies, (c) filtered Raman soliton spectra with represented pulse durations.

In spite of experimentally reached SSFS in the TWL fiber is up to 2.5 μm with low input energy in order of several tens of pJ, we believe, the developed fiber laser system can be used for Raman soliton obtaining in the spectral range beyond 3 μm. As one can see from Fig. 2, the higher input energy, the longer Raman soliton central wavelength is. So, for spectral range extending, it is necessary to operate with higher energies and shorter fiber pieces (due to mid IR loss). We have simulated propagation of 150-fs pump pulses at 2 μm with increased energies using the verified numerical model. Figure 3 demonstrates the central wavelengths and energies of the most red-shifted solitons vs propagation distance respectively. The SSFS up to 3.6 μm for input energy of 1 nJ is limited by high mid-IR loss and large anomalous group velocity dispersion.

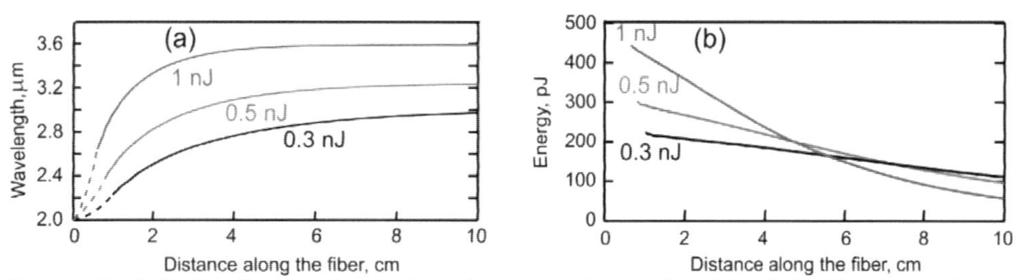

Figure 3. Calculated central wavelengths (a) and energies (b) of the Raman solitons versus propagation distance. The input pump-pulse energies are indicated.

3. Conclusion

The fiber laser source generating high-quality pulses with a spectral sech-shape and Fourier transform-limited duration in order of 100 fs tunable in the range 2-2.5 μm is presented. It is based on Raman soliton self-frequency shifting in the self-made suspended-core microstructured TeO_2-WO_3-La_2O_3 glass fiber pumped by hybrid Er/Tm system. Experimental and theoretical studies, which are in a good agreement, of nonlinear pulse dynamics in the tellurite fiber are demonstrated. The verified numerical model is used to show Raman soliton shift in the range beyond 3 μm for increased pump energy.

Acknowledgments

This work was partly supported by the Russian Foundation for Basic Research through grants 14-22-02076, 14-02-00537, 15-32-20641, 15-03-08324, 15-43-02185, and by Program of fundamental research of the Presidium of RAS.

References

[1] Cheng T, Kanou Y, Asano K, Deng D, Liao M, Matsumoto M, Misumi T, Suzuki T and Ohishi Y 2014 *Appl. Phys. Lett* **104** 121911

[2] Belal M, Xu L, Horak P, Shen L, Feng X, Ettabib M, Richardson D J, Petropoulos P and Price J H V 2015 *Opt. Lett.* **40** 2237

[3] Cheng T, Zhang L, Xue X, Deng D, Suzuki T and Ohishi Y 2015 *Opt. Express* **23** 4125

[4] Thapa R, Rhonehouse D, Nguyen D, Wiersma K, Smith C, Zong J and Chavez-Pirson A 2013 *Proc. of SPIE* **8898** 889808

[5] Ebendorff-Heidepriem H, Kuan K, Oermann M R, Knight K and Monro T M 2012 *Opt. Mater. Express* **2** 432

[6] Koptev M Yu, Anashkina E A, Andrianov A V, Muravyev S V and Kim A V 2014 *Opt. Lett.* **39** 2008

[7] Dorofeev V, Moiseev A, Kraev I, Motorin S, Churbanov M, Kosolapov A F, Dianov E, Plotnichenko V and Philippovskiy D 2012 Tungstate-tellurite glass fibers for spectral region up to 3 μm *in Advanced Photonics Congress, OSA Technical Digest (online)* paper JM5A

[8] Snyder A W, Love J 1983 *Optical Waveguide Theory* (Springer Science & Business Media).

[9] Agrawal G P 2013 *Nonlinear Fiber Optics* (Academic Press, Elsevier 5th ed.)

[10] Anashkina E A, Andrianov A V, Koptev M Yu, Muravyev S V and Kim A V 2014 *IEEE J. of Sel. Top. in Quan. Elect.* **20** 7600608

[11] Yan X, Qin G, Liao M, Suzuki T and Ohishi Y 2011 *J. Opt. Soc. Am. B* **28** 1831

Terahertz emission during interaction of ultrashort laser pulses with gas cluster beam

A V Balakin[1], A V Borodin[1,2], M S Dzhidzhoev[1], V M Gorgienko[1], M N Esaulkov[2], I A Zhvaniya[1], N A Kuzechkin[2], I A Ozheredov, A Yu Sidorov[1], P M Solyankin[1], A P Shkurinov[1,2]

[1]Faculty of Physics and International Laser Center, M.V. Lomonosov Moscow State University, Moscow, 119991, Russia

[2] Institute on Laser and Information Technologies of the Russian Academy of Sciences, Shatura, 140700, Russia

Abstract. We present the results of experimental study of terahertz (THz) generation in gas cluster beam excited by intense femtosecond laser pulses. Cluster beam was produced by partial condensation of pure Ar and mixtures CF_2Cl_2+He, Ar+He during their expansion through a conical nozzle into vacuum. There were used two excitation schemes in our experiments: single color and two color (fundamental frequency mixed with its second harmonic). We have studied how THz signal scales with various control parameters such as laser pulse duration, gas backing pressure and laser pulse energy. Simultaneously we measured intensity of X-Ray emission which originates from laser-cluster interaction. We found that in a single color scheme energy of THz pulses from Ar cluster beam strongly decreases in the region of minimum laser pulse duration while X-Ray power is maximal under these conditions. Both in single- and two color excitation regimes THz signal demonstrated growth without saturation with increasing of optical pulse energy up to its peak value of 25 mJ.

1. Introduction.

Clusters are aggregations of atoms or molecules consisting of $10...10^7$ particles [1]. Interaction of intense laser pulses with clusters is now of great interest due to the possibility of usage in various practical applications and fundamental problems of behavior matter under intense laser fields. There were observed various nonlinear phenomena during interaction of intense laser pulses with clusters: optical harmonic generation [2], self-focusing [3], X-Ray emission [4-6], stimulated Raman scattering [7]. One of the most convenient methods to produce clusters is partial condensation of gas flow during its adiabatic expansion through a nozzle [8]. Cluster beam, obtained by this technique, combines advantages of solid and gaseous media. Cluster inherits high local density and this implies high value of nonlinear response. At the same time cluster beams do not expose to ablation and renew their properties before each act of interaction with the laser pulse. It was shown theoretically and experimentally [6, 9, 10] that absorption of laser radiation in cluster beam can reach high values (up to 95%) which is related to linear (Mie) and nonlinear resonance interactions. Resonance absorption of pulse energy results in efficient production of X-Rays and fast charged particles.

In some recent articles [11-13] intense THz generation in cluster beam excited by ultrashort laser pulses was reported. Authors of [12] observed more than two orders of magnitude enhancement of THz pulse intensity in Ar cluster jet compared that in gaseous Ar with equal average atomic density. Generation of THz pulses in laser produced plasma of gases and solids was widely studied in past two decades [14-17]. Plasma source allows to obtain intense and broadband terahertz pulses. Maximum bandwidth of subpicosecond pulse can reach up to ~100 THz [16] and is limited only by laser pulse duration. In work [14] two color scheme was introduced where the fundamental laser frequency at ω is mixed with its second harmonic at 2ω. Two-color scheme allows to increase THz yield by 2…3 orders of magnitude and at the present time this scheme is widely used for THz generation in gaseous media. Nevertheless, optical to terahertz conversion efficiency in laser-plasma generation method is still low. Also there was observed saturation of THz yield in a two-color scheme at high excitant pulse energy which originates from THz absorption in dense plasma [17]. Cluster target seems to be attractive to solve these problems. As for the possible mechanism of THz emission in clusters, authors of the article [13] proposed that low-frequency radiation originates from time-varying electrical quadrupoles which are produced by ponderomotive induced charge separation in plasma. Ponderomotive force arises from radial gradient of the electric field in Gaussian laser beam and its value is proportional to ∇E^2. Authors of [13] related the enhancement of THz intensity to high local plasma density in clusters. However, complete explanation of nature of low-frequency emission from the clustered plasma is a matter of research.

The subject of the present work is experimental study of THz generation in gas cluster jets both in single- and two color excitation regimes. Simultaneously we study X-Ray emission from clustered plasma to obtain information about laser-cluster interaction.

2. Experimental setup.

The scheme of our experiment is shown if in figure 1. CPA laser system with Ti-Sapphire multipass amplifier was used as the source of high-energy optical pulses for the experiment. Our laser system delivered pulses with maximum energy of 25 mJ at repetition rate of 10 Hz. The central wavelength was 810 nm, the beam diameter 1.5 cm and the quality factor M^2=1.6. Pulse duration could be changed in a range of 50…1000 fs. Laser radiation was focused into a cluster jet by lens L1 (F = 20 cm) at the distance 6 mm below the nozzle edge. Peak laser intensity in a focal plane was estimated to be ~10^{17} W/cm^2 in vacuum. Nanosecond and picosecond contrast of laser pulse was ~10^8 and ~10^7 respectively. As it follows from our evaluation, influence of the prepulse on the laser-cluster interaction can be neglected in our experiments. Entrance window of the vacuum chamber and focusing lens L1 were made of MgF_2 which has low value of nonlinear refraction index and therefore the self-interaction effects were minimized. Second harmonic generation in a two-color scheme was obtained by 300 um-thick I-type BBO crystal. Nonlinear crystal was placed inside the vacuum chamber between the entrance window and the laser beam focus.

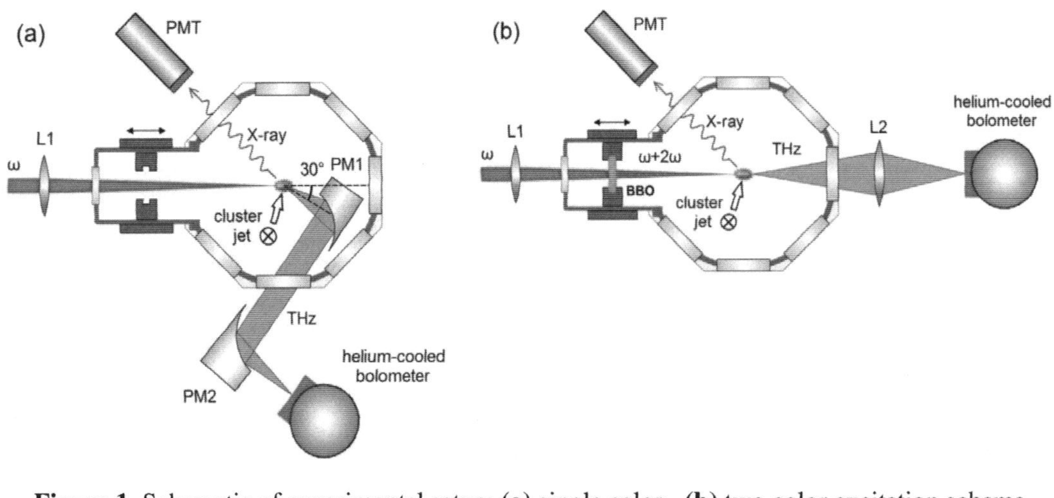

Figure 1. Schematic of experimental setup: **(a)** single color-, **(b)** two color excitation scheme.

For cluster beam production we used conical nozzle with input diameter d_{in} = 0.75 mm, output diameter d_{out} = 4 mm, half opening angle α = 5° and 20 mm length. The nozzle was connected to the high-pressure chamber with a pulsed electromagnetic valve which operated at repetition rate 1.25 Hz and was synchronized with the laser pulse. Delay time between laser pulse and the opening of the valve could be tuned in a timing module to provide the ability to control and optimize cluster formation process. Opening duration of the valve was 400 us to reach stationary jet flow. Working pressure in vacuum chamber was not above 5 mTorr. For cluster beam production we used pure Ar gas and mixtures: CF_2Cl_2+He, Ar+He with various partial concentrations of the components in mixture. Maximum value of gas backing pressure was 4 MPa. Vacuum chamber was equipped with polypropylene windows with transmission band 0…3.5 THz. Terahertz pulse energy was measured with a liquid helium - cooled bolometer (LN-6C type, Infrared Laboratories, Inc). The input THz filter of the bolometer transmitted radiation in the spectral range of 0.1…1.5 THz. In a single-color scheme THz beam was collected and focused by pair of off-axis parabolic mirrors: PM1 (F = 5 cm, D = 2.4 cm) and PM2 (F = 15 cm, D = 5 cm). Optical axis of PM2 was placed at 30° angle from the laser propagation direction in horizontal plane. There was found previously that spatial structure of THz pattern which is generated in cluster jet has four-lobed structure [13]. In our experiments PM1 collected THz radiation in one of the significant peaks of power angular distribution which is located at ~30° direction to laser beam axis. In a two color scheme THz beam was collected by TPX lens L2 (F = 10 cm, D = 5 cm) in forward direction. X-Ray emission of clustered plasma was measured by photomultiplier tube (PMT) equipped with NaJ scintillator. Beryllium window of vacuum chamber transmitted X-Ray radiation with quanta energy above 2 keV. PMT was installed in front of the beryllium window at the angle of 135° from laser propagation direction.

3. Results and discussion.

Energy dependences obtained in a single- and two color excitation regimes are shown in the figure 2 and figure 3 respectively. We have observed that THz radiation intensity scales by a power function ("ω + 2ω" scheme) or lineary ("ω" scheme) with increasing of laser pulse energy up to maximum available pulse energy of 25 mJ. Note that THz signal on graphs in figure 1 and figure 2 cannot be directly compared due to the difference in conditions of THz beams collection. Also we found that in a two color scheme usage of mixture Ar + He at concentration ratio (1:10) results in ~10-fold enhancement of THz yield compared to pure Ar jet.

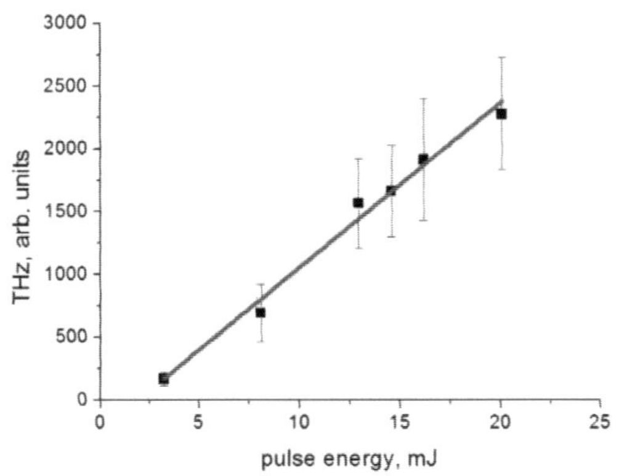

Figure 2. Energy dependence in a single color scheme, Ar cluster jet (backing pressure 2 MPa)

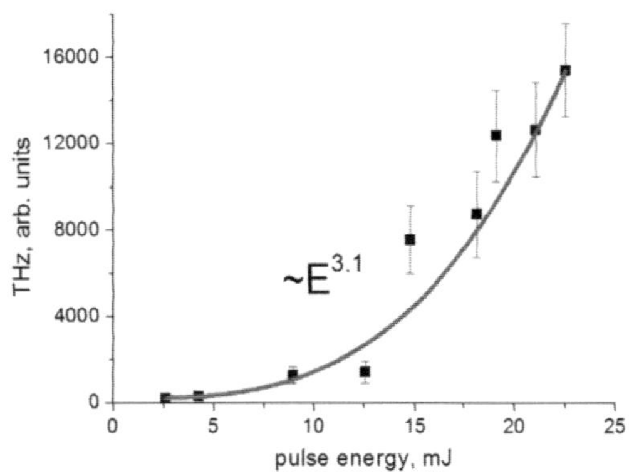

Figure 3. Energy dependence in a two color scheme, Ar + He (1:10) cluster jet (backing pressure 1.8 MPa)

In a single color scheme we observed an interesting phenomenon while we studied influence of laser pulse duration on THz emission intensity. Compressor of our laser system produced positively or negatively chirped pulse with variable duration τ which was defined by value of displacement of diffraction grating. Terahertz and X-Ray intensity scaling with pulse duration τ for positively and negatively chirping are shown in a Fig. 4 (argon cluster beam at a backing pressure of 2 MPa).

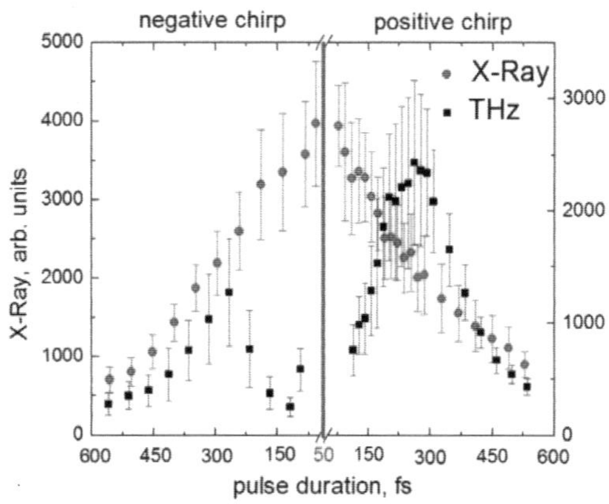

Figure 4. THz and X-Ray yield from Ar cluster jet (backing pressure 2 MPa) as a function of laser pulse duration for positively/negatively chirping. Laser pulse energy E = 22.7 mJ.

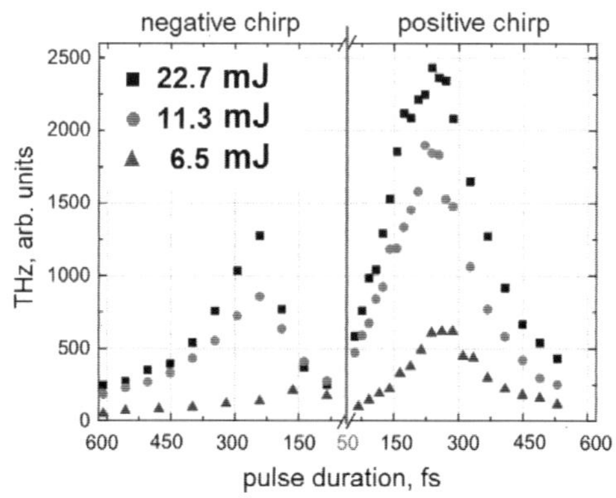

Figure 5. THz yield from Ar cluster jet (backing pressure 2 MPa) as a function of laser pulse duration for positively/ negatively chirping at different pulse energies.

As can be seen, THz yield strongly decreases in the region of minimum τ whereas X-Ray intensity demonstrates maximum values under these conditions. We suggested that THz radiation can be absorbed in the dense cluster plasma since laser intensity takes the maximum values in the region of minimum pulse duration. For further investigation we carried out similar measurement at different laser pulse energies (figure 5). Nevertheless, all curves obtained with various laser pulse energies had the same shape. Influence of pulse duration on X-Ray yield in laser-cluster interaction was studied earlier. X-Ray power scaling with pulse duration, represented in figure 4, was observed in [6] and was explained by nonlinear resonance absorption process which involves energetic electron oscillations back and forth through the cluster.

Also we have studied terahertz generation in molecular clusters of Freon gas (CF_2Cl_2) using mixture CF_2Cl_2 + He at concentration ratio (1:9). Adiabatic condensation of such mixture leads to formation of large clusters with higher atomic density compared to noble gas clusters. There was shown earlier that usage of molecule clusters allows to sufficiently increase X-Ray emission [5, 6]. Peak amplitude of THz signal from CF_2Cl_2 + He jet in our experiment was of the same order that from Ar jet. Figure 6 demonstrates influence of laser pulse width on THz and X-Ray generation in CF_2Cl_2 clusters (at mixture backing pressure of 25 atm and laser pulse energy 22.7 mJ). As can be seen, THz signal decreases in the region of minimum τ values as well as in the case of Ar clusters. Power of X-Ray emission also decreases since laser pulse has the shortest duration and this differs from previous experiment with Ar cluster jet (figure 4). A possible reason of this phenomenon may be self-interaction effects which occur inside cluster jet excited by intense laser pulses. The self-interaction of laser radiation is strong in (CF_2Cl_2 + He) cluster beam since molecular cluster acquires high internal density of electrons during ionization [5].

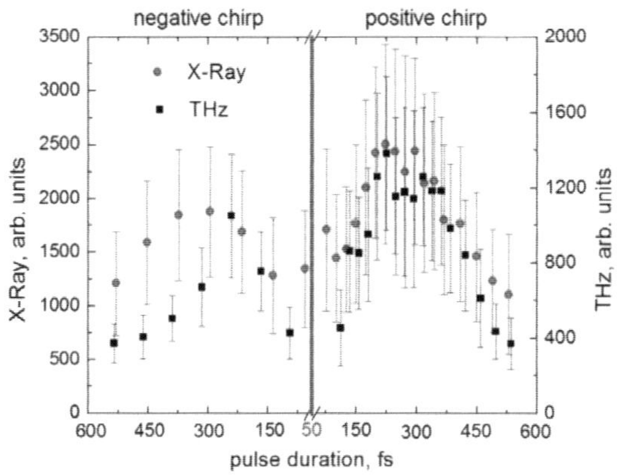

Figure 6. THz and X-Ray yield from CF_2Cl_2 + He (1:9) cluster jet (backing pressure 25 bar) as a function of laser pulse width for positively/negatively chirping. Laser pulse energy E = 22.7 mJ.

4. Conclusion.

We have experimentally investigated terahertz generation in cluster beam excited by single- and two color intense laser pulses. We found that amplitude of THz signal reaches minimum values when laser pulse has the shortest duration and this effect occurred both in atomic (Ar) and molecular (CF_2Cl_2 + He) cluster beams in a single color scheme. Scaling of terahertz radiation power with pulse duration represents dynamics of laser-cluster interaction and there is an actual problem to find theoretical explanation of the observed phenomenon. Both in single- and two color excitation regimes we observed growth of THz yield without saturation with increasing of laser pulse energy up to its maximum value of 25 mJ. We found that in a two color scheme usage of mixture Ar + He at concentration ratio (1:10) results in ~10-fold enhancement of THz pulse energy compared that in pure Ar cluster jet.

Acknowledgements

This work was supported by the Russian Foundation for the Basic Research (Grant №14-22-01098).

References.

[1] Pauly H 2000 *Atom, Molecule and Cluster Beams, Vol. 2: Cluster Beams, Fast and Slow Beams, Accessory Equipment, and Applications* (New York: Springer).
[2] Donnelly T D, Ditmire T, Neuman K, Perry M D and Falcone R W 1996 *Phys. Rev. Lett.* **76** 2472
[3] Alexeev I, Antonsen T M, Kim K Y and Milchberg H M 2003 *Phys. Rev. Lett.* **90** 103402
[4] Gordienko V M, Dzhidzhoev M S, Zhvaniya I A, Platonenko V T, Trubnikov D N and Fedorov D O 2013 *Europ. Phys. J. D* **67** 55
[5] Gordienko V M, Dzhidzhoev M S, Zhvaniya I A, Platonenko V T, Trubnikov D N and Fedorov D O, 2014 *Laser Phys. Lett.* **11** 036003
[6] Chen L M, Liu F, Wang W M, Kando M, Mao J Y, Zhang L, Ma J L, Li Y T, Bulanov S V, Tajima T, Kato Y, Sheng Z M, Wei Z Y and Zhang J 2010 *Phys. Rev. Lett.* **104** 215004
[7] Tiwari P K and Tripathi V K 2004 *Phys. Plasmas* **11** 1674
[8] Hagena O F and Obert W 1972 *J. of Chem. Phys.* **56** 1793
[9] Taguchi T, Antonsen T M and Milchberg H M 2004 *Phys. Rev. Lett.* **92** 2005003
[10] Milchberg H M, McNaught S J and Parra E 2001 *Phys. Rev. E* **64** 056402
[11] Nagashima T, Hirayama H, Shibuya K, Hangyo M, Hashida M, Tokita S and Sakabe S 2009 *Opt. Express* **17** 8807
[12] Jahangiri F, Hashida M, Nagashima T, Tokita S, Hangyo M and Sakabe S 2011 *Appl. Phys. Lett.* **99** 261503

[13] Jahangiri F, Hashida M, Nagashima T, Tokita S, Hangyo M, and Sakabe S 2013 *Appl. Phys. Lett.* **102**, 191106
[14] Cook D J, Hochshtrasser R M 2000 *Opt. Lett.* **25** 1210
[15] Li C, Zhou M L, Ding W J, Du F, Liu F, Li Y T, Wang W M, Sheng ZM, Ma JL, Chen LM, Lu X, Dong Q L, Wang Z H, Lou Z, Shi S C, Wei Z Y, Zhang J 2011 *Phys. Rev. E* **84** 036405
[16] Thomson M D, Blank V, Roskos H G 2010 *Opt. Express* **18** 23173
[17] Thomson M D, Kress M, Loeffler T, Roskos H G 2007 *Laser Photon. Rev.* **1** 349

Influence of convection on the stimulated concentration light scattering

Burkhanov[1] I.S., Krivokhizha[1] S.V., Chaikov[1,2] L.L.
[1]P.N. Lebedev Physical Institute of the Russian Academy of Sciences, Moscow, Russia.
[2]National Research Nuclear University MEPhI, Moscow, Russia.

E-mail: ilyaburkhanov@gmail.com

Abstract. A non-linear growth of the scattering intensity and the frequency shift of the spectral lines of scattered light close to the half-width of the spontaneous scattering in the back scattering of light in the suspensions of latex nanoparticles in water were found. It proves that we observed a stimulated scattering of light on the particle concentration variations. Influence of convection is taken into account using Doppler measurements of fluid flow.

1. Introduction

Each type of spontaneous light scattering has a corresponding type of stimulated scattering (SS). SS attributes are a non-linear increase in the scattered light intensity under increasing of input pump laser radiation, and frequency shift in the spectral line of scattered light, or change of the shift for SBS, STS. In case of absence of frequency shift of the spontaneous scattering line, the value of the SS line shift, in accordance with a steady state model, should be equal to half width of a spectral line of the corresponding spontaneous scattering. An interference grating of intensity is created by interference of the exciting and back scattered waves. It results in the refractive index grating, at which SS occurs [1].

The attempts to study stimulated concentration light scattering (SCoLS) - analogue of spontaneous concentration light scattering (SpCoLS) on concentration fluctuations were undertaken repeatedly in gas mixtures [2-4], in aqueous solutions [5], in the gold nanotubes suspensions [6-8].

Application of correlation spectroscopy and of continuous laser allowed us to observe the frequency shift of SCoLS on particles in liquid in contrast to previous works [6-8].

Frequency shift (the cosine component of the correlation function) and a non-linear increase in intensity SCoLS were measured first time in suspensions of silicon nanoparticles in oil, diamond in water [9] and latex nanoparticles in water [10]. It was found that the cosine component amplitude and the ratio of the scattering intensity I_{sca} to the power of the input laser beam P_{las} increased with laser power P_{las} increasing i.e. I_{sca} depends on P_{las} as a square or even more sharply.

All these facts give reason to believe that we are dealing with stimulated light scattering on variations of the particles concentration in liquid, i.e, SCoLS, analogous with the complex phenomena occurring when a STLS or SSRLW take place [11].

2. Application of dynamic light scattering to study of stimulated light scattering

In the process of stimulated concentration light scattering the interference of exciting and scattered waves creates the interference grating of intensity. Maxima (or minima) of the grating drow the particles of suspension in [12, 13]. It results in creation of the grating of particle concentration and refractive index.

$$N(z,t) = N_o + \delta N(t)\exp(-i\Omega t + i|\vec{k}_L - \vec{k}_S|z) \quad (1)$$

where $N(z,t)$, N_0 are local and mean particle concentration, δN (t) - the amplitude of concentration grating, z - coordinate, $\boldsymbol{k_L}$ and $\boldsymbol{k_S}$ - wave vectors of the exciting and scattered light.

Due to the half-width of the unshifted line of light spontaneously scattered by particles in the fluid is Dq^2, where D - diffusion coefficient, and $q = |\boldsymbol{k_s} - \boldsymbol{k_L}|$ - scattering wave vector, the SS process should lead to the appearance in the scattered light spectrum of the lines shifted relative to the excitation light to the same value $\Omega_{st} = Dq^2$, similar to STS, where $\Omega_{STS}=\chi q^2$.

$$\Omega_{st} = Dq^2 = \frac{k_b T}{6\pi\eta R}\left(\frac{4\pi n}{\lambda_{532}}\sin\left(\frac{\theta}{2}\right)\right)^2 = \frac{k_b T}{6\pi\eta R}\left(\frac{4\pi n}{\lambda_{532}}\right)^2 \quad (2)$$

where k_B-Boltzmann's constant, T - temperature, η - viscosity of the medium, R - radius of the particles, n - refractive index of the medium, λ_{532} – wavelength of green laser.

Note that the grating of concentration (1) moves with the velocity $V_{gr} = \Omega_{st}/q$.

For particle of 480 nm radius in water the theoretical calculation of SS shift from (2) gives values of 478 ± 44 sec^{-1} or 76 ± 7 Hz. The shift of such a quantity can be observed by correlation spectroscopy and can not be detected in experiments with setups using Fabry-Perot interferometers.

If the scattered light reaching the square photodetector has two lines, which are shifted relative to each other on the value of Ω, then the cosine component will appear in the correlation function of the scattered light due to the frequency beats. Analogous to [14], we can derive an expression for the intencity autocorrelation function $G^{(I)}(\tau)$ (ICF) of the scattering intensity in case of beats between the patch of the exciting light (I_h) and a line of stimulated scattering (I_{st}) in the presence of the spontaneous scattering (I_{sp}) [10]:

$$G^{(I)}(\tau) = I_{sp}^2\exp(-2\Gamma_{sp}\tau) + I_{st}^2\exp(-2\Gamma_{st}\tau) + 2I_{sp}I_h\exp(-\Gamma_{sp}\tau) + 2I_{st}I_h\exp(-\Gamma_{st}\tau)\cos(\Omega\tau) + A\langle I\rangle^2 \quad (3)$$

where Γ_{sp} is $G^{(I)}(\tau)$ the half-width of the spontaneous scattering line, Γ_{st} - half-width of the stimulated scattering line. In the case of plane waves and steady state $\Omega = \Omega_{st} = \Gamma_{sp} = Dq^2$. In the case of a stimulated process, the intensity of shifted line increase exponentially with laser pump increase, and at low intensities it should be square-low.

3. Samples and experimental setup

For samples of latex preparation we used monodisperse suspension of spherical latex particles of radius r = 480nm in water produced by company FSUE "NIISK" with a narrow size distribution and with an initial concentration of 6% by weight. We diluted the suspension by dust-free double-distilled water to a volume concentration of polystyrene $Cv = 7 \times 10^{-4}\%$.

Fig. 1. shows a scheme for registration of SCoLS. The light source was a continuous solid state laser with a wavelength λ = 532 nm.

The laser beam, having passed through polarizer P_3, which allows us to adjust the power of the exciting radiation, was focused by lens (Lens, f =15 mm) on the illuminating fiber IF input. The end of optical fiber probe includes the output of the illumination fiber IF and input of the collecting fiber CF. These ones are parallel and placed in direct vicinity to one another. The axes of the optical fibers are at a distance 0.3 mm. The probe immersed in the cuvette containing the suspension. Collecting fiber CF of probe collected light from the scattering volume. The scattered light from the collecting fiber, passing through the polarizer P_2 and a system for spatial coherence providing, SCS, arrived to the cathode of a photomultiplier. The PM pulses were sent to a correlator for obtaining SS ICF. SCS

International Conference of Young Scientists and Specialists "Optics-2015" IOP Publishing
Journal of Physics: Conference Series **735** (2016) 012022 doi:10.1088/1742-6596/735/1/012022

system consisted of aperture D_a, cathode diaphragms D_s, and a lens O. Light fiber system required almost no adjustment, except of the input of illuminating light to optical fiber and alignment of the collecting fiber end image to the diaphragm D_s. It was necessary only to make sure that the light from the fiber probe don't fall on the cell wall.

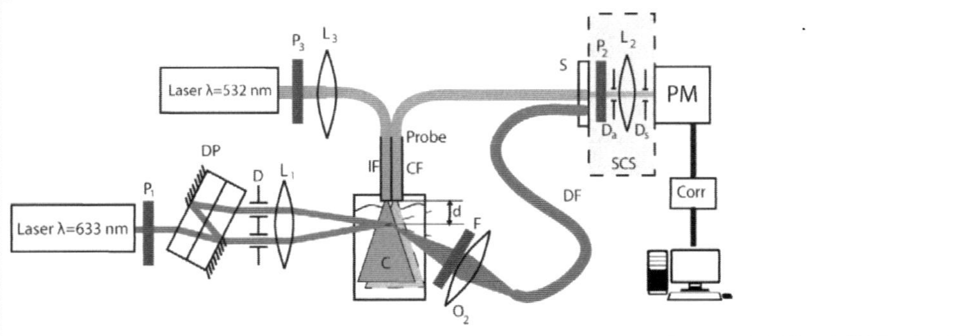

Figure 1. Optical scheme with light fiber probe. Laser - continuous lasers with λ = 532 and 633 nm; P_i - polarizers; L_i - lenses, Probe - a fiber optic probe, IF - illuminating fiber, C - cuvette, CF - collecting fiber, S - fibers switch, SCS - system for spatial coherence providing, D_a - aperture diaphragm, D_s - cathodic diaphragm (before PM), PM - photomultiplier, Corr. - Correlator, DP - beam doubler 50/50%, D - double diaphragm, F - red filter, DF - optical fiber transmitting the Doppler signal, d - distance between the end of optical fiber probe and the intersection of two red beams, d ~ 2.5 mm.

To determine and measure the role of convection velocity, we introduce to the optical circuit He-Ne laser beam, vertical doubler DP and the focus lens L1 (f=10 sm), composing typical two-beam Doppler circuit, as described in [15]. The light scattered at a small angle relative to the axis of the He-Ne laser in a horizontal plane focused by the lens O2 through a red filter F into the fiber DF comes in SCS, to the cathode of a photomultiplier PM, and PM counts passes to correlator for ICF accumulation. It was easy to move from measurement of SS to vertical velocity measurement by switching of fibers CF and DF at the entrance of the SCS system.

Cosine period of ICF of scattered light of two He-Ne laser beams gives us convective flow velocities V_c:

$$V_c = \frac{\Omega_c \lambda_{532}}{4\pi n} = \frac{\lambda_{633}}{2T_{633}\sin(\alpha/2)} \tag{4}$$

where Λ is a period of Bragg grating, λ_{633} – the wavelength of He-Ne laser, T_{633} - cosine period of ICF obtained through the fiber DF, α - the angle of convergence of beams in the air.

The velocity of convective flow is measured at the distance of about 2.5 mm from the end of the fiber probe. This distance is estimated to be the most effective for SS registration from distance dependence of the overlapping area of the cones of illumination region of IF and CF vision field and from the dependence $I_{sp} \sim 1/r^2$.

4. Results and discussion

Fig. 2. shows an increase in the amplitude of the cosinusoidal component of the correlation function with an increase of pump power for latex nanoparticles suspension with radius r = 480 nm., and Fig. 3, 4. shows the scattered light intensity increasing with pump power increasing for the particles. These two facts indicate the presence of SS.

In the experiment we have simultaneously measured total SS spectral shifts Ω_{exp} under covection influence and velocities of convection flow V_c when the fiber probe was directed up and down. The values of the convection velocity V_c was calculated from the period of the cosinusoidal ICF component T_{633} of light collected through the optical fiber DF, according to (3). From the value of the

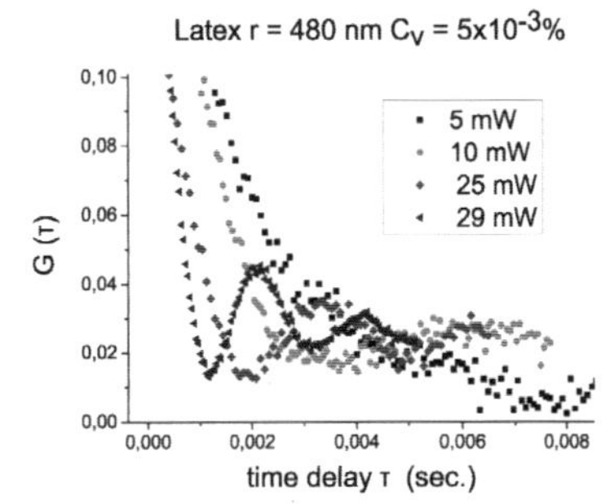

Figure 2. The correlation function of the light scattered by the 480 nm latex suspension when the power of the exciting radiation at the output of the illuminating fiber is P_{las} = 5, 10, 15, 29 mW.

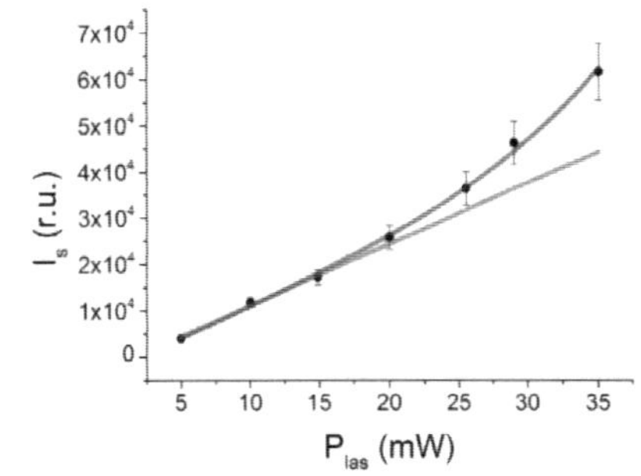

Figure 3. Dependence of the intensity of scattering in the latex suspension on the pump power P_{las}

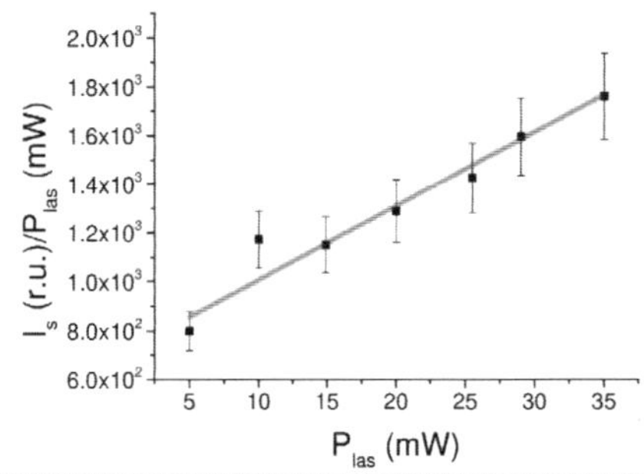

Figure 4. The ratio of the scattering intensity and the pump power P_{las} on the pump power P_{las} for the latex suspension [11].

period of ICF cosinusoidal component T_{532} of light collected through the optical fiber CF, we calculated the total velocity of the concentration SS grating V_{exp}:

$$V_{\exp} = \frac{\Omega_{st}\lambda_{532}}{4\pi n} = \frac{\lambda_{532}}{2T_{532}n} \tag{5}$$

Fig. 5,6. represents the dependences of the velocity of convective flow V_c and the total concentration SS grating velocity V_{exp} on the power of the exciting beam. One can see that for the position of the fiber probe "down" $V_{exp} < V_c$. This indicates that in such position of fiber the speed of the wave concentration V_{gr} relative to the fluid is directed from the fiber and opposite to the velocity of convection flow V_c is more then V_{exp}, and $V_{exp} = V_c - V_{gr}^{DOWN}$ indicating that SS friquency shift is Stokes. For the position "up" $V_{exp} > V_c$, therefore $V_{exp} = V_c + V_{gr}^{UP}$, velocities V_c and V_{gr} are directed in the same direction and have be added, so as shift SS is also Stokes. This shift is defined as:

$$\Omega_{\exp} = \frac{4\pi n V_{gr}}{\lambda_{532}} \tag{6}$$

Fig. 7 shows clean SS spectral shift, calculated from V_{gr} by (5). For the pump power more than 15 mW (when there is SS) the mean SS friquency shift is $\Omega_{st(exp)} = 643 \pm 55$ s^{-1}. The discrepancy between this experimental value and theoretical one $\Omega_{st} = 478 \pm 44$ s^{-1} is of about 12%.

International Conference of Young Scientists and Specialists "Optics-2015" IOP Publishing
Journal of Physics: Conference Series **735** (2016) 012022 doi:10.1088/1742-6596/735/1/012022

5. Conclusion

So we believe that we have detected presence of SS on the particle concentration variation in the liquid and measured the frequency shift of the spectral line of the SS, earlier [11] in a steady state regime, and now in condition of the convection presence.

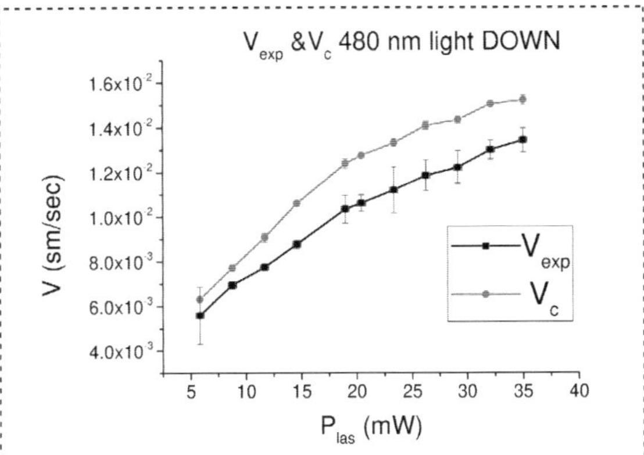

Figure 5. The dependence of the total velocity of the concentration SS grating V_{exp} and convection flow velocity V_c on the power of the exciting beam at the position of the light fiber probe down

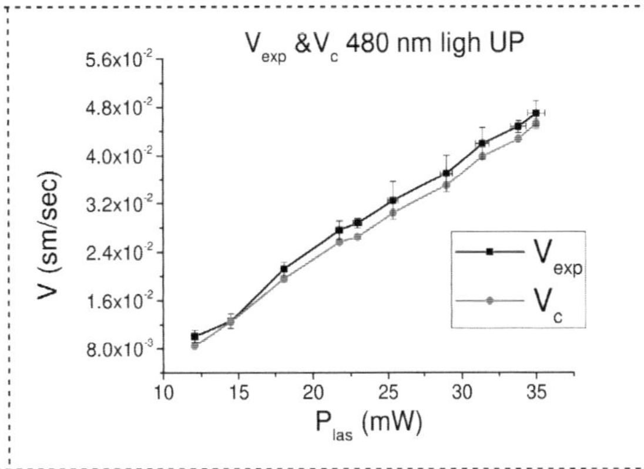

Figure 6. The dependence of the total velocity of the concentration SS grating V_{exp} and convection flow velocity V_c on the power of the exciting beam at the position of the light fiber probe up

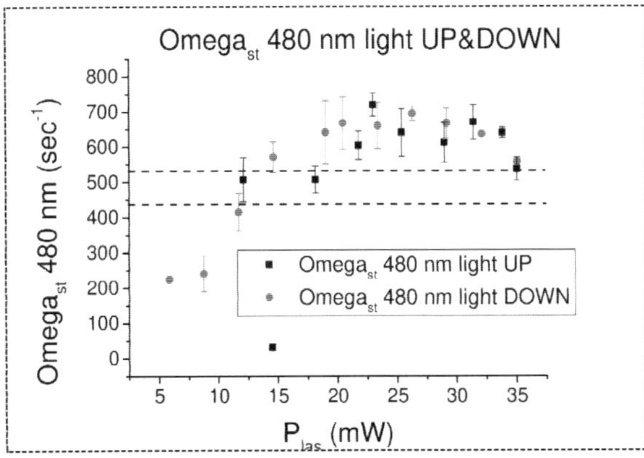

Figure 7. The values of the SS spectral shift obtained from formula (5) under increasing in the pump power. Theoretical limits are denoted by shaded lines.

6. References

[1] Boyd R.W 2003 *Nonlinear Optics* Academic Press **409**
[2] Aref'ev I M, Morozov V 1969 *JETP Letters* **9** 269
[3] Lowdermilk W H, Bloembergen N 1972 *Phys. Rev.* A **5** 1423
[4] Bloembergen N, Lowdermilk W H, Matsuoka M, Wang C S 1971 *Phys. Rev.* A **3** 404
[5] Davydov M A, Shipilov K F 1990 *JETP Letters* **52** 153
[6] He Guang S, Ken-Tye Yong, Jing Zhu, and Prasad P N 2012 *Phys. Rev.* A **85** 043839
[7] He Guang S, Wing-Cheung Law et al. 2012 *Appl. Phys. Lett.* **101** 011110
[8] He Guang S, Wing-Cheung Law et al. 2013 *J. Of Chem. Phys.* **138** 024202
[9] Burkhanov I S, Chaikov L L 2012 *Bull. Lebedev Phys.* Inst **39** 77
[10] Burkhanov I S, Chaikov L L 2013 *Proc. SPIE* 9066 906610
[11] Starunov V S, Fabelinskii I L 1970 *Sov. Phys. Usp.* **13** 428

[12] Afanas'ev A A, Rubinov A N et al. 2007 *Optics and Spectroscopy* **102** 106
[13] Zemanek *et al* 2002 *J. Opt. Soc. Am.* A **19** 1025
[14] Palberg T, Reiber H, K¨oller T at.al. *arXiv*:0811.2321v1
[15] Krivokhizha S V, Lapteva E S, Chaikov L L et al. 2011 *Bull. Lebedev Phys. Inst* **38** 111

Journal of Physics: Conference Series **735** (2016) 012023 doi:10.1088/1742-6596/735/1/012023

Demonstration of vortical beams spectral stability formed in non-zero diffraction orders

S.A. Degtyarev[1,2], A.P. Porfirev[1,2], S.N. Khonina[1,2], S.V. Karpeev[1,2]
[1]Image Processing Systems Institute, 151 Molodogvardejskaya street, 443001 Samara, Russia
[2]Samara State Aerospace University, 34 Moskovskoye shosse, 443086 Samara, Russia

E-mail: sealek@gmail.com

Abstract. We provide a comparative analysis of vortical beams forming by a spiral phase plate and a binary fork diffraction grating. Intensity distributions formed by a diffractive optical element (DOE) have distortions whereas DOE's relief height is not corresponds to the wavelength of an incident beam. These distortions are numerically studied in this work. Different vortical beams are experimentally created and studied. The natural research is conducted with a spatial light modulator and with binary DOE. We experimentally show that fork binary DOEs are resistant to small deviation of incident beam wavelength and vortical beams are formed without any significant distortions.

1. Introduction

A vortical beam former is an in-demand diffractive optical element [1-3]. For this problem, a balance between an energy efficiency and accuracy of required intensity distribution is very critical. This balancing is very common for DOE manufacturing. The reason of these troubles is close tolerance requirements for a phase transmission function. But as usual high accuracy cannot be achieved because of the manufacturing specificity.

Spiral phase plate is one of the simplest devices of singular optics. It was theoretically offered in 1984 [4]. For the first time the binary phase plate was manufactured in 1992 [5]. To this time multilevel DOE manufacturing has been a significant problem. Nevertheless, DOEs cannot be replaced with spatial beams modulators (SLM) because of weak diffraction efficiency of SLM. So a multilevel phase is often coded into a binary-level relief.

If we carefully consider two main ways of phase modulation with DOE, we can find that these methods have absolutely different requirements to the accuracy of manufacturing. Thus, the element for zero-order diffraction phase modulation should have a relief, each part of that produces certain required phase delay.

For vortical phase modulation DOE relief is a helical surface. Integrating the phase of the field around a path enclosing a vortex yields an integer multiple of 2π. However, fabricating inaccuracies of a helical surface causes a similar deviation of formed phase transmission, in contrast to diffractive lenses, which manufacturing deviation only decreases diffraction efficiency but quality degree of point spread function remains on the same level.

However, modern technology level allows us to avoid the problem of close tolerance DOE manufacturing requirements. Obviously, another unavoidable trouble of phase modulation can be a deviation of incident light from the monochromatic wave. This problem plays a main role in data

transmission multiplexing [6-8], because the pulse character of utilized light causes a widening of the spectrum.
All of these reasons make a DOE extremely difficult to use for optical vortices forming in zero diffraction order. But this problem can be solved with using a DOE, which works in non-zero order of diffraction. The fundamental distinguishing difference is that phase modulation is produced by grating bars shifting, instead of phase delaying by optical thickness whereas a DOE works in zero-order. Technologically, optical thickness controlling is more difficult than grating bars shifting because the bars have a constant height. But we should pay with diffraction efficiency for the simpler manufacturing of non-zero DOE, which can provide 40% of efficiency in the plus and minus first order of diffraction.
Moreover, fortunately, related to spectrum widening errors cause only a decreasing the energy efficiency. Indeed, formed wave front shape is kept in good quality whereas incident beam wavelength deviates from the main meaning. Wavelength deviation causes some scale changing because of changing of diffraction angles. However, vortical beams possess unique quality to be invariant to scale-changing. It means that the phase of the vortical beam remains a vortical character whereas the wavelength deviates, just the amplitude distribution changes.

2. Numerical simulation

In this section we show the results of numerical simulation for evaluating of distortions, which arise whereas a DOE is illuminated with deviated wavelength λ. We should mention that the thickness of element's relief corresponds to the main value of the wavelength λ_0.

$$h(x, y) = \frac{\operatorname{mod}_{2\pi}[\psi(x, y)]}{2\pi(n-1)} \lambda_0, \tag{1}$$

where $\psi(x, y)$ is phase of the element, n is refractive index of DOE's material.
For the simulation we use integral transform, which describes the diffraction on thin DOE with a spherical lens:

$$G(u,v) = \frac{2\pi}{\lambda f} \iint_{\Omega} A(x, y) g(x, y) \exp\left[-\frac{i2\pi}{\lambda f}(xu + yv)\right] dxdy, \tag{2}$$

where $A(x,y)$ is an incident beam with wavelength λ, $g(x,y)$ is a complex transmission function of the DOE, f is focal length. The incident beam has Gaussian energy distribution.
We study to types of diffraction optical elements for optical vortices forming. The first type is a multilevel spiral phase plate [5, 9], which is used for optical vortex generation in zero diffraction order. The binary spiral phase plate was considered in the paper [10].
Technology of fabricating multilevel DOE requires is very difficult. In some cases they can be replaced by spatial light modulators whereas diffraction efficiency does not play any significant role.
A spiral phase plate which produces m-order optical vortex, has following complex transmission function:

$$g_m^v(x, y) = \exp\left[im \operatorname{arctg}\left(\frac{y}{x}\right)\right]. \tag{3}$$

Another type of DOE is a fork diffractive grating [11, 12] which forms two optical vortices with topological charges of ±1 in ±1 orders of diffraction respectively. Complex transmission function of the fork grating is

$$g_m^b(x, y) = \exp\left\{ i \arg\left[\cos\left\{ax + m \operatorname{arctg}\left(\frac{y}{x}\right)\right\}\right]\right\}. \tag{4}$$

where a is a factor corresponding to the carrying frequency of the grating.
Gaussian beam transverse intensity distribution (figure 1a), the phase of the spiral phase plate for order m=1 (figure 1b), and the phase of the fork grating for order m=1 (figure 1b) are shown in the figure 1.

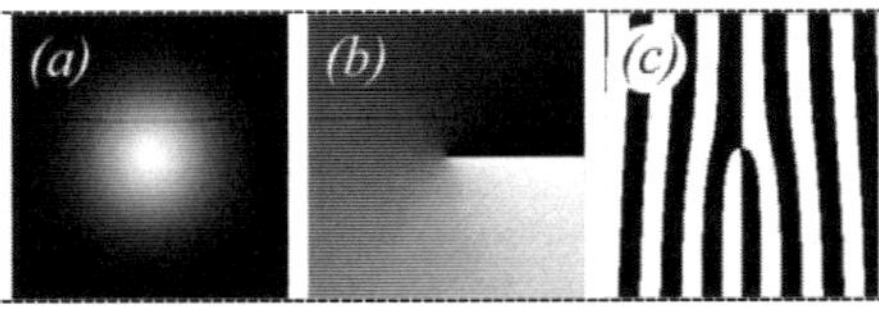

Figure 1. Gaussian beam transverse intensity distribution (a), phase of the spiral phase plate for order m=1 (white color means 2π radian, black color means 0 radian) (b), and phase of the fork grating for order m=1 (white color means π radian, black color means 0 radian) (c).

If the DOE with relief thickness $h(x,y)$ is illuminated by an incident beam with the wavelength λ, suppose that the phase of the incident beam is changed as follow:

$$\psi_{\lambda}(x, y) = \frac{\lambda_0}{\lambda} \psi_{\lambda_0}(x, y). \tag{5}$$

In table 1 we show the results of simulation of vortical beams forming with phase helical plates for m=1,2,3, in case of the incident beam has different wavelength.
From the results we can see that if the incident beam wavelength λ deviates from the base wavelength λ_0 (which corresponds to the element thickness), formed beam has very bad quality, and actually, it is not a vortical beam. Moreover, beams are formed with worse quality for greater phase topological charge orders.

Table 1. Vortical beams forming by means of helical phase plates with orders m=1, 2, 4. The incident Gaussian beam has different wavelengths (negative intensity distributions are shown).

Wavelength	m=1	m=2	m=4
λ=432 nm	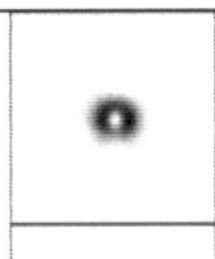	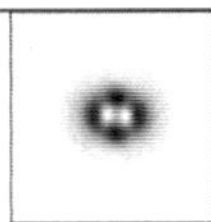	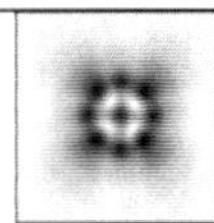
λ_0=532 nm	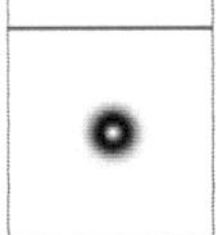	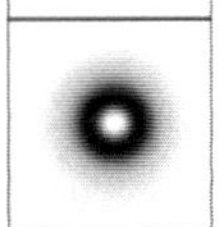	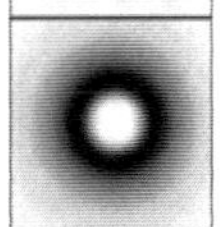
λ=632 nm	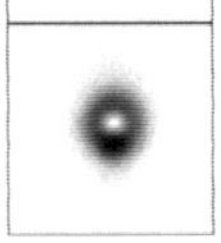	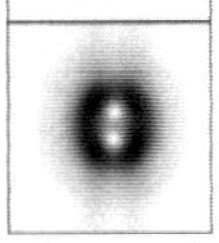	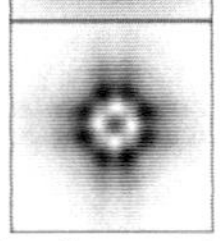

Notice, that wavelength deviation causes separating of m-order vortical phase singularity into m shifted first-ordered vortices. The same situation was mentioned in the works [11, 13] for adjustment errors of DOE and the incident beam. Similar picture arise whereas spiral phase plates of fractional orders are used in [14] and if focusing element has periodical angle dependence as well as in [15].
In table 2 we show the results of simulation of vortical beams forming with fork diffractive gratings in ±1 orders for m=2, 4, in case of the incident beam has different wavelength.
From the tables 1, 2 we can see that incident beam wavelength deviation from the base meaning does not cause the destroying formed vortical beam. Moreover, produced optical vortex quality does not depend on singularity topological charge m.
The main negative effect lies in that fact that the greater the departure of the incident wavelength from the base meaning, the greater the central peak. It means that only diffraction efficiency becomes less, but vortical beam structure does not undergo disturbance.

In addition, we should notice, that wavelength mismatch results to shifting of diffraction orders in the focal plane. Obviously, in that case, a diffraction grating displays its dispersive behavior.

Table 2. Vortical beams forming with fork diffractive gratings in ±1 orders for *m*=2, 4. The incident Gaussian beam has different wavelengths (negative intensity distributions are shown).

Wavelength	*m*=2	*m*=4
λ=533 nm	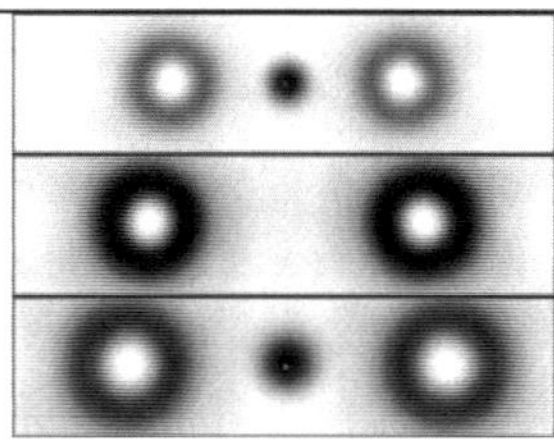	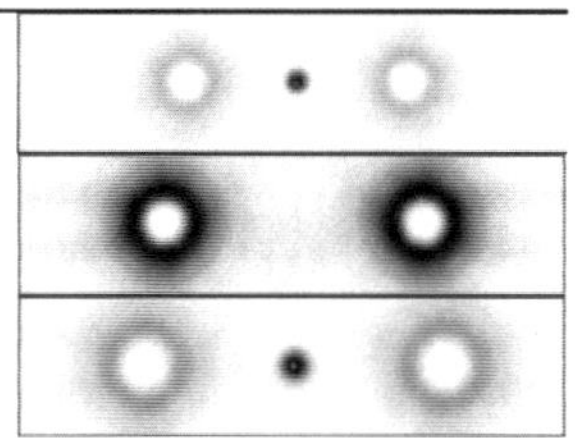
λ₀=633 nm		
λ=733nm		

3. Experimental study

For experimental studying vortical beams forming we have provided two series of experiments. In the first series we investigate vortical beams producing with spiral phase plates. The spiral phase plate was obtained with the spatial light modulator SLM PLUTO-VIS. Incident laser beam has the wavelength of 532 nm. The display of the spatial beam modulator represents the phase distributions of spiral phase plates with different relief thickness. After illuminating the display by laser we take photographs of formed vortical beams intensity distributions. DOE phases is calculated with equation (6). In that way, we investigate DOE relief thickness influence to quality of formed vortical beams. In table 3 we show experimentally received intensity distributions of vortical beams with topological singularity charges of 1, 2 and 4. We vary the relief thickness of the DOE. We can see quite good agreement between experimentally received distributions and simulation results from table 1.

Table 3. Experimental forming of vortical beams with topological singularity charges of 1, 2 and 4, which are formed with the SLM representing spiral phase plates. Laser wavelength is 532 nm.

Wavelength	*m*=1	*m*=2	*m*=4
λ=432 nm	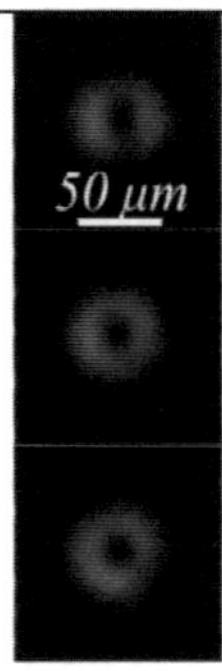		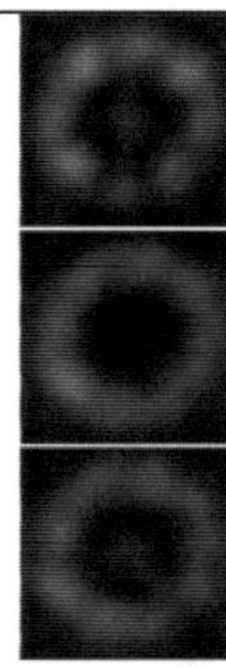
λ₀=532 nm			
λ=632 nm			

Meanwhile, we investigate forming of ±2 and ±4-orders vortical beams by a multi-order DOE. The relief thickness is about 690 nm. According to the expression (1) initial wavelength should be 633 nm because of refractive index of silica glass is 1.46. For demonstration of DOE resistance to spectrum shifting we use laser EKSPLA NT200 with tunable wavelength. In table 4 we show experimentally received intensity distributions of vortical beams with topological singularity charges of ±2 and ±4, which are formed with silica glass multi-order DOE. We vary the wavelength of incident light from 533 nm to 733 nm.

Table 4. Experimental forming of vortical beams with topological singularity charges of 2 and 4 in the +1 diffractive order. Initial laser wavelength is varying from 533 nm to 733 nm.

Wavelength	*m*=2	*m*=4
λ=533 nm	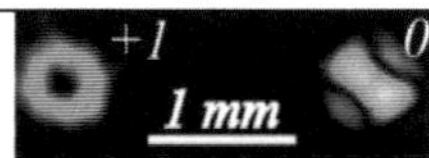	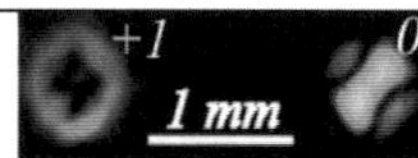
λ_0=633 nm	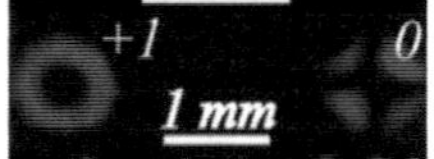	
λ=733nm		

4. Conclusions

We provide the comparative analysis of vortical beams forming by spiral phase plates and binary fork diffraction grating. We show that spiral phase plate does not work whereas relief thickness does not correspond to the incident beam wavelength. The greater the vortex order, the worse the quality of formed beam. However, fork-type multi-order diffractive optical elements which produce optical vortices in ±1-orders are resistant to spectrum shifting. Incident beam wavelength deviations only cause energy losing and shifting of image in the focal plane. Experimental study with the SLM and the binary fork-type multi-order DOE verifies received numerical simulation results.

Acknowledgments

This work was supported by the Ministry of Education and Science of the Russian Federation and also RFBR grants 13-07-00266 and 14-07-31079 mol_a.

References

[1] Soskin M S and Vasnetsov M V 2001 *Progress in Optics* **42** 219-76
[2] Padgett M J, Courtial J, and Allen L 2004 *Phys. Today* **57** 35-40
[3] Rockstuhl C, Salt M, and Herzing H P 2004 *J. Opt. A* **6** S271
[4] Berezny A E, Prokhorov A M, Sisakyan I N and Soifer V A 1984 *Sov. Phys: Doklady* **274** 802
[5] Khonina S N, Kotlyar V V, Shinkaryev M V, Soifer V A, Uspleniev G V 1992 *J. Mod. Opt.* **39** 1147-54
[6] Khonina S N, Kazanskiy N L and Soifer V A 2012 Optical vortices in a fiber: mode division multiplexing and multimode self-imaging *Recent progress in optical fiber research* ed M Yasin, SW Harun, H Arof (Croatia: INTECH)
[7] Bozinovic N, Yue Y, Ren Y, Tur M, Kristensen P, Huang H, Willner A E, Ramachandran S 2013 *Science* **340** 1545-48
[8] Huang H et al 2014 *Opt. Lett.* **39** 197-200
[9] Kotlyar V V, Almazov A A, Khonina S N, Soifer V A, Elfstrom H, Turunen J 2005 *J. Opt. Soc. Am. A* **22** 849-61
[10] Heckenberg N R, McDuff R, Smith C P, White A G 1992 *Opt. Lett.* **17**, 221-23
[11] Bazhenov V Yu, Soskin M S, Vasnetsov M V 1992 *J. Mod. Opt.* **39** 985-90
[12] Khonina S N, Balalayev S A, Skidanov R V, Kotlyar V V, Paivanranta B, Turunen J 2009 *J. Opt. A: Pure Appl. Opt.* **11** 065702
[13] Kotlyar V V, Almazov A A, Khonina S N, Soifer V A, Elfstrom H, Turunen J 2005 *J. Opt. Soc. Am. A* **22** 849-61
[14] Zhang N, Davis J A, Moreno I, Lin J, Moh K-J, Cottrell D M, Yuan X 2010 *Appl. Opt.* **49** 2456
[15] Khonina S N, Ustinov A V 2012 *Comp. Opt.* **36** 72-79

Numerical simulation of interaction of few-cycle pulses counter-propagating in the optical fiber

L S Konev[1], **Yu A Shpolyanskiy**[1]
Department of photonics and optical information
ITMO University, Saint-Petersburg, Russia

E-mail: leonides.konev@gmail.com

Abstract. Interaction of few-cycle pulses counter-propagating in an optical fiber is studied numerically via solution of equations for bi-directional fields that are equivalent to the full scalar wave equation. It is simulated how 3-cycle optical pulse of the Ti:Sa laser is gained as it propagates through the field of the pulse on the second harmonic with higher intensity in the telecommunication type single-mode optical fiber. Rise of sixths harmonic is also observable under considered conditions.

1. Introduction

Qualitative and quantitative description of a few-cycle pulse evolution is one of the most relevant challenges for the modern nonlinear optics. Extremely short durations of such pulses allow for concentration of high peak intensities with a lower amount of the input power, which in turn leads to gain in efficiency of nonlinear light-matter interactions boosting all sorts of nonlinear phenomena. In the linear regime low-intensity pulses propagate independently from each other, so that the resulting field is simply the superposition of those pulses. High-intensity pulses affect each other via nonlinear response of the propagation medium so that the resulting field differs from the simple superposition of electromagnetic waves at the given point in space and time. Some investigations on characteristics of these differences are presented in [1, 2]. However the articles imply quite a few analytical simplifications and do not contain straightforward numerical solution for the equations describing evolution of the counter-propagating pulses.

2. Set of equations

A set of equations for forward and backward waves can be utilized for description of evolution of few-cycle pulses counter-propagating in the optical fiber [3]:

$$\partial_z G_\pm = \pm ik(\omega)G_\pm \pm \frac{1}{2} ik(\omega)N_\omega(E_+ + E_-), \qquad (1)$$

$$N_\omega(E) = 4\pi F[P_{NL}(E)]/n^2(\omega), \qquad (2)$$

where z is the coordinate along the axis of the fiber, $G_\pm = F[E_\pm]$ is the spectral density of the fields of the forward and backward waves $E_\pm$, F is Fourier transform, t is time, ω is frequency, i is the imaginary unit, $k(\omega) = \omega n(\omega)/c$ is the wavenumber, $n(\omega)$ is the refractive

[1] To whom any correspondence should be addressed.

index, c is the speed of light, N_ω is the nonlinear operator in the frequency domain and P_{NL} is the nonlinear response of the medium.

3. Medium parameters

Frequency dependency of refractive index of the silica glass within the range of transparency can be expressed as power series of frequency [4]:

$$n(\omega) = N_0 + ac\omega^2 - \frac{bc}{\omega^2}, \tag{3}$$

where $N_0 = 1.4508$, $a = 2.7401\cdot 10^{-44}$ $s^3\cdot cm^{-1}$, $b = 3.9437\cdot 10^{17}$ $s^{-1}\cdot cm^{-1}$.

Nonlinear response can be described with phenomenological model suggested by Platonenko and Khokhlov [5]:

$$\begin{cases} P_{NL} = \chi_3^e E^3 + \chi_3^{ev} RE \\ \dfrac{\partial^2 R}{\partial t^2} + \dfrac{1}{T_v}\dfrac{\partial R}{\partial t} + \omega_v^2 R = \gamma E^2 \end{cases}, \tag{4}$$

where χ_3^e and χ_3^{ev} are the cubic nonlinear susceptibility of electronic and electronic-vibrational nature, respectively, R is the amplitude of the molecular oscillations, T_v is the relaxation time, ω_v is the Stokes frequency and γ is the damping coefficient.

4. Results of simulations

Equations (1) allow us to represent two counter-propagating pulses as forward and backward waves, respectively. Therefore it seems convenient to investigate how a high-intense pulse influences the field of counter-propagating pulse with lower intensity by numerical solution of equations (1). Within such approach initial field distribution of the high-intensity pulse was represented as the field of the backward wave with Gaussian profile and central frequency of 390 nm (the second harmonic of Ti:Sa laser) and duration 7.8 fs (6 field cycles), and pulse with the lower intensity was represented as the field of the forward wave of Gaussian form with central frequency 780 nm (the first harmonic of Ti:Sa laser) and duration of 7.8 fs (3 field cycles). If initial time delay between the pulses is 50 fs, then the distance in optical waveguide, which is enough for the pulses to propagate through each other with interaction and even accumulate some time delay is about 10 μm.

Since the forward and backward waves are interrelated via the nonlinear term in equations (1), a non-zero forward wave gives rise to non-zero backward and vice versa. Such behavior is independent of initial conditions, hence to exclude peculiarities of the directional fields separation it is worth considering the total field $E = E_+ + E_-$ and compare it between linear and nonlinear regimes. In order to do this comparison we run three numerical experiments. For the first it was assumed that the forward wave is absent, but the intensity of the backward wave is $I = 2\cdot 10^{13}$ W/cm^2. For the second the initial field of the backward wave was set to zero, while the initial amplitude of the forward wave was 5 times lower than that of the backward wave in the first experiment. And finally we considered the initial distributions with both waves presented and amplitudes of backward and forward waves taken from the first and second experiments, respectively. The sum of resulting fields from the first and second experiments $E_1 + E_2$ corresponds to independent propagation of pulses, while the total field E_3 of the third experiment is the result of their interaction. The spectral density of the difference of these fields $G^{diff} = G_3 - (G_1 + G_2)$ depicted in figure 1 illustrates the effect of nonlinear interaction between counter-propagating pulses.

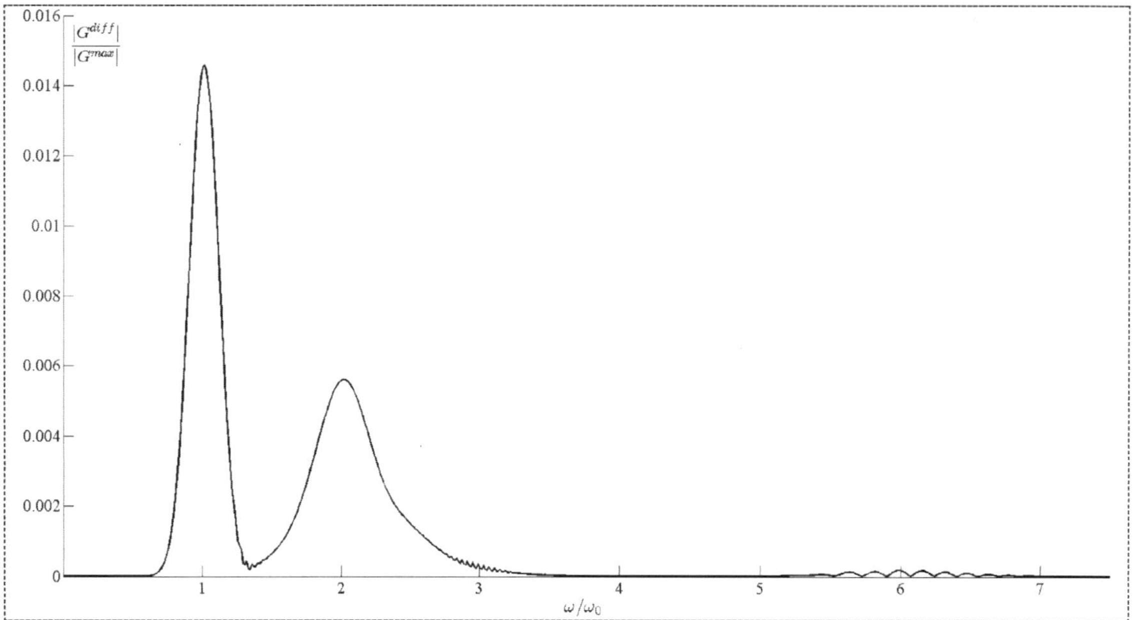

Figure 1. The spectral density of the difference of the resulting fields in non-linear and linear regimes.

It is seen that the interaction leads to the gain of waves on basic frequencies, and the wave on the first harmonic of Ti:Sa laser corresponding to the pulse with the lower intensity is affected more than the one on the second harmonic corresponding to the pulse with the higher intensity. Higher impact of the more intense pulse on the weaker seems justified. Besides one can observe the gain of the sixths harmonic, which corresponds to the triple frequency of the intense backward wave on the second harmonic. The amplitude of the spectral density G^{diff} is in the order of 0.015 of the initial amplitude of the intense pulse.

5. References

[1] Buyanovskaya E M and Kozlov S A 2007 *JETP Lett* **86** 349
[2] Buyanovskaya E M and Kozlov S A 2010 *Sci. and Tech. Journ. of Inf. Tech., Mech. And Opt.* **66** 23
[3] Kinsler P, Radnor S B P and New G H C 2005 *Phys. Rev. A* **72** 063807
[4] Bespalov V G, Kozlov S A, Shpolyanskiy Yu A and Walmsley I A 2002 *Phys. Rev. A* **66** 013811
[5] Platonenko V T, Stamenov K V and Khokhlov R V 1966 *Sov. Phys. JEPT* **22** 827

Journal of Physics: Conference Series **735** (2016) 012025 doi:10.1088/1742-6596/735/1/012025

Entanglement between two qubits induced by thermal field

E K Bashkirov[1] **and M S Mastyugin**
Physics Department, Samara State University, 1 Academician Pavlov Str. 443011, Russia

E-mail: bash@samsu.ru

Abstract. We have investigated the influence of dipole-dipole interaction and initial atomic coherence on atomic entanglement dynamics of two qubits. We have considered a model, in which only one atom couples to a quantum electromagnetic field in a cavity, since one of them can move around the cavity. We have shown that the entanglement arises for all pure atomic state even when both atoms are initially in the excited states. We have also derived that degree of entanglement is enhanced in the presence of the atomic coherence.

1. Introduction

Entanglement is a major resource for many fundamental applications in quantum information science. It is important to implement reliable methods to generate entangled states between systems. In order to function optimally, these applications require maximally entangled states. Several methods of creating entanglement have been proposed involving trapped and cooled ions or neutral atoms in cavities, superconducting circuits, spins in solids etc. [1]. In order to function optimally various applications require maximally entangled states. Because of decoherence, which is generally related to noise, there is great difficulty in generating and keeping the integrity of a pure entangled states. Although the interaction between the environment and quantum systems can lead to decoherence, it may also be associated with the formation of non-classical effects such as entanglement. Thus, understanding and investigating entanglement of mixed states becomes one of the actual problems of quantum information. Kim et al. [2] have investigated the atom-atom entanglement in a system of two identical two-level atoms with one-photon transitions induced by a single-mode thermal field. They showed that a chaotic field with minimal information can entangle atoms which were prepared initially in a separable state. Zhang directly generalizes Kim's study to the case when the atoms are slightly detuned from the thermal field, and study how the detuning would affect atom-atom entanglement [3]. The entanglement between two identical two-level atoms through nonlinear two-photon interaction with one-mode thermal field has been studied by Zhou et al. [4]. They showed that atom-atom entanglement induced by nonlinear interaction is larger than that induced by linear interaction. In [5] has discovered that two atoms can be entangled also through nonlinear nondegenerate two-photon interaction with two-mode thermal field. The influence of dipole-dipole interaction on entanglement between two cubits induced by one-mode and two-mode thermal field has been investigated in [6],[7].

The problem of creating or controlling the atomic entanglement is greatly related to the atomic coherence of population between different levels. Hu et al [8] have shown that the entanglement between two atoms induced by one-mode or two-mode thermal field can be manipulated by changing

[1] To whom any correspondence should be addressed.

the initial parameters of the atoms, such as the superposition coefficients and the relative phases of the initial atomic coherent state and the mean photon number of the cavity field. They have also discovered that entanglement may be greatly enhanced due to dipole-dipole interaction in the presence of the atomic coherence.

The practical applications in quantum information processing require engineering entangled atoms so this expects operable atoms which can be moved to distance without losing of information. Recently many schemes have been proposed to realize the engineering entangled atoms [9]. Guo and co-authors [10] have proposed a simple scheme to realize an easily engineered two-atom entangled state. The advantage of this scheme is only one atom is trapped in a cavity and the other one can be spatially moved freely outside the cavity. But the authors have investigated the atom-atom entanglement only for vacuum initial cavity field. In this paper we study the entanglement dynamics for model supposed in [10] for thermal cavity field taking into account the initial atomic coherence. Let's note that the model under consideration can be most simply realised for system of two superconducting qubits, one of which interacts with the LC superconducting circuit or microwave coplanar cavity. For superconducting qubits the effective dipole-dipole interaction may be much greater than for natural atoms [1].

2. Model and it exact solution

We consider two identical two-level atoms and one-mode quantum electromagnetic cavity field. The first atom is trapped in a lossless microcavity and resonantly interacts with the cavity field of the frequency ω. The second atom lies beside the first atom out of the cavity. We assume that the distance between atoms can compare with a wavelength on working transition. In this case the dipole-dipole interaction should be included. The interaction Hamiltonian of this system can be written as

$$(H = \hbar g(\sigma_1^+ a + a^+ \sigma_1^-) + \hbar J(\sigma_1^+ \sigma_2^- + \sigma_1^- \sigma_2^+), \tag{1}$$

where, $\sigma_i^+ = |+\rangle_{ii}\langle -|$, and $\sigma_i^- = |-\rangle_{ii}\langle +|$ are the transition operators between the excited $|+\rangle_i$ and the ground $|-\rangle_i$ states in the ith atom, a^+ and a are the creation and the annihilation operators of photons of the cavity mode, g is the coupling constant between atom and the cavity field and J is the coupling constant of the dipole interaction between the atoms and $|+\rangle$ and $|-\rangle$ are the excited and the ground states of a single two-level atom. The two-atom wave function can be expressed as a combination of state vectors of the form $|v_1, v_2\rangle = |v_1||v_2\rangle$, where $v_1, v_2 = +, -$.

We consider that the initial state of each atoms is a coherent superposition of the two levels, that is,

$$|\Psi_1(0)\rangle = \cos\theta_1 |+\rangle_1 + e^{i\varphi_1} \sin\theta_1 |-\rangle_1,$$

$$|\Psi_2(0)\rangle = \cos\theta_2 |+\rangle_2 + e^{i\varphi_2} \sin\theta_2 |-\rangle_2.$$

Here θ_1 and θ_2 denote the amplitudes of the polarized atoms, and φ_1 and φ_2 are relative phases of two atoms, respectively.

The initial cavity mode state are assumed to be the thermal one-mode state

$$\rho_F(0) = \sum_n p_n |n\rangle\langle n|.$$

The weight functions are

$$p_n = \frac{\bar{n}^n}{(1+\bar{n})^{n+1}},$$

where $\bar{n}$ is the mean photon number in the i-th cavity mode

$$\bar{n} = (\exp[\hbar\omega_i / k_B T] - 1]^{-1},$$

And k_B is the Boltzmann constant and T is the equilibrium cavity temperature.

The initial density matrix of the whole system is

$$\rho(0)=\rho_A(0)\rho_F(0)=\sum_n p_n\,|\Psi_1(0)\rangle|\Psi_2(0)\rangle\langle\Psi_1(0)|\langle\Psi_2(0)||n\rangle\langle n|. \qquad (2)$$

Before considering the dynamics of the system for thermal initial cavity field, it is straightforward to first study the case when the trapped two-level atom interacts with Fock state. Suppose that the excitation number of the atom-field system is n ($n \geq 0$) the evolution of the system is confined in the subspace $|-,-,n+2\rangle, |+,-,n+1\rangle, |-,+,n+1\rangle, |+,+,n\rangle$. on this basis, the Eigen functions of Hamiltonian of the system (1) can be written as [10]

$$|\Phi_{in}\rangle=\xi_{in}(X_{i1n}|-,-,n+2\rangle+X_{i2n}|+,-,n+1\rangle+$$

$$+X_{i3n}|-,+,n+1\rangle+X_{i4n}|+,+,n\rangle) \quad (i=1,2,3,4),$$

where

$$\xi_{in}=1/\sqrt{|X_{i1n}|^2+|X_{i2n}|^2+|X_{i3n}|^2+|X_{i4n}|^2}$$

and

$$X_{i1n}=1, \quad X_{i2n}=(-1)^{i+1}\frac{S_n}{\sqrt{n+2}},$$

$$X_{i3n}=\frac{S_n^2-(n+2)}{(\sqrt{n+2})\alpha}, \quad X_{i4n}=(-1)^{i+1}\frac{S_n(S_n^2-(n+2)-\alpha^2)}{(\sqrt{n+1}\sqrt{n+2})\alpha}.$$

Here $S_n = A_n$ for odd i and $S_n = B_n$ for even i. The corresponding Eigen value is

$$E_{1n}/\hbar=A_n, \qquad E_{2n}/\hbar=-A_n, \qquad E_{3n}/\hbar=B_n, \qquad E_{4n}/\hbar=-B_n,$$

where

$$A_n=\sqrt{W_n+V_n}/2, \quad B_n=\sqrt{W_n-V_n}/2$$

and

$$W_n=4n+6+2\alpha^2, \quad V_n=2\sqrt{4(n+1)\alpha^2+(\alpha^2+1)}, \quad \alpha=J/g.$$

To derive the full dynamics of our model one can consider also the basis states $|-,-,1\rangle, |+,-,0\rangle, |-,+,0\rangle$. In this basis the Eigen functions and Eigen values of the Hamiltonian (1) are

$$|\varphi_1\rangle=(\alpha^2/\Omega)[|-,-,1\rangle-(1/\alpha)|-,+,0\rangle], \qquad E_1=0;$$

$$|\varphi_2\rangle=(1/\sqrt{2})[(1/\Omega)|-,-,1\rangle+|+,-,0\rangle+(\alpha/\Omega)|-,+,0\rangle], \qquad E_2/\hbar=\Omega;$$

$$|\varphi_3\rangle=(1/\sqrt{2})[-(1/\Omega)|-,-,1\rangle+|+,-,0\rangle-(\alpha/\Omega)|-,+,0\rangle], \qquad E_2/\hbar=-\Omega,$$

where $\Omega=\sqrt{1+\alpha^2}$.

At last, the Hamiltonian has one more Eigen function

$$\varphi_0=|-,-,0\rangle.$$

This corresponds to energy $E_0=0$.

Using the

$$|\Psi(t)\rangle=e^{-iHt/\hbar}|\Psi(0)\rangle=\sum_i C_i(0)e^{-iE_it/\hbar}|\Phi_i\rangle,$$

where the coefficients $C_i(0)$ are determined by initial conditions, one can derive the time evolution of the initial states $|-,-,n\rangle$, $|+,-,n\rangle$, $|-,+,n\rangle$ and $|+,+,n\rangle$. As a result, collecting the gained expressions, we can find a temporary density matrix for an initial state (2). Taking a partial trace over the heat bath variable one can obtain that the reduced atomic density operator in the two-atom basis $|+,+\rangle$, $|+,-\rangle$, $|-,+\rangle$, $|-,-\rangle$ evolves to

$$\rho_A(t) = \begin{pmatrix} \rho_{11} & \rho_{12} & \rho_{13} & \rho_{14} \\ \rho_{12}^* & \rho_{22} & \rho_{23} & \rho_{24} \\ \rho_{13}^* & \rho_{23}^* & \rho_{33} & \rho_{34} \\ \rho_{14}^* & \rho_{24}^* & \rho_{34}^* & \rho_{44} \end{pmatrix}. \quad (3)$$

The elements of matrix (3) are not shown because they have too cumbersome form.

For two-qubit system described by the density operator (3), a measure of entanglement or negativity can be defined in terms of the negative eigenvalues μ_i^- of partial transpose of the reduced atomic density matrix $\rho_A^{T_1}$ [11], [12]. The partial transpose of the reduced atomic density matrix (3) can be written in the form

$$\rho_A^{T_1}(t) = \begin{pmatrix} \rho_{11} & \rho_{12} & \rho_{13}^* & \rho_{23}^* \\ \rho_{12}^* & \rho_{22} & \rho_{14}^* & \rho_{24}^* \\ \rho_{13} & \rho_{14} & \rho_{33} & \rho_{34} \\ \rho_{23} & \rho_{24} & \rho_{34}^* & \rho_{44} \end{pmatrix}.$$

and the negativity is

$$\varepsilon = -2\sum \mu_i^- . \quad (4)$$

When $\varepsilon = 0$ two qubits are separable and $\varepsilon > 0$ means the atom-atom entanglement. The case $\varepsilon = 1$ indicates maximum entanglement.

3. Results and discussion

Using the numerical of calculations of entanglement parameter (4) for initial atomic states (2) we have obtained, firstly, that entanglement can be induced by thermal field for all initial atomic state. The more significant result is that the atomic entanglement may be induced by thermal field when both the atoms are prepared in their excited states. By contrast, the atomic entanglement does not induced by thermal field when both atoms are trapped in cavity and prepared in their excited states [2-8]. Secondly, the maximum degree of entanglement is enhanced owing to the atomic coherence. With increasing of the mean photon number the value of atom–atom negativity decreases. But for coherent states this decreasing is much sharper. Such entanglement behaviour essentially differs from that for model when both atoms are trapped in cavity and interact with a thermal field [8]. In the last case the entanglement is greatly enhanced due to the initial atomic coherence for thermal cavity with large mean photon numbers. The increasing of the strength of the dipole-dipole interaction leads to increasing of the degree of entanglement. In Fig 1 we show time behaviour of the negativity for incoherent initial atomic state $|+,+\rangle$ (a) and coherent initial atomic state $|\Psi_1(0)\rangle = 1/\sqrt{2}(|+\rangle_1 + |-\rangle_1)$, $|\Psi_2(0)\rangle = 1/\sqrt{2}(|+\rangle_2 + |-\rangle_2)$ for the model with mean photon number $\bar{n} = 0.1$ and the strengh of the dipole-dipole interaction $\alpha = 0.5$.

4. Conclusions

We have investigated the effect of the atomic coherence on the entanglement of two dipole coupled two-level atoms when only one atom is trapped in a lossless cavity and interacts with one-mode thermal field, and the other one can be spatially moved freely outside the cavity. We have shown that thermal field can produce atom-atom entanglement for all pure initial atomic states. The results also show that the atom-atom entanglement can be controlled by changing the system parameters, such as the amplitudes of the polarized atoms, the mean photon numbers of thermal field, and the strength of the dipole interaction (or distance between atoms).

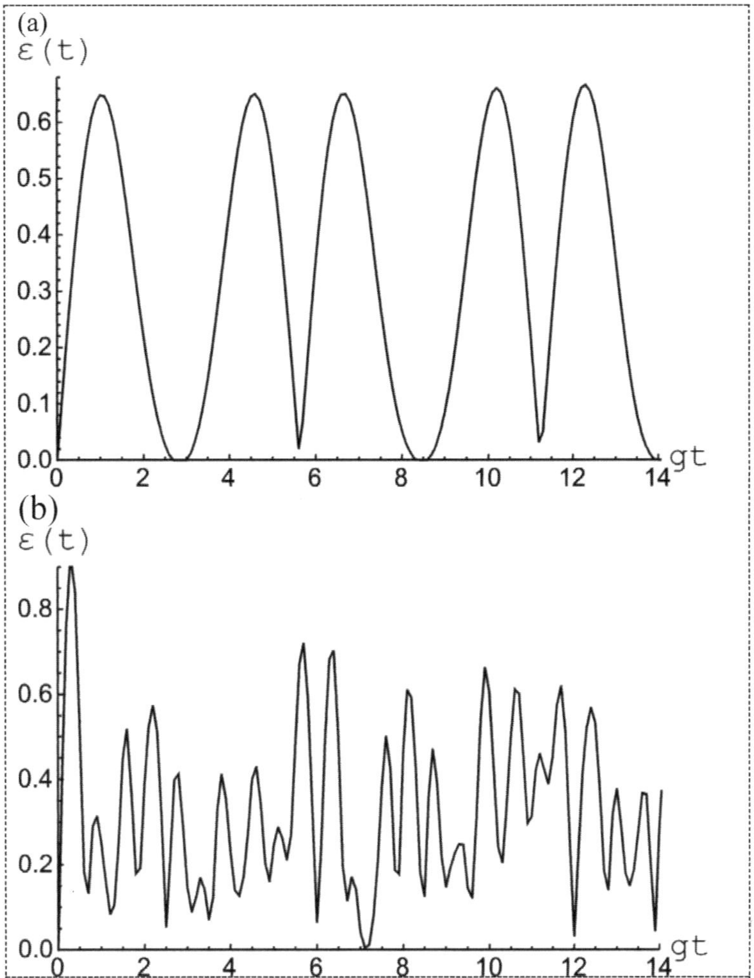

Figure 1. The negativity vs. gt for the incoherent $\theta_1 = \theta_1 = 0$ (a) and coherent $\theta_1 = \theta_1 = \pi/4$ (b). The model parameters are $\bar{n} = 0.1$ and $\alpha = 0.5$.

5. References

[1] Buluta I, Ashhab S and Nori F 2011 *Rep.Prog. Phys.* **74** 104401
[2] Kim M S, Lee J, D. Ahn D and Knight P L 2002 *Phys. Rev.* A. ***65*** 040101
[3] Zhang B 2010 *Optics Commun.* **283** 4676
[4] Zhou L and Song H S 2002 *J. Opt.* B **4** 425
[5] Bashkirov E K 2006 Laser Phys. Lett. **3** 145
[6] Aguiar L S, Munhoz P P, Vidiella-Barranco A and Roversi J A 2005 *J. Opt.* B **7** S769
[7] Bashkirov E K and Stupatskaya M P 2009 Laser Phys. **19** 525
[8] Hu Y H, Fang M F , Jiang C L and Zeng K 2008 *Chin. Phys*. **17** 1784.
[9] Bashkirov E K and Nikiforova Y A 2012 Computer Optics **36** 468 (In Russian)
[10] Guo Y Q, Cao H J and Song H S 2005 ArXiv:quant-ph/0509142 7 pp
[11] Peres A 1996 *Phys. Rev. Lett.* **77** 1413
[12] Horodecki R, Horodecki M and Horodecki P 1996 *Phys. Lett*. A **223** 333

Journal of Physics: Conference Series **735** (2016) 012026 doi:10.1088/1742-6596/735/1/012026

Dynamics for atoms successively passing a cavity in the presence of the initial atomic entanglement

E K Bashkirov[1] and M S Mastyugin
[1]Physics Department, Samara State University, 1 Academician Pavlov Str. 443011, Russia

E-mail: bash@samsu.ru

Abstract. We investigated the dynamics of entanglement between atoms successively passing a cavity and interact with a thermal cavity field. We considered the situation when atoms are initially prepared in a pure entangled state. Using the exact solution of density matrix evolution we calculated the negativity for different values of cavity mean photon numbers. We shown the possibility to save the initial atomic entanglement even for a thermal cavity field with relatively high temperature.

1. Introduction

Entanglement is not only one of the most surprising features of quantum theory, but also provides an important resource for various quantum information processes such as quantum information, quantum communication, and quantum cryptography. Therefore, great efforts have been made to investigate entanglement characterization, entanglement control, and entanglement production in different systems. One of the particular interest schemes in which entanglement can be created is a system containing two two-level atoms, since they can represent two qubits, the building blocks of the quantum gates that are essential to implement quantum protocols in quantum information processing. Two-atom entangled states have been demonstrated experimentally using ultra cold trap ions, impurity spins in solids, superconducting circuits and cavity quantum electrodynamics schemes (QED) [1]. Cavity QED has been used to generate the atom-atom and atom-photon entanglement. The entanglement between two initially independent atoms successively passing the vacuum cavity has been demonstrated by S. Haroche et al. [2]. The entanglement procedure involves the resonant coupling, one by one, of the atoms to a high Q microwave superconducting cavity. The atoms, prepared in circular Rydberg states, exchange a single photon in the cavity and become entangled by this indirect interaction. The effect has been demonstrated with pairs of atoms separated by centimetric distances. Liao et al. [3] investigated the entanglement properties of two atoms passing through a cavity one after another when the field is initially in thermal field. They found that the phenomenon of sudden birth of entanglement occurs in some certain conditions and the entanglement between two atoms can be created even if the two atoms are initially in excited states. Yan [4] also investigated the entanglement properties of two atoms successive passing a cavity with Fock or thermal field but she especially focused on the case that when two atoms are initially in an entangled state. In this paper we examined entanglement properties of two atoms successive passing a thermal field when two atoms are initially in another type of Bell entangled atomic state.

[1] To whom any correspondence should be addressed.

2. Model and it exact solution

Our system consists of two separate identical two-level atoms passing through a cavity one after another with different velocities. We suppose that atoms interact resonantly with cavity field via one-photon transitions. The interaction Hamiltonian of this system can be written as

$$H = \hbar g(a^{+}\sigma^{-} + \sigma^{+}a) \tag{1}$$

where, $\sigma^{+} = |+\rangle\langle -|$ and $\sigma^{-} = |-\rangle\langle +|$ are the transition operators between the excited $|+\rangle$ and the ground $|-\rangle$ states in the two-level atom, a^{+} and a are the creation and the annihilation operators of photons of the cavity mode, g is the coupling constant between atom and the cavity. The two-atom wave function can be expressed as a combination of state vectors of the form $|v_1, v_2\rangle = |v_1\rangle|v_2\rangle$, where $v_1, v_2 = +, -$.

We first examine that the cavity is initially in a Fock state $|n\rangle$ containing n photons and two spatially separated two-level atoms are initially prepared in a pure entangled state of the following form

$$|\Psi(0)\rangle_A = \cos\theta\,|+,+\rangle + \sin\theta\,|-,-\rangle, \tag{2}$$

which pass through the cavity one after the other. Here θ is the amplitude.

Using the Hamiltonian (1) and the initial conditions (2) one can obtain the time dependent wave function for considered system at the moment τ when the first atom leaves the cavity

$$|\Psi_{1n}(\tau)\rangle = Y_{1n}(\tau)\,|+,+,n\rangle + Y_{2n}(\tau)\,|+,-,n-1\rangle + Y_{3n}(\tau)\,|-,+,n+1\rangle + Y_{4n}(\tau)\,|-,-,n\rangle, \tag{3}$$

where

$$Y_{1n}(\tau) = \cos\theta\cos(\sqrt{n+1}g\tau),$$
$$Y_{2n}(\tau) = -i\sin\theta\sin(\sqrt{n}g\tau),$$
$$Y_{3n}(\tau) = -i\cos\theta\sin(\sqrt{n+1}g\tau),$$
$$Y_{4n}(\tau) = \sin\theta\cos(\sqrt{n}g\tau).$$

The wave function (3) on the other hand is the initial state of the system prior to entering into the cavity of the second atom. At the moment t when the second atom leaves the cavity the time-dependent wave function takes the form (for $n \geq 1$)

$$|\Psi_{2n}(t)\rangle = X_{1n}(t)\,|+,+,n\rangle + X_{2n}(t)\,|+,+,n-2\rangle + X_{3n}(t)\,|+,-,n-1\rangle + X_{4n}(t)\,|+,-,n+1\rangle +$$
$$+ X_{5n}(t)\,|-,+,n+1\rangle + X_{6n}(t)\,|-,+,n-1\rangle + X_{7n}(t)\,|-,-,n\rangle + X_{8n}(t)\,|-,-,n+2\rangle, \tag{4}$$

where

$$X_{1n}(t) = \cos\theta\cos(\sqrt{n+1}g\tau)\cos(\sqrt{n+1}gt),$$
$$X_{2n}(t) = \sin\theta\sin(\sqrt{n}g\tau)\cos(\sqrt{n-1}gt),$$
$$X_{4n}(t) = -i\cos\theta\cos(\sqrt{n+1}g\tau)\sin(\sqrt{n+1}gt),$$
$$X_{5n}(t) = -i\cos\theta\sin(\sqrt{n+1}g\tau)\cos(\sqrt{n+2}gt),$$
$$X_{6n}(t) = -i\sin\theta\cos(\sqrt{n}g\tau)\cos(\sqrt{n}gt),$$
$$X_{7n}(t) = \sin\theta\cos(\sqrt{n}g\tau)\cos(\sqrt{n}gt),$$
$$X_{8n}(t) = \cos\theta\sin(\sqrt{n+1}g\tau)\sin(\sqrt{n+2}gt).$$

For the case $n = 0$ the wave-function can be derived as

$$|\Psi_2(t)\rangle = Z_1(t)|+,+,0\rangle + Z_2(t)|+,-,1\rangle + Z_3(t)|-,+,1\rangle + Z_4(t)|-,-,0\rangle + Z_5(t)|-,-,2\rangle, \quad (5)$$

where

$$Z_1(t) = \cos\theta\cos(g\tau)\cos(gt), \quad Z_2(t) = -i\cos\theta\sin(g\tau)\sin(gt),$$

$$Z_3(t) = -i\cos\theta\sin(g\tau)\cos(\sqrt{2}gt), \quad Z_4(t) = \sin\theta, \quad Z_5(t) = \cos\theta\sin(g\tau)\sin(\sqrt{2}gt).$$

To investigate the atom-atom entanglement we calculate negativity. The reduced atomic density matrix can be derived by tracing the density matrix of the whole system on the field variables. As a result one can obtain for atomic density matrix

$$\rho_{at}(\tau,t) = \begin{pmatrix} \rho_{11} & \rho_{12} & \rho_{13} & \rho_{14} \\ \rho_{12}^* & \rho_{22} & \rho_{23} & \rho_{24} \\ \rho_{13}^* & \rho_{23}^* & \rho_{33} & \rho_{34} \\ \rho_{14}^* & \rho_{24}^* & \rho_{34}^* & \rho_{44} \end{pmatrix}, \quad (6)$$

where for the case $n \geq 1$

$$\rho_{11} = |X_{1n}(t)|^2 + |X_{2n}(t)|^2, \quad \rho_{12} = 0, \quad \rho_{13} = 0, \quad \rho_{14} = X_{1n}(t)X_{7n}(t)^*,$$

$$\rho_{22} = |X_{3n}(t)|^2 + |X_{4n}(t)|^2, \quad \rho_{23} = X_{3n}(t)X_{6n}(t)^* + X_{4n}(t)X_{5n}(t)^*,$$

$$\rho_{24} = 0, \quad \rho_{33} = |X_{5n}(t)|^2 + |X_{6n}(t)|^2, \quad \rho_{34} = 0, \quad \rho_{44} = |X_{7n}(t)|^2 + |X_{8n}(t)|^2 .$$

For two-qubit system described by the density operator (6), a measure of entanglement or negativity can be defined in terms of the negative eigenvalues μ_i^- of partial transpose of the reduced atomic density matrix $\rho_A^{T_1}$ [5], [6]. The partial transpose of the reduced atomic density matrix (6) can be written in the form

$$\rho_{at}^{T_1}(\tau,t) = \begin{pmatrix} \rho_{11} & \rho_{12} & \rho_{13}^* & \rho_{23}^* \\ \rho_{12}^* & \rho_{22} & \rho_{14}^* & \rho_{24}^* \\ \rho_{13} & \rho_{14} & \rho_{33} & \rho_{34} \\ \rho_{23} & \rho_{24} & \rho_{34}^* & \rho_{44} \end{pmatrix}. \quad (7)$$

and the negativity is

$$\varepsilon = -2\sum \mu_i^- . \quad (8)$$

When $\varepsilon = 0$ two qubits are separable, and $\varepsilon > 0$ means the atom-atom entanglement. The case $\varepsilon = 1$ indicates maximum entanglement.

The partial transpose of the reduced atomic density matrix (7) has only two eigenvalues that can be negative

$$\mu_1 = (1/2)\Big[|X_{1n}|^2 + |X_{2n}|^2 + |X_{7n}|^2 + |X_{8n}|^2 -$$
$$-\sqrt{(|X_{1n}|^2 + |X_{2n}|^2 - |X_{7n}|^2 - |X_{8n}|^2)^2 + 4|X_{6n}X_{3n}^* + X_{5n}X_{4n}^*|^2}\Big], \quad (9)$$
$$\mu_2 = (1/2)\Big[|X_{3n}|^2 + |X_{4n}|^2 + |X_{5n}|^2 + |X_{6n}|^2 -$$
$$-\sqrt{(|X_{3n}|^2 + |X_{4n}|^2 - |X_{5n}|^2 - |X_{6n}|^2)^2 + 4|X_{1n}|^2|X_{7n}|^2}\Big].$$

Using formulae (9) one can calculate the negativity (9). Analogously we can derive negativity for the case when $n = 0$.

Now we consider the case in which two initially entangled atoms successively pass a thermal field, which is the most easily available radiation field. This initial one-mode thermal field can be written as

$$\rho_F(0) = \sum_n p_n |n\rangle\langle n|.$$

The weight functions are

$$p_n = \frac{\bar{n}^n}{(1+\bar{n})^{n+1}},$$

where $\bar{n}$ is the mean photon number in the cavity mode $\bar{n} = (\exp[\hbar\omega_i / k_B T]-1]^{-1}$, k_B is the Boltzmann constant and T is the equilibrium cavity temperature.

For the thermal field the matrix element of (7) have the form

$$\rho_{11} = \sum_{n=1} p_n(|X_{1n}(t)|^2 + |X_{2n}(t)|^2 + p_0 |Z_1|^2, \quad \rho_{12} = 0, \quad \rho_{13} = 0,$$

$$\rho_{14} = \sum_{n=1} p_n X_{1n}(t) X_{7n}(t)^* + p_0 Z_1(t) Z_4(t)^*, \quad \rho_{22} = \sum_{n=1} p_n(|X_{3n}(t)|^2 + |X_{4n}(t)|^2) + p_0 |Z_1(t)|^2,$$

$$\rho_{23} = \sum_{n=1} p_n (X_{3n}(t) X_{6n}(t)^* + X_{4n}(t) X_{5n}(t)^* + p_0 Z_2(t) Z_3(t)^*$$

$$\rho_{24} = 0, \quad \rho_{33} = \sum_{n=1} p_n(|X_{5n}(t)|^2 + |X_{6n}(t)|^2 + p_0 |Z_3(t)|^2,$$

$$\rho_{34} = 0, \quad \rho_{44} = \sum_{n=1} p_n |X_{7n}(t)|^2 + |X_{8n}(t)|^2 + p_0(|Z_4|^2 + |Z_5|^2).$$

For the thermal field the partial transpose of the reduced atomic density matrix (7) also has only two eigenvalues that can be negative. But these have too cumbersome form to present in the paper.

3. Results and discussion

The results of the numerical of calculations of entanglement parameter (8) for initial atomic states (2) ant thermal fields with different mean photons numbers are shown in the Fig.1. The curves were obtained under the assumption that $t = 2\tau$ as in [2]. One can see from the Fig.1 that for vacuum cavity state the initial maximum degree of atomic entanglement restore for the moments $t_k = 2\pi k \ (k = 1,2,...)$. Thermal field destroys the initial atomic quantum correlations only for certain time of the atom-field interaction. For other moments the partial degree of atom-atom entanglement is restored, even for relatively high cavity temperatures. to increasing of the degree of entanglement.

4. Conclusions

Thus, we investigated the dynamics of entanglement of two two-level atoms successively passing the thermal cavity and resonantly interacting with one-mode quantum cavity electromagnetic field through one-photon transitions. The atoms are assumed to be initially prepared in the entangled state. We derived the exact expression for the reduced atomic density matrix and calculated the analytical expression for negativity. Our numerical results reveal that the initial atomic entanglement can partially restore its original value for a finite interval of time, even for relatively high cavity temperatures.

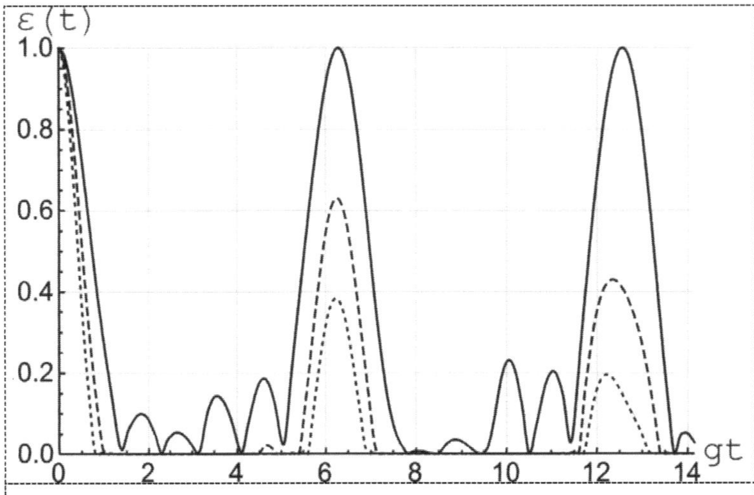

Figure 1. The negativity vs. *gt* for entangled initial atomic state (2) with $\theta = \pi/4$. The thermal mean photon numbers are $\bar{n} = 0 (T = 0)$ (solid), $(T = 2.4\ K)$ (dashed) and $\bar{n} = 1 (T = 3.8\ K)$ (dotted).

5. References

[1] Buluta I, Ashhab S and Nori F 2011 Rep.Prog. Phys. **74** 104401
[2] Hagley E, Maitre X, Nogues G, Wunderlich C, Brune M, Raimond J M and Haroche S. 1997 Phys. Rev. Lett. **79** 1
[3] Liao Q, Fang G, Ahmad M A and Liu S 2011 Optics Commun. **384** 201
[4] Yan X Q 2009 Chaos, Solitons and Fractals **41** 1645
[5] Peres A 1996 Phys. Rev. Lett. **77** 1413
[6] Horodecki R, Horodecki M and Horodecki P 1996 Phys. Lett. A **223** 333

Journal of Physics: Conference Series **735** (2016) 012027 doi:10.1088/1742-6596/735/1/012027

Phase quantization of diffractive optical elements for the formation of predetermined symmetric light distributions

A P Porfirev
Samara State Aerospace University, 34 Moskovskoye shosse, 443086 Samara, Russia
Image Processing Systems Institute, 151 Molodogvardejskaya street, 443001 Samara, Russia

E-mail: lporfirev@rambler.ru

Abstract. A method for the effective formation of symmetric light fields with a given phase distribution by binary diffractive optical elements is proposed. The design method is based on two-level phase quantisation of multi-level diffractive optical elements that form a basic distribution of light field. Numerical simulation of the formation of predetermined light fields is performed. The possibility of forming such light fields by a spatial light modulator is experimentally demonstrated.

1. Introduction

Nowadays, formation of light fields with a predetermined spatial structure is widely used in various applications of biophotonics and micromechanics [1-3]. Optical beams with a predetermined amplitude-phase distribution are used in passive transporting [4,5], sorting [6,7] and rotating of micro- and nano-objects [8,9]. Formation of light beams with special intensity distributions helps to reduce the probability of thermal damage to the biological micro-objects trapped in the laser beam [10]. It is important to achieve effective formation of such light fields, as this will reduce the cost of operations through the use of lower-cost lasers.

To solve the problem of the formation of predetermined light fields with complex spatial structures, spatial light modulators (SLM) are commonly used [11]. Despite the advantages for dynamic control of light field distribution, most commercial SLM have a number of limitations: high costs, low energy efficiency and the possibility of damage to the modulator display using high-power lasers.

The use of diffractive optical elements (DOEs) can increase the efficiency of forming light fields. Furthermore, in many applications it can reduce the cost of the final toolkit.

This paper presents a method of calculating of DOE phase forming light field that have a symmetry and a predetermined distribution phase. These elements have two-level relief. This simplifies the process of manufacture. These light fields can be used for biophotonics applications, for example, for passive sorting and transport of microscopic objects.

Nowadays, DOEs with a continuous profile are used for the formation of light fields with predetermined amplitude-phase distributions. Manufacturing of DOEs with a two-level profile (binary DOEs) is a cheaper and less complicated technological process. In addition, DOEs with continuous profile are more sensitive to manufacturing errors [12] and laser wavelength variations [13]. At the same time, a large number of diffraction orders are created when using binary DOEs, thus the energy of the incident beam is divided between these diffraction orders. As a result, the maximum efficiency of the first diffraction order for a binary DOE is about 40% [14]. Formation of a symmetric light field, which is a distribution composed of light fields generated in the +1 and -1 diffraction orders, can theoretically lead to doubling of the efficiency value.

2. Design method

Let τ(x,y) be the phase transmission function multilevel DOE that forms a predetermined amplitude-phase distribution. One of the many currently existing methods (amplitude encoding techniques [15, 16], iterative methods [17]) can be used to calculate this function. Reducing the number of quantisation levels for the phase function in multilevel DOEs reduces the efficiency value. The DOE quantisation operation can be described by the expression:

$$\Phi(\tau(x,y)) = \left[\frac{\tau(x,y)}{\Delta}\right]\Delta \tag{1}$$

where $\Delta = \dfrac{2\pi}{M}$, M is the number of quantisation levels, and [..] is an integer part of number.

Then, the complex transmission function of the DOE is written as

$$T(x,y) = \exp\{i\Phi(\tau(x,y))\}. \tag{2}$$

In case of two-level quantisation (M = 2), N irregular diffraction grating is created. In this case, light fields that are symmetric about the origin are formed in symmetrical diffraction orders (for example, +1 and -1 diffraction orders). Under specific initial parameters of a multilevel binary DOE, quantisation may lead to the light distribution composed of the fields formed in these orders forming a symmetrical light field.

In the simulation, we used the integral transform, which describes the diffraction of a laser beam on a thin optical element supplemented with lens:

$$G(u,v) = \frac{2\pi}{\lambda f}\iint_{\Omega} A(x,y)g(x,y)\exp\left[-\frac{i2\pi}{\lambda f}(xu+yv)\right]dxdy. \tag{3}$$

where $A(x,y)$ is the laser beam distribution, λ is the laser wavelength, $g(x,y)$ is the complex function of the optical element, and f is the focal length.

The simulation results for the case of calculating the binary DOE forming the light distribution in the form of a cross, the letter «S» and the contour of the rectangle are shown in Fig. 1. Fig. 1 also shows the light fields generated by a base multilevel DOE (Fig. 1, top row), which were taken as a basis, as part of a composite light field with symmetry.

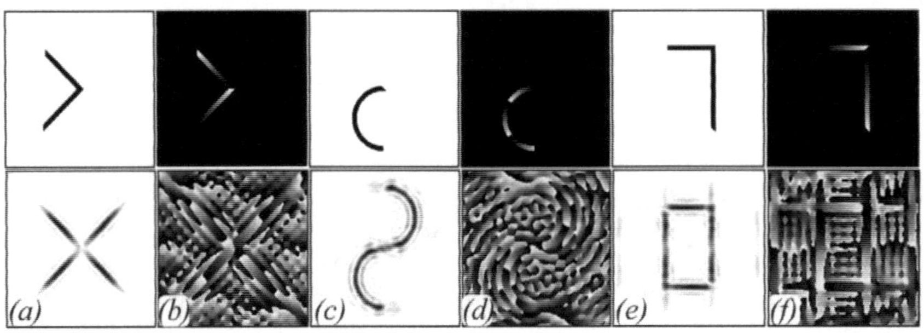

Figure 1. Examples of basic light fields generated by multi-level DOE (top row: (a), (c), (e) intensity distributions; (b), (d), (f) phase distributions). The symmetrical light fields distribution formed by DOE synthesised as a result of quantisation of the original multi-level DOE (bottom row: (a), (c), (e) intensity distributions; (b), (d), (f) phase distributions). For intensity distributions, negative images are presented.

3. Experimental results

To generate the calculated symmetric light fields experimentally, we utilised a SLM PLUTO VIS (1920 × 1080 pixel resolution, 8 μm pixel size). The optical setup is shown in Figure 2. The calculated binary phase displayed at the SLM is illuminated by a laser beam (wavelength of 532 nm). With the phase pattern of 1024 × 1024 pixels used in the experiment, the size of the phase element at the SLM output was approximately equal to 8.2 mm ×8.2 mm. A system composed of a lens L_1 (f_1 = 150 mm) and a lens L_2 (f = 250 mm) was utilised to expand the laser beam incident on the SLM. A system of two lenses L_3, L_4 (f_3 = 350 mm and f_4 = 150 mm) and a diaphragm D was used for high-frequency spatial filtering. A lens L_5 (f = 90 mm) was used to focus the generated light beam. Then, a microobjective *MO* (8×, NA = 0.2) formed an image on the matrix of a CMOS camera LOMO TC-1000 (3856 × 2764-pixel resolution, 1.67 × 1.67 μm pixel size). The zero and first diffraction orders were spatially separated by the superposition of the original phase function and a linear phase mask.

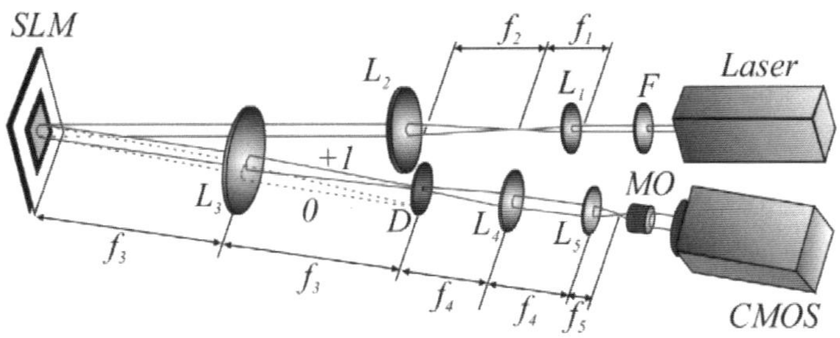

Figure 2. Experimental optical setup: *Laser* is a solid-state laser, L_1, L_2, L_3, L_4, and L_5 are the lenses with focal lengths of f_1 = 150 mm, f_2 = 250 mm, f_3 = 350 mm, f_4 = 150 mm, and f_5 = 90 mm, respectively, *SLM* is a spatial light modulator, D is a diaphragm, *MO* is a microobjective (8×, NA = 0.2), *CMOS* is a video camera (LOMO TC-1000).

Figure 3 shows examples of a binary phase DOE used in the experiment, and the intensity distribution formed in the lens focus. It can be clearly seen that the experimental images are in good agreement with the simulation results.

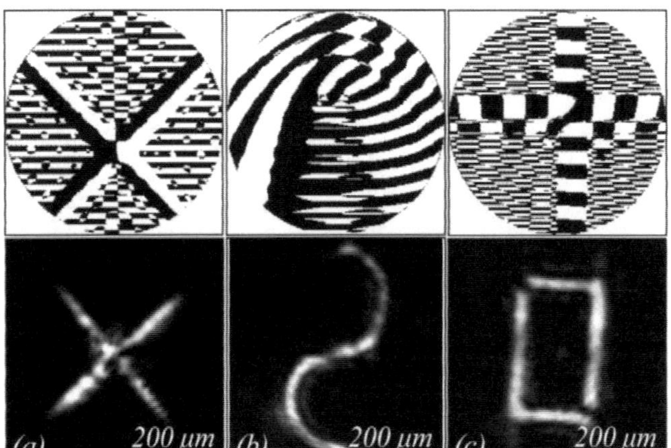

Figire 3. Examples of a binary DOE for the formation of symmetrical predetermined amplitude-phase distributions (top row: black, 0; white, π) and experimentally formed intensity distributions in the focal plane of the lens L_5 (bottom row): (a) the cross, (b) the letter "S", and (c) a rectangle.

4. Conclusions

The method of calculation of DOEs that form a predetermined, symmetrical amplitude-phase distribution is proposed. This method is based on two-level phase quantisation of DOE. Due to the fact that the designed DOE has a binary relief, the technology for their production is significantly simplified. In addition, owing to the formation of the light field as a sum of the light fields formed in the +1 and -1 diffraction orders, the theoretical efficiency is the sum of the efficiency of diffraction orders.

The simulation of the formation of some symmetric fields in the thin optical element approximation is performed. Predetermined symmetrical fields are experimentally formed by a SLM. The experimental results are in qualitative agreement with the simulation results.

Acknowledgments

The research was financially supported by RSF grant No. 14-19-00114.

References

[1] Farre A, van der Horst A, Blab G A, Downing B P B, Forde N R 2010 *J. Biophotonics* **3** 224-33
[2] Svoboda K, Block S M 1994 *Ann. Rev. Biophys. Biomol. Struct.* **23** 247-85
[3] Kim K, Yoon J, Park Y K 2015 *Optica* **2** 343-46
[4] Jesacher A, Maurer C, Schwaighofer A, Bernet S, Ritsch-Marte M 2008 *Opt. Express* **16** 4479-86
[5] Porfirev A P, Skidanov R V 2014 *Opt. Mem. Neural Network* **23** 164-69
[6] Ma B, Yao B, Peng F, Yan S, Lei M, Rupp R 2012 *J. Opt.* **14** 105702
[7] Chapin S C, Germain V, Dufresne E R 2006 *Opt. Express* **14** 13095-100
[8] Morozov A A, Skidanov R V 2013 *Comp. Optics* **37** 203-07
[9] Padgett M, Bowman R 2011 *Nat. Photonics* **5** 343-48
[10] Skidanov R V, Rykov M A 2013 *Comp. Optics* **37** 431-35
[11] Grier D G 2003 *Nature* **424** 810-16
[12] Unno Y 1998 *Appl. Opt.* **37** 3401-07
[13] Davis J A, Chambers J B, Slovick B A, Moreno I 2008 *Appl. Opt.* **24** 4375-80
[14] Cox J A, Werner T, Lee J, Nelson S, Fritz B, and Bergstrom J 1990 *Proc. SPIE* **1211** 116-24
[15] Khonina S N, Balalayev S A, Skidanov R V, Kotlyar V V, Paivanranta B, Turunen J 2009 *J. Opt. A: Pure Appl. Opt.* **11** 065702
[16] Flavin M A, Horner J L 1989 *Appl. Opt.* **28** 1692-96
[17] Abramochkin E G, Afanasiev K N, Volostnikov V G, Korobtsov A V, Kotova S P, Losevsky N N, Mayorova A M, Razueva E V 2008 *Izv. Ros. Akad. Nauk: Seriya Fiz.* **72** 76-79

Journal of Physics: Conference Series **735** (2016) 012028 doi:10.1088/1742-6596/735/1/012028

Features of diffraction and collimation of few-cycle optical beams

Danila Puzyrev

Department of Photonics and Optical Information Technology, ITMO University, Saint-Petersburg 197101, Russia

E-mail: puzurdan@gmail.com

Abstract. The paper reports theoretical spectral and spatial-temporal features of diffraction and collimation of few-cycle optical beams in transparent isotropic linear media. Equation describing asymmetric distribution in temporal structure of electric field in far field region of diffraction, which with increasing distance becomes symmetric is given. It is shown the frequency of maximum spectral density increases with distance from source of radiation along the axis of propagation. The frequency shift is inversely proportional to the initial duration of the wave packet and decreases with increasing of oscillations number.

1. Introduction

The development of methods of generation terahertz radiation made it possible to obtain few-cycle (less than 10 field oscillations) electromagnetic waves [1]. Such radiation is widely applicable in spectroscopy, in drugs and explosive materials detection systems and for medical diagnostics [2]. Changes in electro-magnetic spatial-temporal structures in diffracted radiation lead to task of optimization optical systems of emission, collimation and focusing initially few-cycle waves [3,4].

In this paper some spectral and spatial-temporal features of diffraction and collimation of few-cycle beams are reviewed. We got expression, which describes the dynamics of redistribution of electric field in temporal structure of diffracting wave packet in far field region of diffraction. Also we got analytic formulas describing spatial-temporal structure of collimated few-cycle beams.

2. Diffraction of few-cycle beams

Spectral profile of few-cycle beams is defined by the expression

$$G_0\left(\omega\right) = \frac{\sqrt{\pi}}{2} i^{n+1} \tau_p^n \omega^{n-1} E_0 \exp\left(-\left(\frac{\omega\tau_p}{2}\right)^2\right). \qquad (1)$$

If we use Fourier transform $E\left(t\right) = \frac{1}{2\pi}\int\limits_{-\infty}^{+\infty} G\left(\omega\right)\exp\left(i\omega t\right)d\omega$, we get the spatial-temporal profile of electric field on radiation emitter [5]

$$E_0(t) = i^{2n} 2^{n-2} E_0 \sum_{k=2l, l\in\mathbb{N}_0}^{n-1} \left(\frac{(n-1)! i^{-k}}{2^k (n-1-k)! \left(\frac{k}{2}\right)!}\left(\frac{t}{\tau_p}\right)^{n-1-k}\right) \exp\left(-\left(\frac{t}{\tau_p}\right)^2\right), \qquad (2)$$

where ω is the temporal cyclic frequency of radiation, n is the number of half-cycle electric field oscillation in wave packet, τ_p is the duration on emitter of radiation.

Paraxial dynamics of spatial-temporal spectrum of initial Gaussian beams can be described by the expression [4]

$$G(\omega,x,y,z) = z_R(\omega)\frac{z_R(\omega)+iz}{z_R^2(\omega)+z^2}\exp\left(-\frac{x^2+y^2}{\rho^2}z_R(\omega)\frac{z_R(\omega)+iz}{z_R^2(\omega)+z^2} - ik(\omega)z\right)G_0(\omega), \quad (3)$$

where x,y,z is the Cartesian axes coordinates (z axis coincides with wave propagation direction) $z_R(\omega) = \frac{\rho^2 n(\omega)\omega}{2c}$ is the Rayleigh range, ρ is the transverse beam width, $k(\omega) = \frac{\omega}{c}n(\omega)$ is the wave number, $n(\omega)$ is the refractive index of the medium.

In non-dispersive media, like atmospheric air [6], we can assume dispersive index in form $n(\omega) = N_0$.

Along the z-axis frequency of maximum of temporal spectrum can be received by time derivative of the function (3):

$$\omega_{\max}^{x=0,y=0}(z) = \frac{\sqrt{L^2(n-1) - \frac{z^2}{2} + \sqrt{\left(L^2(n-1) - \frac{z^2}{2}\right)^2 + 2nz^2L^2}}}{\tau_p L} \quad (4)$$

where $L = \frac{\rho^2 N_0}{2c\tau_p}$.

The function (4) is limited on the source of radiation by value

$$\omega_{\max}^{z=0} = \frac{\sqrt{2(n-1)}}{\tau_p}. \quad (5)$$

In far field region of diffraction:

$$\omega_{\max}^{z\to\infty} = \frac{\sqrt{2\,n}}{\tau_p}. \quad (6)$$

Fig.1 shows the dynamics of frequency temporal spectrum maximum along z-axis for different values of n.

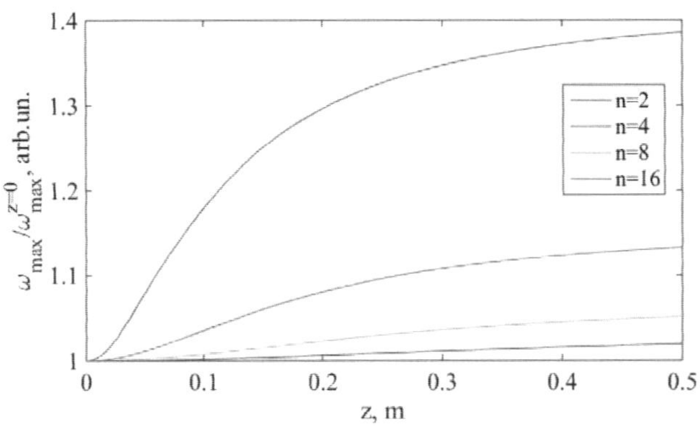

Figure 1. Dynamics of frequency temporal spectrum maximum along z-axis with initial duration $\tau_p = 0.2ps$

Spatial-temporal structure of electric field in far-field region ($z^2 \gg \{z_R(\omega)\}^2_{\max}$ is the spherics condition of the wave front curvature. $\{z_R(\omega)\}^2_{\max} = \frac{\rho^2 N_0 \omega_{\max}}{2c}$, where ω_{max} is maximum of frequency from frequency range where the majority of initial radiation energy is trapped) is defined by Furier transform of the function (3) instead of paper [3], where only initial one cycle beams was reviewed:

$$E\left(t',x,y,z\right)=i^{2n+2}E_0 2^{n-1}\frac{L}{z}\left(\frac{\tau}{\tau}\right)^{n+1}\left(\begin{array}{l}\frac{2L}{z}\frac{\tau_p}{\tau}\sum\limits_{k=2l,l\in\mathbb{N}_0}^{n+1}\left(\frac{(n+1)!i^{-k}}{2^k(n+1-k)!\left(\frac{k}{2}\right)!}\left(\frac{t'}{\tau}\right)^{n+1-k}\right)+\\+\sum\limits_{k=2l,l\in\mathbb{N}_0}^{n}\left(\frac{n!i^{-k}}{2^k(n-k)!\left(\frac{k}{2}\right)!}\left(\frac{t'}{\tau}\right)^{n-k}\right)\end{array}\right)\times\exp\left(-\left(\frac{t'}{\tau}\right)^2\right), \quad (7)$$

where "delay time"

$$t'=t-\frac{N}{c}\left(z+\frac{x^2+y^2}{2z}\right) \quad (8)$$

describes curvature of the wave front. The wave packet duration τ

$$\tau=\tau_p\sqrt{1+\frac{x^2+y^2}{\rho^2}\left(\frac{2L}{z}\right)^2} \quad (9)$$

is directly proportional to distance from z-axis.

Analitic formula (7) matches well with numerical calculations of few-cycle wave diffraction. Fig.2 shows it on example of one-cycle wave diffraction.

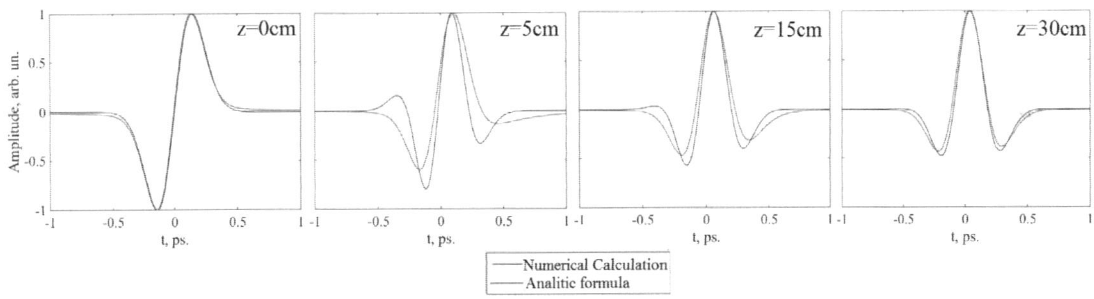

Figure 2. Comparison of numerical calculation with analytic formula (7) on example of along z-axis one-cycle wave diffraction $n=2$; $\tau_p=0.2ps$.

Visualization of formula (7) is shown at fig.3 on example of diffraction of half and one cycle beam.

3. Collimation of few-cycle beams

Collimation mirror with the reflection function $F_{col}(\omega,x,y)=\exp\left(ik(\omega)\left(x^2+y^2\right)/2f_{col}\right)$ is placed at the distance $z_0=f_{col}$ from radiation source. Temporal spectrum of radiation at that distance is transformed to form

$$G_{col}(\omega,x,y,z)=\frac{iz_0\left(z_{col}(\omega)+i(z-z_0)\right)}{\left(z_{col}(\omega)\right)^2+(z-z_0)^2}\exp\left(\begin{array}{l}-\frac{x^2+y^2}{\rho^2}\frac{z_R(\omega)(z_{col}(\omega)+i(z-z_0))}{(z_{col}(\omega))^2+(z-z_0)^2}\\-ik(\omega)(z_0+z)\end{array}\right)G_0(\omega), \quad (10)$$

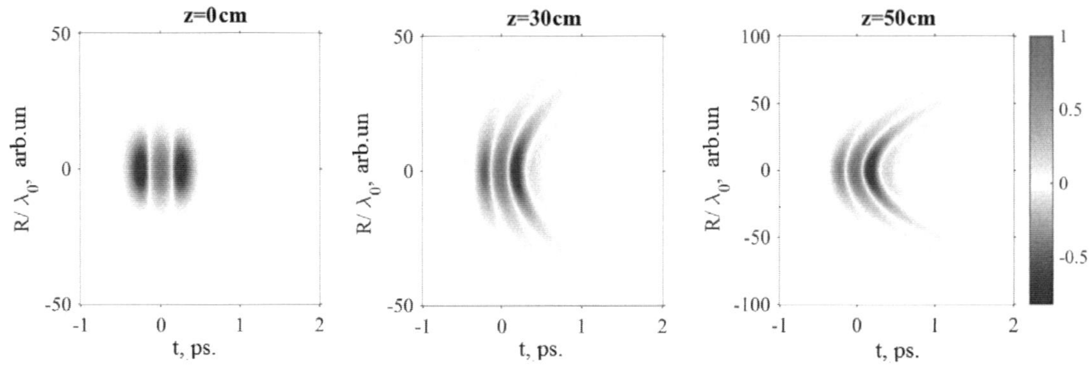

Figure 3. Spatial-temporal structure of electric field at different distance from source of radiation on example of half and one cycle beam. $n = 3; \tau_p = 0.2ps, R = \sqrt{x^2 + y^2}$.

where $z_{col}(\omega) = \frac{z_0^2}{z_R(\omega)}$.

If we know the expression of dynamics of spatial-temporal spectrum in linear isotropic media [4] and Fourier transform

$$G(\omega, x, y, z) = \frac{1}{(2\pi)^2} \int\limits_{-\infty}^{+\infty} \int\limits_{-\infty}^{+\infty} g(\omega, k_x, k_y, z) \exp\left(i\left(k_x x + k_y y\right)\right) dk_x dk_y,$$

we can produce expressions, which describe temporal dynamics of collimated beam's spectrum:

$$G_{col}(\omega, x, y, z) = \frac{iz_0\left(z_{col}(\omega) + i(z - z_0)\right)}{\left(z_{col}(\omega)\right)^2 + (z - z_0)^2} \exp\left(\begin{array}{c} -\frac{x^2+y^2}{\rho^2} \frac{z_R(\omega)(z_{col}(\omega)+i(z-z_0))}{(z_{col}(\omega))^2+(z-z_0)^2} - \\ -ik(\omega)(z_0 + z) \end{array}\right) G_0(\omega), \quad (11)$$

where distance z is counted from collimation mirror. Spatial-temporal structure of collimated wave packet at the distance z_0 from mirror, where the waist of beam is located, can be expressed in the form

$$E_{col}^{z=z_0}(t', x, y, z_0) = i^{2n+2} E_0 \frac{2^{n-1} L}{z_0} \left(\frac{\tau_p}{\tau}\right)^{n+2} \sum_{k=2l, l \in \mathbb{N}_0}^{n} \left(\frac{n!(2i)^{-k}}{(n-k)!\left(\frac{k}{2}\right)!} \left(\frac{t'}{\tau}\right)^{n-k}\right) \exp\left(-\left(\frac{t'}{\tau}\right)^2\right), \quad (12)$$

where duration of the wave packet τ is described by expression

$$\tau = \tau_p \sqrt{1 + \frac{x^2 + y^2}{\rho^2} \left(\frac{2L}{z_0}\right)^2}, \quad (13)$$

which is equivalent to beam's duration at the distance $z0$ from the radiation source. Delay time is

$$t' = t - \frac{N_0}{c}(z_0 + z). \quad (14)$$

shows the flat phase shift at the focus distance from collimation mirror.

In far-field region spatial-temporal structure simplifies to form

$$E_{col}^{far}(t'', x, y, z_0) = -i^{2n} E_0 \frac{L}{z_0} 2^{n-1} \sum_{k=2l, l \in \mathbb{N}_0}^{n-1} \left(\frac{n-1!(2i)^{-k}}{(n-1-k)!\left(\frac{k}{2}\right)!} \left(\frac{t''}{\tau_p}\right)^{n-1-k}\right) \exp\left(-\left(\frac{t''}{\tau_p}\right)^2\right) \quad (15)$$

where curvative of the wave front becomes to be spherical and can be described by expression:

$$t'' = t - \frac{N_0}{c}\left(z_0 + z + \frac{x^2 + y^2}{2(z - z_0)}\right). \quad (16)$$

The example of collimation of few-cycle waves are shown at fig.4

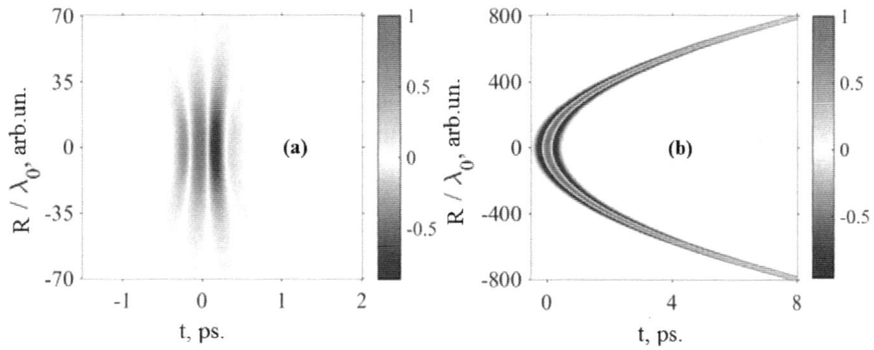

Figure 4. Spatial-temporal structure of electric field at distance $z_0 = f_{col}$ (a) and $z_0 = 100 f_{col}$ (b) from collimation mirror on example of initial half and one cycle beam. $n = 3$; $\tau_p = 0.2ps$; $R = \sqrt{x^2 + y^2}$.

4. Conclusion

The spectral approach was used for the task of few-cycle wave propagation in linear isotropic media. In far-field region diffracted n-half-cycle beam is described by superposition of (n+1)-half-cycle and (n+2)-half-cycle wave packets. It produce asymmetric temporal structure, which with distance becomes symmetric, because (n+2)-half-cycle component decreases faster than (n+1)-half-cycle. The frequency of maximum spectral density increases with distance from source of radiation along the axis of propagation (4). The frequency shift is inversely proportional to the initial duration of the wave packet and decreases with increasing of oscillations number. For collimated few-cycle beams equations, which describes electric field spatial-temporal structure at focus distance from collimation mirror and in far-field region of diffraction are produced.

5. Acknowledgments

This work was financially supported by Government of Russian Federation, Grant 074-U01.

References

[1] Bespalov V G, Gorodetski A A, Denisyuk I Y, Kozlov S A, Krylov V N, Lukomski G V, Petrov N V and Putilin S E 2008 J. Opt. Technol. 75 636-642
[2] Sakhnov S N, Leksutkina E V, Smolyanskaya O A, Usov A V, Parakhuda S E, Grachev Y V and Kozlov S A 2011 Optics and Spectroscopy 111 257-261
[3] Puzyrev D., Drozdov A.A. Collimation and focusing of initially single-cycle paraxial optical beams. Journal of Physics: Conference Series. 2014. Vol. 541. No. 1. pp. 012065.
[4] Ezerskaya A A, Ivanov D V, Kozlov S A and Kivshar Y S 2012 J. Infrared Millim. Terahertz Waves 33 926942
[5] Drozdov A A, Kozlov S A, Sukhorukov A A and Kivshar Y S 2012 Phys. Rev. A 86 053822
[6] Kulya M S, Grachev Y V, Bespalov V G and Kujanpaa V P 2013 Nanosystems: Physics, Chemistry,Mathematics 4 206213

Journal of Physics: Conference Series **735** (2016) 012029 doi:10.1088/1742-6596/735/1/012029

Polarization of two-level atom in a weak polychromatic field

A G Antipov, A A Kalinichev, S A Pulkin, S V Saveleva, A C Sumarokov, S V Uvarova and V I Yakovleva

Saint Petersburg State University, St.Petersburg, Russia

E-mail: usvik2009@yandex.ru

Abstract. Numerical solutions have been obtained for 2-level atom driven by a polychromatic field. Two different methods of analytical solution were proposed for any case of position of the field's components.

1. Introduction

The aim of the present study was to investigate the polarization spectrum of two-level homogeneously broadened atomic system in a weak polychromatic field.

Let the medium of two-level atoms is subjected to the action of quasi-resonant polychromatic field consisted of $(2M+1)$ equidistant monochromatic components:

$$E(t)=\frac{1}{2}\left(\left[E_{s0}+\sum_{m=1}^{M}E_{sm}(e^{im\Delta_s t}+e^{-im\Delta_s t})\right]e^{i\omega_{s0}t}+\kappa.c.\right), \qquad (1)$$

where $\omega_{s0}=(\omega_{2M+1}+\omega_1)/2$ is the median frequency of the polychromatic field, $\Delta_s=\omega_{j+1}-\omega_j$ is the frequency distance between components and is the amplitude of m-th component (Figure 1).

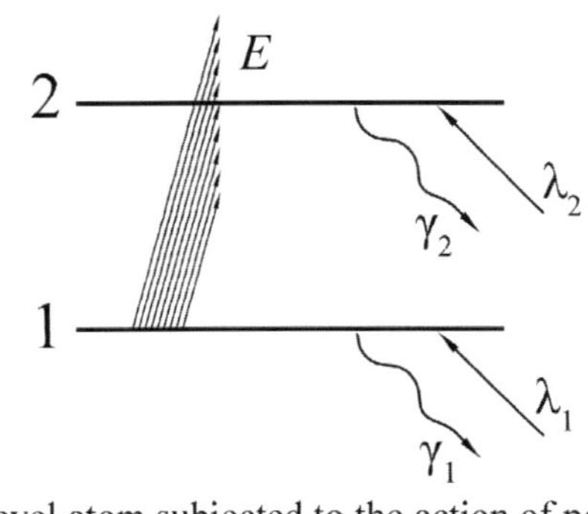

Figure 1. The diagram of two-level atom subjected to the action of polychromatic field *E(t)*.

The density matrix equations for two-level system in quasi-resonant rotating wave approximation have the following form:

$$\frac{d\rho_{12}}{dt} = -(\Gamma + i\delta)\rho_{12} + iV_{21}^{*}N_{12}$$
$$\frac{dN_{12}}{dt} = \lambda_{12} - \gamma N_{12} - 4\,\mathrm{Im}(V_{21}\rho_{12}), \qquad (2)$$

where $\lambda_{12} = \lambda_1 - \lambda_2$ is the level pump difference, $V_{21} = \frac{1}{2}\Omega_{s0} + \sum_{m=1}^{M}\Omega_{sm}\cos m\Delta_s t$ is the matrix element of interacting energy in dipole approximation, $\Omega_{sm} = -\frac{d_{21}E_{sm}}{\hbar}$ is the Rabi frequency, d_{21} is the dipole moment of the transition, $N_{12} = \rho_{11} - \rho_{22}$ is the population difference, $\delta = \omega_{S0} - \omega_{21}$ is the shift from field median frequency to a transition frequency ω_{21}, $\gamma = \gamma_1 = \gamma_2$ and Γ are the longitudial and transverse relaxation constants respectively.

2. Computations

The first way to solve differential equation system (2) is to apply to it direct numerical methods. Runge-Kutta method gives the population difference N_{12} and the non-diagonal element of density matrix as the discrete functions of time (Figure 2). Frequency dependencies can be obtained by Fourier transform.

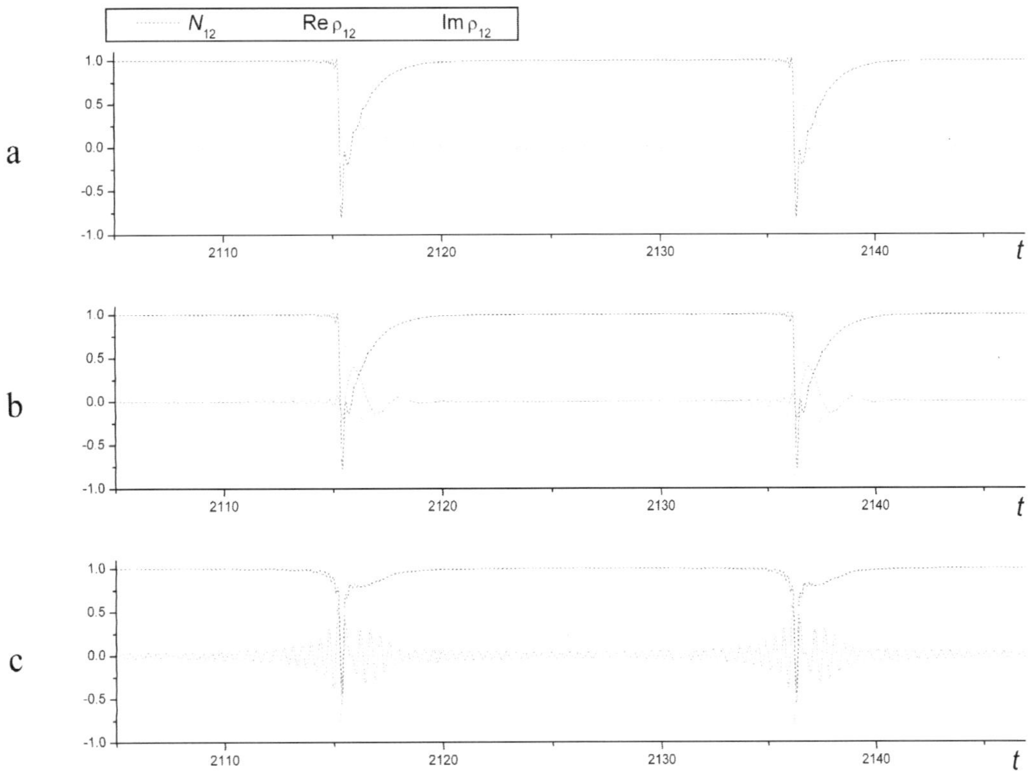

Figure 2. Population difference N_{12}, real and imaginary parts of non-diagonal density matrix element ρ_{12} as functions of time at some values of shift $\delta = \omega_{S0} - \omega_{21}$. $M = 50$, $\Delta_s = 0.3\Gamma$,

$\Omega_{sm} = 0.2\Gamma$, $m = 0, \pm 1, \ldots, \pm M$, $\lambda_{12} = 1$; a) $\delta = 0$, b) $\delta = 3\Gamma$, c) $\delta = 13\Gamma$.

The second way to solve (2) which is close to Stenholm method of chain ratios [2] is to decompose the unknown functions in harmonic basis.

The system of differential equations (2) simplifies for the special symmetric case of zero shift $\delta = 0$ and equal relaxation constants $\gamma = \Gamma$. Introducing new variable $y = N_{12} + i2\,\mathrm{Im}\,\rho_{12}$ one can reduce equation system (2) to a single equation:

$$\frac{dy}{dt} = -(\gamma - i2V_{21})y + \lambda_{12}. \tag{3}$$

Substituting into (3) harmonic decomposition $y(t) = \sum_{j=-\infty}^{+\infty} a_j e^{ij\Delta_s t}$, we arrive to an infinite linear equation system for unknown coefficients $(\gamma + ij\Delta_s)a_j - i\sum_{m=-M}^{M}\Omega_{sm}a_{j-m} = \lambda\delta_{0,j}$, $j = 0, \pm 1, \pm 2, \ldots$, where $\delta_{0,j}$ is the Kronecker symbol.

Due to linear growth of respective matrix diagonal elements as j increases while non-diagonal elements remain the same one can realize a reduction, i.e. to consider finite system of equations instead of infinite one. The values of diagonal elements are defined by Δ_s, while the values of non-diagonal ones are defined by Ω_{sm}, thus at small ratio Ω_{sm}/Δ_s the effectiveness of reduction increases. While coefficients a_j are computed, we can obtain the population difference $N_{12}(t) = \mathrm{Re}\,y(t)$ and the imaginary part of non-diagonal density matrix element $\mathrm{Im}\,\rho_{12}(t) = \mathrm{Im}\,y(t)/2.$ as functions of time.

The polarization is defined by non-diagonal density matrix elements:

$$P(t) = d_{21}\rho_{12}(t) + c.c.$$

Spectral components of polarization oscillating on frequencies are equal to

$$P_{\pm j} = -\frac{d_{21}}{2} < \rho_{12}e^{\pm ij\Delta_s t} >_t. \tag{4}$$

In isotropic medium a direction of polarization vector coincides with the field orientation and the susceptibility is given by

$$\chi_j = \frac{P_j}{E_{sj}} = -\frac{P_j d_{21}}{\Omega_{sj}\hbar}$$

The real and the imaginary parts of susceptibility define the dispersion and the absorption coefficient respectively.

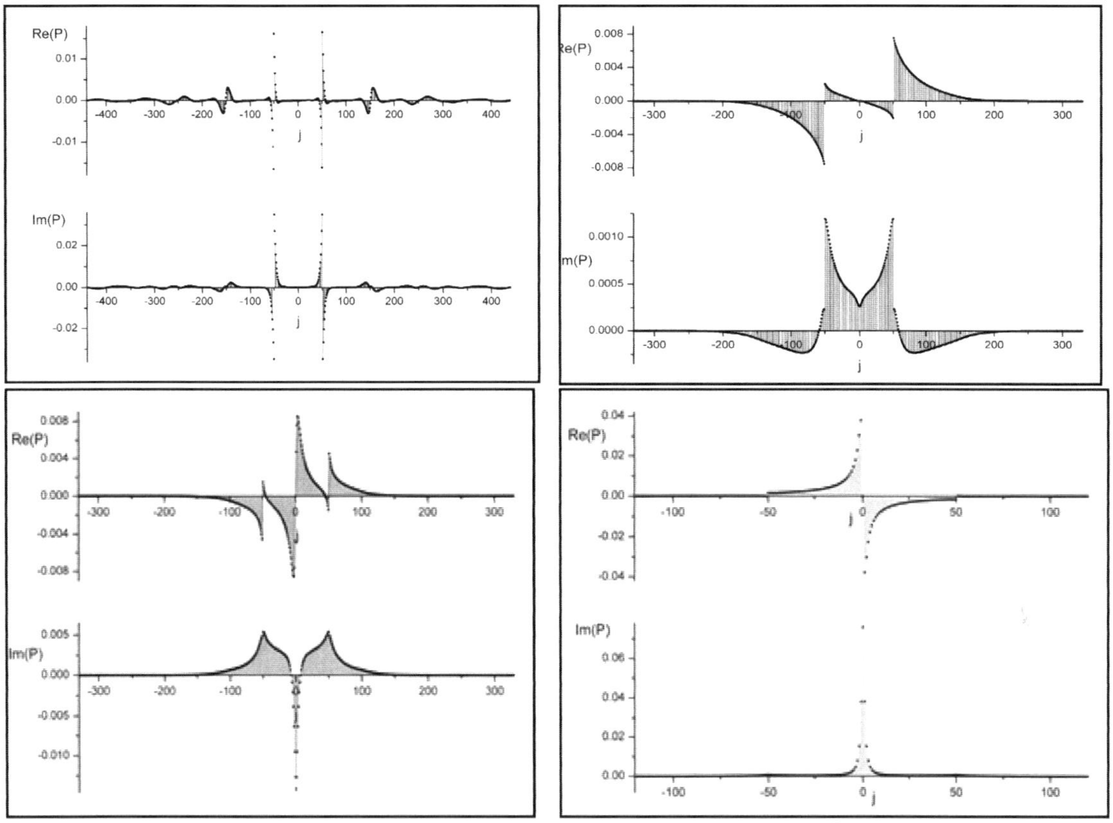

c d

Figure 3. The real and the imaginary parts of polarization as functions of component number at different distances between components Δ_s. $M = 50$, $\Omega_{sm} = 0.2\Gamma$, $m = 0, \pm 1, \ldots, \pm M$, $\lambda_{12} = 1$, $\delta = 0$; a) $\Delta_s = 0.01$, b) $\Delta_s = 0.2$, c) $\Delta_s = 0.3$, d) $\Delta_s = 1$.

In weak fields with small distances between components where Δ_s is less or equal to Ω_{sm}, the imaginary part of polarization have positive values for $-M < j < M$, therefore these components are absorbed (Figure 3 a, b, c). At $|j| > M$ the amplitude of imaginary part of polarization for $M-1$-th component is negative and its absolute value is almost equal to the absolute value of $M+1$-th component, thus new components appear at frequencies where acting field is absent. In spite of the only weak field acts non-linear processes involving new harmonics forming at $|j| > M$ arise.

At some parameters of atom and field the amplification exists on the transition frequency i.e. the imaginary part of polarization is negative.

At large distances between field components $\Delta_s > \Omega_{sm}$ the dependence of polarization imaginary part on component number j looks like a Lorentz contour (Figure 3 d) as in the case of homogeneous broadening and one-component weak field near the absorption line. A dispersion has abnormal characteristics near the transition line and normal characteristics aside of it.

In the common case ($\delta \neq 0$) considering N_{12}, $\mathrm{Re}\,\rho_{12}$, $\mathrm{Im}\,\rho_{12}$ in density matrix equations (2) as vectors in harmonic space we obtain the following operator equation for population difference:

$$\left(\frac{d}{dt}+\gamma+4V\left(\left(\frac{d}{dt}+\Gamma\right)+\delta^2\left(\frac{d}{dt}+\Gamma\right)^{-1}\right)^{-1}V\right)N_{12}=\lambda.$$

Using reduction to solve the last equation numerically we can find N_{12} as decomposition on finite number of harmonics. The imaginary and real parts of non-diagonal element of density matrix are expressed in terms of population difference N_{12} :

$$\operatorname{Im}\rho_{12}=\left(\left(\frac{d}{dt}+\Gamma\right)+\delta^2\left(\frac{d}{dt}+\Gamma\right)^{-1}\right)^{-1}VN_{12}$$

$$\operatorname{Re}\rho_{12}=\delta\left(\frac{d}{dt}+\Gamma\right)^{-1}\left(\left(\frac{d}{dt}+\Gamma\right)+\delta^2\left(\frac{d}{dt}+\Gamma\right)^{-1}\right)^{-1}VN_{12}.$$

Polarization can be computed using (4).

Figures 4 and 5 show the evolution of amplification resonances as the frequency shift δ changes. A projection on the plane $\delta 0 j$ demonstrates the central resonance ($\delta=0$, $j=0$) follows the transition frequency. The obtained numerical data show that at shifts $\delta < M\Delta_S = 15$ the resonance is slightly pulled by the median frequency of acting field. At $\delta > 15$ the resonance on the contrary is slightly pushed away from the transition frequency by the acting field. Near $\delta = M\Delta_S = 15$ while the left margin of acting field spectrum is about the transition frequency one can see a transformation of one amplification resonance into another with a lesser characteristic frequency.

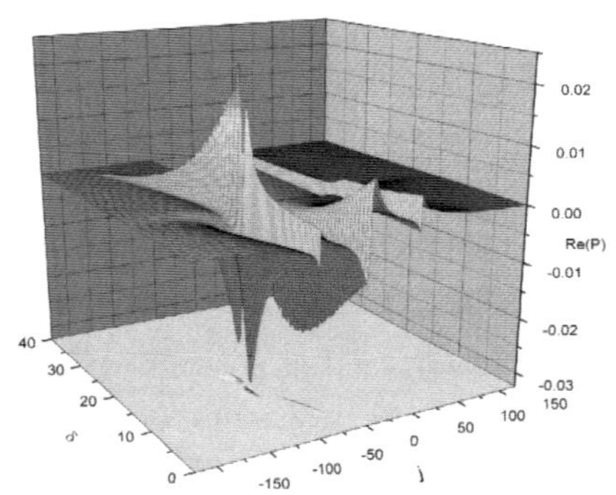

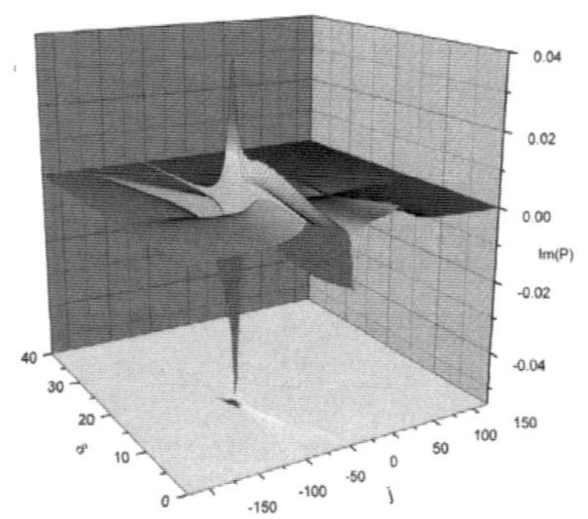

Figure 4. The real $\mathrm{Re}(P)$ and the imaginary $\mathrm{Im}(P)$ parts of polarization as functions of component number and shift δ of transition frequency from the median field frequency. $M=50$, $\Delta_s=0.3\Gamma$, $\Omega_{sm}=0.2\Gamma$, $m=0,\pm1,\ldots,\pm M$, $\lambda_{12}=1$.

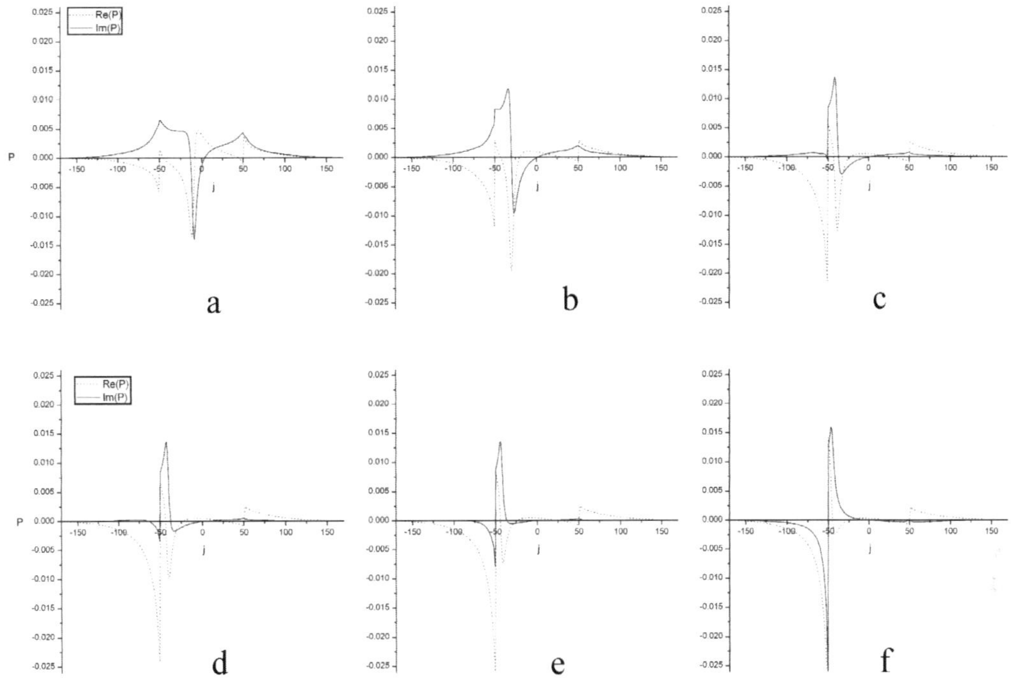

Figure 5. The real and the imaginary parts of polarization as functions of component number j at different shifts δ. $M=50$, $\Delta_s=0.3\Gamma$, $\Omega_{sm}=0.2\Gamma$, $m=0,\pm1,\ldots,\pm M$, $\lambda_{12}=1$; a) $\delta=3$, b) $\delta=9$, c) $\delta=12$, d) $\delta=12.5$, e) $\delta=13$, f) $\delta=14$.

The results obtained by the both considered methods are in good agreement with each other [3].

3. Conclusions

In comb-spectroscopy the nonlinear multi-quantum processes usually are not considered.

We showed that even in weak fields the nonlinear effects take place and this nonlinearity should be taken into account.

When median field frequency does not coincide with the transition frequency the resonance mostly follows the transition frequency. But some frequency pulling or pushing is observed depending on the transition frequency inside the field component bounds or not. This phenomenon can be considered as an evidence of nonlinear interaction of the weak field components.

The results could be used for nonlinear optics and spectroscopy [4].

References

[1] Vitushkin L F a*t al* 1993 *Optika i spectroskopiya* **74** pp. 786-794
[2] Stanholm S 1984 *Foundations of laser spectroscopy* (New York: Wiley)
[3] Antipov A G, Pulkin S A, Sumarokov A S, Uvarova S V, Yakovleva V I, *Optics and Spectroscopy* **118** pp. 945–948
[4] Pulkin S A *at al* 2015 *Russian Journal of Physical Chemistry* B **9** pp. 571–573

Journal of Physics: Conference Series **735** (2016) 012030 doi:10.1088/1742-6596/735/1/012030

Light beam diffraction on inhomogeneous holographic photonic PDLC structures under the influence of spatially non-uniform electric field

A O Semkin[1], **S N Sharangovich**[1,2]
[1]Tomsk state university of control systems and radioelectronics, 40 Lenina av., Tomsk, Russia

E-mail: shr@tusur.ru

Abstract. In this work the theoretical model of two-dimensional Bragg diffraction of quasi-monochromatic light beams on amplitude- and phase- inhomogeneous holographic photonic PDLC structures under the impact of spatially non-uniform electric field is proposed. The self-consistent solutions for the light diffraction on PDLC structure with uniform amplitude and quasi-quadratic profiles are obtained for the case of influence of linearly varying electric field. The possibility to compensate the PDLC structure response inhomogeneity by the impact of non-unifrom external field is shown.

1. Introduction

Photonic structures with inhomogeneous amplitude and phase profiles have a wide range of possible applications. One of the practical problems which can be solved - improving the efficiency of light input to the polymer optical fiber due to a wider angular aperture of such structures. The solution to this problem using chirped diffraction gratings in photopolymer materials has been discussed in detail previously [1]. Development of optical data transmission technologies requires the creation of a fully optical manageable elements. By administering a certain quantity of liquid crystal (LC) molecules to the photopolymer material it is possible to obtain a composite material, called polymer-dispersed liquid crystal (PDLC) which allows us to holographically form the photonic structure (PS), whose optical properties can be controlled by an external electric field. The peculiarity of this material is the liquid crystal "droplets" formation of in a polymer medium [2].

In the case of recording holographic amplitude- and phase- inhomogeneous PS in PDLC the impact of the control field would distort diffraction characteristics of PS, so the problem of optimizing the response of PS to an external field is actual now. The aim of this study is to investigate the influence of spatially non-uniform electric field on the diffraction characteristics of phase-inhomogeneous PDLC PS with linearly changing period (chirped diffraction gratings).

2. Theoretical model

The presence of the LC molecules in the investigated environment causes anisotropy of its optical properties. This statement allows us to use the results of [3.4] to solve the problem by replacing the temperature-induced optical inhomogeneity [4] by the electrically-induced. Similarly to [4] we will

[2] To whom any correspondence should be addressed.

consider a two-dimensional Bragg diffraction of quasi-monochromatic beam $E_p(\mathbf{r})$, corresponding to an extra-ordinary wave, on inhomogeneous photonic structure (figure 1a). The direction of a periodic perturbation of the dielectric constant is given by the vector $\mathbf{K}$, which is due to inhomogeneity of PS represents a function of the coordinates $\mathbf{K}=\mathbf{K}(\mathbf{r})$. The spatial inhomogeneity of the external electric field is set by a special topology of the electrode structure (figure 1a). As an illustration, figure 1b shows a linearly varying function of the field at the aperture of the beam w.

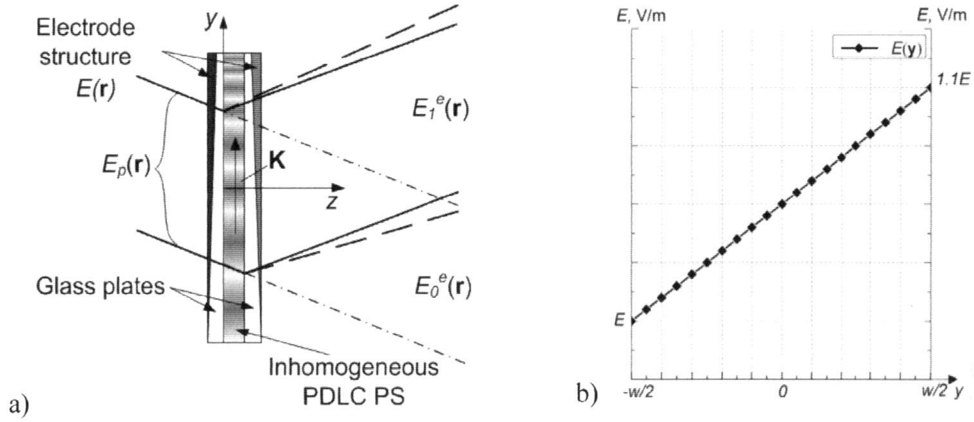

Figure 1. Diffraction geometry (a) and function of external electric field (b)

The perturbation of the sample dielectric tensor because of its smallness in relation to $\hat{\varepsilon}_0$ we will present as the following sum:

$$\hat{\varepsilon}(\mathbf{r}) = \hat{\varepsilon}_0 + \Delta\hat{\varepsilon}(\mathbf{r}) + \Delta\hat{\varepsilon}_e(\mathbf{r}), \quad (1)$$

where $\hat{\varepsilon}_0$ – unperturbed dielectric tensor; $\Delta\hat{\varepsilon}_a(\mathbf{r})$ – periodical perturbation, caused by the inhomogeneous PS recording process; $\Delta\hat{\varepsilon}_e(\mathbf{r})$ – smooth perturbation, caused by the impact of spatially non-uniform electric field.

$$\Delta\hat{\varepsilon}(\mathbf{r}) = 0.5\Delta\hat{\varepsilon}[U_0 U_m(\mathbf{r})\exp[i\cdot\varphi(\mathbf{r})]], \quad (2)$$

where U_0, $U_m(\mathbf{r})$ – amplitude of perturbation and normalized amplitude profile; $\varphi(\mathbf{r})$ – phase profile, which in the case of PS with varying period can be represented as $\varphi(\mathbf{r})=\mathbf{K}(\mathbf{r})\cdot\mathbf{r}$; $\Delta\hat{\varepsilon}$ – perturbation of $\hat{\varepsilon}_0$, corresponding to $U_0=1$; $\mathbf{r}$ – radius-vector.

Changing of the dielectric constant caused by the impact of external field is due to the reorientation of LC director in the electric field under the action of Fredericks effect [2,5,6]. As in the considered medium the statistical distribution of the LC molecules preferred orientation in the "droplets" is observed [2] we will introduce the corresponding changing of the dielectric constant in the form of statistically averaged tensor:

$$\langle\Delta\hat{\varepsilon}_e(\mathbf{r},E)\rangle = -\Delta\varepsilon\int_0^{\pi}\int_{-1}^{1}\mathbf{C}(\mathbf{r},E)\mathbf{C}(\mathbf{r},E)p(\alpha)q(\phi)d\alpha d\phi, \quad (3)$$

where $\mathbf{C}(\mathbf{r},E)$ – spatial distribution of LC director orientation; $\langle...\rangle$ means statistical averaging.

In (3) we introduced the Gaussian functions of director distribution in the quasi-ellipsoidal LC "droplet" [5]:

$$p(\alpha) = A\exp\left[-(\alpha-\overline{\alpha})^2/2\sigma_\alpha^2\right],\ q(\phi) = B\exp\left[-(\phi-\overline{\phi})^2/2\sigma_\phi^2\right] \tag{4}$$

where α, ϕ – orientation angles of LC molecules in the "droplet"; $\overline{\alpha}$, $\overline{\phi}$ – averages; σ_α, σ_ϕ – standard deviations.

LC director orientation changing in each spatial point under the influence of an electric field will be given by the angle of declination from its initial direction. As the considered medium is a set of LC "droplets" suspended in the polymer, to determine the distribution of the "droplets" rotation angle in the sample we will use the theoretical model, developed in [2]:

$$\gamma(\mathbf{r},E) = \psi(\mathbf{r}) + \frac{1}{2}\cdot arctg\left[\frac{\sin[2\vartheta_0(\mathbf{r})]}{e^2+\cos[2\vartheta_0(\mathbf{r})]}\right], \tag{5}$$

where $\psi(\mathbf{r})$ – orientation angle of external electric field vector $\mathbf{E}$; $\vartheta_0(\mathbf{r}) = \gamma_0 - \psi(\mathbf{r})$ – initial angle between $\mathbf{E}$ and $\mathbf{C}$ vectors; $e = |\mathbf{E}|R\sqrt{\Delta\overline{\varepsilon}/K(5.7\delta^2+2.1\lambda)}$ – parameter, which characterizes the impact of electric field on bipolar LC droplet; R – radius of the droplet; δ – eccentricity of the droplet; $\lambda = RW_a/K$ – parameter of surface adhesion; W_a – azimuth surface adhesion coefficient; K – Frank's elasticity coefficient; $\Delta\overline{\varepsilon}$ – effective dielectric anisotropy of bipolar droplet.

Critical electric field at which the Fredericks effect will occur is given by [2]:

$$Ec = \frac{1}{R}\cdot\left(\frac{5.7K\delta^2+2.1W_aR}{\Delta\overline{\varepsilon}}\right)^{1/2}. \tag{6}$$

In the considered case of Bragg diffraction of the extraordinary waves on grating (1), the total optical field in the sample can be represented as [4]:

$$E_s(\mathbf{r},E) = \sum_{i=0,1}\mathbf{e}_i E_i^e(\mathbf{r},E)\exp\left[i\left(\omega_i t - \int \mathbf{k}_i\cdot d\mathbf{r}\right)\right] \tag{7}$$

where $\mathbf{e}_i$, $\mathbf{k}_i$, ω_i – polarization vectors, wave vectors and frequencies of transmitted $E_0^e(\mathbf{r},E)$ and diffracted $E_1^e(\mathbf{r},E)$ beams.

Amplitude distributions $E_{0,1}^e(\mathbf{r},E)$ can be determined by solving the system of coupled waves equations (CWE) [4]:

$$\mathbf{N}_{r0}^e\cdot\nabla E_0^e(\mathbf{r},E) = -iC_1^e(\mathbf{r},E)U_m^*(\mathbf{r})E_1^e(\mathbf{r},E)\exp(+i\int_0^r \Delta K(\mathbf{r},E)dr), \tag{8}$$

$$\mathbf{N}_{r1}^e\cdot\nabla E_1^e(\mathbf{r},E) = -iC_0^e(\mathbf{r},E)U_m(\mathbf{r})E_0^e(\mathbf{r},E)\exp(-i\int_0^r \Delta K(\mathbf{r},E)dr), \tag{9}$$

where $C_i^e(\mathbf{r},E) = \frac{1}{4}\frac{\omega\cdot U_0}{c_c n_i^e(\mathbf{r},E)}\mathbf{e}_1^e(\mathbf{r})\cdot\Delta\hat{\varepsilon}(\mathbf{r},E)\cdot\mathbf{e}_0^e(\mathbf{r})$ – amplitude coupling coefficients; $\mathbf{N}_{r0,1}^e$ – group normals; c_c – speed of light; $\Delta K(\mathbf{r})$ – length of spatially non-uniform phase mismatch vector $\Delta\mathbf{K}(\mathbf{r},E) = \mathbf{k}_0(\mathbf{r},E) - \mathbf{k}_1(\mathbf{r},E) + \mathbf{K}(\mathbf{r})$, characterizing the diffraction geometry changing by the impact of external electric field and PS period varying. Parameter $\Delta K(\mathbf{r})$ can be determined as in [4].

Refraction indexes $n_{0,1}^e(\mathbf{r},E)$ for the diffracted light beams are given by:

$$n_i^e(\mathbf{r},E) = n_o n_e \left[n_e^2 \cdot \sin^2(\gamma(\mathbf{r},E) \pm \theta_i^e(\mathbf{r})) + n_o^2 \cdot \cos^2(\gamma(\mathbf{r},E) \pm \theta_i^e(\mathbf{r}))\right]^{-1/2}, \tag{10}$$

where $\theta_i^e(\mathbf{r})$ – angles of incidence and diffraction of extra-ordinary waves.

Solution of the diffraction problem we will seek for a uniform amplitude profile $U_m(\mathbf{r}) = 1$ and quasi-quadratic phase front $\varphi(\mathbf{r})$. We will consider the case when the $\nabla\varphi(\mathbf{r})$ is directed along the grating vector $\mathbf{K}$ (figure 1a), which corresponds to the chirped PS. In this case, the phase profile of the PS included in (2) we can represent as a Taylor series:

$$\varphi(\mathbf{r}) = \varphi_0 + \nabla\varphi \cdot \mathbf{r} + 0.5\nabla^2\varphi \cdot \mathbf{r}^2, \tag{11}$$

where $\nabla\varphi = K_0$ – average length of vector $\mathbf{K}$; $0.5\nabla^2\varphi$ – deviation of $\mathbf{K}$ length, caused by the inhomogeneity of PS phase profile.

Substituting (11) to (2), considering a uniform amplitude profile, we will find the solution of CWE system (8)-(9) in analogy with [4] in aperture coordinates ξ_0, ξ_1, without taking into account the attenuation of light. Then the distribution of the diffracted light field amplitude at the output of the PDLC sample can be written as follows:

$$E_1(\eta) = -i\frac{C_1 U_0 L}{2\upsilon_1}\int_{-1}^{+1}\left\{\exp\left[\delta m(1-y) + \delta^2 n(1-y^2)\right]\left[\Phi\left(\frac{d}{a}, 1; a\delta^2\frac{\upsilon_1}{\upsilon_0}(1-y)^z\right)E_0\left(\frac{\delta(1-y)}{\upsilon_0} - \frac{\eta}{\upsilon_1}\right)\right]\right\}dy, \tag{12}$$

where $\Phi(a,c,z)$ – confluent hypergeometric function of the first kind; L – sample thickness; the remaining symbols are shown in [4].

3. Numerical simulations and discussions

For the analysis of the obtained solution and diffraction field amplitude dependencies on external field impact we will consider the incidence of a plane monochromatic wave with unit amplitude on the inhomogeneous PS. For this, we will introduce the relative phase mismatch $\Delta = f(\delta\theta)$, where $\delta\theta = \theta - \theta_б$ – the deviation of the plane wave components incidence angle from the Bragg angle. Figure 2 shows the results of numerical simulations of $|E_1(\Delta,E)|^2$ dependencies on the relative phase mismatch and the external field amplitude for amplitude-uniform photonic structures with different quadratic phase profiles (figure 2a) in the absence of external influences and on the different values of the external impact (Fig. 2b) for the PS with phase profile with $0.5\nabla^2\varphi = 2.5\cdot 10^5$. Wavelength of the incident light is $\lambda = 1.55$ um.

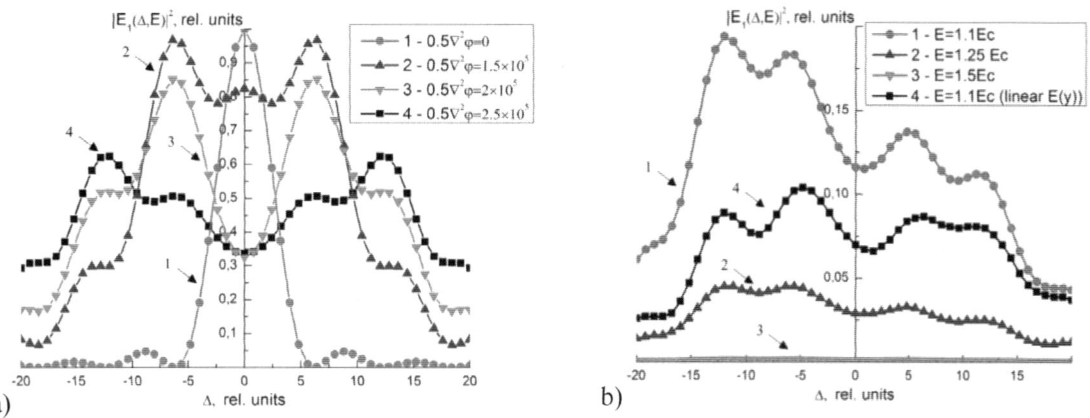

a) b)

Figure 2. Numerical simulations results

As it can be seen from figure 2b, the impact of the external field distorts the PS diffraction characteristics. The impact of the linear non-uniform (on the incident beam aperture) electric field (function is shown in figure 1b) partially solves this problem. The amplitude of the diffraction field at the output of PS under the influence of linear non-uniform external field is shown in figure 2b, curve 4. A more careful selection of function of spatial distribution of the external field $E(\mathbf{r})$ can lead to the full dependence $|E_1(\Delta, E)|^2$ restoration to the original form.

4. Conclusion

Thus, the possibility to compensate the response inhomogeneity of phase inhomogeneous photonic structure holographically formed in the PDLC by the impact of spatially non-unifrom electric field is shown in this work.

Acknowledgments

The work is performed as a base part of Government Task of Russian Ministry of Education for 2015 (project № 2014/225).

References

[1] Dovolnov E A, Sharangovich S N, Sheridan J T 2005 *Photorefractive Effects, Materials, and Devices 2005 (PR05), OSA Trends in Optics and Photonics Series (TOPS)* **99** 337
[2] Afonin O A, Nazvanov V F 1998 *Technical physics letters* **24** 451
[3] Kushnarev I N, Sharangovich S N 1992 *Technical physics* **62** №1 171
[4] Kushnarev I N, Sharangovich S N 1993 *Technical physics* **63** №2 24
[5] Dovolnov E A, Sharangovich S N 2008 *Izvestiya vuzov. Fizika* **51** № 9/3 171
[6] Sonin A S *Introduction to the physics of liquid crystals* (Science Press, Moscow) 1983

Journal of Physics: Conference Series **735** (2016) 012031 doi:10.1088/1742-6596/735/1/012031

Fluorescent standards for photodynamic therapy

N Belko[1,3], **S Kavalenka**[1] **and M Samtsov**[2]

[1] Faculty of Physics, Belarusian State University, Minsk, Belarus
[2] Sevchenko Institute of Applied Physical Problems of Belarusian State University, Minsk, Belarus

E-mail: nikita.belko@gmail.com

Abstract. Photodynamic therapy is an evolving technique for treatment of various oncological diseases. This method employs photosensitizers – species that lead to death of tumor cells after the photoactivation. For further development and novel applications of photodynamic therapy new photosensitizers are required. After synthesis of a new photosensitizer it is important to know its concentration in different biological tissues after its administration and distribution. The concentration is frequently measured by the extraction method, which has some disadvantages, e.g. it requires many biological test subjects that are euthanized during the measurement. We propose to measure the photosensitizer concentration in tissue by its fluorescence. For this purpose fluorescent standards were developed. The standards are robust and simple to produce; their fluorescence signal does not change with time. The fluorescence intensity of fluorescent standards seems to depend linearly on the dye concentration. A set of standards thus allow the calibration of a spectrometer. Finally, the photosensitizer concentration can be determined by the fluorescence intensity after comparing the corresponding spectrum with spectra of the set of fluorescent standards. A biological test subject is not euthanized during this kind of experiment. We hope this more humane technique can be used in future instead of the extraction method.

1. Introduction

Photodynamic therapy (PDT) is a clinically approved, minimally invasive technique for cancer treatment [1]. This technique requires a photosensitizer (PS) – a compound that absorbs the light and, consequently, initiates a photophysical reaction. This reaction results in the production of highly reactive species that damage nearby cells. A photosensitizer should thus accumulate selectively in malignant cells and have an absorption band in the optical window of tissue (i.e. in the region between 600 and 1200 nm, where tissue is transparent).

However, there is not an ideal PS for PDT. For this reason chemists are searching for new PSs that have better properties. Before some newly synthesized PS is clinically approved, its properties must be investigated in numerous experiments on biological objects (e.g. mice, rats). Specifically, it is important to know the concentration of PS in various biological tissues. The concentration is conventionally measured by the extraction method, which is time demanding and requires plenty of biological test subjects.

Alternatively, the concentration of PS can be measured by its fluorescence signal, since all PSs are fluorescent species. We call this technique for the concentration measurement the fluorescent method.

[3] To whom any correspondence should be addressed.

Although the fluorescence of PSs is commonly used for tumor imaging [2], exact values of the concentration are not important in this case.

In contrast, the fluorescent method allows us to determine the exact value of the PS concentration in tissue by its fluorescence, i.e. by using a spectrometer. All spectrometers are plagued by instabilities and should be calibrated before experiments. Consequently, we need objects with the fluorescence signal that does not change over a long period of time. This kind of objects we call fluorescent standards.

Our goal is to create fluorescent standards that allow us to measure the concentration of PS in biological tissue by a spectrometer (i.e. without extraction). The fluorescence of these objects is required to be similar to the fluorescence of the studied PS in tissue. In addition, fluorescent standards should be stable and simple to produce.

2. Methods

A tricarbocyanine dye (TD) was chosen as a fluorescent component of fluorescent standards (see the structural formula in figure 1). The dye belongs to the class of polymethine dyes, which are promising for use as PSs for PDT [3]. The fluorescence of TD can be efficiently excited at 680 nm, whereas widely used for PDT porphyrins should be excited at 630 nm [1]. TD has therefore an advantage over porphyrins when treating tumors of high thickness, since light with a longer wavelength penetrates more deeply through tissue.

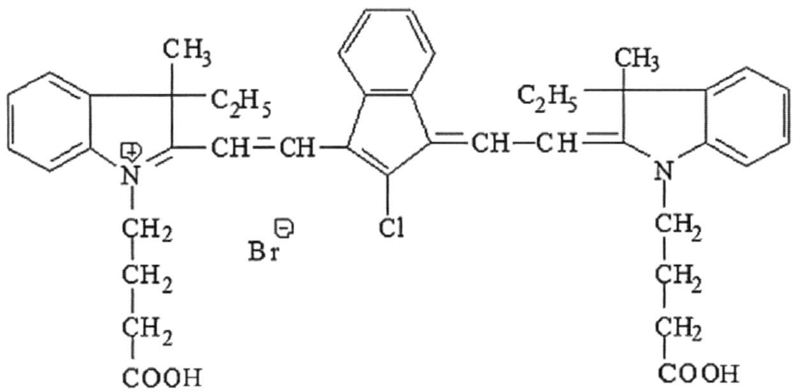

Figure 1. The structural formula of the utilized tricarbocyanine dye.

Disks of filter paper, which consists of cellulose, were chosen as a substrate for the dye. Dye solution was dropped on filter paper (the volume of droplets was 20 microliters), and then solvent evaporated at room temperature. Consequently, dye molecules were in nonpolar environment of cellulose fibers. Ethanol and acetone were used as solvents. The utilized dye is water-insoluble to prevent effects of air humidity.

The fluorescence was excited by a semiconductor laser at 680 nm corresponding to the mentioned above absorption peak of TD. To eliminate scattered laser light a colored glass filter was used when registering the fluorescence of the dye. The fluorescence signal was registered by a spectrometer with the Toshiba TCD1304A CCD linear image sensor.

3. Results and Discussion

The fluorescence of developed fluorescent standards can be observed in the range ca 700-900 nm, the peak of the spectrum is located at 757 nm (figure 2). The fluorescence spectrum has the same shape, when the dye concentration in the initial solution belongs to the range from 10^{-8} to 10^{-6} M. Lower dye concentration results in too low signal-to-noise ratio, whereas higher concentration leads to the deformation of the fluorescence spectrum due to the reabsorption. The spectrum does not seem to change with time.

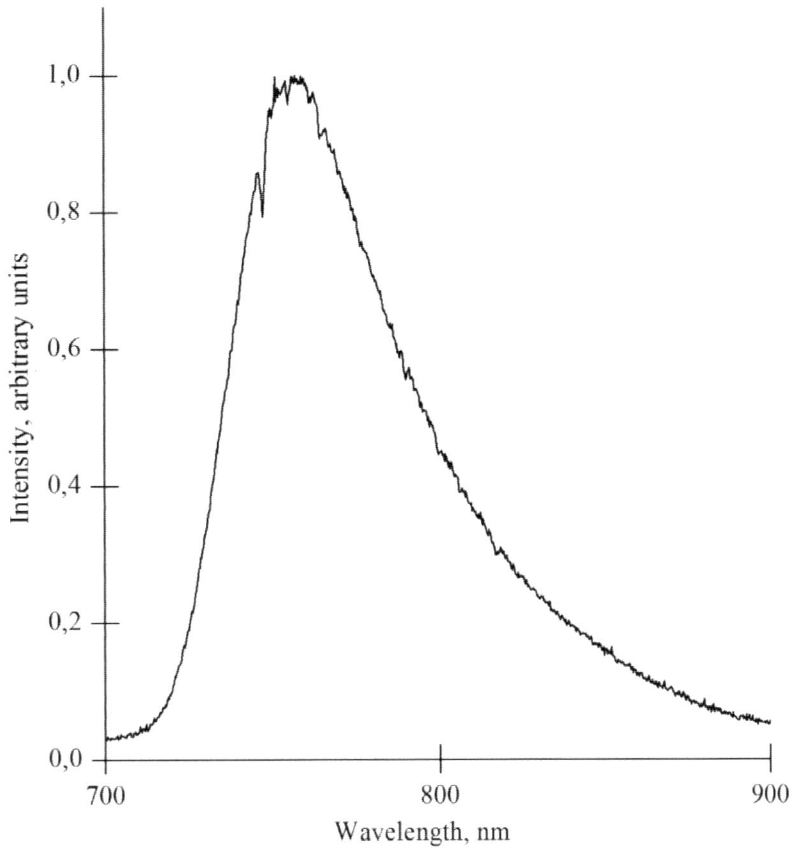

Figure 2. The fluorescence spectrum (normalized) of one of fluorescent standards.

The dye is distributed over fluorescent standards uniformly after the evaporation of solvent. The standard deviation of the peak intensity in the fluorescence spectra of different points of a fluorescent standard was 5 %.

The dye concentration on fluorescent standards can be multiplied by dropping of several solution droplets (after drying of previous droplets). It means the spectrum of a fluorescent standard with n droplets of dye solution with the concentration c_0 is almost identical to the spectrum of a fluorescent standard with one droplet of dye solution with the concentration nc_0. It was proved for $n=2$, $n=5$, $n=10$. For example, the maximal deviation between two corresponding spectra for $n=2$ was only 2 %.

The fluorescence intensity of fluorescent standards appears to depend linearly on the number of dropped dye droplets (figure 3), i.e. on the concentration of TD according to the previous result. Deviations from the linear dependence can be explained by the incomplete uniformity of the dye distribution and filter paper.

We believe this result can be useful when calibrating a spectrometer. For the correctly calibrated spectrometer the dependence of the fluorescence intensity on the dye concentration must be linear. Taking this fact into account, the calibration can be conducted after measuring of this dependence by the utilized spectrometer. For convenient calibration a set of fluorescent standards with different dye concentrations is required.

The spectrum of fluorescent standards is similar to the spectra of TD in different tissues of a rat (figure 4). The TD concentration in some tissue of a living biological test subject can therefore be measured by the fluorescence signal of the dye in the tissue. To do this the corresponding spectrum is compared with the spectra of a set of fluorescent standards. In this case the biological test subject is not euthanized. The fluorescence of tissues of the living test subject can be excited and registered by optical fibers after administration and distribution of TD.

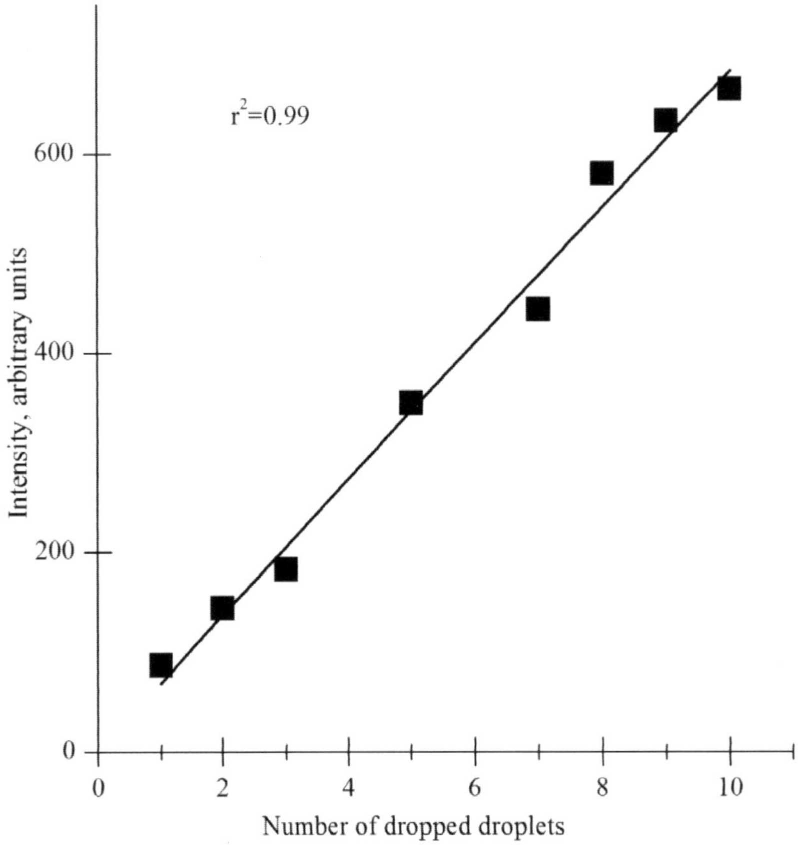

Figure 3. The dependence of the fluorescence intensity of fluorescent standards on the tricarbocyanine dye concentration.

This technique for concentration measurements is significantly simpler than the conventional extraction method, it requires less time and few biological objects. Moreover, the micro-environment of dye molecules does not change during measurements, whereas it does change when using the extraction method. The extraction method has therefore a disadvantage, since the change of the micro-environment can probably bias the measured values.

4. Summary

Photodynamic therapy is an efficient method for cancer treatment. A key component of photodynamic therapy is the photosensitizer. However, no photosensitizer is ideal, and new species are steadily synthesized. Properties of new photosensitizers must be carefully studied in experiments on biological object before clinical approval. In particular, it is important to know the distribution of photosensitizer in different biological tissues.

Robust, simply producible fluorescent standards for photodynamic therapy were developed. The standards can be used for the measurements of the photosensitizer concentration in various tissues by its fluorescence, i.e. by a spectrometer. A set of fluorescent standards allows the calibration of the spectrometer, which is important, since spectrometric facilities are subject to instabilities. Consequently, the photosensitizer concentration can be determined by comparing the spectrum of the tissue with spectra of the set of fluorescent standards.

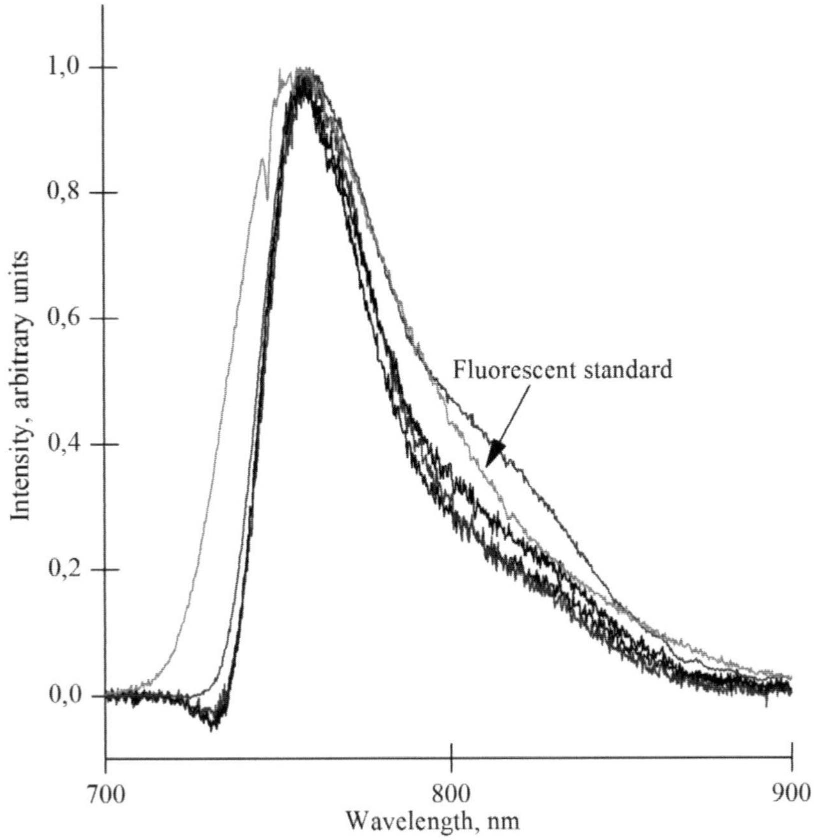

Figure 4. Normalized fluorescence spectra of tricarbocyanine dye in tissues of a rat (skin, brain, heart, liver, and kidney) and of a fluorescent standard.

This technique, i.e. the fluorescent method, is simple, not time-demanding, requires a little amount of biological objects, and thus can be efficiently used instead of the extraction method. We hope the fluorescence method will be widely employed in the future, since few biological objects are required and are not euthanized during the measurement.

References

[1] Agostinis P *et al* 2011 *CA-Cancer J. Clin.* **61** pp 250–81
[2] Celli J P, Spring B Q, Rizvi I, Evans C L, Samkoe K S, Verma S, Pogue B W and Hasan T 2010 *Chem. Rev.* **110** pp 2795–838
[3] Samtsov M P, Voropai E S, Mel'nikov D G, Lyashenko L S, Lugovskii A A and Istomin Yu P 2010 *Journal of Applied Spectroscopy* **77** pp 406–12

Journal of Physics: Conference Series **735** (2016) 012032 doi:10.1088/1742-6596/735/1/012032

Effects of femtosecond laser radiation on the skin

P Yu Rogov, V G Bespalov
ITMO University, Optics and Photonics Department, 49 Kronverksky Pr.
St. Petersburg, Russia

E-mail: RogovPU@niuitmo.ru

Abstract. A mathematical model of linear and nonlinear processes is presented occurring under the influence of femtosecond laser radiation on the skin. There was held an analysis and the numerical solution of an equation system describing the dynamics of the electron and phonon subsystems were received. The results can be used to determine the maximum permissible levels of energy generated by femtosecond laser systems and the establishment of Russian laser safety standards for femtosecond laser systems.

1. Introduction

Currently femtosecond laser systems (FLS) are used in scientific applications and in medical technology [1, 2]. However, despite the rapid development of femtosecond technologies and their usage in biology and medicine, currently in the Russian Federation there are no standards for safe levels of energy intense femtosecond laser pulses. It should be noted that the mechanism of interaction of intense femtosecond laser radiation with matter is significantly different from the mechanism of interaction of pulses of longer duration since the exposure time is less than the time required for excitation of the phonon subsystem [3]. In addition, the power density that occurs when using the FLS results in a variety of non-linear processes, such as self-focusing, two- and three-photon absorption, multiphoton ionization and in high energy density leads to an optical breakdown [4]. Particularly relevant the process is when femtosecond radiation affects the biological tissue, as an intense femtosecond pulses can be a significant danger to the skin, eyes and other organs [5].

In this paper we consider the linear and nonlinear effects of femtosecond laser radiation on the skin and a mathematical model that describes the approximate exposure to individual processes. In the future the model will be used to determine the maximum permissible energy levels of FLS.

2. Mechanism of action of femtosecond radiation on the skin

Human skin is optically transparent turbid medium (there are both the absorption and scattering). At the same time the skin is a heterogeneous structure consisting of the inclusion of various types and sizes (blood vessels, hair follicles, and so forth.), that complicates the understanding of the processes taking place under the influence of laser radiation on the skin. The main components of the skin are water (70%) and proteins (27%). The main structural protein of body is collagen, which is about 70% of the dry weight of skin. Moreover, the water molecules can be divided into two separate classes: the free molecules and water molecules in the bound state, involved in protein compounds (in triple helical collagen group represents about 500 molecules of water) [6]. Thus, in a first approximation, the skin can be considered water with protein inclusions by electrodynamics properties close to semiconductors.

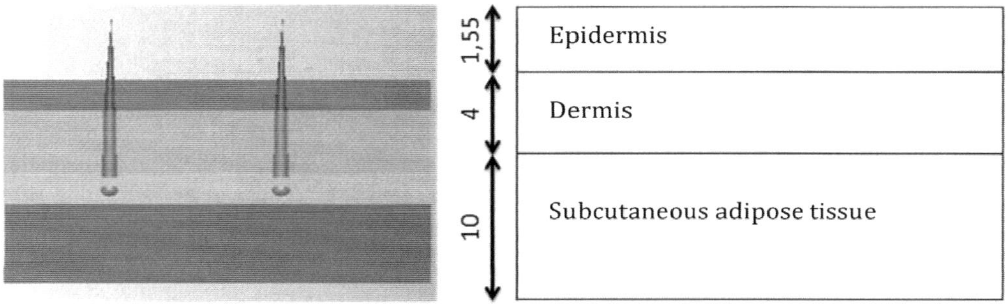

Figure 1. Model of human skin

Table 1. Optical (λ= 800 nm) and thermal parameters used in the model for calculating skin laser heating of the skin [7]

Tissue layer	d, mm	μ_a, 1/mm	μ_s, 1/mm	g	n	C, J/gK	k, W/mK
Epidermis	0,03-2	0,083	21,37	0,85	1,5	3,05	0,29
Dermis	0,5-5	0,026	9,67	0,86	1,5	3,52	0,41
Subcutaneous adipose tissue	>10	0,025	12,39	0,86	1,5	2,3	0,2

Where d - depth of the layer, μ_a- absorption coefficient, μ_s- scattering coefficient, g - mean cosine of the scattering angle, n - refractive index of the medium, C – heat capacity, k - the coefficient of thermal conductivity.

For evaluation we chose a wavelength of 800 nm for the most common FLS Ti: sapphire. The duration of the pulses in the system may vary from 10 to 150 fs. Irradiation of ultrashort laser pulses results in damage that cannot be explained by the mechanism of the heat fusion [8]. Such changes may be caused by the excitation of valent electrons with laser pulses [9, 10]. There are 500 water molecules for a triple helical collagen group. Thus the molecules of water are used as the transmission link between absorbing energy from the laser and exciting vibrational (phonon) mode of collagen (relaxation time is about 3 ps) [11, 12].

3. Mathematical model

In this study we used balanced system, the dynamic equation of electron-phonon subsystem. In the model atoms interact through semi-empirical potentials and electronic degrees of freedom are not taken into account. Due to the influence of the laser radiation non-equilibrium carriers is generated, which are described by the integral concentration [13]. We can neglect the process of Auger-recombination and impact ionization as it has small effect.

Yakovlev, E. B. proposed the idea of the definition of the relationship between the potentials of the atoms and the state of the electron subsystem, allowing to simulate the effects of heat when to femtosecond pulses expose on the dielectric material. [14]. Kinetic model of photo-excited semiconductor structure is taking into account external electron emission is a system of differential equations with the appropriate boundary and initial conditions. Differential Bouguer-Lambert law most accurately determines the distribution of laser intensity $J\ (z,t)$ within a solid (the z axis is directed into). Below there is a system of equations of thermal conductivity in one-dimensional approximation describing the dynamics of the electron and phonon subsystems:

$$\frac{\partial T_e}{\partial t} = \alpha_e \frac{\partial^2 T_e}{\partial z^2} - \frac{1}{C_e \tau_{ep}}(T_e - T_a) + \frac{\alpha_e h\nu}{C_e} J(t,z) \qquad (1.1)$$

$$\frac{\partial T_a}{\partial t} = \alpha_a \frac{\partial^2 T_a}{\partial z^2} + \frac{1}{C_a \tau_{ep}}(T_e - T_a) \qquad (1.2)$$

where $C_e = \frac{\pi^2 k_B^2 N_e}{2E_F}$; $J(t,z)$ – distribution of the intensity of the laser radiation within a solid, C_e - electron gas heat capacity, C_a - atomic heat capacity, τ_{ep} - time of electron-phonon relaxation (3 pc), T_e - the electron temperature, T_a - the temperature of the atoms, α_e and α_a - thermal diffusivity; E_F - Fermi energy; k_B - Boltzmann constant.

To simulate the effects of a femtosecond laser radiation on the skin, we chose the following boundary conditions:

$$T_e|_{\,t=0} = T_e|_{\,t=0} = T_0 \qquad (2.1)$$

$$\left.\frac{\partial T_e}{\partial z}\right|_{z=0} = \left.\frac{\partial T_e}{\partial z}\right|_{z=L} = \left.\frac{\partial T_a}{\partial z}\right|_{z=0} = \left.\frac{\partial T_a}{\partial z}\right|_{z=L} = 0 \quad (4.2)$$

Since the pulse width is negligible ($t \ll \frac{x^2}{a_e}$, $t \ll \frac{x^2}{a_a}$, where x - a spatial representation of the pulse, and the thermal diffusivity of biological tissue: α_a = 1,54 × 10–3 см^{2}с [15], а $\alpha_e < \alpha_a$), we can put $a_e \frac{\partial^2 T_e}{\partial z^2}$, $a_a \frac{\partial^2 T_e}{\partial z^2} = 0$, and $\frac{a_e h\nu}{C_e} \cdot J(t,z)$ can be replaced by a function $\Gamma\delta(z - vt)$ describes a heat source due to the absorption of radiation in the environment. Solution can be obtained in the form of:

$$T_e(t) - T_e(0) = \int_0^t (A + (1 - A)\exp(-\lambda(t - t'))) \cdot \Gamma\delta(z - vt')dt' \qquad (3.1)$$

$$T_a(t) - T_a(0) = \frac{1}{\tau_{ep} \cdot C_e} \int_0^t (1 - \exp(-\lambda(t - t'))) \cdot \Gamma\delta(z - vt')dt' \qquad (3.2)$$

where $A = \frac{1}{C_e^2} + \frac{1}{C_e + C_a}$, $\lambda = \frac{1}{\tau_{ep}}(\frac{1}{C_e} + \frac{1}{C_a})$.

4. Results

With the help of numerical simulation propagation of radiation in the skin tissues was obtained the data of the intensity distribution in depth:

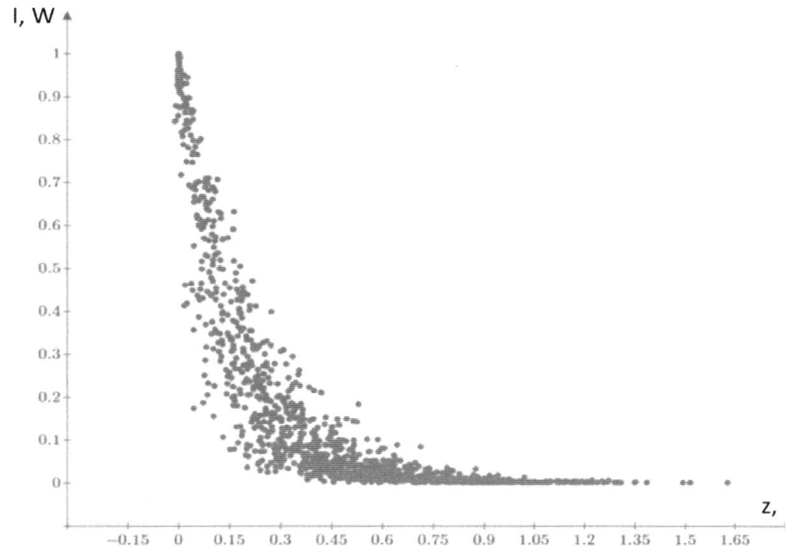

Figure 2. The intensity of the radiation on the depth of penetration (at λ = 800 nm).

Figure 2 shows that at a depth of 0.3 mm the radiation intensity is less than 30% of the initial level. Based on these data, we assume that most of the radiation is absorbed in a thin layer of the epidermis.

The dynamics of the electron subsystem can be described by the energy distribution. Figure 3 shows the energy distribution of the electrons as a function of time after the pulse: status of the electronic subsystem immediately after exposure, after 1 ps (some of the electrons relaxed after transferring energy to the phonon subsystem), more than 3 ps (most of the electrons transfered energy to the atomic subsystem, the whole system tends to thermodynamic equilibrium).

The dynamics of the electron subsystem can be described by the energy distribution:

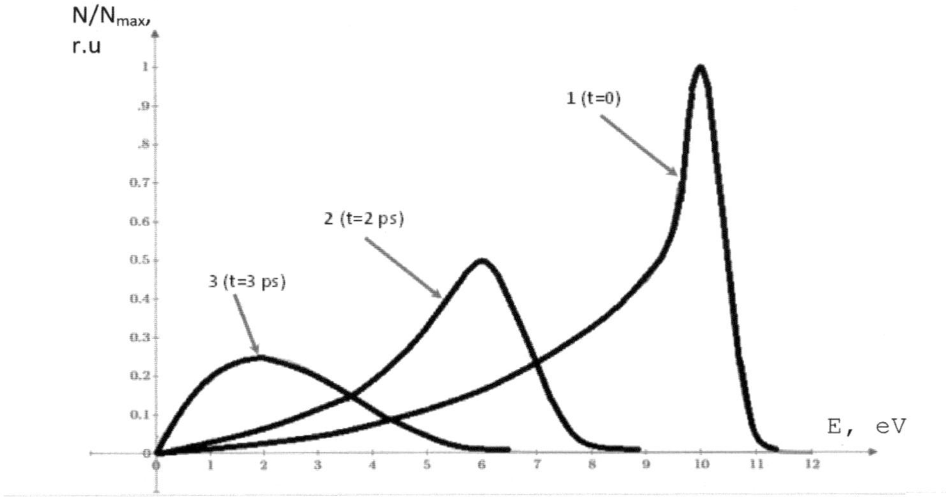

Figure 3. The distribution of the electron energy depending on the time: 1 - energy transfer to the electron subsystem, 2 and 3 - partial relaxation of electrons and atomic energy transfer subsystem.

Thus the dynamics of electron-phonon subsystem can be represented as:

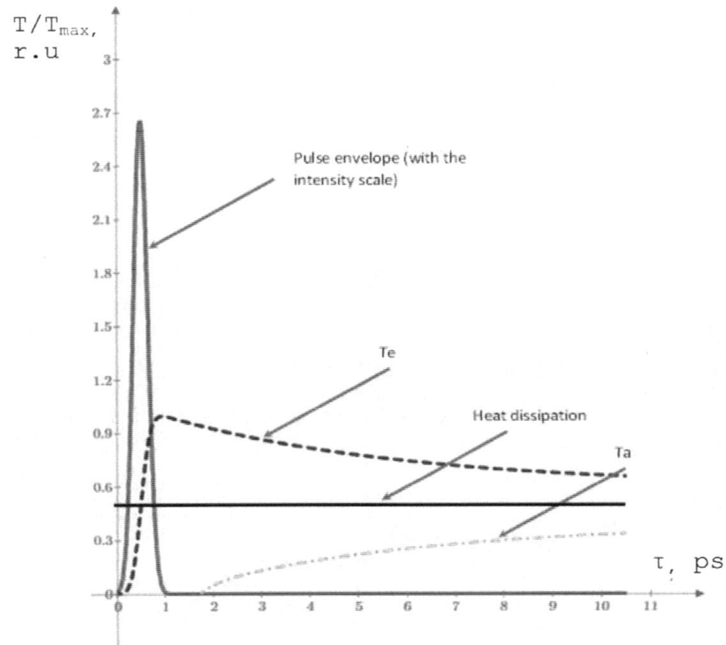

Figure 4. Thermodynamics of system, where T_e is the dynamics of the electron gas temperature, T_a is the dynamics of the crystal lattice temperature.

The process of exposure to the skin can be divided into three stages: the first - during the action of a femtosecond pulse multiphoton excitation of water molecules. The ionization energy in this process is 6.5 eV, so for the ionization of one water molecule about 5 photons (800 nm) are required, which is the reason for the decrease of the quantum efficiency of the process ($\leq 20\%$); at the same time there is the process of impact ionization. As a result by the end of the laser pulse $N_{max} \approx \eta \cdot E/\hbar w$ (E is energy of the pulse) of the electrons will be initiated in the upper ionized water molecules with more energy $U_a = 6.5$ eV. The second stage occurs after the action of a femtosecond pulse and continues until the full transfer of electronic energy to phonon subsystem, thus there is a cooling of the electron gas and heating the collagen. The third stage - the heat distribution by bulk.

5. Discussion

The maximum temperature of the electron gas is achieved at the end of the pulse; the phonon subsystem can be heated to a maximum temperature in a time longer than the duration of the pulse. Thus femtosecond radiation exposure on the skin is the process of heat transferring, extended in time to far greater time, than the duration of the pulse. From this system of equations the time at which the temperature of the crystal lattice of the gas reaches a maximum is calculated by the formula [13]:

$$t_{max} \approx \frac{C_e}{\beta} ln(1 + \frac{\beta C_a}{ka^2 C_e})$$

where β - the heat transfer coefficient between the electronic and lattice subsystems; ci, ce - the heat capacity of the crystal lattice and the electron gas; k - the lattice thermal conductivity. However, if we believe that the radiation is absorbed by a thin layer, the maximum temperature will be almost identical under the influence of 10 fs and 150 fs pulses with the same energy. Thus the three stages of exposure for short times can be set independently (since the time of the pulse is much shorter than the time of establishing the equilibrium temperature) between the atomic lattice and the electron gas which in its turn is negligibly compared with the time characterizing the heat dissipation. A more detailed analysis of the system of equations is needed in terms of the intensity distribution in the volume of the illuminated object and keeping multiphoton effects which depend on the intensity.

6. Conclusion

A mathematical model of femtosecond radiation as it passes through the skin was presented. With the help of numerical simulation (Monte-Carlo method), the dependence of the radiation intensity on the depth of penetration was measured. The solution of the equations was made by analytical methods describing the electron balance between electronic and atomic subsystem, while establishing thermal equilibrium. The mechanism of femtosecond radiation exposure on the skin was described.

Acknowledgements

Results of this work were obtained within the framework of the state order №3.1675.2014/K of Ministry of Education and Science of the Russian Federation.

Thanks to Sergey Chivilikhin for consultations on mathematical questions and substantial contribution to the paper.

7. References

[1] Akhmanov SA, 1988 *Optics femtosecond laser pulses* (Moscow: Nauka)
[2] Rulliere, C. 2005 *Femtosecond laser pulses* (Springer, Ed. Rulliere)
[3] Yilbas, B. S., and A. F. M. Arif. 2005 *Laser short-pulse heating with time-varying intensity and thermal stress development in the lattice subsystem.*(Proceedings of the Institution of Mechanical Engineers, Part C: Journal of Mechanical Engineering Science) pp 73-81
[4] Habbema, Louis, 2012 *Minimally invasive non-thermal laser technology using laser-induced optical breakdown for skin rejuvenation* (Journal of biophotonics) pp. 194-199.
[5] GOST 50723-94 1995 *Laser Safety. General safety requirements for the design and operation of laser products* (M.: Publishing Standards) 34c.

[6] Sriramoju V., Alfano R. R. 2015 *In vivo studies of ultrafast near-infrared laser tissue bonding and wound healing* (Journal of biomedical optics)

[7] Pushkareva AE (2008) *Methods of mathematical modeling of biological tissues in optics: the manual* (St. Petersburg: ITMO)

[8] Fredriksson, Ingemar 2008 *Optical microcirculatory skin model: assessed by Monte Carlo simulations paired with in vivo laser Doppler flowmetry* (Journal of Biomedical Optics)

[9] P. Stampfli and K. H. Bennemann 1990 *Theory of the instability of the diamond structure of Si, Ge, and C induced by a dense electron–hole plasma (*Phys. Rev. B42)

[10] P. Stampfli and K. H. Bennemann 1992 *Dynamical theory of the laser-induced instability of silicon* (Phys. Rev. B46)

[11] M. F. Kropman and H. J. Bakker 2001 *Dynamics of water molecules in aqueous solvation shells* (Science 291(5511)).

[12] M. F. Kropman, H. K. Nienhuys, and H. J. Bakker 2014 *Real-time meas-urement of the orientation dynamics of aqueous solvation shells in bulk liquid water* (Phys. Rev. Lett. 88(7)).

[13] Lipp V. P. et al. 2014 *On the interatomic interaction potential that describes bond weakening in classical molecular-dynamic modeling* (Journal of Optical Technology) pp 254-255.

[14] Dyukin R. V. et al. 2011 *Dynamics of the permittivity of a semiconductor acted on by a femtosecond laser* (Journal of Optical Technology) pp. 558-562.

[15] Serebryakov VA et al. 2015 *Laser mid-infrared spectral range for precision surgery* (Opt// Opt.. - V. 82)

Theoretical analysis of gyrotropy and absorption of terahertz electromagnetic waves in layer of DNA molecules

A Semenova[11], **V Vaks**[1,2,3]
[1]Institute for Physics of Microstructures, Nizhny Novgorod, Russia
[2]Lobachevsky State University, Nizhny Novgorod, Russia
[3]ITMO University, Saint-Petersburg, Russia

E-mail: semenanna@yahoo.com

Abstract. Certain type of low-frequency DNA molecular oscillations was analysed within the self-consistent phonon approximation. There were calculated dispersion relationship, exiting the oscillations by electromagnetic wave and corresponding contribution to the absorption spectrum of ensemble of parallel DNA molecules. The dependence of the DNA spectral characteristics on the length and period of the DNA duplex structure is revealed. The method of experimental check of obtained results is suggested. If the described model is confirmed by experiment, the obtained results available to reconstruct the length and duplex period of the DNA in a sample by its absorption spectrum.

1. Introduction

Chemical properties and functions of big complex bio molecules, such as DNA and proteins, are determined by their spatial structure – so-called conformation. Conformational state of many biological molecules can change without any chemical reaction, just by thermal motion. So investigation of conformations needs non-invasive methods. Electron and atomic force microscopy, traditionally used for the molecular visualisation, works only with adsorbed molecules, but adsorption can influence on the conformation itself [1]. Roentgen structural analysis excels at spatial resolution but this method works well only with molecular crystals but not solutions of molecules.
The terahertz (THz) spectroscopy is considered as perspective method to get information on structure and dynamics of biological molecules [2,3,4,5,6]. THz spectroscopy enables measurements of bio molecules *in situ*, in aqueous solution. Instead of roentgen, THz irradiation is non-ionizing and is suspected to be non-invasive for bio molecules [2,3,4,5,6], at least if the power of the irradiation lies below some threshold. Unfortunately, today there are no well established methods of interpreting molecular spectra in the THz frequency range. This work is aimed at simulation of dependence of DNA absorption spectra on the length and the period of the DNA duplex. If the experimental data confirm the model predictions, the algorithm of interpreting the absorption spectra is suggested.

[1] To whom any correspondence should be addressed.

2. Model description.

The dynamics of the DNA molecule was analysed within the self consistent phonon approximation [7,8]. The DNA molecule was described as a chain of coupled oscillators having the same partial frequency. Each oscillator presents a pair of nucleotides which have opposite charges, different masses and can oscillate along the line perpendicular to the molecules axis. Each nucleotide is springily connected with its neighbors (as shown on Fig. 1). A direction of oscillations is not the same for all nucleotide pairs but it turns by the same small angle for each next oscillator which is used to simulate the duplex structure of DNA molecule. The DNA layer was simulated as parallel duplexes placed in isotropic and spatially uniform dielectric medium.

The two Cartesian coordinate systems were used. The equations of motion of each nucleotide were write in a 'local' coordinates, where the axis x_n is parallel to the hydrogen bond in the nth base pair (as shown on Fig. 2) and the axis z_n is perpendicular to the plane of the nth base pair. The 'global' coordinates correspond to the DNA film: the z axis is parallel to the helix axes of DNA molecules and the axes x and y lie in the plane of the film.

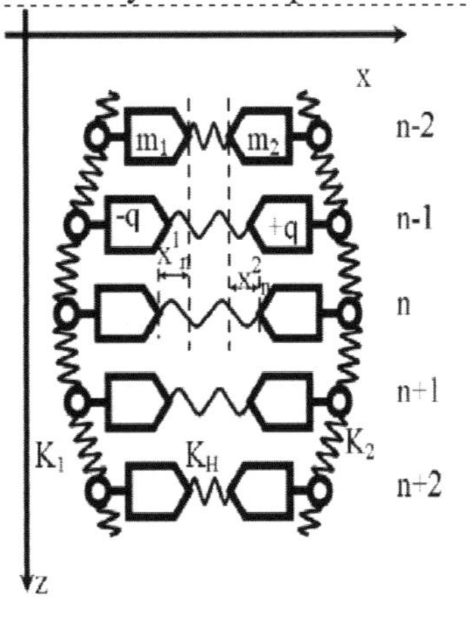

Figure 1. The model of the DNA molecule. Helix structure is omitted on the picture.

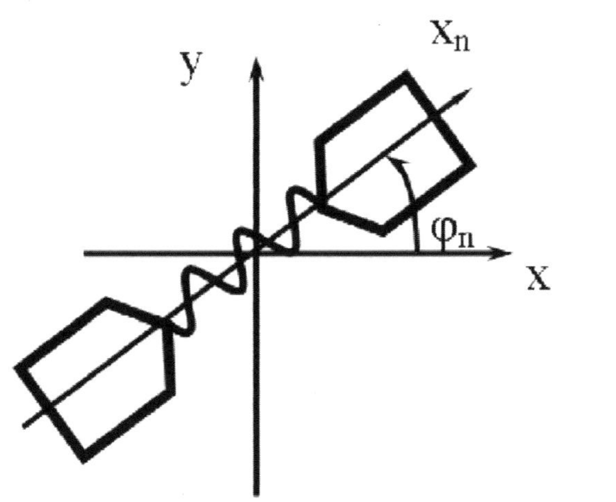

Figure 2. The local and global Cartesian coordinates in the model.

An interaction of DNA molecule with electromagnetic field was considered as interaction of charged nucleotides with plane spatially uniform monochromatic electromagnetic wave, propagating along the DNA helix axis. The equations of motion are given below (1):

$$m_{1,2}\frac{\mathrm{d}x_n^{1,2}}{\mathrm{d}t} = \pm K_H\left(x_n^2 - x_n^1\right) - K_{1,2}\left(2x_n^{1,2} - x_{n-1}^{1,2} - x_{n+1}^{1,2}\right) - m_{1,2}\gamma\frac{\mathrm{d}x_n^{1,2}}{\mathrm{d}t} \mp qE_{xn}e^{ikz_n - i\omega t} \quad (1)$$

where $x_n^{1,2}$ and $m_{1,2}$ are coordinates and masses of nucleotides, K_H and $K_{1,2}$ are rigidities of hydrogen and chemical bonds correspondingly, γ is damping coefficient and $E_{xn}e^{ikz-i\omega t}$ is the projection of the electric field vector on the axis x_n in the local coordinates. Indexes 1 and 2 correspond to the "left" and "right" strand of the DNA molecule on the Fig. 1, index n corresponds to the number of base pair.

The masses of nucleotides were calculated from their chemical structure while rigidities of bonds were adjustable parameters.

3. The simulation results.

3.1. Eigen modes.

There are two types of eigen modes of the DNA model (1): acoustic and optic (fig.3). The acoustic modes correspond to in-phase oscillations of nucleotides in each bas pair while the optic modes correspond to anti-phase oscillations. The frequencies of eigen modes is written below (2):

$$\omega_{\kappa a,o}{}^2 = 0.5\left(\omega_{H1}{}^2 + \omega_{H2}{}^2\right) + 2\left(\omega_1{}^2 + \omega_2{}^2\right)\sin^2(\pi\kappa/2N) \mp \mp\left[\left(0.5\left(\omega_{H1}{}^2 - \omega_{H2}{}^2\right) + 2\left(\omega_1{}^2 - \omega_2{}^2\right)\sin^2(\pi\kappa/2N)\right)^2 + \omega_{H1}{}^2\omega_{H2}{}^2\right]^{1/2} \quad (2)$$

where $\omega_{H1,2}{}^2 = K_H/m_{1,2}$; $\omega_{1,2}{}^2 = K_{1,2}/m_{1,2}$; N is the number of base pairs (b.p.) in the DNA molecule and κ is an integer number, determining the spatial structure of oscillations (3). For the κ eigen mode the spatial structure of oscillations is written below:

$$x_n^{1,2} = A^{1,2}\cos\left(\frac{\pi\kappa}{N}(n+1/2)\right)e^{-i\omega t} \quad (3)$$

where $A^{1,2}$ is amplitudes of eigen modes.

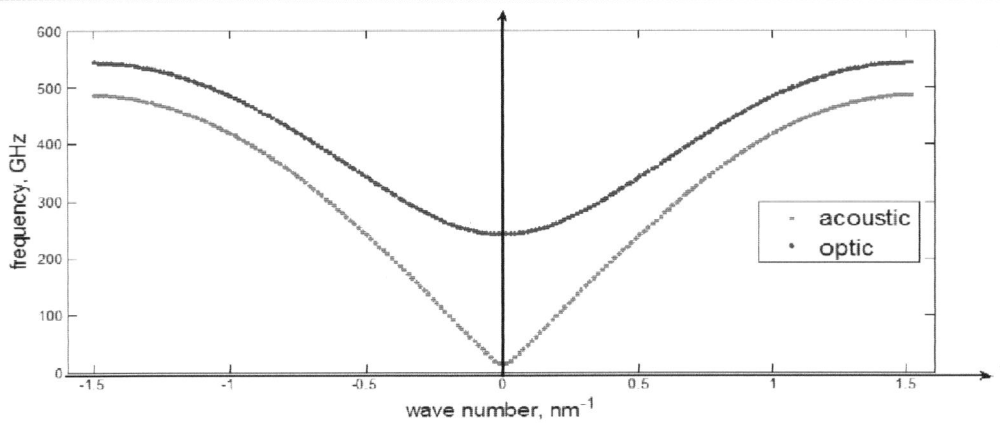

Figure 3. Dispersion relationship for both types of eigen modes of the DNA. The lowest frequency of the acoustic branch is equal to 0; the lowest frequency of the optic branch is determined by the properties of hydrogen bonds

The stationary solution of the equation (1) is written below (2):

$$p_\kappa(\omega) = \frac{qE_\kappa}{m_1 m_2}\frac{4(K_1+K_2)\sin^2(\pi\kappa/2N) - (\omega^2 + i\gamma\omega)(m_1+m_2)}{\left((\omega^2+i\gamma\omega) - \omega_{\kappa a}{}^2\right)\left((\omega^2+i\gamma\omega) - \omega_{\kappa o}{}^2\right)} \quad (4)$$

where $p_\kappa(\omega)$ and E_κ is coefficients of Fourier finite transformation (5):

$$p_n = \sum_{\kappa=0}^{N-1} p_\kappa(\omega)\cos\left(\frac{\pi\kappa(n+1/2)}{N}\right)e^{-i\omega t}$$
$$E_{xn}e^{ikz} = E_0\cos(\phi_n)e^{ikz} = \sum_{\kappa=0}^{N-1} E_\kappa \cos\left(\frac{\pi\kappa(n+1/2)}{N}\right) \quad (5)$$

where p_n is a projection on the nth base pair dipole moment on the axis x_n and ϕ_n is the angle between the electric field vector and the nth nucleotide pair.

Journal of Physics: Conference Series **735** (2016) 012033 doi:10.1088/1742-6596/735/1/012033

The absorption of the DNA-containing media is proportional to the electrical field work on the excitation the molecular oscillations (6):

$$abs \propto \sum_{n=0}^{N-1} E_{xn} p_n^*$$
$$abs \propto \sum_{\kappa=0}^{N-1} \frac{q|E_\kappa|^2}{m_1 m_2} \frac{4(K_1+K_2)\sin^2(\pi\kappa/2N) - (\omega^2+i\gamma\omega)(m_1+m_2)}{\left((\omega^2+i\gamma\omega)-\omega_{\kappa a}{}^2\right)\left((\omega^2+i\gamma\omega)-\omega_{\kappa o}{}^2\right)} \tag{6}$$

where *abs* is absorption coefficient.

There are two factors, determining the $p_\kappa(\omega)$: frequency and spatial structure. First, $|p_\kappa(\omega)|^2$ is maximum if the frequency of the wave is close to the eigen frequencies $\omega_{\kappa a,o}$. Second, E_κ and correspondingly $p_\kappa(\omega)$ strongly depend on the number κ. For the regular DNA duplex with the period N_{hel} E_κ can be estimated as (6):

$$|E_\kappa|^2 \propto \begin{cases} \left(1+(\kappa-\kappa_{hel})^2\right)^{-1}, & |\kappa-\kappa_{hel}| << N \\ 1/N^2, \ |\kappa-\kappa_{hel}| \propto N \end{cases} \tag{6}$$

where $\kappa_{hel} = \frac{2N}{N_{hel}} + kaN$ and a is the distance between neighbor nucleotide pairs in the DNA.

The absorption spectra of the DNA molecule calculated in the context of the presented model are shown on the fig. 4 (a, b). The lines of absorption correspond to the frequencies of the eigen modes. The frequency of the maximum absorption depends on the helix period of DNA molecule.

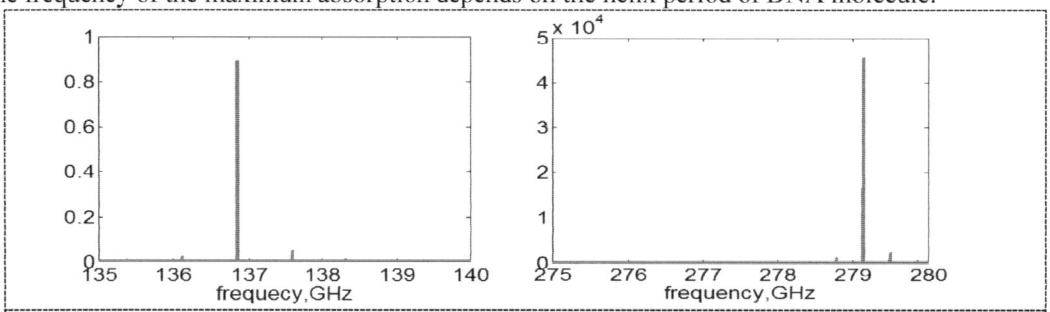

Figure 4a. Absorption of the DNA molecule with the length 1002 b.p. and helix period 11 b.p.

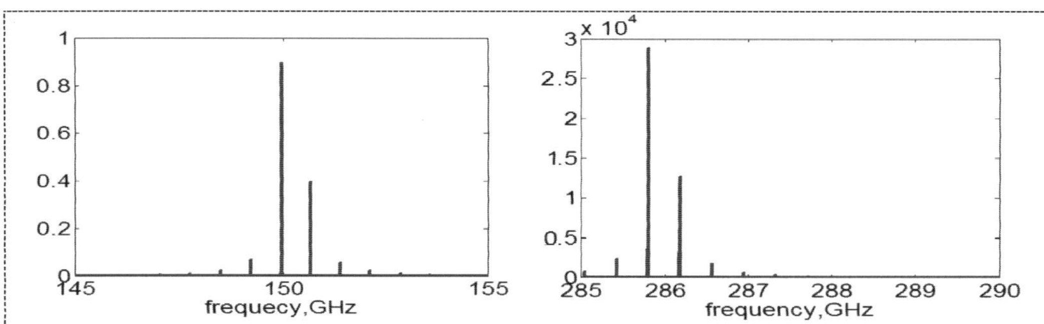

Figure 4b Absorption of the DNA with the length 1002 b.p. and helix period 10 b.p.

4. Conclusion

In the self-consistent phonon approximation the eigen modes and absorption spectra of the layer of identical parallel DNA molecules were calculated. The absorption spectra contain two series of lines corresponding to the acoustic and optic branches of the dispersion relationship. Each series presents one big central (major) line and few satellites. The frequency of the major line in the both series depends on the DNA helix period while the distance between the major line and its neighbour satellites depends on the length of the DNA molecules. It's interesting, that the distance between the lines in the absorption spectra doesn't depend on the helix structure of the DNA molecule and *vice versa*, the frequency of the major lines doesn't depend on the length of the molecule according to the model.

These results can be simply checked by comparing absorption spectra of DNA molecules with different length or different helix structure.

The results can be important for practical applications. It is possible to-determine the helix structure of the DNA molecules in the sample from the THz absorption spectra. If somebody measures the both major absorption lines, predicted by the model, and two satellites for each major line, he can calculate such parameters as K_H, $K_{1,2}$, N and N_{hel} from the frequencies of the measured lines. Also somebody can put together absorption spectra of DNA molecules alter by length but with the same duplex structure (or *vice versa*) and then compare the spectrum of an unknown sample with these standard spectra. If the frequency of the mayor line in the unknown sample is similar with the standard sample, the helix structures of the both samples are identical. Also if the distances between lines in the absorption spectra of the samples are equal, the length of the DNA molecules in the samples are equal.

Acknowledgments
I would like to acknowledge Dr. Smolyanskaya O.A. (ITMO University) for useful discussion. This work was supported by the RFBR fund (Grant 15-32-51049, young-science) and Government of Russian Federation (Grant 074 – U01).

5. References

[1] Klinov D V, Martynkina L P, Yurchenko V Yu et al. 2003 *Russian Journal of Bioorganic Chemistry* **29** (4) 397-402
[2] Zhang X-C 2002 *Physics In Medicine & Biology* **47** 3667-3677
[3] Vaks V L 2012 *J. Infrared Milli Terahertz Waves* **33** 43-53
[4] Vaks V L 2014 *Journal of Physics: Conference series* **486** 012002 1-2
[5] Strepitov E A, Khodzitsky M K, Smolyanskaya O A et al. 2014 *Progress in Electromagnetic Research Symposium* 1707-1710.
[6] Yablokov A A, Anfert'ev V A, Vaks V L et al. 2015 *IEEE Transaction on Terahertz Science and Technology* **5** (5) 845-851
[7] Dauxois T, Peyrerd M, Bishop A R 1993 *Physical Review E* **47** (1) R44-R47
[8] Yakushevich L V, Savin A V, Manevitch L I 2002 *Physical Review E* **66** 016601-14

Journal of Physics: Conference Series **735** (2016) 012034 doi:10.1088/1742-6596/735/1/012034

Femtosecond scalpel-optical tweezers: efficient tool for assisted hatching and trophectoderm biopsy

D S Sitnikov[1,3], **I V Ilina**[1], **Yu V Khramova**[1,2], **M A Filatov**[2] **and M L Semenova**[1,2]
[1]Joint Institute for High Temperatures of the Russian Academy of Sciences, Moscow, Russia,
[2]Department of Biology, M.V. Lomonosov Moscow State University, Moscow, Russia,

E-mail: sitnik.ds@gmail.com

Abstract. Ultrashort laser pulses have enabled highly precise and delicate processing of biological specimens. We present the results of using femtosecond (fs) laser pulses for dissection of zona pellucida (*ZP*) in mouse embryos during assisted hatching procedure and for trophectoderm biopsy as well. We studied the effects of application of fs laser radiation in the infrared (1028 nm) and visible (514 nm) wavelength ranges. Laser irradiation parameters were optimized so as not to compromise the viability of the treated embryos. Embryo biopsy was carried out in late-stage mouse preimplantation embryos. Femtosecond laser pulses were applied to detach the desired amount of trophectoderm cells from the blastocyst, while the optical tweezers trapped the cells and moved them out of the embryo. The parameters of laser radiation were optimized so as to efficiently perform embryo biopsy and preserve the viability of the treated embryos. The thermal effects can be significantly lower when fs lasers are used as compared to CW or long-pulse lasers. It is crucial when dealing with living cells or organisms.

1. Introduction

It is hard to imagine that modern medicine and biology can operate without high-tech equipment and lasers, in particular. Advances in the field of laser technologies make it possible to perform rather challenging and time-consuming manipulations on cellular and even subcellular levels with unprecedented accuracy and selectivity. In recent decades, lasers have been used not only for research purposes in the fields of embryology and developmental biology, but also in clinical practice, for example, in various medical centers for reproductive medicine. The most interesting and challenging task is to apply laser sources in assisted reproduction techniques. This implies performing various micromanipulations on human oocytes and/or embryos within a laboratory environment that result in successful fertilization, normal developmental physiology of the preimplantation embryo, high implantation rates and, finally, birth of a healthy baby.

Laser technologies have rapidly developed in recent years and a variety of laser systems have been employed to perform different microsurgical procedures on embryos in nearly noncontact mode. Lasers have been used not only for assisted hatching, but also for sperm immobilization and selection, and for embryo biopsy for preimplantation genetic diagnosis as well. Laser technologies have also

[3] To whom any correspondence should be addressed.

been used in research laboratories for stem cell derivation or cellular microsurgery (in particular, for oocyte enucleation). In earlier papers it has been shown that nanosecond lasers may be successfully applied for embryo microsurgery [1–2]. At the present time picosecond [3–4] and femtosecond (fs) lasers [5–9] are mainly used. It should be pointed out that femtosecond lasers have several advantages as compared to CW lasers or long-pulsed lasers. The main mechanism of microsurgery with femtosecond laser pulses is based on nonlinear absorption [10]. According to the fact that absorption of laser radiation and heating outside the area of laser focal spot are negligible, extremely high accuracy of laser microsurgery can be achieved with a minimal risk of damaging living cells and tissues.

2. Materials and Methods

2.1. Embryo collection and culture

All procedures performed in studies involving animals were in accordance with the bioethics standards of Biology department of the M.V. Lomonosov Moscow State University. Female mice of the BALB/cJ strain aged 1.5–2 months were paired with males late in the evening. Females were checked the following morning for a presence of a copulation plug. The day of plug detection was counted as embryonic day 0.5 (E0.5).

Zygotes were collected from fresh oviducts according to standard protocol [11]. Cumulus cells were removed by exposure to hyaluronidase solution (27 ME/ml, Cook). Embryos were then washed several times in fresh M2 (Millipore) medium and kept in culture in M16 (Millipore) medium in 4-well culture dishes under standard incubation conditions (37 ºC, 5 % CO2, 98 % humidity).

Embryos at E.2.5–E.3.5 were used in laser-assisted hatching procedure and at E.2.5–E.3.5 — for trophectoderm biopsy. Embryos were transferred to fresh drops of M2 medium in a 35 mm glass-bottom Petri dishes (170 um thick, 5 embryos per dish) under mineral oil balanced with the medium. Experimental dish was placed on the motorized microscope stage and covered with a compact on-stage thermostat to maintain a temperature of 37°C. After laser-assisted manipulations, embryos were transferred from the Petri dish into 4-well dish and filled with M16 (Millipore) culture medium. Embryos were then cultured under standard incubation conditions for 2 days.

2.2. Experimental Groups

To examine the influence of laser irradiation parameters on the embryos viability all embryos were divided into 3 groups: 1) a group of intact control (embryos cultured in vitro and not subjected to any manipulations and laser exposure), 2) a group of control embryos that were manipulated in the same way as the embryos from the experimental group except laser treatment, 3) a group of laser-treated embryos (femtosecond laser pulses were applied to dissect *ZP*). Embryos from both the second and the third groups (after all the manipulations were complete) and the intact control embryos were cultured under similar conditions.

2.3. Experimental set-up

Detailed information about scheme of the combined system "Femtosecond laser scalpel-optical tweezers" is given in [13]. The system consists of the inverted microscope Olympus IX-71 and two laser sources – CW and femtosecond lasers applied as optical tweezers and laser scalpel respectively. For laser scalpel, experimental setup allows implementation of both the fundamental frequency of laser radiation (wavelength of 1028 nm) and the second harmonic (wavelength of 514 nm). A 20× (NA=0.5) microscope lens was used to focus the laser beam on the zona pellucida to carry out assisted hatching procedure; laser spot diameter was ~ 3 microns for the fundamental frequency and about 2 microns for the second harmonic. For trophectoderm biopsy, a 40× (NA=0.8) microscope lens was used; laser spot diameter was of the order of 1 micron for the second harmonic

3. Zona pellucida drilling with femtosecond laser pulses

Assisted hatching can be performed by thinning a large area of the zona pellucida, partially dissecting (creating an incomplete hole) or drilling completely through the zona (creating a hole 10–25 um in diameter). A purpose of this work was to find an optimum range of laser irradiation parameters required for efficient *ZP* dissection that does not compromise viability of the treated embryos and their preimplantation development. One of the most important parameter is the energy (or intensity) of laser pulses. For both wavelengths (1028 nm and 514 nm) we have determined the acceptable range of laser pulse energies required for successful *ZP* dissection.

Dissection of the zona pellucida was performed by application of laser pulses along a freehand or straight line defined by an operator in the system control software. The microscope was focused at the so-called "equatorial plane" of the embryo. A freehand line was drawn in this plane and exposed to femtosecond laser pulses with 2.5 kHz repetition rate. The movement of X-Y microscope stage resulted in movement of the embryo relative to the focused laser beam with a speed of 0.01 mm/s.

Energy threshold value of laser pulse for assisted hatching procedure appeared to be $E_\omega \approx 140$ nJ for the fundamental frequency and $E_{2\omega} \approx 47$ nJ for the second harmonic. Thereafter we gradually increased laser pulse energy. After a certain value further energy increase resulted in an optical breakdown of the medium followed by formation of cavitation bubble. This caused embryo displacement relative to its initial position due to bubble expansion and shockwave formation. Such phenomena appeared at $E^{\omega}_{max} = 390$ nJ и $E^{2\omega}_{max} = 130$ nJ for the wavelengths of 1028 nm and 514 nm respectively.

Figure 1 demonstrates a successful dissection of *ZP* at a stage of development of 3.5 day by femtosecond laser pulses with the wavelength of 514 nm. A path irradiated with fs laser pulses is marked by a solid curve.

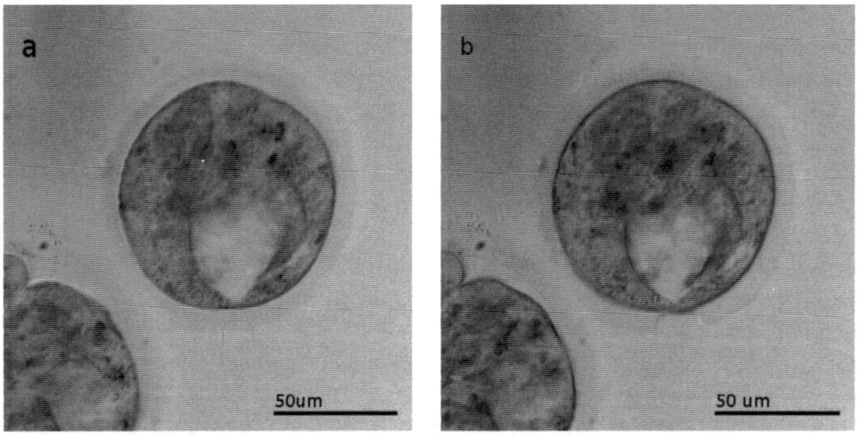

Figure 1. *ZP* dissection mouse embryos by means of femtosecond laser pulses.
(a) Embryo before the exposure. (b) Dissected *ZP* after laser treatment.

After determining the acceptable energy range of laser pulses causing ZP dissection, an optimal laser pulse energy was found in terms of viability and developmental rates of laser-treated embryos. Four energy values were chosen for the second harmonic of laser radiation between $E^{2\omega}_{tr} = 47$ nJ and $E^{2\omega}_{max} = 112$ nJ. The maximum energy $E^{2\omega}_{max} = 112$ nJ was chosen to be 15% less than the value of $E^{2\omega}_{max} = 130$ nJ obtained above to avoid embryo displacement (due to shock wave formation in the medium) during laser treatment. A total of 100 embryos were divided into 6 groups (10 embryos per control group, 20 embryos per each experimental group): two control and four experimental groups. In the experimental groups of embryos *ZP* dissection was performed by applying laser pulses with the following energies: group 1 – 47 nJ, group 2 – 66 nJ, group 3 – 83 nJ, and group 4 – 112 nJ.

Viability and developmental rates of embryos subjected to laser-assisted *ZP* drilling with fs laser pulses at the wavelength of 1028 nm were also investigated. A total of 40 embryos were divided into two experimental groups. In the first and the second experimental group of embryos, zona pellucida was drilled by applying femtosecond laser pulses with energies of 250 nJ and 320 nJ respectively.

At the same time the analysis of embryos development in the groups of intact (Control 1) and parallel (Control 2) controls was performed. Data on blastocyst formation and hatching rates are presented in table 1.

Table 1. Data on blastocyst formation rate and hatching rate in experimental group treated with laser pulses

	λ = 514 nm Energy (nJ)				λ = 1028 nm Energy (nJ)	
	47	66	83	112	250	320
No. of embryos for assisted hatching	10	10	10	10	10	10
Blastocyst formation rate, %	100	100	100	100	100	100
Totally hatched embryos, %	100	100	90	100	90	100

As can be seen from the table, the majority of embryos has developed to the blastocyst stage (> 90%) and has successfully hatched in almost all embryo groups. There were no considerable differences in hatching rate for both experimental and control groups. We suggest that although no differences were found in terms of assisted hatching efficiency for the two wavelengths applied, using laser radiation at the wavelength of 514 nm would minimize thermal influence on embryos because of the less pulse energy required to cause *ZP* drilling.

4. Trophectoderm biopsy using femtosecond laser scalpel

Trophectoderm biopsy procedure was performed on embryos at E.4.5 — the natural hatching stage. At this stage an embryo leaves *ZP* and a portion of the trophectoderm cells outside the shell can be cut by laser pulses. One needs to monitor constantly the state of embryos since the first signs of hatching to succeed in trophectoderm biopsy. The goals is not to miss a moment of hatching, when a specified number of cells is outside the *ZP*. Then a series of single femtosecond laser pulses is applied to remove several trophectoderm cells. As a rule, it is required to apply several (2–4) laser actions in order to fully detach trophectoderm cells. We performed this procedure using the second harmonic of laser scalpel radiation.

It is important not only to remove trophectoderm cells from the embryo, but also to preserve the viability of the last one. Thus, our goal was to determine the optimal laser pulse energy in terms of embryos survival. We took four values of laser energy between threshold and maximum energy values: 100 nJ, 140 nJ, 190 nJ, and 220 nJ respectively. The error in energy determination in the experiment was ± 5%. A total of 80 embryos were divided into four groups (20 embryos per group). In each group biopsy of trophectoderm cells was performed by application of laser pulses with the prescribed energy. Assessment of the viability of embryos was performed the next day according to the percentage of blastocysts finished their hatching process, as well as to staining the cells with fluorescent dyes *Hoechst* 33258 (Sigma) and *Propidium iodide* (Sigma). At the same time we controlled the survival rates in the first and second control groups. The experimental results are summarized in table 2.

Table 2. Survival rates of embryos in experimental group (after trophectoderm biopsy with femtosecond laser pulses with the wavelength of 514 nm).

	Energy (nJ)			
	100	140	190	220
No. of embryos for trophectoderm biopsy	20	20	20	20
Totally hatched embryos, %	100	90	70	10

As can be seen, laser-assisted trophectoderm biopsy is a kind of balance between maximum efficiency of cell detachment (at high energy) and maximum survival rate (at low energy) in the range of acceptable laser pulse energies. From this point of view, energies of femtosecond laser pulses lying in the range of 140-190 nJ seem to be optimal. Energy value should be selected individually for each embryo.Figure 2 demonstrates the case, when 3 laser actions are applied to detach the trophectoderm cells. The position of action of laser scalpel is indicated by "+" marker. Dashed line represents the boundaries of the *ZP*. Laser tweezers was used to hold the dissected cells.

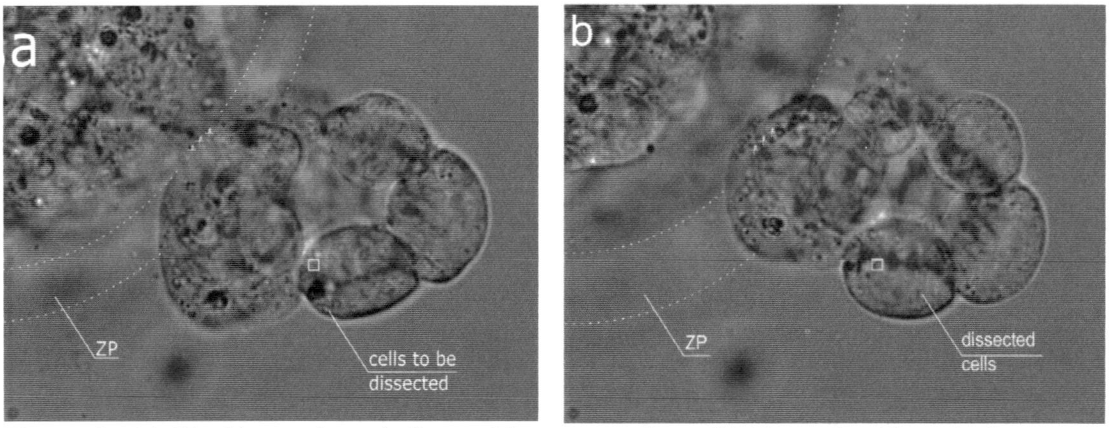

Figure 2. (a) Hatching embryo before a biopsy procedure. (b) Trapped with optical tweezers trophectoderm cells right after the dissection. + Markers indicate the places where laser shots are applied; □ marker indicates the position of optical trap

Thus we demonstrated efficient assisted hatching and trophectoderm biopsy procedures performed by the femtosecond scalpel-optical tweezers.

Acknowledgments

This work was supported by the Ministry of Education and Science of the Russian Federation, project No. 14.604.21.0026, RFMEFI60414X0026.

References

[1] Clement-Sengewald A, Schutze K, Ashkin A, Palma G A, Kerlen G and Brem G 1996 *J. Assist. Reprod. Genet.* **13** 259

[2] Clement-Sengewald A, Buchholz T, Schutze K, Berg U and Berg F 2002 *J. Assist Reprod Genet.* **19** 183

[3] Karmenyan A V, Shakhbazyan A K, Sviridova-Chailakhyan T A, Krivokharchenko A S, Chiou A E and Chailakhyan L M 2009 *Mol. Reprod. Dev.* **76** 975

[4] Krivokharchenko A, Karmenyan A, Sarkisov O, Bader M, Chiou A and Shakhbazyan A 2012 *PLos ONE* 7 e50029

[5] Torres-Mapa M L, Antkowiak M, Cizmarova H, Ferrier DEK., Dholakia K and Gunn-Moore F J 2011 *Biomed Opt Express* **2** 1564

[6] Rakityanskiy M M, Ilina I V, Sitnikov D S, Ovchinnikov A V, Agranat M B, Khramova Y V and Semenova M L. 2011 *Eur. Biophys. J. Biophys. Lett.* **40** S-237

[7] Ilina I V, Rakityanskiy M M, Sitnikov D S, Ovchinnikov A V, Agranat M B, Khramova Yu V and Semenova M L 2012 *AIP Conf. Proc.* **1464** 560

[9] Torres-Mapa M L, Gardner J, Bradburn H, King J, Dholakia K and Gunn-Moore F J 2013 *Proc. SPIE* **8611** 861104

[10] Il'ina I V, Sitnikov D S, Ovchinnikov A V, Agranat M B, Khramova Yu V and Semenova M L 2012 *Proc. SPIE* **8427** 84270S1

[11] Vogel A, Noack J, Huttman G and Paltauf G 2005 *Appl. Phys. B.* **81** 1015

[12] Hogan B, Beddington R, Costantini F and Lacy E. 1994 Manipulation the Mouse Embryo: A laboratory manual. 2-nd edition (New York: Cold Spring Harbor Laboratory Press) pp 136-145

[13] Sitnikov D S, Ovchinnikov A V, Ilina I V, Chefonov O V and Agranat M B 2014 *High Temp.* **52** 803

Journal of Physics: Conference Series **735** (2016) 012035 doi:10.1088/1742-6596/735/1/012035

Possibility of determination of the level of antioxidants in human body using spectroscopic methods

E Timofeeva[1, 2], **E Gorbunova** [1]
[1] ITMO University (Saint Petersburg National Research University of Information Technologies, Mechanics and Optics), Saint Petersburg, Russia

E-mail: elyatimofeeva@gmail.com

Abstract. In this work, the processes of antioxidant defence against aggressive free radicals in human body were investigated theoretically; and the existing methods of diagnosis of oxidative stress and disturbance of antioxidant activity were reviewed. Also, the kinetics of free radical reactions in the oxidation of luminol and interaction antioxidants (such as chlorophyll in the multicomponent system of plant's leaves and ubiquinone) with the UV radiation were investigated experimentally by spectroscopic method. The results showed that this method is effective for recording the luminescence of antioxidants, free radicals, chemiluminescent reactions and fluorescence. In addition these results reveal new opportunities for the study of the antioxidant activity and antioxidant balance in a multicomponent system by allocating features of the individual components in spectral composition. A creation of quality control method for drugs, that are required for oxidative stress diagnosis, is a promising direction in the development of given work.

1. Introduction

Evidently, an oxidative damage in cell structures is a basis of most mechanisms that lead to health diseases and senescence of human body. The main factor for such cell damage is oxygen. Oxygen permeates into the human body and, before turning into water, oxygen molecule becomes an aggressive oxidizing agent in the form of hydrogen peroxide or oxygen radical. This oxidizing agent can react with molecules, which are susceptible to oxidants and free radicals, under the influence of various interfering factors and deviations from the normal functioning of the body. Aggressive forms of oxygen molecules are triggers for a chain reaction of lipid oxidation. That leads to damages of the cell membranes and destruction of the cell. Antioxidants fully or partially suppress the activity of free radicals, while maintaining the integrity of cells and tissues, but prevent from the process of phagocytosis, in which a greater concentration of toxic oxygen is required [1]. In the process of phagocytosis (which was discovered in the 19th century by Ilya Mechnikov) specialized white blood cells (phagocytes) destroy morbific bacteria using a high concentration of oxygen free radicals.
It can easily be said that direct analysis of antioxidant activity allows diagnostics of diseases connected with the damage of tissues and organs structure, as well as genetic material, and is a topical issue of clinical diagnosis. Besides medical intervention is required in the presence of excess of normal antioxidants concentration in human body because of probable loss in defence mechanisms

[2] To whom any correspondence should be addressed.

against pathogenic bacteria (due to the lack of free radicals which were neutralized by antioxidants). Therefore, the problem of antioxidant studying requires to develop several ways of research to analyse directly the balance of antioxidants and free radicals, to control drugs for the treatment of oxidative stress, to analyse the mechanisms of antioxidants interaction with free radicals by spectroscopic method and to find promising antioxidants on the basis of knowledge of their interactions with radiation of a special spectral composition.

2. Problem description

The imbalance of antioxidants diagnosis in modern clinical diagnostics can be based only on the testimony that has an indirect relation to the antioxidant status. Therefore direct indicators of antioxidant status of the patient and high accuracy of diagnostic methods are required (taking into account the direct impact of the diagnosis on the action of physician). Immunoassay methods provide data on the reaction products of antioxidants or free radicals in the blood plasma. However, cell damage by free radicals most takes place directly in the tissues and organs. So body's tissue samples (that may be under adversely oxidative influences) are to be selected as analysis objects.
Moreover the work with biological objects has to be considered as an analysis of multi-component system, which is a determining factor in choosing a method of diagnosis in the absence of a preliminary chemical intervention.

3. Current methods

There are many methods for the study of antioxidants:

- Volumetric – the study of the antioxidant activity by volumetric absorption of oxygen;
- Photometric – the determination of the optical density for colorimetric analysis;
- Biological marking – the implementation of fluorescent markers in a living organism;
- Electrochemical – measurement of voltammetric characteristics of cathode reduced oxygen in a solvent, as it is known that the electrochemical activity correlates with the ability to inhibit lipid peroxidation;
- Radio spectroscopy or EPR-spectroscopy is the most direct method for the study of free radicals, but it is not applicable to dynamic systems;
- Chemiluminescent – is currently the most promising because it has significant advantages over the other: high accuracy and safety for the patient and the laboratory assistant [2].

4. Materials and methods

4.1. Investigation the kinetics of free radical reactions

First of all it was decided to study the free radical reaction with the help of reaction of luminol transformation to 3-aminophthalate, because it is well-studied by chemiluminescent method (Figure 1). This test is a part of the laboratory practical training program in the education of specialists in the medical field. This reaction is chosen deliberately with the aim of the study the kinetic of free radical reaction not by chemiluminometer but using spectroscopic method. It was necessary to compare the accuracy of the spectroscopic method for determining the concentration of the reactants by the energy from luminescence of radical reaction.

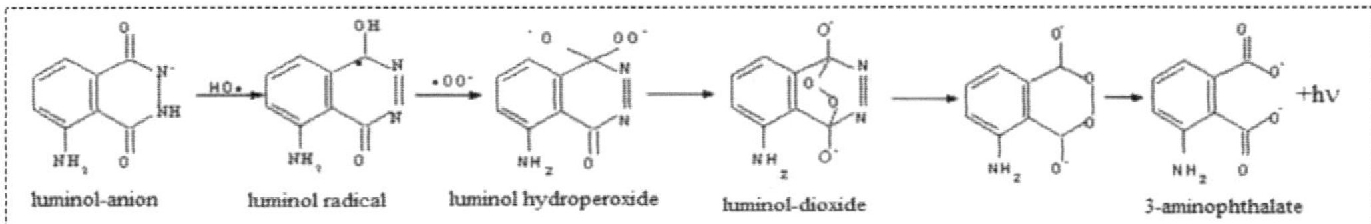

Figure 1. Reaction scheme of luminol transformation to 3-aminophthalate

The luminol chemiluminescent reaction with kalio heksacianoferatas (III) was selected as an object for studying the kinetics of the oxidation process. The luminescence of reaction was registered with the help of spectrometer USB4000 of Ocean Optics firm. Software for recording and processing the results of experiments were developed in LabVIEW environment. During the experiment, the obtained data was compared with the known model of the chemical reaction [3] and a match with the model of the behaviour of chemiluminescence reaction was found. Also, luminescence wavelength of the chemiluminescent reaction was defined.
As a result, the conclusion about the possibilities of studying the kinetics of the chemiluminescent reaction by spectroscopic method was made.

4.2. Identification of obstacles at this point
Following problems for direct analysis were detected during the research:

- Study of multi-component system;
- At the moment, there is no opportunity to avoid a chemical treatment to the sample;
- Instability of free radicals for their study by spectroscopic method.

The review of proposed solutions of this situation and a selection of samples for testing were made. So it was decided to study directly antioxidants (that are presented in the human body) to isolate their features from the rest of the components of blood plasma or tissue samples. Free radicals cannot be studied directly in this method of analysis due to their short lifetime, so the individual study plan will be developed for them.
The realization of the experimental setup as a whole depends on the subject and the research goals: working with chemicals or biological materials; register the luminescence or the fluorescence caused by ultraviolet light. These facts suggest that the evaluation of the total antioxidant status in human body requires a complex analysis of biological samples of patients. Nevertheless, the work is done in stages for following methodology development and design of the device for the complex analysis of biological samples and the diagnosis of oxidative stress.

4.3. Investigation of plants as sources of antioxidants
The experimental study of UV radiation influence to the plant leaves were carried out along with the analysis of the chemiluminescent reaction.
Despite the fact that the leaves of the plants are multi-component system, it was decided to carry out experiments to study how UV radiation effects on the leaves of plants as sources of antioxidants for humans [4].
The optical properties of chlorophyll in the leaves of plants were investigated in the theory before the experiment. The special equipment (Figure 2) for the experiment was prepared.

Figure 2. Experimental setup for investigation of plants as sources of antioxidants. It comprises a radiation source DT-Mini-2-GS of Ocean Optics firm (1), a fibre optic probe (2), a spectrometer USB4000 Ocean Optics firm (3), a data processing unit (4), a cardboard case (with a hole for the probe) for the object (5) and objects of the study (6).

The radiation source DT-Mini-2-GT contains the Halogen and the Deuterium lamps. Preliminary experiments have shown that for the study of fluorescence should be used the Deuterium lamp.
Figure 3 shows the one of results of fluorescence study of plant leaves.

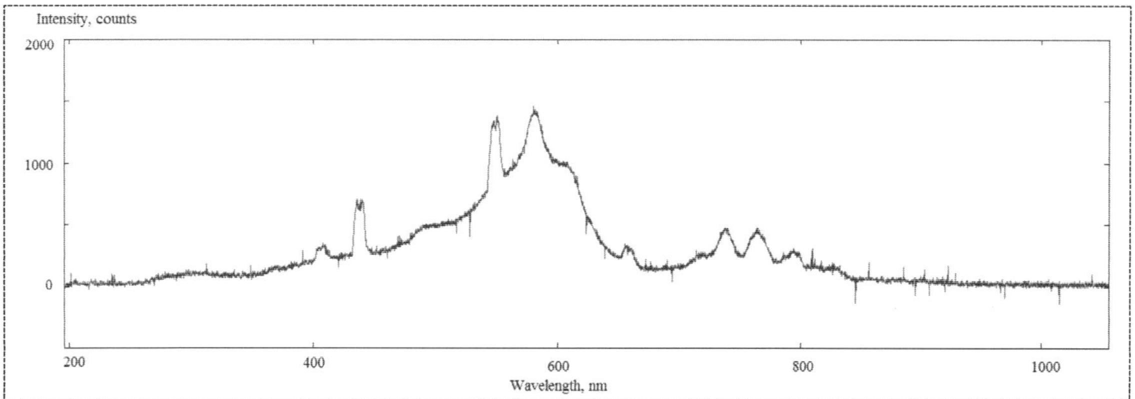

Figure 3. Fluorescence emission spectrum of a plant leaves under radiation of Deuterium lamp.

The results of experiments on plant leaves irradiation by Deuterium lamp showed that the spectroscopic method allows the registration of chlorophyll fluorescence and photosynthesis process of plants. A comparison of different plant species and plants growing in different conditions was made and the match of the spectral composition of fluorescence response was found.
As a result, it may be said that the spectral method allow us to divide the signals from the various components during the analysis and processing of the plant leaves fluorescence under known fluorescent responses from the each individual component of the plants leaves.

4.4. Investigation of the optical parameters of antioxidants used to treat patients
The ubiquinone was selected for the investigation as a substance insufficiently explored by optical methods. Ubiquinone is known in the pharmaceutical industry as a coenzyme q10. It is currently widespread in the dietary supplement form, which is called "anti-aging". And it was first used for the treatment of cardiovascular diseases in 1965. This antioxidant actually inhibits cell senescence. But an excess of antioxidants can slow down the normal metabolism and lead to disease, oxidative stress feedback, in which process the leukocytes are not provided with enough free radicals to fight against pathogenic bacteria. Therefore it is necessary to study not only the quantity of products of chemiluminescent reaction, but antioxidants themselves.
Experiments with ubiquinone were based on the phenomenon of fluorescence coenzyme solutions in oil. The registration of the fluorescence was carried out by the spectrometer. UV LEDs with a narrow spectral composition of the maxims at 280 and 355 nm were chosen as fluorescence excitation sources. Figures 4 and 5 represent the effects of UV irradiation of different wavelengths (280 and 325 nm) to ubiquinone solution and solvent.
The results of studying the optical properties of the ubiquinone showed that the spectroscopic method can be applied to the fluorescence investigation of this antioxidant. Also the data obtained are dependent on the luminescence excitation source and on the solvent, if the solution is irradiated.

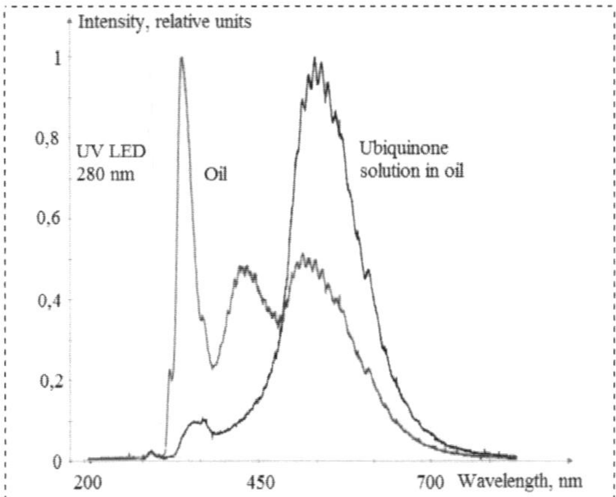

Figure 4. Fluorescence emission spectra of oil and ubiquinone solution in oil under radiation of UV LED with maximum at 280nm.

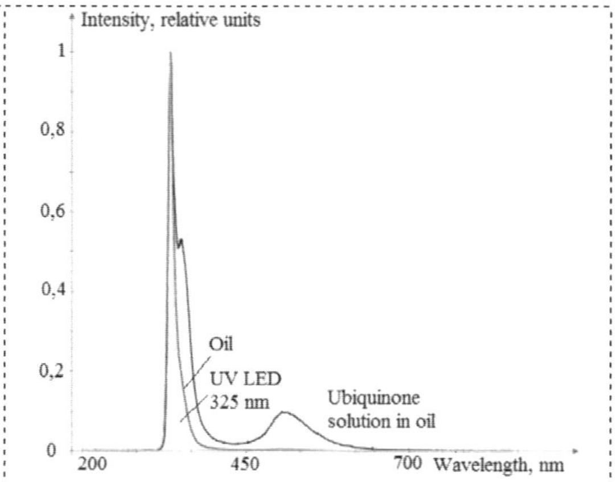

Figure 5. Fluorescence emission spectra of oil and ubiquinone solution in oil under radiation of UV LED with maximum at 325nm.

5. Results and discussion

The results of these studies show the possibility of antioxidant research by spectroscopic methods. Also the prospects of this method in clinical diagnosis of possible antioxidant imbalances (on the basis of the known normalized antioxidant activity and in the early stages of the disease) were proved. Spectroscopic method allows us to separate the fluorescent response of the various components, has a sufficiently high sensitivity and allows analysis of the multicomponent system, highlighting the characteristics of each component on the background of the whole system.

6. Conclusions

The research results at this stage allowed us to make a choice between the directions of future research and to identify entirely new research problems:

- The study of the relation between optical density of ubiquinone solutions and the ubiquinone concentrations in different solvents;
- The research of colour characteristics of ubiquinone solutions (for possible use in quality control of drugs, which are based on this coenzyme);
- The study of antioxidants in association with their pro-oxidant counterparts (ubiquinone-ubiquinol);
- Organization of parallel researches with the existing biochemical analysis of antioxidants.

7. Acknowledgments

This work was financially supported by Ministry of Science and Education of the Russian Federation (GOSZADANIE 2014/190).

8. References

[1] Severin E S 2004 *Biochemistry* pp 294-296, 491, 616-635, 666
[2] Abramova Zh I, Oksengendler G I 1985 *Human and antioxidant substances* pp 34-36
[3] Vladimirov Yu A 2001 *Soros Educational Journal* **7** pp 16-23.
[4] Schulman S G 1976 *Modern fluorescence spectroscopy* **2** 36.

DEVELOPMENT OF FIBER OPTIC SENSOR FOR FLUID FLOW OF ASTRONAUTS' LIFE-SUPPORT SYSTEM

E.A. Shachneva[1], T.I. Murashkina[2]

[1]Penza State Technological University, Russia, Penza
[2]Penza State University, Russia, Penza

E-mails:e_shashneva@mail.ru and timurashkina@mail.ru

Abstract. This paper proposes a fiber optic sensor consumption (volume, speed) of liquids in life-support systems of astronauts, as well as offers a simple method and apparatus for reproducing the parameters of fluid flow needed in research, yustiovke and adjusting the optical sensor system.

1. Introduction

Space medicine in its historical development has gone from a simulation of space flight in the laboratory and during the flights animals on rockets and satellites to research related to the long flight of space stations and international flight crews. The main problem in the early stages of development of space medicine has focused on the question of the principal possibility of human space flight and problem solving creating systems that ensure the presence of man in the cabin of the spacecraft during orbital flight [1].

In unusual conditions of space flight (vacuum, radiant heat, ionizing radiation), a person must be in a closed sealed compartment of the spacecraft. In the habitable compartments is necessary to create conditions for normal life and work of man. These conditions must be maintained during the entire flight, feeding into the bay substances for human consumption and removing the products of its life. The onboard systems of the spacecraft (spacecraft) that solve these problems, called life-support systems (LSS) (Figure 1) [2].

Currently, we are talking about improving the life support systems of astronauts in an increasingly hard impact of external factors on the astronaut on a long flight.

2. Materials and methods

The authors propose new methods and apparatus for measuring fluid flow and reproduction based on the use of fiber optic elements.

The proposed method for reproducing liquid flow parameters is known that through the pipe cross-section is passed the liquid flow parameters are determined by the change in liquid level in the containers at the inlet or outlet pipe for a fixed period of time.

Figure 1 shows a simplified structural diagram of the device for the implementation of the process parameters of liquid media playback.

The apparatus includes the following elements for playback parameters pumped liquid conduit 1 through which flows a liquid container 2 with a scale calibrated with the height of the liquid level in the values of capacity to drain the scale 3 is also calibrated with the height of the liquid level in the values of hoses 4 and 5, a pump 6, a timer 7 is connected to the starter pump. The pump 6 is placed at the bottom of a transparent container with a liquid 2, the level of which is determined by the scale. The ends of the hoses 4 and 5 are omitted in the container 2 and 3, respectively. The other ends of the hoses 4 and 5 are sealed with different angles to the conduit 1.

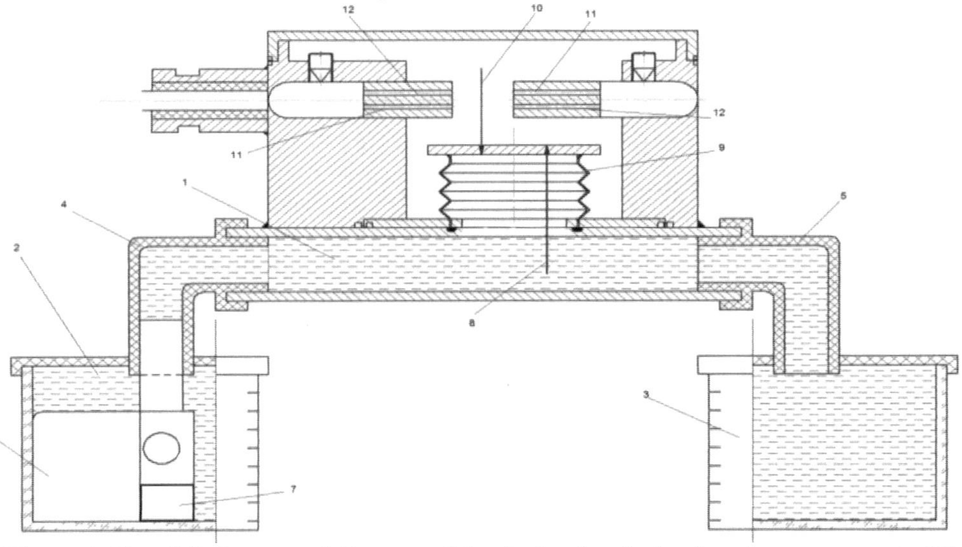

Figure 1. Simplified structural diagram of the device for playback and measurement of liquid media

The apparatus includes the following elements for measuring parameters of fluid flow: receptive element 8 having one end secured to the blind end 9 inside the bellows, and the other free end - located in the conduit 1 in the liquid flow zone; plate 10 with two reflecting surfaces, one end of which is also fixed to the outside of the blind end of the bellows 9 and the other free end - located at a certain distance with respect of lead 11 and outlet 12 of optical fiber [5] (Figure 1).

The device works as follows: when the pump 6, the timer starts counting time $t_1 = 0$, when you turn off stops at time t_2, operation time is defined as $\Delta t = t_2 - t_1$. Before the measurements is fixed liquid H_1 level in the tank 2. At the end of measurement is fixed at the time level H_2.

3. Conclusion

The proposed method and apparatus for reproducing and measurement of fluid flows (volume, velocity, flow rate) will simplify the adjustment and adjustment of the optical system of optical fiber flow sensor used in life support systems of astronauts, to improve their performance.

4. References

[1] Malozemov V, Roznov VF Pravetsky VN Life support systems crew of the aircraft. Publisher "Engineering", 1986

[2] Serebryakov V Basics of designing life-support systems of crew of spacecraft. Publisher "Engineering", 1983

A thermal lens response of the two components liquid in a thin film cell

V I Ivanov[1], G D Ivanova[1]
Department of Physics and Theoretical mechanics, Far Eastern State Transport University, Khabarovsk, Seryshev street, 47, Russia

Abstract. It was proposed a new thermal lens scheme with a thin layer of cell thickness which is significantly less than the size of the beam. As a result the exact analytical expression for the thermal lens response is achieved, taking into account the thermal lens in the windows of the cell.

1. Introduction

The thermal lens method is widely used in spectrometry and thermo-optical diagnostics of materials [1-3]. The application of this method in the two-component liquid has its own peculiarities, because in addition to normal thermal response associated with thermal expansion of environment here it may arise resulting from flows of concentration phenomenon of thermal diffusion (effect of Soret) [2-5]. This component concentration redistribution results a corresponding change in the refractive index (and absorption) of the liquid. We propose a thermal lens scheme with thin-layer cell in which the cell thickness is significantly less than the size of the beam.

2. A thermal lens response in the thin optical cell with two-component liquid

We consider the one-beam scheme of the thermal lens signal measurement (fig. 1). Let's the absorption coefficient α is determined entirely by one component with mass concentration C ($\alpha = \beta C$, где $\beta = (\partial\alpha/\partial C)$ – constant) and liquid is located in a thin cell with thickness d_0.

We have for a Gaussian beam distribution in a plane perpendicular to the optical axis z:

$$I = I_0\left(1+\left(\frac{\lambda z}{\pi r_0^2}\right)^2\right)^{-1}\exp(-r^2(z)/r_1^2(z)), \tag{1}$$

the radius of the beam $r_1(z) = r_0\left[1+\left(\frac{\lambda z}{\pi r_0^2}\right)^2\right]^{0.5}$, r – the distance from the axis of the beam, λ - radiation wavelength, r_0 - the radius of the beam in the beam waist, I_0 - intensity of radiation on the axis in the plane in the beam waist, $I(r,z)$ - radiation intensity ($\alpha d_0 << 1$).

[1] E-mail: ivanov@festu.khv.ru

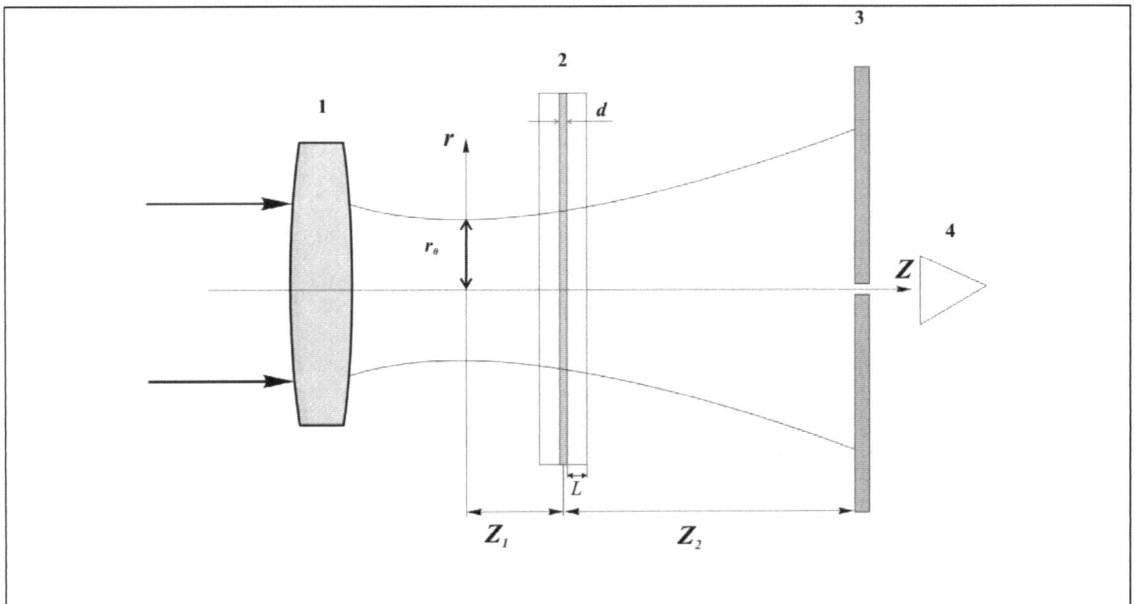

Figure1. The scheme of the thermal lens signal measurement: 1- input lens, 2- optical cell, 3- screen, 4-photodetector

We assume small thicknesses of liquid layer d and cell window L ($d, L << r_1$). In this case the radial heat flow is negligible and we get a one-dimensional thermal task:

$$c_m \rho_m dT_m / dt = -\chi_m d^2 T_m / dz^2 + \alpha I(r,z), \qquad (2)$$

were c_m, ρ_m - specific heat capacity and density environment, T_m - liquid temperature, χ_m - heat conductivity coefficient,.

The similarly thermal task is considered challenge for the temperature in the window of the cell T_w:

$$c_w \rho_w dT_w / dt = -\chi_w d^2 T_w / dz^2, \qquad (3)$$

where c_w, ρ_w, χ_w - thermo-physical window material parameters.

Boundary conditions on the borders are next:

$$\chi_w (dT_w / dz)_{z=L+d/2} = \gamma (T_z - T_0). \qquad (4)$$

$$T_m(d/2) = T_w(d/2). \qquad (5)$$

$$\chi_m (dT_m / dz)_{z=d/2} = \chi_w (dT_w / dz)_{z=d/2}. \qquad (6)$$

where γ, T_0 - the convective heat transfer coefficient and ambient temperature accordingly, $T_z = T_w(L + d/2)$.

In the stationary mode we have the following system solution for the distribution of temperature in the liquid T_m^s and the cell window T_w^s:

$$T_m^s = T_0 + T\left(d/2\right) + \alpha I\left(r\right)\chi_m^{-1}\left\{\left(d/2\right)^2 - z^2\right\}, \tag{7}$$

$$T_w^s = T_0 + T\left(d/2\right) + \alpha d I\left(r\right)\chi_w^{-1}\left\{\left(d/2\right) - z\right\}, \tag{8}$$

$$T\left(d/2\right) = \alpha I\left(r\right) d\left(\gamma^{-1} - L\chi_w^{-1}\right). \tag{9}$$

Now let's consider the effect of thermal diffusion. Since the diffusion processes are much orders of magnitude slower than heat transfer processes, we can assume that thermal diffusion of absorbing component occurs in stationary temperature field, which is determined by formulas (7-9), but with a concentration-dependent absorption coefficient.

Balanced system of equations for the concentration of absorbing particles C is follows:

$$dC/dt = -\nabla J\ , \tag{10}$$

$$J = -D_{21}C\nabla_r T - D_{22}\nabla_r C\ , \tag{11}$$

here: J – concentration flow, D_{22} – the diffusion coefficient of absorbing particles, D_{21} – thermal diffusion coefficient. For layer thicknesses $d_0 << L$ we can neglect the temperature change along o the thickness of cell and take it equal $T(r, z=0)$. In the set mode ($\left(dC/dt\right)=0$) we have from (10-11) for the stationary concentration C:

$$-D_{21}C\nabla_r T - D_{22}\nabla_r C = 0 \tag{12}$$

The concentration change of absorbing components is small compared to primary one - $C_1 = C - C_0$, where $C_1 << C_0$. Then we get the equation for C_1:

$$-\Delta_{thd}C_0\nabla_r I - \nabla_r C_1 = 0\ , \tag{13}$$

here we denoted the next values:

$$\Delta_{thd} = \alpha_0 d(L\chi_0^{-1} + \gamma^{-1} + d\chi_c^{-1}/2)D_{21}D_{22}^{-1}\ ,\quad \alpha_0 = \beta C_0 .$$

Than we get from (13):

$$C_1 = -C_0\Delta_{thd}I_0\ . \tag{14}$$

We use the term for the lens transparency cell to calculate the thermal lens signal [1]:

$$\Theta = 1 - \frac{2\left(z_1/l_0\right)\Phi_n\left(0\right)}{\left(1 + z_1^2/l_0^2\right)\left(1 + 3z_1^2/l_0^2\right)}, \tag{15}$$

where $l_0 = \pi r_0^2/\lambda$, $\Phi_{nl}\left(0\right)$ - nonlinear phase in optical cell on the beam axis. The latter value includes the nonlinear phases in the liquid and in the windows of the cell:

$$\Delta\Phi_{nl}^m\left(0\right) = 2k\int_0^{d/2}\left(\frac{\partial n_m}{\partial T}\right)\Delta T_m^s\left(z, r=0\right)dz\ . \tag{16}$$

$$\Delta\Phi_{nl}^{w}(0)=2k\int_{d/2}^{(d/2+L)}\left(\frac{\partial n_w}{\partial T}\right)\Delta T_w^s(z,r=0)dz\,, \tag{17}$$

where $\left(\frac{\partial n_m}{\partial T}\right)$ и $\left(\frac{\partial n_w}{\partial T}\right)$ are constant coefficients for nonlinear medium and material of the window respectively, $k=2\pi/\lambda$ - wave vector of radiation.

Using (15-17), we get:

$$\Delta\Phi_{nl}^{m}(0)=\alpha_0 kI_0 d^2\left\{\gamma^{-1}-L\chi_w^{-1}+d\chi_m^{-1}/6\right\}\left(\frac{\partial n_m}{\partial T}\right)(1-\Delta_{thd}I_0). \tag{18}$$

$$\Delta\Phi_{nl}^{w}(0)=\alpha_0 kI_0 dL\left\{\gamma^{-1}-L\chi_w^{-1}+\chi_m^{-1}(L+d)\right\}\left(\frac{\partial n_w}{\partial T}\right)(1-\Delta_{thd}I_0). \tag{19}$$

Finally we have an expression for the stationary lens transparency value:

$$\Theta(t=\infty)=1-\frac{2(z_1/l_0)\left[\Delta\Phi_{nl}^{m}(0)+\Delta\Phi_{nl}^{w}(0)\right]}{\left(1+z_1^2/l_0^2\right)\left(1+3z_1^2/l_0^2\right)}. \tag{20}$$

The resulting expression allows to calculate the influence of thermal diffusion on the value of the stationary thermal lens response in the optical cells with two-component liquid.

3. Conclusions

Thus it is shown that thermal lens response in the two-component liquid contains additional contribution due to thermodiffusion concentration change of absorbing components. The magnitude of this contribution can be quite large and it have a different sign for different liquids, depending on the sign of the coefficient of thermal diffusion. Secondly, the amount of thermal lens response will influence and change layer transmission. Thus a self-induced absorption coefficient modulation must be taken into account in the thermal lens spectroscopy analyses of the multi-component liquids [1-3]. Received expressions can be used in the experimental determination of the heat and mass transfer coefficients in such multi-component materials [4-6].

4. References

[1] Cabrera H, Sira E, Rahn K, García-Sucre M 2009 *Appl. Phys. Let.* **94** 051103
[2] Hoffmann B, Kohler W, Krekhova M 2003 *J. of Chem. Phys.* **118** 3237
[3] Ivanov V I 2006 *Thermo-induced Mechanisms of Dynamic Holograms Recording* (Khabarovsk: FESTU).
[4] Giglio M, Vendramini A. 1974 *Appl. Phys. Lett.* **25** 555-557
[5] Vicary L 2002 *Philos. Mag. B.* **82** 447-452
[6] Okishev K N, Ivanov V I, Kliment'ev S V, Kuzin A A, Livashvili A I 2010 *Opt. Atm. and Ocean.* **23** 106-107

Journal of Physics: Conference Series **735** (2016) 012038 doi:10.1088/1742-6596/735/1/012038

Selectivity control of photosensitivity Au-AlGaN structures

M Yu Andreev[1], S A Tarasov[1], I A Lamkin[1], I I Mikhailov[1], A V Solomonov[1], S Yu Kurin[2]

[1]Saint-Petersburg Electrotechnical University "LETI", Prof. Popova 5, St. Petersburg 197376, Russia

[2]GaN-Crystals Ltd., 27 Engels Ave, St. Petersburg 194156, Russia,

E-mail: ialamkin@mail.ru

Abstract. The Created selective photosensitive structures based on the Schottky barrier Au-AlGaN are for different ranges of the ultraviolet region of the spectrum. The methods that has been shown here, of controlling the spectrum of photosensitivity, are due to the effect of the wide windows and the effect of above-barrier transport in the structures Au-AlGaN.

1. Introduction

Photosensitive structures and devices, at the present time, are widely using in almost all areas of human activity [1], [2]. In recent years, photodetectors, which are sensitive in the ultraviolet spectral range, are becoming more popular and requested for a number of military and civilian applications [3]. Systems of UV location is one of the most relevant examples of the use of such devices, that capable of effectively detecting apparatuses for various purposes by the ultraviolet radiation component of reactive jet, including in outer space [4]. Controlling the UV levels can be used for peaceful applications in disinfection of water and some medical procedures; environmental monitoring, including the studies of the ozone layer and finding of the ozone holes; building of security systems, in particular, flame detectors and pollution; and a number of other scientific and economic tasks. For most of these applications are required detectors, that are insensitive to radiation in the visible range of the spectrum, the so-called apparently blind photodetectors, while in some cases - solarblind photodetectors, that are capable of correctly detecting the UV light in the conditions of strong backlight by the sunlight [5]. Without the use of expensive optical filters, it can be implemented only by using the wide-band semiconductor materials with a band gap of 3 eV or more [6]. On the other hand for the successful implementation of such detectors in production, it is necessary to develop a technology of their building within which it is possible to achieve high parameters of devices at preservation of their low cost [7]. In addition, one of the current highly relevant branches of development photodetectors is the creation of selective photodiodes. Using the effect of the wide window allows you to create photodiodes with a short-wavelength of boundary photosensitivity in a given region. In the structure, a wideband window is a semiconductor layer, with the smallest band gap, through which light passes. The use of structures on the basis of Schottky barrier one of the promising options, which should enable the creation of highly efficient photodetectors by relatively simple and inexpensive methods [8].

2. Sample and experimental technique

$Al_xGa_{1-x}N$ epitaxial layers were grown by hydride vapour phase epitaxy (HVPE) technique [9]. The top doped n-$Al_xGa_{1-x}N$ layer had a carrier concentration of 10^{17} cm^{-3}. Metal layers were deposited by vacuum thermal evaporation at a residual gas pressure of 10^{-5} Torr. Au was used as a rectifying contact material. Al with Ti underlayers were used to form Ohmic contacts [10]. Prior to metal deposition the structures were cleaned by potassium hydroxide (KOH) during 2 min and washed with distilled water. For good metal adhesion the structures were heated to a temperature of 300°C during deposition. Structure characteristics were measured using a set-up based on a monochromator with a diffraction grating and a xenon lamp as a light source.

Journal of Physics: Conference Series **735** (2016) 012038 doi:10.1088/1742-6596/735/1/012038

3. Experimental results and discussions

The main focus in the work was given to the creation of the effect of the selective sensitivity in grown structures without needing for using optical filters, as well as studies the possibility of controlling this effect. When illuminated by the side of metal, spectrum of sensitivity of the photodetector has a broadband characteristic with a half-width greater than 40 nm.

On the fig. 1a shown the structure on the basis of epitaxial layer AlGaN, with proportion of AlN equaled to 0.1, that forming a Schottky barrier with Au.

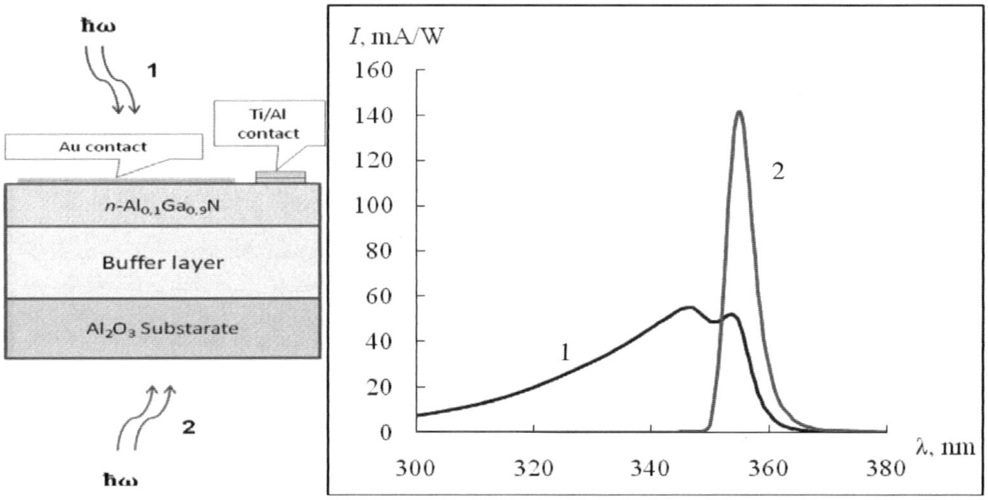

(a) (b)

Figure 1. An example of the structure of the photodiode Au-$Al_{0.1}Ga_{0.9}N$ and the measured spectra: a) structure of the photodetector, b) spectrum of photosensitivity.

When illuminating this structure from the back side of the solid solution layer AlGaN with proportion of AlN equaled to 0.06, It serves as the wide windows (Fig. 2b). Thus, it is possible to create a selective photodetector with high photosensitivity at 362 nm.

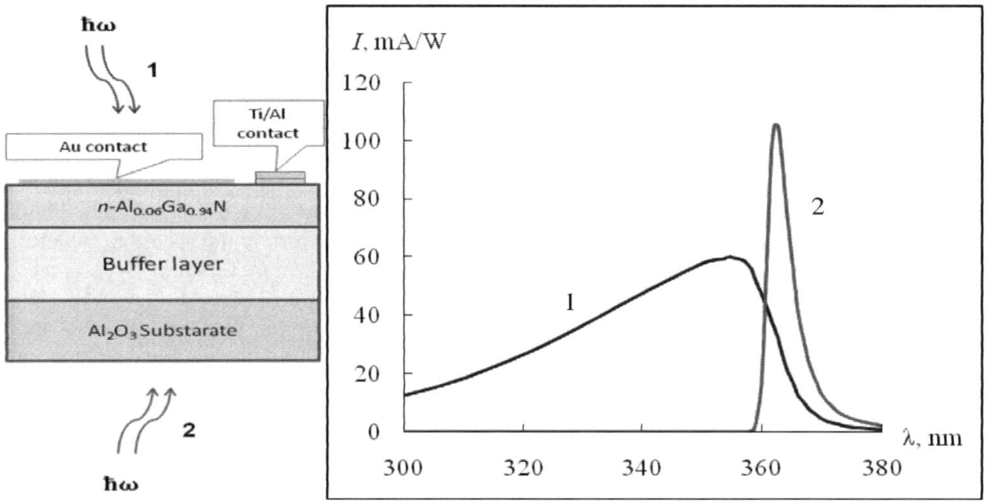

(a) (b)

Figure 2. An example of the structure of the photodiode Au-$Al_{0.06}Ga_{0.94}N$ and the measured spectra: a) structure of the photodetector, b) spectrum of photosensitivity.

It is possible to controlling the sensitivity from two sides, by using an additional layer in the composition of the photodetector (Fig. 3a). GaN is used in this structure, as the wide-gap window layer, which displaces the short-wave boundary of the photosensitivity at wavelength of about 360 nm. The long-wavelength boundary is determined by the high of the potential barrier contact Au-AlGaN with a proportion of AlN equaled to 0.1.

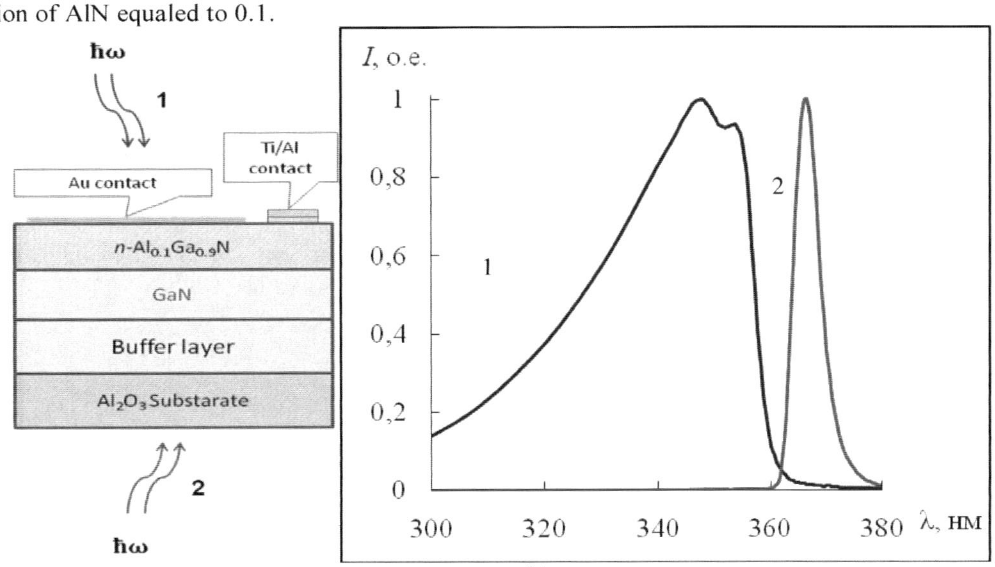

Figure 3. An example of the structure of the photodiode Au-$Al_{0.1}Ga_{0.9}N$-GaN and the measured spectra: a) structure of the photodetector, b) spectrum of photosensitivity.

Although in this case the maximum of the photosensitivity is reduced, but this structure makes it possible to demonstrate how the combination of the effects of a wide-window and above-barrier transfer allows controlling the sensitivity range of the photodetector. By changing the composition of the solid solution, that performing the role of the wide windows, lead to changing the range of the photosensitivity in the shortwave region of the spectrum.

Thus, here it was demonstrated the possibility of controlling the selectivity of sensitivity structures based on Au-AlGaN (Fig. 4).

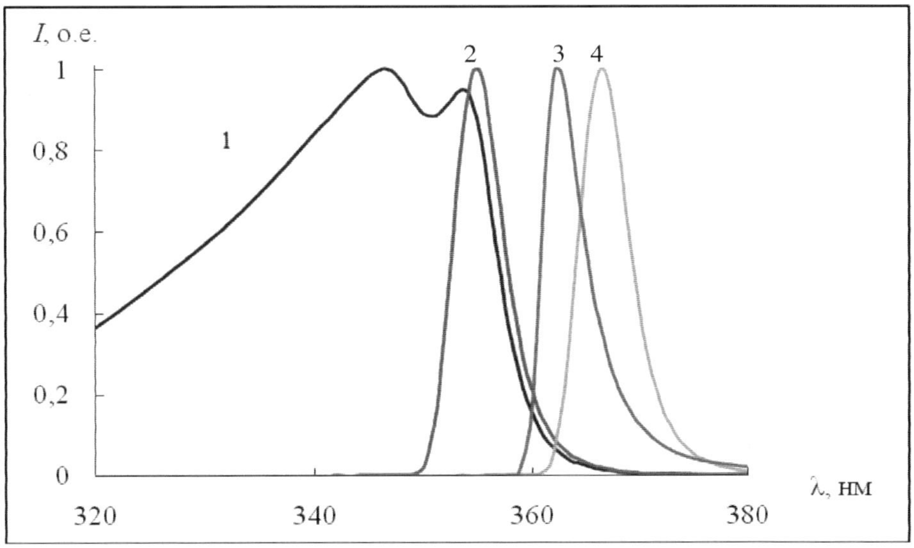

Figure 4. Spectral characteristics of photodiodes Au-AlGaN.

Selective photodetectors were obtained with the following parameters: half-width of 5-6 nm, 351-373 nm range of photosensitivity with a maximum at 355 nm, 362 nm, 366 nm and a sensitivity to 140 mA / W.

4. Conclusion

Selective photosensitive structures based on solid solution AlGaN were created. On the basis of structure with the upper epitaxial layer AlGaN and with proportion of AlN equaled to 0.1, that forming a Schottky barrier with Au, a selective photodiode with a maximum at 355 nm photosensitivity and a half-width 6 nm has been created. Reducing the proportion of AlN to 0.06 in the sold solution AlGaN caused the shifting of the photosensitivity maximum of the selective detector for a wavelength 362 nm, reducing the half-width of 5 nm. Adding an extra layer of GaN, performing the role of the wide windows in the structure Au-$Al_{0.1}Ga_{0.9}N$, also caused to shifting the maximum sensitivity. The maximum of spectral characteristics of selective photodiode was at 366 nm. It has been shown that changing the composition of the solid solution, which performing the role of the wide window, change the range of the photosensitivity in the shortwave region of the spectrum.

Acknowledgment
The work was supported by the Ministry of Education of the Russian Federation in the framework of the project of the state task in the field of scientific activity, project number 16.1307.2014K.

References

[1] Kung P, Yasan A, McClintock R, Darvish S R, Mi K, Razeghi M Future of $Al_xGa_{1-x}N$ materials and device technology for ultraviolet photodetectors, *Proc. SPIE*, vol. 4650, p. 199-206, 2002.
[2] Helava H I, Menkovich E A, Kurin S Yu, Antipov A A, Roenkov A D, Barash I S, Tarasov S A, Shmidt N M, Makarov Yu N, Lamkin I A UV LEDs for high-current operation *Journal of Physics: Conference Series* vol. 461, p. 012028, 2013.
[3] Lamkin I, Tarasov S Ultraviolet photodiodes based on AlGaN solid solutions *Journal of Physics: Conference Series* vol. 461, p. 012025, 2013.
[4] Ulmer M, Razeghi M, Bigan E, Ultra-Violet Detectors for Astrophysics, Present and Future, *Proc. SPIE.* vol. 2397, p. 210-216, 1995.
[5] Petrov A A, Tarasov S A, Lamkin I A, Menkovich E A, Solomonov A V, Kurin S Yu Research of the solar-blind and visible-blind photodetectors, based on the AlGaN solid solutions *Journal of Physics: Conference Series*. vol. 572, p. 012063, 2014.
[6] Menkovich E A, Tarasov S A, Lamkin I A Luminescence of nanostructures based on semiconductor nitrides *Functional Materials* vol. 19, № 2, p. 233, 2012.
[7] Kurin S, Antipov A, Barash I, Roenkov A, Usikov A, Helava H, Ratnikov V, Shmidt N, Sakharov A, Tarasov S, Menkovich E, Lamkin I, Papchenko B, Makarov Y Characterization of HVPE-grown UV LED heterostructures *Physica status solidi (C)* №3, p. 813–816, 2014
[8] Rumyantsev S L, Pala N, Shur M S, Gaska R, Levinshtein M E Low-frequency noise in $Al_{0.4}Ga_{0.6}N$-based Schottky barrier photodetectors *Applied Physics Letters* vol. 79, (6), p. 866-858, 2001.
[9] Barash I, Roenkov A, Kurin S, Antipov A, Helava H, Tarasov S, Menkovich E, Lamkin I, Makarov Yu. CHVPE growth of AlGaN-based UV LEDS *Physica Status Solidi (C) Current Topics in Solid State Physics* vol. 10, № 3, p. 289-293, 2013.
[10] Lamkin I A, Tarasov S A, Petrov A A, Menkovich E A, Solomonov A V, Kurin S Yu Investigation into the processes of atom redistribution encountered during of the formation of metal layers on the surface of aluminum gallium nitride *Journal of Physics: Conference Series* vol. 572, p. 012062, 2014.

Journal of Physics: Conference Series **735** (2016) 012039 doi:10.1088/1742-6596/735/1/012039

Problem analysis of image processing in two-axis autocollimator

A Nogin, I Konyakhin

Chair of Optical-Electronic Devices and Systems, ITMO University, 49 Kronverksky Pr. St. Petersburg, 197101, Russia

E-mail: rujusted@gmail.com

Abstract. The article deals with image processing algorithms in the analysis plane of an angle measuring two-axis autocollimator, which uses a reflector in the form of a quadrangular pyramid. This algorithm uses Hough transform and the method of weighted summation. The proposed algorithm can reduce the area of nonoperability and open up new possibilities for this class of devices.

1. Collimation scheme as the angle measuring system

Angle measurement is widely used in enterprises of mechanical engineering and optical industry to measure the deviations from the straightness guide, deviation from planarity faceplate and surface plate, configuration tools and machines, the analysis of vibration and temperature changes, as well as measuring of angles of optical wedges. Optoelectronic systems allow non-contact, quickly and high-precision solution of the given tasks[1]. There are two schemes of constructing angle measuring systems - collimation and autocollimation.

To use the first scheme, the collimation, it is necessary to place the collimator on base object, and the registration system of spatial position the beam emitted from the collimator on the controlled object.

In the second scheme, autocollimation, the controlled object is a passive reflector, and on the base object is located autocollimator. The autocollimator is both a collimator with an eyepiece attached to it for lighting grid and registering its reflection from the reflector on the object. This system is the most preferred because it does not require a power supply at the control point. It simplifies and reduce the cost of installation the entire system.[2,3]

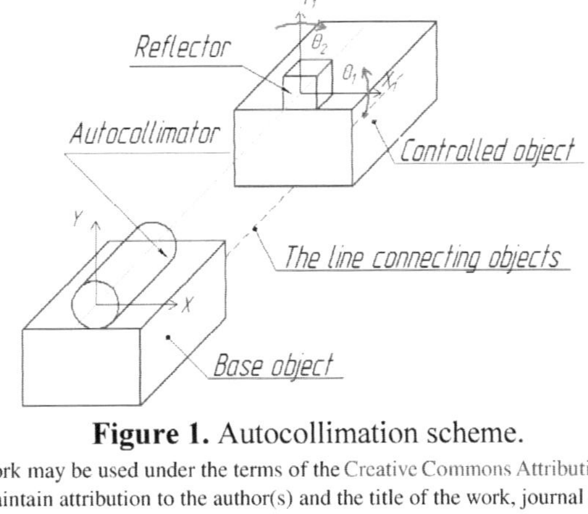

Figure 1. Autocollimation scheme.

2. Analysis of the image plane

Autocollimator consists of emitting optic-electronics channel and a receiver with a microprocessor. Emitting channel generates an optical beam that falls on the reflector. Optical electronic receiver meant for registration of optical radiation from the control element and measuring its parameters, which determine rotation angles of the object. When the reflector element rotates on angles Θ_1 and Θ_2, the reflected beam has deviated from the original direction. As a result the image shifts on the matrix photo-receiver of the autocollimator. The microprocessor calculates video frames from the matrix photo-receiver and determines the shift of the image. Angular coordinates Θ_1 and Θ_2, of the reflector, are determined as a result of this processing

In the process of selecting a base object, some of the key parameters are the size of the autocollimator and the distance to the control object. One way to solve this problem is to use a special reflective element instead of an ordinary flat mirror. The reflector in the form of quadrangular pyramids can be used as the special reflective element. This control element allows to increase the working distance angle measuring system and reduce the diameter of the objective to receive the beam.

A characteristic feature is that the opposed faces comprise a single working surface, and for this reason, the parallel beam incident on the refractive face at the reflection, is divided into two beams, each of which is formed a corresponding equivalent prism BR-180. By rotating the pyramid relative to one axis, each reflected beam in the plane of the aperture the objective will move only along this axis while the other coordinates of its position will not change.

The initial stage of the calculation of the angles Θ_1 and Θ_2 is to determine the center of the marks registered by the system. Using the control elements different from flat mirror creates a situation when in the image plane will mark overlap with each other. In this connection, the question arises what should be the distance between the centers of the marks in order to save serviceability and accuracy of the measurement system

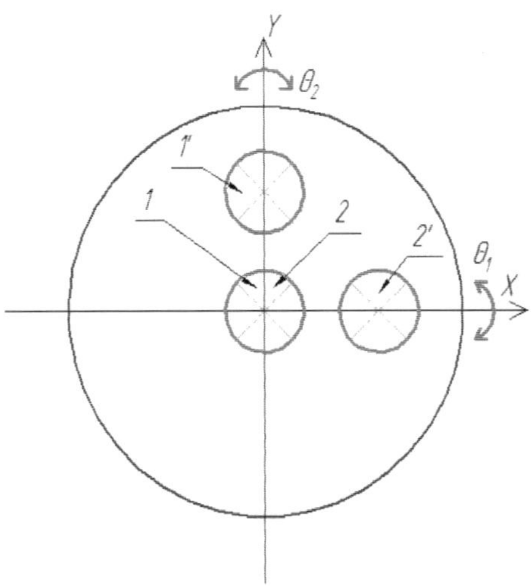

Figure 2. The images on the photodetector matrix: 1, 2 – before the rotation of the reflector, 1 'and 2' - after the rotation.

3. Algorithm for measurement of coordinates in the overlapping marks

To study the possible solution of this problem has been developed and implemented a model for the processing of overlapping arrays of irradiance in technology MatLab. This model has allowed to investigate the influence marks overlapping on the measuring accuracy of the coordinates.

In general, the idea of the algorithm consists in preliminary find the most probable centers of marks and their subsequent a clarification by the weighted summation method. To minimize the effect marks overlapping decided to use one of the key features of the control element, namely the fact that each mark can move only along one axis.

Table 1. The model parameters.

Parameter	Value
The size of the matrix	128*128 pixels
The radius of the mark	12 pixels
Signal-To-Noise Ratio	17
The law of distribution of noise	Normal

3.1. The preparatory stage of the algorithm

The first stage is to prepare the image to be processed. At this stage, the image processing by the circular averaging filter with a window size equal to the diameter of a circle plus one pixel. Then apply a threshold filtering that allows to cut noise. After filtering threshold is necessary to apply a Gaussian filter as the image becomes unusable for processing by a weighted summation method.[3,4]

Hough transform is used to find the most probable centres of marks. This method is based on an assessment of the normal orientation of the voting contour points.

The first step of the process is finding edge pixels surrounding the perimeter of the object. For this purpose used evaluation of the amplitude and direction of the gradient vector. Voting contour point is considered in terms of high modulus gradient.

The second step for each edge pixel is used position estimate and the orientation of the circuit in order to evaluate the center of a circular object of radius R by the movement over a distance R from the edge of the pixel in the direction normal to the contour (i.e. in the direction of the gradient vector).

If this operation is repeated for each edge pixel will be found a variety of positions alleged points of the center, which can be averaged to determine the exact location of the center.[6]

It should be noted that the coordinates of the center of the marks obtained by Hough transform has a strong dependence on the degree of marks overlapping and signal-to-noise ratio.

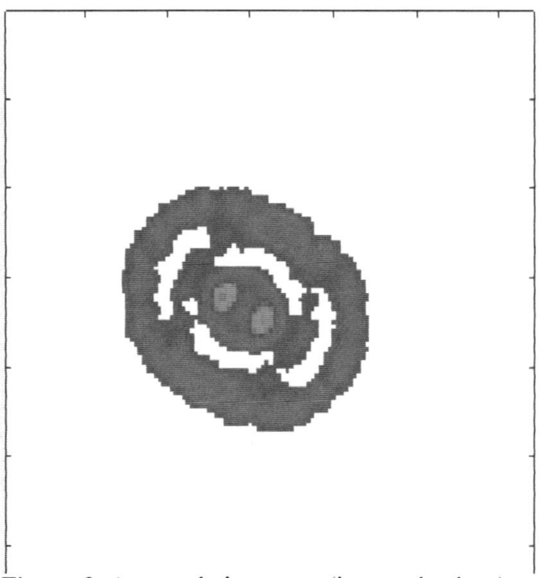

Figure 3. Accumulation array (inverted colour).

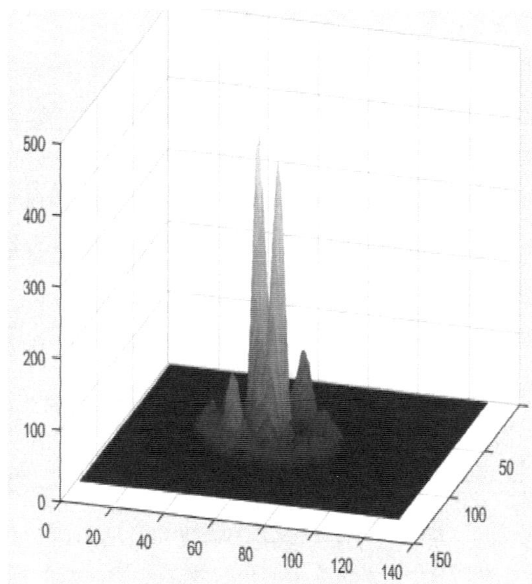

Figure 4. 3D view of the accumulation array (inverted colour).

Journal of Physics: Conference Series **735** (2016) 012039 doi:10.1088/1742-6596/735/1/012039

3.2. The main stage of the algorithm

On the main stage, the definition of so-called "zones of interest" i.e. in what quarters mark do not overlap. The definition of "zones of interest" based on the approximate coordinates of the center. After determining the zone of interest, there are an allocation disjoint parts of each mark into its own array. This array is restored and then use the weighted summation method for determining the centers of marks. Since the mark is moved along only one of its coordinate axes, one already known and the problem reduces to the determination of only the second coordinate.

The result of the program is to display the result in the image plane with an indication of the measured centers of coordinates and the distance between the centers.

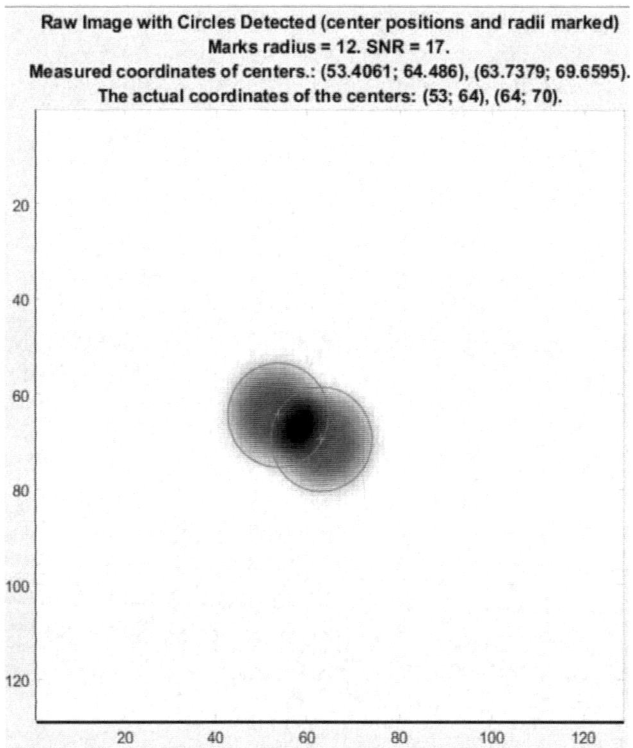

Figure 5. The result of the program. (Inverted colour).

4. Capabilities of the algorithm and the results of research its usage

4.1. Estimation of measurement error

During research of the algorithm using the model was evaluated errors. An analysis of the data determined that positioning accuracy is dependent only on the distance between the centers of the marks and the degree of the image overlapping. Parameters such as the size of the matrix, the size of marks and other factors on the accuracy of coordinate measurement is not affected. This algorithm is unusable when the distance between the centers of marks less than half the radius of the marks.

To estimate the errors were constructing graphs below. From this graph, it is clear that takes place to be a large error in the interval 1.1-1.6 radii in the deviation of the measured *X* coordinate from the real.

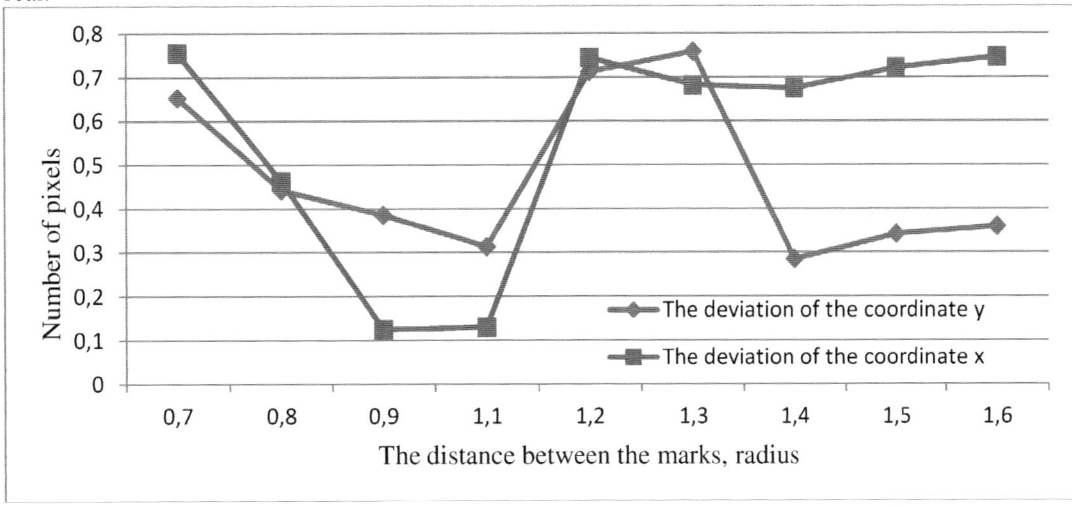

Figure 6. Measurement errors for overlapping marks.

The research also found that when the distance between the marks more than two radii error does not exceed half a pixel.

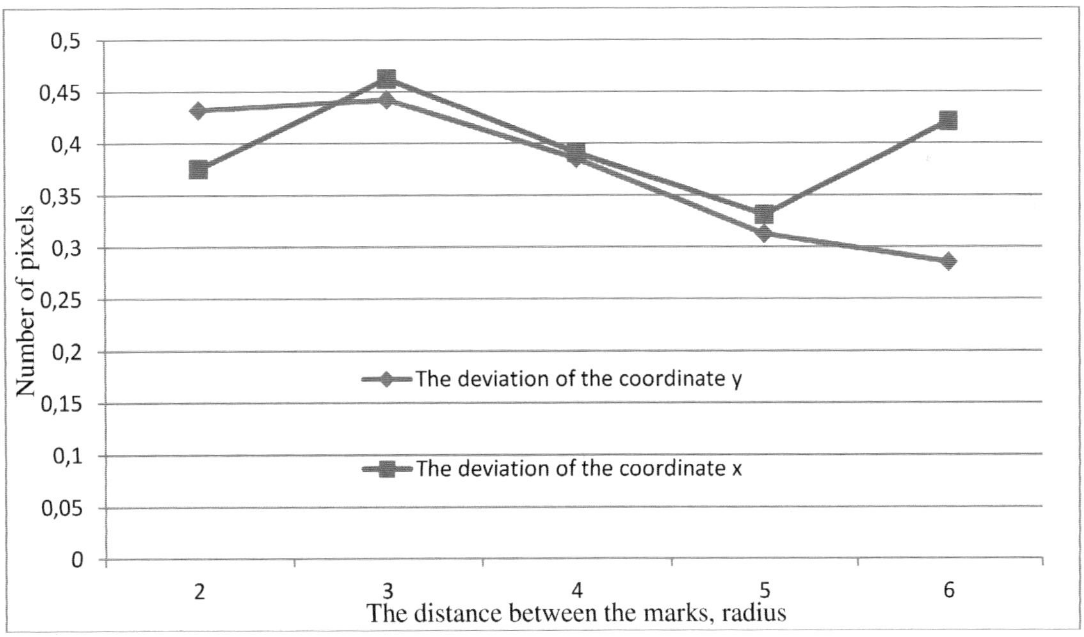

Figure 7. Measurement errors for nonoverlapping marks.

4.2. Functional testing of the algorithm

The algorithm was tested for functional three-axis autocollimator with the tetrahedral reflector. Despite the fact that the algorithm is not optimized for a given reflector were detected correct marks and measured their centers. The algorithm has also been able to ignore the parasitic image. When we crossing marks algorithm also retained its capacity for work and was able to detect the mark and measure their position and ignore the parasitic image. The experiment was to evaluate the accuracy of the algorithm will be held in subsequent experiments.

5. Summary

In conclusion, it is necessary to highlight the main advantages of the algorithm: a small area of malfunction, an error less than half a pixel at a distance least two radii and less than 1 pixel in the other cases. Also, worth noting is the possibility of the algorithm to work with any number of marks and high noise immunity.

The algorithm allows to use autocollimation scheme where it was previously not possible. The algorithm after improvement can use not only in the angle measuring systems but also in medicine and biology.

6. Acknowledgments

This work was financially supported by Government of Russian Federation, Grant 074-U01.

References

[1] T Turgalieva, I Konyakhin, 2013, Research of autocollimating angular deformation measurement system for large-size objects control, *Proc. SPIE* **8788, 878832**

[2] I Konyakhin, T Kopylova, A Konyakhin, 2012 Optic-electronic autocollimation sensor for measurement of the three-axis angular deformation of industry objects, *Proc. SPIE* **8439, 84391N**

[3] I Konyakhin, T Kopylova, A Konyakhin and A Smekhov, 2013, Optic-electronic systems for measurement the three-dimension angular deformation of axles at the millimeter wave range radiotelescope, **Proc. SPIE** *8759, 87593E*

[4] D.H. Ballard, 1981, *Generalizing the Hough transform to detect arbitrary shapes*, **13(2),** pp.111-112,.

[5] L.A.F Fernandes, M.M. Oliveira, Real-time line detection through an improved Hough transform voting scheme, №41, pp.299-314, (2008).

[6] R. Gonzalez, R. Woods, S. Eddins, 2010, *Digital Image Processing Using MATLAB*

Journal of Physics: Conference Series **735** (2016) 012040 doi:10.1088/1742-6596/735/1/012040

An investigation of the influence of residual amplitude modulation in phase electro-optic modulator on the signal of fiber-optic gyroscope

D A Pogorelaya, M A Smolovik, V E Strigalev, A S Aleynik, I G Deyneka
ITMO University, 197101, Kronverskiy pr. 49, Saint-Petersburg, Russian Federation

E-mail: pdaria@mail.ru

Abstract. The investigation is devoted to residual amplitude modulation (RAM) of phase electro-optic modulator, which guides are made in $LiNbO_3$ crystal by Ti diffusion technology. An analysis is presented that shows influence of RAM on the signal of fiber-optic gyroscope. The RAM compensation method is offered.

1. Introduction

The principle of fiber-optic gyroscope (FOG) is based on Sagnac effect. Two beams from a laser are injected into the same fiber but in opposite directions. Due to the Sagnac effect, the beam travelling against the rotation experiences a slightly shorter path delay than the other beam [1]. The resulting differential phase shift is measured through interferometry, and it is proportional to the angular velocity.

A minimal configuration of FOG [2] is presented in Figure 1.

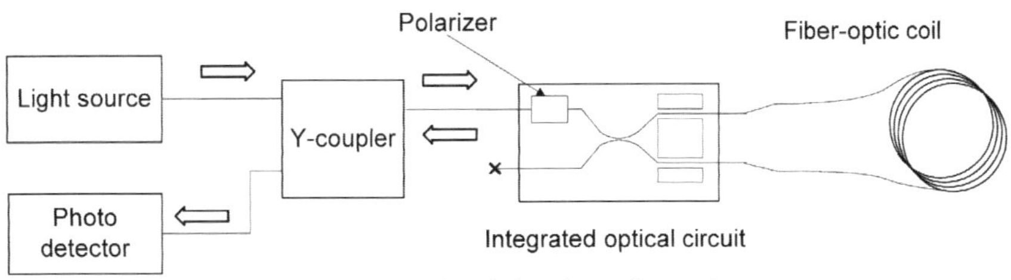

Figure 1. FOG minimal configuration.

When the optic system is rotating in inertial space with angular velocity Ω, differential phase shift is expressed as:

$$\Delta\varphi_r = (2\pi LD / \lambda_0 c)\Omega,$$

where: λ_0 – the wavelength of light in vacuum, D – the diameter of fiber-optic coil, L – the length of optic fiber, Ω – the detected angular velocity [3].

The intensity of the output signal of the interferometer is changing by cosine law depending on phase difference $\Delta\varphi_r$, defined by angular velocity:

$$I(\Delta\varphi_r) = I_1 + I_2 + 2\sqrt{I_1 I_2}\cos(\Delta\varphi_r),$$

where I_1 and I_2 – intensities of the interfering waves.

In order to achieve a higher sensitivity the signal is shifted by the phase modulator in the operating point where $I(\Delta\varphi)$ is maximal. Then the signal of the interference is expressed as:

Journal of Physics: Conference Series 735 (2016) 012040 doi:10.1088/1742-6596/735/1/012040

$$I(\Delta\varphi_r) = I_1 + I_2 + 2\sqrt{I_1 I_2}\cos(\Delta\varphi_m + \Delta\varphi_r), \tag{1}$$

where $\Delta\varphi_m$ – the phase shift, provided by phase modulator.

One of the most widespread methods of maintaining the operating point to quadrature is using of an electro-optic phase modulator. The most often used electro-optic phase modulators are modulators on $LiNbO_3$ crystal because of high linearity of the electro-optic effect and high operation speed.

But $LiNbO_3$ phase modulators have a significant shortcoming, which is expressed in residual amplitude modulation (RAM). As a result a light beam receives both phase shift and intensity changes [4]. As RAM magnitude increases so the phase shift provided by modulator becomes less well defined [5]. So lower accuracy of fiber-optic interferometry sensors is a consequence of RAM. Therefore it is necessary to reduce RAM for increasing FOG accuracy.

2. RAM measures

RAM is evaluated by coefficient which is expressed as:

$$k_{RAM} = \Delta I / I_m,$$

where: ΔI – the magnitude of oscillation of interferometry signal caused by RAM, I_m – the medium value of signal intensity without modulation.

The integrated optical circuit made in $LiNbO_3$ crystal by Ti diffusion technology is presented in Figure 2.

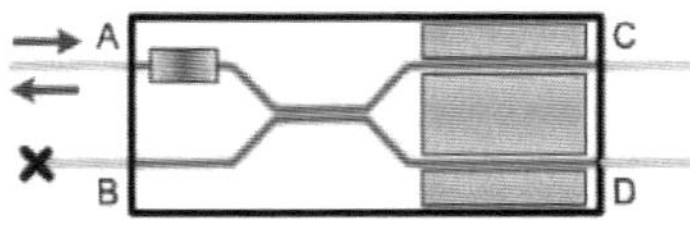

Figure 2. Integrated optical circuit.

The integrated optical circuit consists of the integrated optical polarizer in the arm A, X-splitter and the electro-optic phase modulator with three electrodes. After the polarizer the light beam is divided on two similar beams by X-splitter and propagates to phase modulator where the phase of both beams is modulated.

The $k_{RAM}(U)$ has been measured by the next way: the voltage has been applied to electrodes of the phase modulator, and output signals of each arm (C and D) have been detected synchronously with the applying voltage. The resulting RAM coefficients in the upper arm of modulator $k^U_{RAM}(U)$ and lower one $k^L_{RAM}(U)$ are shown in Figure 3.

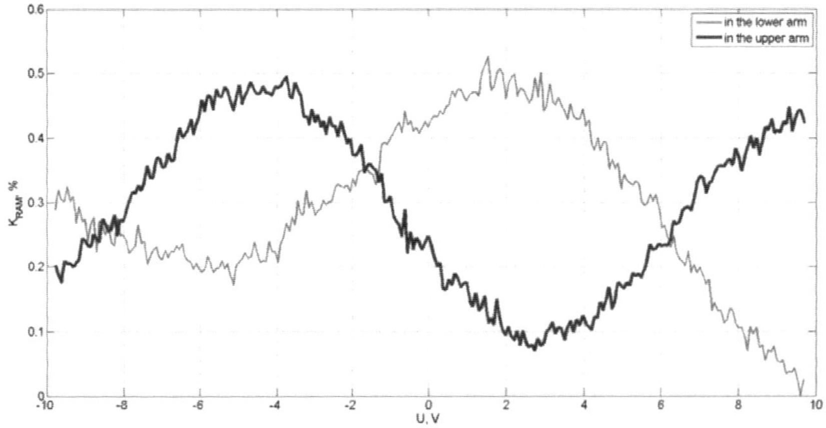

Figure 3. RAM coefficient dependence on applied voltage.

Journal of Physics: Conference Series **735** (2016) 012040 doi:10.1088/1742-6596/735/1/012040

3. RAM influence on FOG signal

Measured $k^U_{RAM}(U)$ and $k^L_{RAM}(U)$ were entered in the program model of the FOG signal processing. Taking the RAM into account the intensities of two interfering beams from equation (1):

$$I_1 = \frac{I_0}{2}\left[1 - k^U_{RAM}(U_1)\right]\left[1 - k^L_{RAM}(U_2)\right], \tag{2}$$

$$I_2 = \frac{I_0}{2}\left[1 - k^L_{RAM}(U_1)\right]\left[1 - k^U_{RAM}(U_2)\right]. \tag{3}$$

where I_0 - the intensity of optical radiance of interfering waves.

In order to simplify the next text it is suggested to enter the replacement:

$$k_U(U) = 1 - k^U_{\mathrm{RAM}}(U),$$

$$k_L(U) = 1 - k^L_{RAM}(U).$$

Then equation (2) and equation (3) are expressed as:

$$I_1 = \frac{I_0}{2} k_U(U_1) k_L(U_2),$$

$$I_2 = \frac{I_0}{2} k_L(U_1) k_U(U_2).$$

In order to describe the influence of RAM on the FOG signal it is suggested to consider the differential signal of photo detector which is using for forming the signal of angular velocity:

$$\Delta I = I - I_{pr},$$

where: I – the current signal of photo detector, I_{pr} – the previous signal of photo detector.

$$I = I_1 + I_2 + 2\sqrt{I_1 I_2}\cos(\Delta\varphi) = \frac{I_0}{2}[k_U(U_1)k_L(U_2) + k_L(U_1)k_U(U_2)]$$
$$+ I_0\sqrt{k_U(U_1)k_L(U_2)k_L(U_1)k_U(U_2)}\cos(\Delta\varphi),$$

$$I_{pr} = I_{1pr} + I_{2pr} + 2\sqrt{I_{1pr} I_{2pr}}\cos(\Delta\varphi_{pr}) = \frac{I_0}{2}[k_U(U_2)k_L(U_3) + k_L(U_2)k_U(U_3)]$$
$$+ I_0\sqrt{k_U(U_2)k_L(U_3)k_L(U_2)k_U(U_3)}\cos(\Delta\varphi_{pr}).$$

We use the FOG signal processing method based on quadrature modulation with a feedback loop which compensates the phase shift provided by the angular velocity [3]. In case when the feedback compensates the phase shift completely, the resulting phase shift under the cosine function is $\pi/2$ and $\cos(\pi/2) = 0$. So it is possible to neglect the second term of the expression for the intensity of output signal of the photo detector. Then the differential signal of the photo detector is expressed as:

$$\Delta I = I - I_{pr} = \frac{I_0}{2}[k_U(U_1)k_L(U_2) + k_L(U_1)k_U(U_2) - k_U(U_2)k_L(U_3) - k_L(U_2)k_U(U_3)]$$
$$= \frac{I_0}{2}\{k_L(U_2)[k_U(U_1) - k_U(U_3)] + k_U(U_2)[k_L(U_1) - k_L(U_3)]\}. \tag{4}$$

Due to equation (4) the error in the differential signal of the photo detector is conditioned by the difference of amplitude coefficients $k_U(U_1)$ and $k_U(U_3)$, and $k_L(U_1)$ and $k_L(U_3)$. Consequently the error of the FOG output signal depends on the slope of $k_U(U)$ and $k_L(U)$.

4. RAM compensation method

There is offered an algorithmic method of compensation of the RAM influence on the FOG signal. It is possible to adjust the signal of the photo detector on the correction factor. In case when the feedback compensates the phase shift provided by the angular velocity completely, the output signal of the photodetector can be expressed as:

$$I = I_1 + I_2 = (I_0/2)\cdot[k_U(U_1)k_L(U_2) + k_L(U_1)k_U(U_2)]. \qquad (5)$$

Due to equation (5) the adjusted signal of the photodetector is expressed as:

$$I' = 2I\cdot[k_U(U_1)k_L(U_2) + k_L(U_1)k_U(U_2)]^{-1}.$$

This method of the signal adjusting was added into the program model of the FOG signal processing. The comparison of the FOG output signal with averaging over 1 sec of algorithm with the RAM compensation method and without it was made by computer simulation. Results are presented in Figure 4.

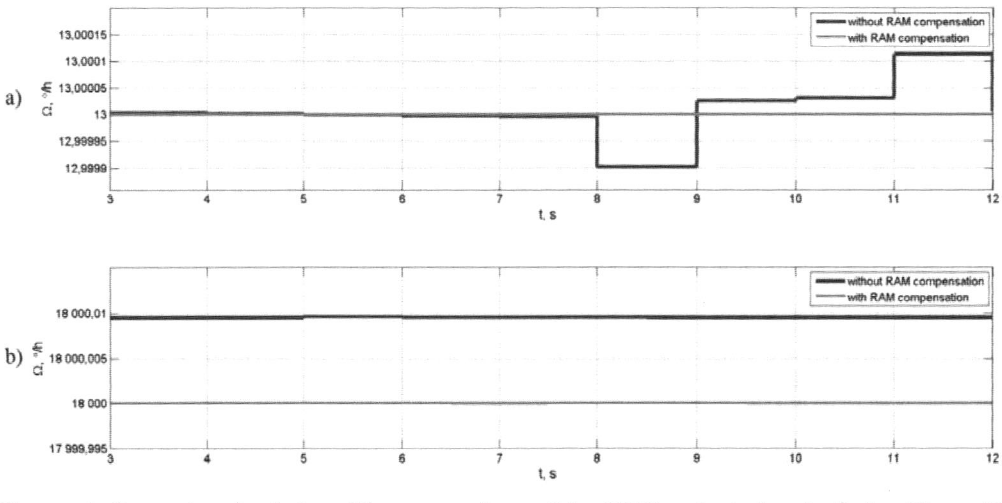

Figure 4. Computer simulation. The comparison of the FOG output signal of algorithm with the RAM compensation method and without it: a) in case the angular velocity is 13 °/h, b) in case the angular velocity is 18000 °/h.

The comparison of algorithms shows that described RAM compensation method promotes a decreasing of the standard deviation and promotes a decreasing of signal`s bias in case of high angular velocities.

5. Conclusion

This paper is devoted to investigation of residual amplitude modulation (RAM) of phase electro-optic modulator, which guides are made in $LiNbO_3$ crystal by Ti diffusion technology. The investigation shows that RAM promotes increasing of standard deviation and bias of the FOG`s output signal. The RAM compensation method, offered in the paper, allows to decrease the angular rate signal`s standard deviation and bias.

References

[1] Sagnac G 1913 *Comptes Rendus L'ether lumineux demontre par l'effet du vent relatif d'ether dans un interferometre en rotation uniforme* **157** 708
[2] Merlo S, Norgia M, Donati S 2000 *Handbook of Fibre Optic Sensing* (New York) pp 331-348
[3] Lefevre H C 2014 *The Fiber Optic Gyroscope* (London: Artech House)
[4] Ishibashi C, Ye J, Hall J L 2002 Quantum Electronics and Laser science Conference 91
[5] Sathian J, Jaatinen E 2013 *Optics express* **21** 12309-17

Development of microprojection system of mixed and augmented reality

M S Rudakova
Saint-Petersburg State University of Informational Technologies, Mechanics and Optics, 197191, 49 Kronverksky prospect, Saint-Petersburg, Russian Federation

e-mail: RudakovaMS@bk.ru

Abstract. The paper deals with the problems of development and designing of microprojection system of mixed and augmented reality designed so that an observer could see the information of the microdisplay and the surrounding space as the background at the same time. The combiner on planar waveguides screens based on the composite structure of the prism elements was developed. In this work different results of the simulation with the TracePro software are considered and also the main problems encountered in the development of such systems are considered.

1. Introduction

In the last decade in patent literature and articles a lot of information about the input image systems in the observer's eye from the reflective microdisplays can be found. Such systems consist of the image generator (most often it's the microdisplay with the illumination system if the display is not self-luminous), the lens that forms an image at infinity and the combiner which creates the required field of view and ensures the formation of images in the form of a raster-character information on the retina at the background of the surrounding space [1]. Field of application of such systems is quite wide: civil and military avionics, training systems, automobile production, etc.

The present work is devoted to the development of new methods of calculation and design of microprojection systems of mixed and augmented reality.

Among the patent literature the most common method to provide the operation of the combiner is the use of holographic optical elements. However, the reflectivity of the holographic combiners strongly depends on the incidence angle and wavelength, thus limiting the development, primarily color displays [2]. Moreover, most software can not realize the full and correct operation with holographic elements, because there are no exact calculation models and algorithms for solving the ray tracing from the object plane to the image plane. The relevance of the work related to the solution of these problems and is dedicated to the development of calculation models of microprojection display systems.

2. Theoretical model

One of the ways to build combiners is that the radiation is delivered to the observer's eye through the prism elements, the surface of which has a reflective/transmissive coating. This prismatic structure of the display allows to increase the size of the pupil area. However, the combiner of this type has some disadvantages. For example, an increasing of the pupil area requires an increasing of the size of the combiner. This leads to an increasing of the number of components of the combiner prism elements. And that presents some technical difficulties and increases the cost of production. But this solution provides a significant gain in overall dimensions.

The figure 1 shows a radiation transport scheme: image from microdisplay 1 enters the optical system 2, and then on the screen 3-8, and transfers to the observation plane 9. The main structural elements of the screen 3-8 are two crossed prismatic blocks: block 3, 4, 5, referred to as a multiplier of the horizontal field and block 6, 7, 8 – multiplier of the vertical field. The number of elements in multipliers can be different depending on the thickness of the device d and the required field of view.

Radiation enters multiples through input prism 3, 6, whose surface is formed as a mirror [3]. The inclination angle of the input prisms in the multipliers is equal to α.

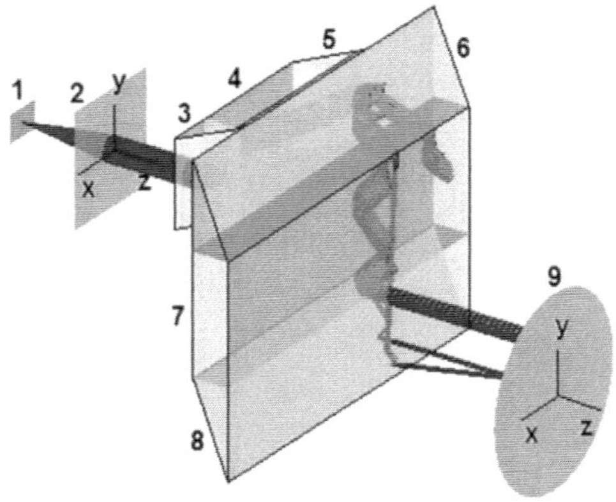

Figure 1. Planar waveguide screen

Were obtained analytical expressions that allow to modify the system according to the changing conditions of input radiation, the value of the field angle and allow to change system parameters such as the thickness and inclination angle of prisms.

The angle of total internal reflection α_{tir} defined by equitation (1):

$$\alpha_{tir} = \arcsin\left(\frac{n_1}{n_2}\right), \tag{1}$$

where n_1 - the index of refraction of the medium from which the light falls (n_1=1 for air); n_2 - refractive index of the medium in which the radiation is incident (the refractive index of the material multiplier).

In order the total internal reflection of radiation on the edges occurs, it is necessary that the angle of incidence β was greater than the angle of total internal reflection α_{tir}.

Entering radiation to the multiplier can be implemented in different ways: using a diffraction grating, a semitransparent mirror, a reflecting mirror, and so on. In this case, we assume that the radiation is introduced through the input prism one surface of which a mirror. The angle of incidence on the faces of the multiplier will be, in this case, $\beta = 2\alpha'$:

All layers in the structure are at a distance from each other, the structure is called the period T. If you have a field angle ω between the layers, period must be equal to $T = b + c$, where b - the height of the prism, and c - the distance from the end of one layer to the next (figure 2).

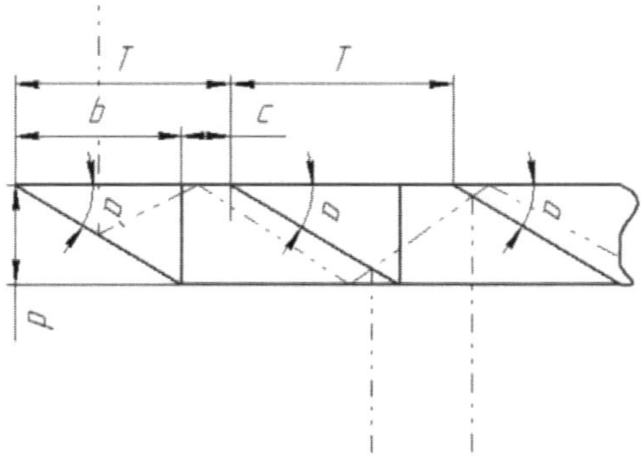

Figure 2. The period T of the structure

The value of c (as in equitation 2) defines the interval at which rays will be reflected.

$$c = d \cdot \mathrm{tg}(\omega_2), \quad (2)$$

where d – thickness of the multiplier; ω_2 - angle of refraction at the transition from a medium with a refractive index n_1 (in this case, from the air) in a medium with a refractive index n_2 (refractive index of the material of the prism). If the radiation falls directly on the input prism, the angle ω_2 is the angle of Snell's law (equitation 3).

$$\omega_2 = \arcsin(\frac{n_1 \cdot \sin\omega}{n_2}). \quad (3)$$

The last issue is the provision of necessary reflection and transmission of layers. It can be done by applying a special coating on each profiled layer. That is, for a system of three layers is enough that the first layer reflects 33.3%, the second - 50%, and the third - 100%. According to the equitation 4 can be calculated reflection coefficient ρ_i of the layer depending on the overall reflectivity and the number of layers in the multiplier.

$$\rho_i = \rho_0 \frac{1}{(N - i + 1)}, \quad (4)$$

where ρ_0 - reflectance multiplier; N - number of layers; i - layer number (i = *1, 2 ... N*).

3. Simulations in TracePro software

On the basis of analytical expressions in TracePro software were simulated several variants to evaluate image quality and system performance. Analysis of the distribution of power density on the screen was carried out on the brightness of the image in the pupil area, uniformity of illumination of the image, the presence of blind spots and losses in the optical path.

Table 1 shows the variants of multipliers to be modeled in the software TracePro. The initial number of beams is 119401, the radiation power P=1W. The wavelength λ=0,534 μm. The refractive index of the material of multipliers n_c=1,65. Horizontal field of view of ± 11.3 °, vertical field of view of ± 8.5 °. The thickness of the device in all cases d=4 mm and the number of layers n=3. Rays emerging from the multiplier register on the screen, with a diameter of 50 mm at a distance of 28 mm from the last surface of the multiplier. The analysis was conducted also on the distribution in the pupil area, which was located at the same distance from the device (multiplier) as the screen.

Table 1. Simulation variants

Parameters	Var. 1	Var. 2	Var. 3	Var. 4
		Multiplier of horizontal field		
The angle of the input prism α', °	45	30	25	30
Tilt angle α, °	45	30	25	30
Period T_1^a/T_2^b, mm	4/4.47	6.93/7.4	8.57/9.05	6.93/7.4
		Multiplier of vertical field		
The angle of the input prism α', °	45	30	25	25
Tilt angle α, °	45	30	25	25
Period T_1^a/T_2^b, mm	4/4.36	6.93/7.28	8.57/8.94	8.57/8.94

[a] *T1* – period between the layers with a parallel ray path.
[b] *T2* – period between the layers in the presence of a field angle ω.

The radiation power of 1 W, it can be assumed that the screen brightness microdisplay B=2100 nt when the value of the emitting area $S = \pi r^2$, that when a beam diameter of 4 mm yields S = 12.56 mm^2 and total aperture angle of 14,4°. We take the value of the pupil of the observer equal to d = 5 mm, which corresponds to the area S'= 19.625 mm^2, and the angular field of 22.6° x17°. The transmittance τ is calculated as the ratio of the number of the transmitted rays to the number of the incident rays.

The brightness of the resulting image B' is calculated by the equitation 5:

$$B' = \frac{\tau B S \Omega}{S' \Omega'}, \tag{5}$$

where τ – transmittance in pupil area; B - the estimated brightness of the screen; S - area of the emitting area; $\boldsymbol{\Omega}$ - full aperture angle of the lens of the microdisplay; S'- the area of the pupil area; Ω '- the angular field of the pupil.

One of the plots that we obtain in our research presented in the figure 2. Here presented the simulation of variant 4. The figure 3 shows, that we have optical lose about 31%, good brightness, sufficiently good uniformity of illumination in horizontal plane and no blind spots. In vertical plane is a redistribution of the intensity if illumination. On the whole obtained results meet the requirements.

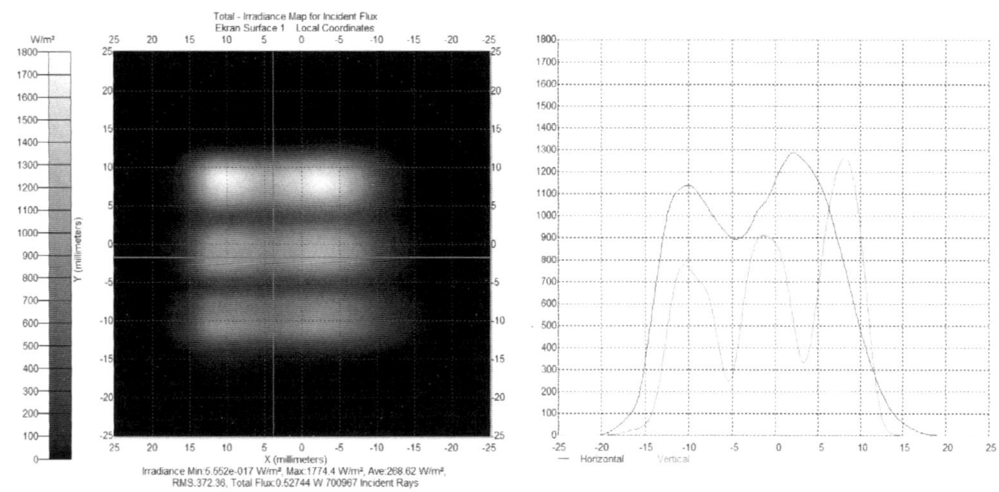

Figure 3. Distribution of power density on the screen

The figure 4 shows the distribution of power density in pupil area. The displacement of the pupil center relative to the center of the screen was: x =2 mm, y =-1.5 mm. The transmittance for the pupil: τ = 25420/119401 = 0.21 (or 21%). The brightness of the image is B'= 152.4 nt.

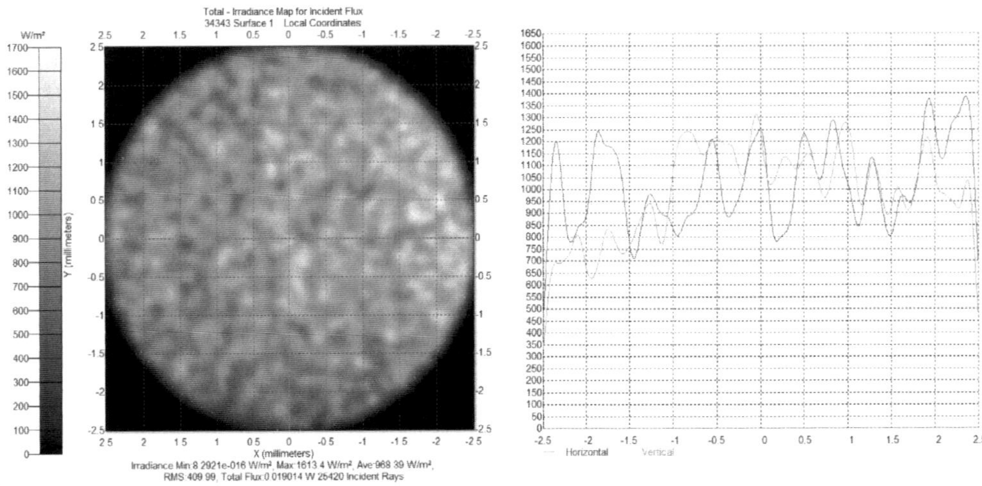

Figure 4. Distribution of power density in pupil area

The system considered efficient if the brightness of the image in the pupil area was not lower than 50 nt and the path loss don't exceed 40%. All of the variants satisfy this condition, but in the distribution of power density are some distortions such as wrong color image, phantom glare, «blind» spots, that can be connected to the frustrated total internal reflection and polarization.

In this case also issues related to the propagation of polarized radiation in the system were discussed. It is necessary to define system performance and the cause of the problems associated with distortion in image.

4. Conclusion

It was found, that the system works and provides replication of the light rays that fill the exit pupil of the collimator lens. Variants that can be considered more suitable for further study and analysis have been identified. In general the fulfilled work matches to the aim, i.e. the calculation model for combiner on planar waveguide screens was developed. In prospect is planned to implement a prototype of the combiner.

References

[1] Vasilev V N, Grimm V A, Romanova G E, Smirnov S A, Bakholdin A V and Grishina N Y 2014 Designing and researching of the virtual display system based on the prism elements *Proc. SPIE* **9131**

[2] Vasilev V N, Bakholdin A V, Grimm V A, Romanova G E and Smirnov S A 2013 Optical devices of virtual displays *J. of Opt. Tech.* **80** 17-24

[3] Grimm V A, Rudakova M S and Smirnov S A 2013 Microprojection systems of mixed and augmented reality on planar waveguide screens *Izv. Vyssh. Uch. Zav. Priborostr.* **56** 84-89

Journal of Physics: Conference Series **735** (2016) 012042 doi:10.1088/1742-6596/735/1/012042

A study on reducing of the focal spot size using spatial filtering and phase apodization

D Savelyev[1,2]
[1] Samara State Aerospace University, 34, Moskovskoye Shosse, Samara, Russia, 443086
[2] Image Processing Systems Institute of the RAS, 151, Molodogvardeyskay st., Samara, Russia, 443001

E-mail: dmitrey.savelyev@yandex.ru

Abstract. The resizing of focal spot using an annular diaphragm with phase jump in the scalar paraxial case was investigated numerically. Due to the apodization optical system, we succeeded in reducing the size of the focal spot in 2.3 times.

1. Introduction

One of the drawbacks of the conventional light microscope - beams out of focus, which reduce image contrast. If in the scheme of optical microscope to introduce a special aperture located in the intermediate image plane, it will miss the only light rays that come from a very small area of the object. Aperture will act as a spatial filter. The smaller the diameter of the aperture then is the smaller the area.

It is well known that one way the formation of narrow beams of extended focal plane is to use a narrow annular aperture blocking radiation in the central portion of the lens [1]. Modifications of this method in the full-aperture pupil apodization function using amplitude-phase and pure phase distributions are considered in [2-4]. Reducing the size of the focal spot, usually accompanied by a redistribution of power from the central portion to the side lobes, which is consistent with the theory Toraldo di France [5]. In [6,7] numerically investigated reducing the size of the focal spot in the radial polarization using binary annular element. It was shown that due to making the phase jump and simultaneously broadening the annular aperture can form a focal spot and size of this spot does not exceed the limit corresponding to a narrow annular aperture, and the intensity is increased by almost 6 times.

In this paper numerically investigate the influence of the size of the size of annular aperture with a radial phase jump to the formed the focal spot in the scalar paraxial case. The scalar case corresponds to uniform polarization and focusing with a small numerical aperture will not lead to strengthening of the longitudinal component of the electric field which broadens in this case, the focal spot [8,9].

2. Spatial filtering and phase apodization

In this paper, to focus limited plane wave (wavelength $\lambda = 0.633$ micrometres) with a lens radius R and the focal length f (numerical aperture $NA \approx f/R$) used transformation of Fourier-Bessel. The intensity in the focal plane in this case is:

$$I(\rho) = \left| \frac{2\pi}{\lambda f} \int_0^R J_0\left(\frac{2\pi\rho r}{\lambda f}\right) r\,\mathrm{d}r \right|^2 = \left| \frac{R}{\rho} J_1\left(\frac{2\pi\rho R}{\lambda f}\right) \right|^2 . \quad (1)$$

The radius of the central spot in the focusing (1) without apodization is determined by the first zero of the Bessel function of the first order ($J_1(\gamma_{11}) = 0,\ \gamma_{11} = 3.83$):

$$\rho_{lens} = \frac{3.83\lambda f}{2\pi R} = \frac{0.61\lambda f}{R} \approx 0.61\lambda/NA . \quad (2)$$

If limited the exit pupil to the narrow annular aperture with a relative width Δ, we obtain [6]:

$$I_\Delta(\rho) = \left| \frac{2\pi}{\lambda f} \int_{R(1-\Delta)}^{R} J_0\left(\frac{kr\rho}{f}\right) r\,\mathrm{d}r \right|^2 \approx \left| \frac{2\pi R^2 \Delta}{\lambda f} J_0\left(\frac{k\rho R}{f}\right) \right|^2 . \quad (3)$$

Then the radius of the focal spot is determined by the first zero of the Bessel function of zero order ($J_0(\gamma_{01}) = 0,\ \gamma_{01} = 2.405$):

$$\rho_{ring} = \frac{2.4\lambda f}{2\pi R} = \frac{0.38\lambda f}{R} \approx 0.38\lambda/NA , \quad (4)$$

that is, 1.6 times less than without apodization.
The additional introduction of a radial phase jump can further reduce the size of focal spot [6]:

$$\rho_{\min}^{jump} = \frac{0.22\lambda f}{R} \approx 0.22\lambda/NA , \quad (5)$$

Thus, theoretically, the amplitude-phase apodization reduces the size of the focal spot of 2.77 times, but this comes at the expense of a significant loss of energy. To estimate the losses, perform the numerical simulation.
The calculation parameters: the radius of the lens is 1 mm (equal to the outer radius of the diaphragm), the focus is 100 mm (NA = 0,01), r_1 - the inner radius of the diaphragm, r_2 - the radius of the phase jump on π radians. Figure 1 shows the schema of using installation.

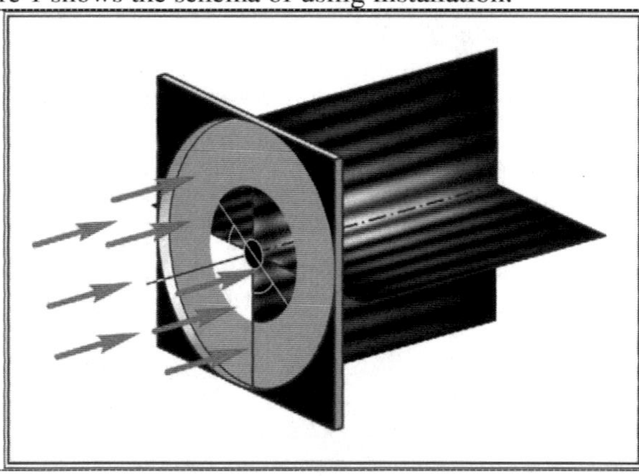

Figure 1. Optical scheme.

To evaluate the effectiveness type the following expression:

$$\varepsilon = \frac{I}{I_0}, \tag{5}$$

I - a value of intensity at the focus using the spatial filtering and phase apodization, I_0 - maximum value of the focus (without apodization).

Table 1 shows the cross-section intensity pattern obtained in the focal region, and input values of the phase and amplitude of the diaphragm (the dotted line in the pictures shows the amplitude size of the inner radius of the aperture).

Table 1. Change the radius of the diaphragm and the phase jump (picture 2 mm×2mm).

The radius of the aperture, $r_1 \backslash r_2$, мм	without apodization	0,2 \ 0,6	0,4 \ 0,7	0,6 \ 0,8	0,7 \ 0,85	0,76 \ 0,88	0,8 \ 0,9
Input, amplitude	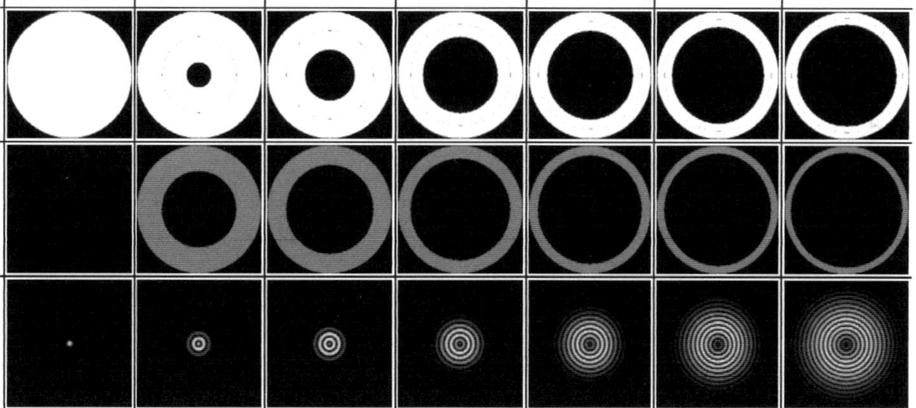						
Input, phase							
The focus, intensity							
FWHM, mm	0,0326	0,0208	0,0192	0,0159	0,015	0,0144	0,014
ε	1	0,102	$3{,}2\cdot10^{-2}$	$6{,}1\cdot10^{-3}$	$1{,}9\cdot10^{-3}$	$7{,}8\cdot10^{-4}$	$3{,}6\cdot10^{-4}$

As seen from table 1, the decrease in the radius of the phase ring and amplitude ring reduces the size of the focal spot, which is accompanied by the growth of side lobes. The lowest value was obtained when $r_1 = 0.8$ mm, $r_2 = 0.9$ mm and made FWHM = 0.014 mm, it is better theoretical value 2.3 times. However, in this case, the intensity value is 0.036% of the peak without using a aperture. It looks acceptable result for $r_1 = 0.6$ mm, $r_2 = 0.8$ mm. Then, the intensity value is about 0.6% of the maximum value when the size of focal spot of 0.0159 mm.

On figure 2 is a graph of intensity for the lowest value of FWHM.

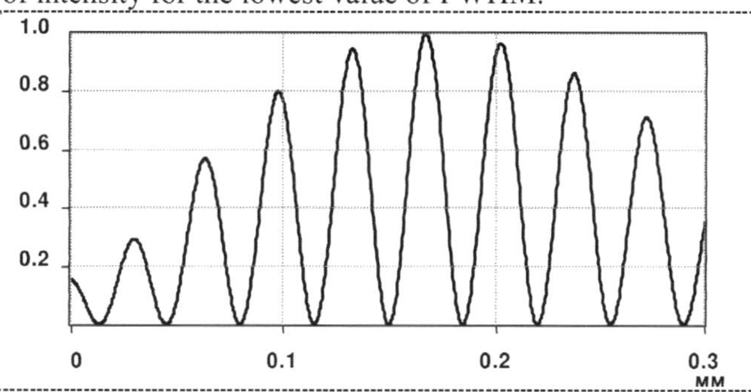

Figure 2. Graph intensities (normalized) for the case of $r_1 = 0{,}8$ mm, $r_2 = 0{,}9$ mm.

As seen from the graph, the total intensity of the picture there are significant side lobes, yet they are commensurate with the height of the central peak, in contrast to super-oscillating functions [10].

3. Conclusion

Thus, in the numerically shown that the focus is uniformly polarized plane wave at a low numerical aperture lens can be used the annular diaphragm with the phase jump. Thus it was possible to achieve reduction of the size of the focal spot is 2.3 times as compared with the theoretical value (the minimum value of FWHM = 0,014 mm) but in this case an increase in the intensity of the side lobe in a significant drop in intensity at the focus.

Acknowledgments

The work was financially supported by the Russian Foundation for Basic Research (grant 14-07-31079 mol_a) and by the Ministry of Education and Science of Russian Federation.

4. References

[1] Sheppard C J R 2004 *Appl. Opt.* **43** 4322
[2] Wang H, Shi L, Lukyanchuk B, Sheppard C, C T Chong 2008 *Nat. Photonics* **2** 501
[3] Khonina S N, Volotovsky S G 2010 *Comp. Opt.* **34** 58
[4] Khonina S N 2013 *Opt. Engineering* **52** 091711
[5] Toraldo di Francia G 1969 *J. Opt. Soc. Am.* **59** 799
[6] Khonina S N, Ustinov A V 2012 *Comp. Opt.* **36** 219
[7] Khonina S N, Ustinov A V 2013 *J. of Engineering (Hindawi Publishing Corporation)* ID 512971, 8
[8] Khonina S N, Savelyev D A 2013 *J. of Experimental and Theoretical Physics* **117** 623
[9] Khonina S N, Karpeev S V, Alferov S V, Savelyev D A, Laukkanen J, Turunen J 2013 *J. of Optics* **15** 085704
10. Huang F M, Zheludev N I 2009 *Nano Lett.* **9** 1249

Design of a new multiapertured diffraction interferometer scheme for small angular displacement control

Yu Sokolov[1]
[1] ITMO University, 49, Kronverksky Pr., St. Petersburg, 197101, Russia

E-mail: sokolov-juri@yandex.ru

Abstract. The compact diffraction circuit controlling small angular movements with small errors (2 ") and implementation simplicity is considered. The mathematical model describing the radiation intensity distribution in a receiver is obtained. The experiment results (at TU Ilmenau, Germany) confirmed the basic theoretical principles described in the paper.

1. Introduction

At present, various optoelectronic devices and sensors are widely used for positioning and carrying out angular measurements. The bitmap sensors and autocollimators are used to solve many angular measurement problems [1]. The advantages of photoelectric autocollimators are their small measurement errors (0.01 - 1)" [2,3], as well as high speed and the possibility to automate the measurements. Their disadvantages are high cost and large dimensions of the devices, as it is necessary to use long lenses f = (300-1000) mm. Compared to them, the coherent optical diagrams for controlling angular values are characterized by the same accuracy and greater construction simplicity.

But there are some drawbacks limiting their application areas: significantly smaller measuring range with comparable sensitivity (diffraction control circuit) [4]; structure complexity with large measuring range (interferometric control circuits) [1]; specific construction making autocollimation measurements difficult and having high sensitivity and large measurement range used as an angle measurement circuit sensor [5]. In this connection, the development and research of new control diffraction methods without the aforementioned drawbacks is of interest.

The scheme of measuring small angular displacements is researched in this paper. It is based on the analysis of the interference pattern formed by a multi-aperturated diffraction interferometer. Unlike other interferometer circuits dividing the grating interferometer radiation beam, the interfering wave fronts are formed by diffraction elements, such as slotted apertures. This approach simplifies the interferometer construction and its alignment reducing its size and price.

The scheme researched in paper is presented in figure 1. Here, the plane-parallel glass plate (2) with a reflective coating, in which the same rectangular holes are made, is mounted in front of the lens (1). Using an autocollimating mirror (3), the plate is illuminated by an oblique collimated radiation beam from a laser source (4). The interference pattern superposition from each of three pairs of slotted apertures and the diffraction pattern from a single aperture is observed in the receiver plane (5). Sensitivity to the autocollimation mirror rotation angle at all points in one interference pattern is constant. At the same time, sensitivity of the interference pattern formed by a pair of axial apertures (having zero lateral displacement from the vertical dash-dotted line $v1 = v2 = 0$ mm) is minimal. It is

[1] To whom any correspondence should be addressed.

determined by the distance between the apertures along the dash-dotted line and the incidence radiation angle. Sensitivity of the off-axis interference pattern (from a pair of apertures positioned on a sloping dash-dotted line with the value $v1 = 0$; $v4 \neq 0$) is increased with the aperture lateral offset from the system axis $v4$ (figure 1). It is not dependent on the incidence radiation angle and the prism-plate distance (6). Having a higher sensitivity and measurement range, the circuit is more compact compared to that without any prism and with one pair of apertures researched in [6].

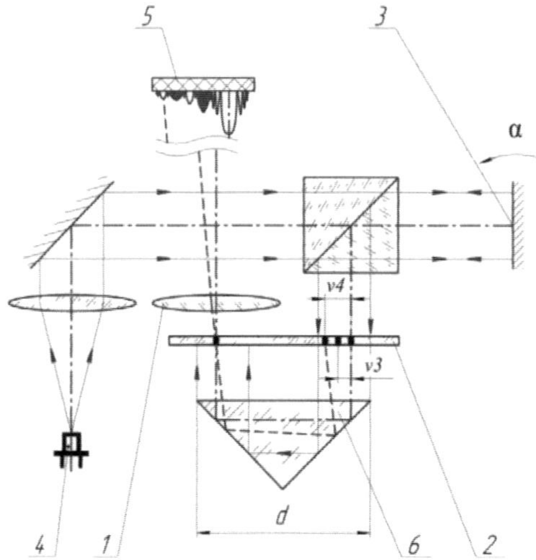

Figure 1. The multiaperture diffraction autocollimator circuit.

The mathematical expression describing two-dimensional radiation distribution in the thin lens focal plane at collimated radiation beam diffraction in several apertures disposed on either side of a plane-parallel glass plate, which thickness is equal to the axial ray stroke geometric length in the prism d (figure 1), has been derived in studying the circuit and assessing its sensitivity. The engineering calculation method of the basic circuit parameters has been developed to evaluate the measurement range of the diffraction autocollimator and its dimensions.

As it is shown by the calculation results, the scheme resolution capability is about 1.5"; the measurement range varies from -2000" to +3000" with the circuit's overall dimensions equal to (200 x 135) mm with the beam radiation diameter $D_{cl.ap.} = 27{,}5$ mm; the wavelength $\lambda = 0.405$ micron; the controlled mirror distance $L_{meas.} = 200$ mm; the beam splitter width $D_{b.s.} \approx 43$ mm; the axial ray stroke length in the prism $d \approx 65$ mm; the aperture sizes $a = 11$ micron; the off-axis aperture maximum displacement $v4 = 8$ mm. and the lens focus $f \approx 106$ mm.

2. Experimental

To verify the obtained theoretical propositions and formulas, two glass masks with different sizes and mutual aperture arrangements («Compugraphics Jena GmbH») were developed and the experimental installation was assembled at Ilmenau Technical University. The EVS VEI-535 color camera with 2.775 × 2.775 mm. pixel size operating in monochromatic mode was used during the experiments. The photos of interference patterns from the camera and its parameter adjustment were carried out automatically in the MatLAB medium. The sequential two-shot recording with different levels of maximum exposure was taken to neutralize the interference pattern illumination (due to the intensity difference between the diffraction pattern principal and minor maximums) for each mirror rotation angle. The mirror rotation angle was monitored with the help of the Elcomat 3000 autocollimator with 0.1" resolution capability.

The incidence radiation angle (λ = 0.65) effect on the interference patterns was researched at three different plate-prism distances and various transverse distances between the apertures $v3$, $v4$ (figure 1) for the mirror rotation angles from 0 to 2000" in increments ~ 1.7".

3. Discussion

The examples of intensity oscillations graphs, which are obtained from the central point of each interference pattern, are shown in figure 2. The experimental results show that the mathematical model adequately defines the sensitivity of the interference patterns formed by different apertures pairs.

Determination of the mirror rotation angle from the obtained photographs is carried out with a proprietary algorithm. The maximum modulo deviation obtained in the experiment from the autocollimator readings is 2.2" for most sensitive third interference pattern (figure 2, curved line (c), the standard deviation is 1.2".

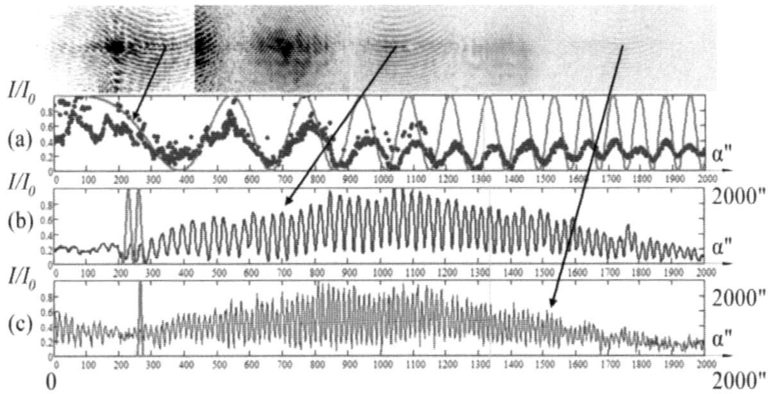

Figure 2. Experimental diffraction pattern with three interference patterns from three pairs of slits (50 x 100) μm (Figure 1.).

4. Conclusions

The study results proved the possibility of developing a new means of small angular value measurements having higher compactness (200 x 135) mm. and the measurement range [-2000"; + 3000"] in comparison with the photoelectric autocollimator with comparable measurement error (1-1.5)". To verify the theoretical propositions, the measuring bench was assembled and the experimental researches were carried out in the laboratory of TU Ilmenau (Germany). The error obtained experimentally was about 2.2". The obtained metrological parameters are not ultimate and they can be made better by improving the method of processing the interference fringes and/or modifying the circuit parameters (for example, by compensation of intensity difference between principal and minor diffraction maximums with the aid of gradient filter). Further researches are directed at developing a compact two-dimensional diffraction autocollimator with improved metrological parameters.

References

[1] Anikst D A, Konstantinovich K M and Mes'kin I V 1987 *Vysokotochnyye Uglovyye Izmereniya* (Moskva, Mashinostroyeniye) (in Russian)
[2] 2014 *The MOELLER-WEDEL Prospectus* (Germany)
[3] 2014 *The TRIOPTICS Prospectus* (Germany)
[4] Nazarov V N and Linkov A E 2002 Diffraction methods for monitoring the geometrical parameters and spatial position of objects *J. Opt. Technol.* **69** 129–34
[5] Arefiev A A, Ivanov A and Kotenok A 1996 Interferometric devices for angular measurements *SPIE* **2775** 189–96
[6] Nazarov V N and Sokolov Yu A 2013 New method of diffraction monitoring of small angular and linear displacements *J. Opt. Technol.* **80** 756–60

Monitoring of quantum mode correlation functions in the presence of pointer state phase relaxation

E Trifanova and A Trifanov
ITMO University, Kronverksky 49, Saint Petersburg, 197101, Russia

E-mail: etrifanova@gmail.com

Abstract. In present work we investigate the process of interaction between two-level atom-pointer and cavity mode to obtain correlation functions of mode quantum state. In particular, we analyze the protocol of indirect photodetection which allow estimations for average values of photon number operator in any order using statistics of atom state detector clicks.

1. Introduction

Weak and indirect measurements are widely used in quantum physics for extracting information about the quantum system with minimum perturbation of its state [1]. The indirect photodetection process of quantum resonator mode implies organization of interaction between this mode and two level atom, which is passed through the cavity. This process is followed by atom state detection. Repetition of this process several times and collection of detector statistical sampling allow to estimate the statistical properties of cavity field quantum state [2].

Actually each finite quantum system is open. The atom and the electromagnetic field interact with the environment and it leads to relaxation of their common state. In connection to the cavity, in our case, it influence the evolution of monitored object (cavity mode). However state relaxation of the atom-pointer leads to distortions of the detection results [3, 4]. Finally it gives an erroneous statistical properties estimation of the mode. Therefore it is highly useful to study detector's nonideality, originated from environment influence on atomic state, and to take this into account in the statistics processing of the atomic state detectors counts.

In this work we present a method for describing the evolution of cavity mode, conditioned by atomic state measurement result in the case of indirect photodetection protocol. Assuming weak excited state relaxation of the atom compared to decoherence of its state in resonance conditions we obtain analytical expressions for superoperators of cavity mode conditional evolution. These allows us to show that averaged values of photon number operator and it grades are proportional to time derivatives of probability to detect atom in its ground state.

2. The model

Let us describe the common evolution of two level atom and cavity mode of electromagnetic field in the presence of environment which influences on the atom and follows to decoherention. In this case the master equation for density matrix $\rho(t)$ has the following (Lindblad) form:

$$i\frac{d}{dt}\rho=\left[H_{\text{int}},\rho\right]+iD_A\rho, \tag{1}$$

where H_{int} is interaction Jaynes - Cummings Hamiltonian in interaction picture:

$$H_{\text{int}} = \Omega\sigma_{+}a\exp(i\Delta t) + \Omega^{*}\sigma_{-}a^{\dagger}\exp(-i\Delta t), \tag{2}$$

D_A is the atomic dissipater characterized the interaction of the atom with the external modes:

$$D_A\rho = \frac{\gamma_{ge}}{2}\left(2\sigma_{-}\rho\sigma_{+} - \sigma_{+}\sigma_{-}\rho - \rho\sigma_{+}\sigma_{-}\right) + \frac{\gamma_{ge}}{2}(c.c.), \tag{3}$$

Ω is the Raby frequency of atomic transition; $\sigma_{+} = |e\rangle\langle g|$, $\sigma_{-} = |g\rangle\langle e|$ are atomic operators, a is photon annihilation operator in the cavity, Δ is atom-field detuning, γ_{ge}, γ_{eg} are coefficients of spontaneous decay and excitation, respectively. For convenience we rewrite Eq. (3) to the from:

$$\begin{aligned} \left(D_A\rho\right)_{gg} &= \gamma_{eg}\rho_{ee} - \gamma_{ge}\rho_{gg}, & \left(D_A\rho\right)_{ge} &= -\rho_{ge}\Gamma, \\ \left(D_A\rho\right)_{ee} &= \gamma_{ge}\rho_{gg} - \gamma_{eg}\rho_{ee}, & \left(D_A\rho\right)_{eg} &= -\rho_{eg}\Gamma, \end{aligned} \tag{5}$$

where Γ has the meaning of atomic phase relaxation (decoherence) rate.

To solve Eq.(1) we represent the density matrix as a decomposition:

$$\rho(t) = \sum_{\mu,\nu\in\{g,e\}} |\mu\rangle\langle\nu| \otimes \rho_{\mu\nu}(t) \tag{4}$$

and assume $\Gamma \gg \Omega$, which is equivalent to condition that coherence relaxation time is much shorter than the typical evolution time of the system [4]. From here it is easy to show that the following approach is applicable:

$$\frac{d}{dt}\left(\rho_{ge}e^{i\Delta t}\right) \approx \frac{d}{dt}\left(\rho_{eg}e^{-i\Delta t}\right) \approx 0, \tag{6}$$

and system (1) reduces to the closed form for diagonal elements of density matrix in the basis of atomic states:

$$\begin{aligned} \dot{\rho}_{gg} &= \kappa\Gamma\left(2a^{\dagger}\rho_{ee}a - \rho_{gg}a^{\dagger}a - a^{\dagger}a\rho_{gg}\right) - i\kappa\Delta\left[a^{\dagger}a, \rho_{gg}\right] + \gamma_{eg}\rho_{ee} - \gamma_{ge}\rho_{gg}, \\ \dot{\rho}_{ee} &= \kappa\Gamma\left(2a\rho_{ee}a^{\dagger} - \rho_{gg}aa^{\dagger} - aa^{\dagger}\rho_{gg}\right) - i\kappa\Delta\left[a^{\dagger}a, \rho_{ee}\right] + \gamma_{ge}\rho_{gg} - \gamma_{eg}\rho_{ee}. \end{aligned} \tag{7}$$

Here $\kappa = |\Omega|^2/\left(\Gamma^2 + \Delta^2\right)$. We will solve this system in *superoperator form*. To this end we introduce the following superoperators:

$$2K_0\rho_F = a^{\dagger}a\rho_F + \rho_F aa^{\dagger}, \; K_{+}\rho_F = a^{\dagger}\rho_F a, \; K_{-}\rho_F = a\rho_F a^{\dagger}, \; N\rho_F = \left[a^{\dagger}a, \rho_F\right], \tag{8}$$

which act on the mode state space and form SU(1,1) algebra. The solution of Eq.(7) may be found in formalism of evolution superoperator. To obtain this we write:

$$\rho(0) = U(t)\rho(0) = \sum_{\mu,\nu\in\{g,e\}} M_{\mu,\nu}(t) \otimes |\mu\rangle\rangle\langle\langle\nu|\rho(0), \tag{9}$$

where we use notations $|\mu\rangle\rangle\langle\langle\mu|\rho = |\mu\rangle\langle\nu|\rho|\nu\rangle\langle\mu|$ and $M_{\mu\nu}$ is conditional evolution superoperator of mode state corresponding to the case then atom prepared in state $|\mu\rangle$ was detected in state $|\nu\rangle$. Rewriting (7) with (8) and (9) in use the following system of differential equation appears:

$$\begin{cases} \dfrac{d}{dt}M_{gg} = -\left(\alpha K_0 + \beta + \gamma_{ge}\right)M_{gg} + \left(\alpha K_{+} + \gamma_{eg}\right)M_{eg}, \\ \dfrac{d}{dt}M_{eg} = -\left(\alpha K_0 - \beta + \gamma_{eg}\right)M_{eg} + \left(\alpha K_{-} + \gamma_{ge}\right)M_{gg}. \end{cases} \tag{10}$$

Here we use the notations $\alpha = 2\kappa\Gamma$, $\beta = -\kappa(i\Delta N + \Gamma)$.

3. Analytical estimations

Assume resonance interaction between atom and cavity mode, that is $\Delta = 0$, and absence of spontaneous relaxation and excitation at the considered time scale, that means $\gamma_{ge} = \gamma_{eg} = 0$. In this case the solution of system (10) can be found in analytical form. Let write system (10) in symmetric form using the substitution $M_{gg} = \exp[-(\alpha K_0 + \beta)t]M'_{gg}$, $M_{eg} = \exp[-(\alpha K_0 - \beta)t]M'_{eg}$. Then

$$\begin{cases} \dfrac{d}{dt}M'_{gg} = \alpha e^{2\beta t} e^{\alpha K_0 t} K_+ e^{-\alpha K_0 t} M'_{eg}, \\ \dfrac{d}{dt}M'_{eg} = \alpha e^{-2\beta t} e^{\alpha K_0 t} K_- e^{-\alpha K_0 t} M'_{gg}. \end{cases} \tag{11}$$

Using operator equalities $e^{\alpha K_0 t} K_+ e^{-\alpha K_0 t} = e^{\alpha t} K_+$, $e^{\alpha K_0 t} K_- e^{-\alpha K_0 t} = e^{-\alpha t} K_-$ [6], and condition $\alpha + 2\beta = 0$ we obtain

$$\begin{cases} \dfrac{d}{dt}M'_{gg} = \alpha K_+ M'_{eg}, \\ \dfrac{d}{dt}M'_{eg} = \alpha K_- M'_{gg}. \end{cases} \tag{12}$$

Superoperator M'_{gg} satisfies the differential equation

$$\frac{d^2}{dt^2}M'_{gg} - \alpha^2 K_+ K_- M'_{gg} = 0, \quad M'_{gg}(0) = I,\ M'_{eg}(0) = 0. \tag{13}$$

The solution of this Cauchy problem (13) is superoperator

$$M'_{gg} = \cosh \alpha\sqrt{K_+ K_-}\,t, \tag{14}$$

wherefrom

$$M_{gg} = e^{-(\alpha K_0 + \beta)t} \cosh \alpha\sqrt{K_+ K_-}\,t. \tag{15}$$

To find the averaged value of photon number operator lets expand (15) in series in powers of t:

$$M_{gg} = 1 - (\alpha K_0 + \beta)t + \frac{1}{2}\left(\alpha^2 K_+ K_- + \alpha^2 K_0^2 + 2\alpha\beta K_0 + \beta^2\right)t^2 + \dots. \tag{16}$$

Let us find the probability of detecting the atom in the ground state if it was prepared in ground state. From (14) we get

$$P_{gg}(t) = \langle M_{gg}(t)\rangle = \mathrm{Tr}(M_{gg}\rho) = 1 - \alpha\langle a^\dagger a\rangle t + \alpha^2\left\langle (a^\dagger a)^2\right\rangle t^2 + \dots. \tag{17}$$

Hence we can write the expressions for photon numbers operator and its square:

$$\langle a^\dagger a\rangle = -\frac{1}{\alpha}\frac{d}{dt}P_{gg}(t)\Big|_{t=0}, \quad \left\langle (a^\dagger a)^2\right\rangle = \frac{1}{2\alpha^2}\frac{d^2}{dt^2}P_{gg}(t)\Big|_{t=0}. \tag{18}$$

4. Discussion

Figure 1 shows the dynamics of the average photon number for initial Fock state with $n = 2$ (a) and coherent state with amplitude equals to $\sqrt{2}$ (b). The black horizontal solid line in both cases corresponds to the analytical estimation (18) of the average photon number at the time of photodetection. Calculations were made for different values of the coefficient of spontaneous excitation (γ_{eg}, first number in the legend) and decay (γ_{ge}, second number in the legend). We can see that curves are grouped in threes according to the same value of γ_{ge}. In the case of the initial Fock state (a) the curves are grouped in pairs with the same value of the difference $\gamma_{ge} - \gamma_{eg}$.

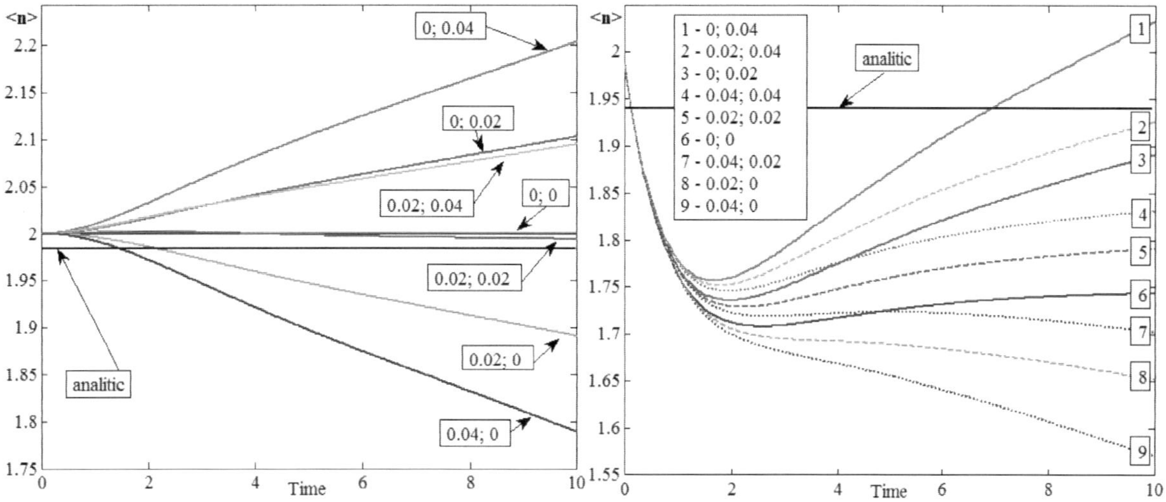

Figure 1. Average photon number for initial Fock state (a) and coherent state (b) of cavity mode. Black horizontal line coresponds to analytical estimation (18).

The conditional number of intracavity photons is cruicially depends on relation between spontaneous excitation and relaxation rates. For the case when the first process dominates we observe increasing in time number of photons while if the second process dominates the number of cavity photons is decreased. The character of dynamics is different in presented cases and this is a problem for the following research. Analytical estimation of mean photon number in the cavity is in good agreement with numerical simulations. Small difference between initial value and observed result may rise from approximate calculation of time derivative of probability and restricted basis of mode state by ten photons.

In conclusion, we suggest a protocol of mode statistical properties monitoring with atom pointer in use. Here we acount phase relaxation of the pointer and obtain conditional dynamics of intracavity mode in this case. Finaly, using exact solution for field state transformers we got analytical expressions for some statistical properties of cavity mode in the case of imperfect measurement procedure.

Acknowledgements
This work was partially financially supported by the Government of the Russian Federation (grant 074-U01), by the Ministry of Education and Science of the Russian Federation (GOSZADANIE 2014/190, Projects No 14.Z50.31.0031 and No. 1.754.2014/K)), by grant MK-5001.2015.1 of the President of the Russian Federation, by ITMO University from centralized resources (grant 415829).

[1] Braginsky B and Khalili F Ya 1992 *Quantum Measurement* (Cambridge) 191 p
[2] Briegel H-J et al. 1994 *Phys Rev A* **49** (4) 2962-2985
[3] Miroshnichenko G P and Trifanov A I 2014 *Phys Scr* **160** 014042(4).
[4] Trifanov A I and Miroshnichenko G P 2013 *Nanosystems: Phys Chem Math* **4** (5) 635–647

Calculation of polarization sensitivity of image sensors

A. Trushkina[1], **V. Ryzhova, V. Korotaev**
Production Editor, *Journal of Physics: Conference Series*,
ITMO University, St. Petersburg

E-mail: trushkens@gmail.com

Abstract. The work deals with the analysis of effect of image sensors structure on the distribution of polarization on its area. It is also about polarization sensitivity calculation with the method based on Jones matrix formalism.

1. Introduction

CCD and CMOS detectors have many applications. Such sensors are in particular used for measurements, but because of complexity of their structure many factors can decrease its accuracy. First of all, many researches of effect of signal-noise ratio on the accuracy are known [1-4]. Relative orientation of polarizers in polarization CCD camera also exerts a strong influence on received signal [5, 6]. As for polarization sensitivity, there are only researches of SeGA detectors known [7]. But polarization sensitivity of general familiar sensors of quartz is no less important.

CCD or CMOS sensors are usually made of isotropic materials, so they are considered not to be sensitive to polarization. However, when a detector is disposed at an angle to the axis of incident light beam, it becomes polarization sensitive. It can be critical in precise measurement performed in polarized light.

Therefore, the main purpose of this research is to find out, how does the polarization sensitivity of specific photosensors affect the result of polarimetric and ellipsometric measurements.

2. Theoretical aspects

When a photosensitive element is disposed at an angle to the axis of incident light beam, the transmittance of boundary between electrode end detector's crystal is defined by

$$T = \frac{n_2 \cos \varepsilon''}{n_1 \cos \varepsilon}\left(t_s^2 \cos^2 \alpha_{in} + t_p^2 \sin^2 \alpha_{in}\right) = T_s^2 \cos^2 \alpha_{in} + T_p^2 \sin^2 \alpha_{in} \quad (1)$$

Here, ε is the incidence angle, ε'' is the refraction angle, T_s and T_p are transmission coefficients for light which is polarized parallel and perpendicular to incidence plane, α_{in} is an azimuth of incident polarization, t_s are t_p the transmission coefficients of orthogonal Jones vector's components, which are given by

$$t_s = \frac{2n_1 \cos \varepsilon}{n_1 \cos \varepsilon + n_2 \cos \varepsilon''}, t_p = \frac{2n_1 \cos \varepsilon}{n_2 \cos \varepsilon + n_1 \cos \varepsilon''} \quad (2)$$

[1] To whom any correspondence should be addressed.

Photocurrent is defined by

$$i(\lambda,\varepsilon,\alpha_{in}) = \Phi\frac{e\eta\beta\lambda}{hc}\left[T_p(\varepsilon)\sin^2\alpha_{in} + T_s(\varepsilon)\cos^2\alpha_{in}\right] \quad (3)$$

where e is the electron charge, η is sensors's quantum efficiency, β is the coefficient of separation of charge carriers couples, h is the Planck constant, c is the speed of light in vacuum, λ is the wavelength.

So the sensitivity of sensors is characterized by two values: to light, polarized parallel (s_p) and perpendicular (s_s) to incidence plane. Normalized dependences of sensitivity on incidence angle are calculated with the following equations:

$$s_p(\varepsilon) = \frac{n_2\cos\varepsilon''}{n_1\cos\varepsilon}t_p^2, s_s(\varepsilon) = \frac{n_2\cos\varepsilon''}{n_1\cos\varepsilon}t_s^2 \quad (4)$$

The main parameter of polarization sensitive sensor is the effective polarization sensitivity, which is the difference between p- and s- components. The signal's amplitude is proportional to this value, thus it depends on refraction and absorbtion indices of photodetector crystal, on quantum efficiency of detector, on incidence angle and light wavelength. Consequently it is advisable to consider the changing of signal magnitude owing to variation of detector orientation. For this purpose it is necessary to calculate the distribution of polarization sensitivity as a function of the light incident angle at each pixel.

3. Method of calculation

The suggested method of calculation is based on well-known Jones matrix formalism [8]. Each pixel of sensor array is concepted as a complex of boundaries between isotropic media. When passing through each one, the light is refracted, so the wave ort changes its orientation, namely it rotates at an angle α_i. To get to new coordinate system, the boundary Jones matrix should be multiplied at the rotation matrix. Since a pixel is concepted as it is shown on the Fig. 1, its Jones matrix is given by

$$\mathbf{M}_{MPD} = \mathbf{M}_R(\alpha_5)\mathbf{M}_4\mathbf{M}_R(\alpha_4)\mathbf{M}_3\mathbf{M}_R(\alpha_3)\mathbf{M}_2\mathbf{M}_R(\alpha_2)\mathbf{M}_1\mathbf{M}_R(\alpha_1) \quad (5)$$

This equation contains matrices between following boundaries:

- $\mathbf{M}_1$ – air-microlens;
- $\mathbf{M}_2$ – microlens-color filter (red, green or blue);
- $\mathbf{M}_3$ – color filter-electrode (Si:H);
- $\mathbf{M}_4$ – electrode-dielectric (SiO_2);
- $\mathbf{M}_5$ – dielectric-photosensor (Si (p/n)).

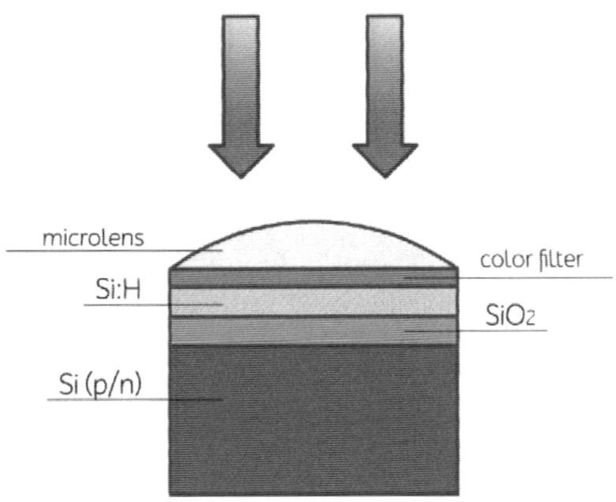

Figure 1. Pixel structure

Variation of this angle changes the transmittance of detectors structure, which accordingly causes the rotation of polarization plane. The value of this rotation is given by

$$\Delta\alpha_{out} = \arctan\frac{\tan(\alpha_1 - \alpha_{in})[1 - (t_p/t_s)]}{1 + (t_p/t_s)\tan^2(\alpha_1 - \alpha_{in})} \quad (6)$$

So, to perform the calculation, it's necessary to know the orientation of coordinate system, in which the light is described, and refraction indices of each aforesaid material. Indices of color filter were calculated from sensitivity specters of researched sensor OV5620 (Fig. 2) and sensitivity spectrum of the same sensor without color filters (Fig. 3) [9, 10].

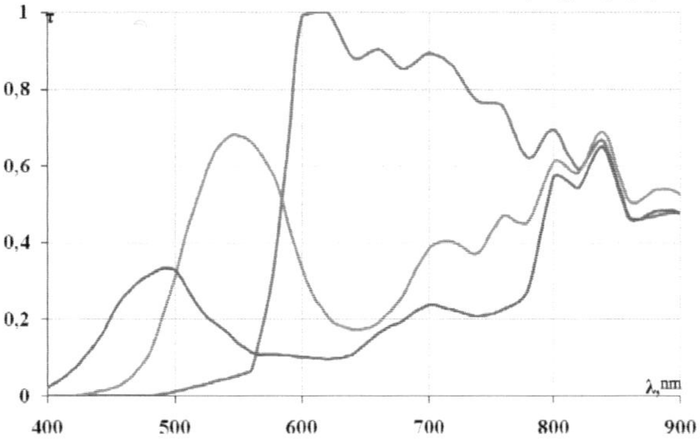

Figure 2. Sensitivity spectrum of OV5620 sensor with color filters

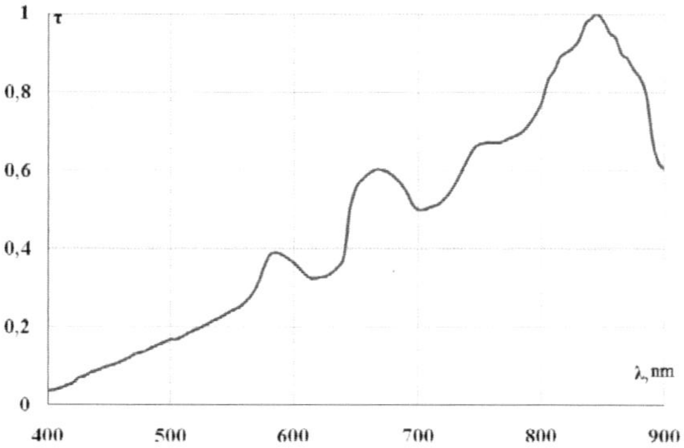

Figure 3. Sensitivity spectrum of OV5620 sensor without color filters

As a result, transmission spectres of filter look like it is shown at the Fig. 4.

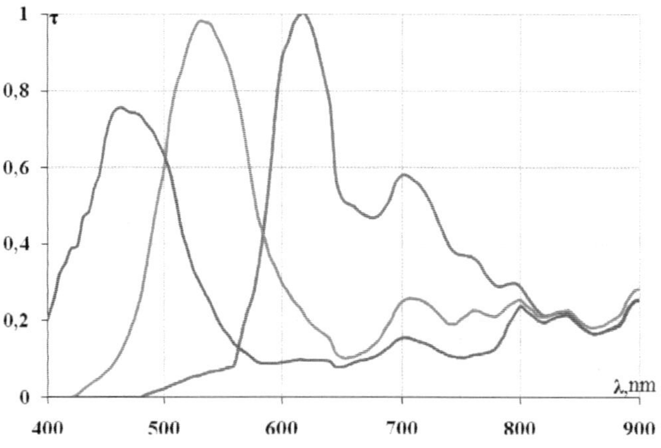

Figure 4. Transmission of color filters

The transmissions coefficients for λ = 532 nm, are: τ_R = 0,057165; τ_G = 0,9813; τ_B = 0,30062. Refraction indices were calculated using Fresnel equations: n_R = 33,987; n_G = 1,038; n_B = 5,653.

4. Results of calculations

The software for calculation was developed in LabVIEW environment. The results of calculations for λ = 532 nm as a functions $\Delta\alpha_{out}(\alpha_{in})$ and $\Delta\alpha_{out}(\varepsilon)$ will be shown at Fig. 5 and 6.

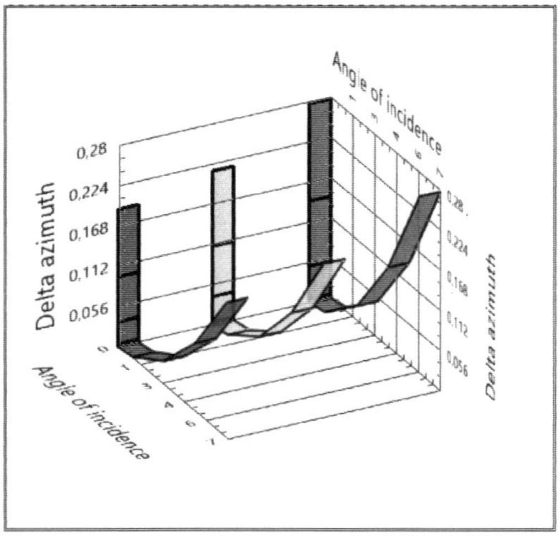

Figure 5. Changing of polarization azimuth depending on the incidence angle ($\alpha_{in} = 45^o$)

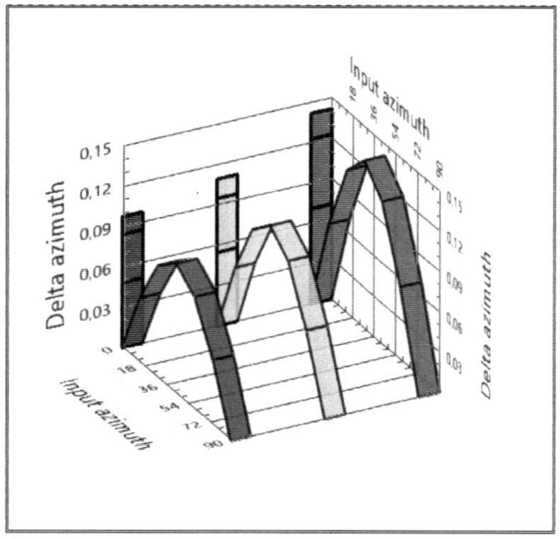

Figure 6. Changing of polarization azimuth depending on the input azimuth ($\varepsilon = 5^o$)

Since the sensitive array is illuminated with parallel light beam, the distribution of output characteristics is uniform and depends only on transmittance of color filters. So, the value of $\Delta\alpha_{out}$ is the same for each red, green or blue pixel.

The *p*- and *s*- components of polarization sensitivity are calculated by equations (4). The dependences $s_p(\varepsilon)$ and $s_s(\varepsilon)$ with $\varepsilon = 0\text{-}7^o$ are shown at the Fig. 7.

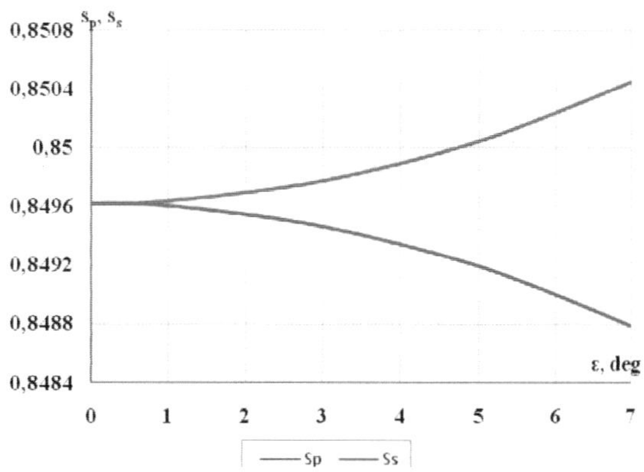

Figure 7. Components of polarization sensitivity

5. Experimental research

Experimental setup for validating the calculation was built by classical Stokes polarimeter scheme [11, 12]. It is shown at the Fig. 8.

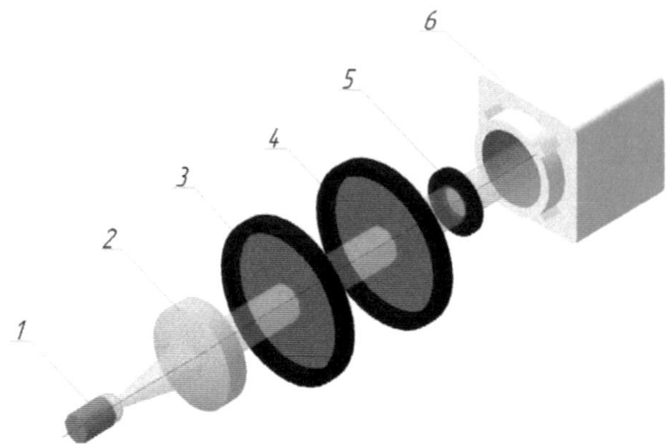

Figure 8. Experimental setup: 1 – LED (λ = 532 nm), 2 – collimating lens, 3 – polarizer, 4 – analyser, 5 – waveplate λ/4, 6 – camera with researched sensor

The comparison of theoretical and experimental results is shown at Fig. 9. Theoretical data are shown by solid lines, experimental data – by dashed lines.

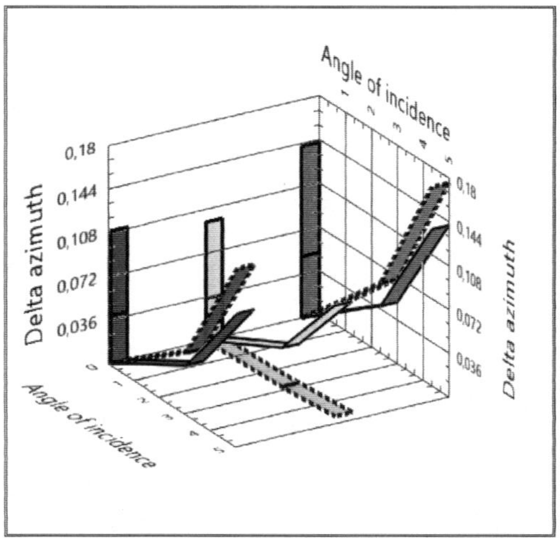

Figure 9. Changing of azimuth gotten by theoretical and experimental methods

This char shows, that the theoretical results for red end green pixel are correct within the error, and the experimental and theoretical results for green pixel are quite different. It is associated with experiment conditions and meathematical model validity. So, it is necessary to improve a mathematical model. Then we will get better results.

6. Conclusions

So, the program for calculation how the structure of detectors affects the polarizaton distribution on its sensitive area was developed basing on the Jones matrix formalism. The obtained results can be used for metrology attestation of specific CCD or CMOS detectors from the viewpoint of polarization sensitivity and also for considering the effect of photodetector on measurements results. It would improve measurements accuracy by its performing in polarized light.

This work was partially financially supported by Government of Russian Federation, Grant 074-U01.

7. References

[1] Chen L, Zhang X, Fedler F, Lin J, Sha D 2009 Signal-to-noise ratio evaluation of a CCD camera *J. Optics & Laser Technology* **41** 574-579

[2] Wang D, Zhang T, Kuang VH 2012 Relationship between the charge-coupled device signal-to-noise ratio and dynamic range with respect to the analog gain *J. Appl. Opt.* **51** 7103-7114

[3] Weiwei F, Yanjun J, Ligang C 2013 The impact of signal–noise ratio on degree of linear polarization measurement *Optik - International Journal for Light and Electron Optics* **124(3)** 192-194

[4] Tyo J S 2002 Design of optimal polarimeters: Maximization of signal-to-noise ratio and minimization of systematic error *J. Appl. Opt.* **41** 619–630

[5] Feng W and Chen L 2010 Relative orientation accuracy analysis of the polarizers in a polarization CCD camera *Optik* **121** 1401-1404

[6] Feng W and Chen L 2010 The impact of the orientation angles uncertainty of instrument polarizers on polarization measurement accuracy *Optik.* **121** 2276-2279

[7] Millera D, Chester A, Moellera V, Starostaa K, Vamana C, Weisshaara D 2007 Linear polarization sensitivity of SeGA detectors *Nuclear Instruments and Methods in Physics Research.* **A 581** 713-718

[8] Панков Э.Д., Коротаев В.В. 1992 Поляризационные угломеры *М.: Недра* 240

[9] Datasheet OV5610 Color CMOS QSXGA (5.17 MPixel) CameraChip with OmniPixel Technology

[10] Müllerová J, S. Jurečka, Šutta 2006 Optical characterization of polysilicon thin films for solar applications *Solar Renewable Energy News Conference* **80(6)** 667-674

[11] Prokopyeva E., Tománek P., Kocová L., Palai-Dany T., Balík Z., Škarvada P., Grmela L 2013 Comparison of optical and electrical investigations of meat ageing *J. Optical Sensors* **8774**

[12] Azzam R M A 1978 Photopolarimetric measurement of the Mueller matrix by Fourier analysis of a single detected signal *J. Optics letters.* **2(6)**

Laser forming micro geometric structures on the surface of roller rolling mill

O S Vasilyev [1 2], J S Ruzankina [1]
[1] ITMO University, 49 Kronverksky Pr., Saint-Petersburg, Russia
[2] Laser centre Ltd, Saint-Petersburg, Russia

E-mail: tesla_90@mail.ru

Abstract. This paper presents a method of metal surface microstructuring by means of radiation of an impulse fiber laser with a scanning system, used for the displacement of the light beam. The topographic relief being presented was modeled in the graphics editor and then was manufactured on the experimental material. We also created a "parameter matrix", which allows to determine the optimal modes of operation of the laser system for treatment of a specific material. The paper describes three stages of the microstructuring process: profiling, cleaning, and polishing, and the corresponding geometrical parameters of the structures manufactured. A method for decreasing the level of relief roughness (for Ra and Rz) was developed.

1. Introduction

It is well known that the physical and chemical properties (mechanical, chemical, optical, etc.) of the surface are largely determined by its microgeometry [1]. Engineering, biomedicine, decorative and applied arts and other branches of industry require microstructuring technology; because it allows to improve certain performance characteristics of the surface (durability, friction factor, adhesiveness, corrosion resistance, electric endurance, etc.). Laser microprocessing is a relatively new method of surface microrelief formation, which is researched and used worldwide in a variety of industries [2]. However, most studies only solve the problem of optimization of a particular surface to solve a specific problem [3]. Besides the solution of each particular problem requires a full cycle of time-consuming research. It seems more promising to us to establish universal regularities of formation of various regular reliefs using laser radiation, which would greatly reduce the time, required for experimental formation of a certain microgeometry. In this paper, we focus on the formation of microstructures of regular geometric shape, with typical values of altitude and lateral settings. [4] The aim of this work is to demonstrate the possibility of creating microgeometric structures on the metal surface and to optimize laser modes, used for their formation. As a result, an interrelation between the parameters of laser treatment and geometrical characteristics of the surface structures is established.

To create surface structures we used a method of laser micro-ablation. [5] When this process is repeated many times, a system of depressions or grooves is formed on the surface, the shape and relative position of which can be changed programmatically. [6]

2. Experimental set.

The experiment was conducted using a high-precision laser marking system with impulse fiber laser and scanning light beam guidance system.

Table 1. Main specifications of the laser

Parameters	Value
Ytterbium fiber laser	
Wavelength (λ)	1064 nm
Average power (P_m)	up to 20 W
Maximum speed (V_{max})	2000 mm/s
Pulse repetition frequency (f)	20-100 KHz
Pulse length (τ)	≤ 200 ns
Beam diameter at the focus	~30 μm
Pulse energy	1 mJ
Processing field	100x100 mm

Initial and direct purpose of the selected laser system is laser marking of metal surfaces. However, its design is universal: the optical fiber supplies laser radiation to the processing area; catadioptric optical system focuses the laser beam onto any area of the processing field; additional vertical positioning system allows to process various surfaces of the item. Mentioned technical characteristics make the system in question multifunctional – the list of its technological capabilities includes marking and cutting of metal, application of oxidation coating, hole cutting and, lastly, the three-axis microstructuring, described in the present work.

3. Creation of a graphic model.

To create graphic models we used the ArtCam software. This software creates a three-dimensional image of the planned surface geometry. Laser microstructuring technology is implemented as laser ablation. In the graphics editor, we create the image in shades of gray. ArtyRay4 software breaks the image into 256 layers. Each layer is processed separately and has its own limits of the image. As a result, a darker parts of the image will be processed more times and therefore have the greatest depth [7].

4. In this paper, it is proposed to carry out the process of structuring in three stages.

The first stage - ablation - 3D-removal of surface layers. The thickness of the removed layer (H) can be up to 300 μm in one pass of the laser beam in the processing area.

The aim of the second stage is to clean the material of the products of material combustion from a metal surface after the ablation process.

The third stage - the polishing of the surface - is characterized by a decrease in surface roughness of the processed material.

Let us estimate the parameters of laser treatment, which are required to implement each of the processes above – ablation, cleaning and polishing. The problem specification satisfies the case of heating with a rapidly moving heat source, because $V_{sc}r_0/a>1$, therefore, to determine the maximum temperature of the processed surface when exposed to a series of short pulses we will use the formula [8]:

$$T = \frac{2q(1-R)\sqrt{a\tau}}{k\sqrt{\pi}}\sqrt{\frac{2ar_0}{V_{sc}}} + T_0, \quad (1)$$

where, q is the power density of laser beam, R is the reflection coefficient, a is the thermal diffusivity of the material, τ is the pulse duration, k is the thermal conductivity of the material, r_0 is the radius of the waist spot, Vsc is the laser beam scanning speed, T is the initial temperature)

4.1 I processing step (ablation)

In this step, the laser beam heats the surface of the metal to a temperature of T_I≈3200°K. During that, a part of the material evaporates in form of vapor and other part is removed in a molten form under the influence of recoil vapors pressure $P_{rt} \approx F_{rt}/S = mW/S\tau$, where W is the pulse energy. The metal in the liquid phase is redistributed over the surface under the influence of P_{return} and the surface tension of the melt. Then it cools, forming a rough surface, and oxidizes. In addition, during the ablation process the vaporized material is accumulated on the surface of the material.

4.2 II processing step (cleaning)

At this stage, the combustion products are removed from the surface. At a temperature of T_{II}≈1600°K the metal is not evaporating yet, while the accumulated large and small particles of the material after the first processing step (contamination) are removed from the surface of the processed material. The condition of $T_{surface}$ less than $T_{evaporation}$ is facilitated by the fact that the focal plane is located above the surface of the processed area.

Cleaning mechanism is as follows: the particles accumulated on the surface are sintered together and stick to the processed surface. Consequently, a porous structure is formed. During the repeated irradiation of such a structure, the upper layers of the material are heated faster and more heavily, due to the reduced heat dissipation from the laser pulse irradiation area. Because of the temperature differences, the sintered structure becomes deformed: a porous structure collapses. Under the influence of the recoil vapor pressure, the residue of porous contamination is removed. If an additional cleaning is needed, this processing step can be repeated.

4.3 III processing step (polishing)

Since the surface temperature is significantly lower than T_I and exceeds $T_{melting}$, a recurring melting of the roughened surface of the processed sample occurs. Due to the surface tension in the liquid phase, the unevenness of the surface is smoothed. Moreover, due to the shorter pulse duration of τ = 4 ns, a smaller amount of metal enters the liquid phase, than when τ = 100 ns.

The greater the exposure time, the greater the volume of the liquid phase and therefore the greater the dispersion of sizes of the processed side. In addition, long time of exposure causes an increase in the depth of the heat influence zone (of the heated layer $h_\lambda = (a\tau)^{1/2}$ (3)), where the chemical and structural changes and the appearance of defects on the surface of the processed area occur. [9]

Table 2. Specific parameters of the laser system during three modes of processing and the parameters of the processed material.

	ablation	cleaning	polishing
P [W]	14	6	15
τ [ns]	100	100	4
V [m/s]	0,6	0,8	1,5
f [KHz]	30	40	100
q [W/m^2]	3,6*10^9	3*10^9	7*10^9
T [K]	3200	1200	1700
Ra [μm]	0,78	0,66	0,26
Rz [μm]	4,14	2,61	1,10
l [μm]	49,9	49,8	50

5. Results of the experiment

The total height of the structure and the density of objects on the surface is determined by the task given.

It was planned to create 3D-microstructure in the form of a cones 'Figure 1' [10] for cold steel rolling, where a highly developed surface is needed for its further protection by means of coating.

Figure 1. Images of a 3D-micro-structure on the surface of Steels 10: a series of 200 cones on the area of 0.25 cm^2

Table 3. Characteristics of microstructures in 'Figure 1'.

	untreated surface	cones
the height of the structure, h	-	195 μm
Diameter, d	-	350 μm
Density, N	-	800 thing/cm^2
Roughness Ra, Rz	0,6 μm 2,77 μm	3,48 μm 13,33 μm
the total thickness of the removed layer, H	-	200 μm
run time, t	-	195 μm

In general, the height of the structures h equals the depth of the total removed layer H, with regard to the surface layers cleaning mode of h <H. Reduction of the roughness can be achieved at the final stage of "polishing" the surface. However, in some cases, III processing step is not required.

It is important to note again that the surface microtopography is set programmatically by imaging in a graphics editor. This makes it possible to manufacture a wide variety of structures: intended for industrial use (the surface frictional sliding wetting liquid collecting oil, etc.), as well as decorative crafts (three-dimensional image of a human face or an animal bas) [11]

6. Conclusion.

In the presented work a surface microstructuring technology was examined using a Stell 10 sample. We obtained a relief of the surface in the form of pyramid raster, cones, hemispheres, cylinders and images. It is essential that by changing the surface topography of the metal we can modify various surface properties: mechanical, optical, electrical, etc., at the same time each material, of course, requires the individual parameters of the laser processing.

It should be noted, that the accuracy and performance of this technology on the laser system, that we used, is limited by several factors: the resolving power of the focusing optics, resolution and beam-scanning field, reasonable processing time, etc.

However, in the form in which it currently exists, this technology can be used in various industrial processes: during the hot and cold embossing in the automotive industry [12], to increase the corrosion resistance of steel in metallurgical industry [13], to increase adhesiveness of the surfaces in the varnish-and-paint operations [14], in the jewelry industry for the production of stamps and seals, in microelectronics for processing of film elements [15], for the manufacture of reflective optical rasters, etc.

The proposed method of forming the microgeometry of surface using a volume-controlled ablation can be used for development of laser-based surface engineering in various industries.

7. References

[1] Schneider J. G. Operational properties of details with regular relief. Leningrad, *"engineering"*, 1982

[2] Veiko V.P. Laser–assisted micro shaping. Proceedings of SPIE, v.4157, *Laser–assisted Microtechnology* (LAM–2000), p. **93–103**, 2001.

[3] B. Medres, Menachem Bamberger. LASER SURFACE TEXTURING. Fricso Ltd, *Technion-Israel Institute of Technology* 2008

[4] Nadeem H. Rizvi, Phil T. Rumsby. New Developments and Applications in the Production of 3D Micro-structures by Laser Micro-machining. Exitech Limited Hanborough Park, Long Hanborough, Oxford OX8 8LH, United Kingdom 2012.

[5] V. P. Veiko, A. A. Kislov, T. Y. Mutin, V. N. Smirnov. Prospects of industrial applications of optical materials. *Scientific and technical journal of information technologies, mechanics and optics,* 2012, № 3 (79), **50-54**

[6] Jin-zhong Lu, Chao-jun Yang, Lei Zhang, Ai-xin Feng, Yin-fang Jiang. Mechanical Properties and Microstructure of Bionic Non-Smooth Stainless Steel Surface by Laser Multiple Processing. *Journal of Bionic Engineering* 6 (2009) **180–185**.

[7] Ralf Knappe. Applications of picosecond lasers and pulse-bursts in precision manufacturing. *LUMERA LASER GmbH*, Opelstr. 10, 67661 Kaiserslautern, Germany 2014.

[8] Veiko VP, Shahno E. A. Collection of problems on laser technologies. SPb – *SPb NRU ITMO* 2007.

[9] V. P. Veiko, E. A. Shahno. Laser technology in problems and examples. the textbook for the course *"Physical and technical bases of laser technology".* SPb 2014

[10] S.G. Scholz, C.A. Griffiths, S.S. Dimov, E.B. Brousseau, G. Lalev, P. Petkov. Manufacturing routes for replicating micro and nano surface structures with bio-mimetic applications. *Manufacturing Engineering Centre, School of Engineering*, Cardiff University, Cardiff, UK. CIRP Journal of Manufacturing Science and Technology 4 (2011) **347–356**.

[11] Manuel Pfeiffer, Andy Engel, Steffen Weibmantel, Stefan Scholze, Guenter Reisse. Microstructuring of Steel and Hard Metal using Femtosecond Laser Pulses. *University of Applied Sciences Mittweida, Technikumplatz* 17, 09648 Mittweida, Germany. Physics Procedia 12 (2011) **60–66.**

[12] Urtsev V. N., Platov S. I., Anzupow V. P. A method of processing rolls. Patent of Russia № 2224822 2004

[13] Marshakov, I. K. Corrosion resistance and Obeschanie brass. Physicochemistry of poverhnosti and protection of materials. center of RAS *"Publishing house "Nauka"* (Moscow) ISSN: 0044-1856 2005, **PP 227-233**

[14] Won-Seock Kim , Il-HanYun , Jung-JuLee , Hee-TaeJung. Evaluation of mechanical interlock effect on adhesion strength of polymer–metal interfaces using micro-patterned surface topography. *International Journal of Adhesion & Adhesives* 30 (2010) **408–417**

[15] A. Schubert, S. Gross, J. Edelmanna, B. Schulz. Laser micro structuring of high-stressed embossing dies. **Physics Procedia** 5 (2010) **261–268**

Journal of Physics: Conference Series **735** (2016) 012047 doi:10.1088/1742-6596/735/1/012047

Laser technologies for three-dimensional polymeric structures fabrication

V. O. Veselov, N. D. Vorzobova and V. G. Bulgakova

National Research University of Information Technologies, Mechanics and Optics, Saint-Petersburg, Russia

Email: vorzobova@mail.ifmo.ru , veselov.vasily@gmail.com

Abstract. The processes of formation volume polymeric structures with periodic and arbitrary three-dimensional configuration and optical properties of the elements on their basis are described.

Laser technologies for three-dimensional polymeric structures are an issue of wide research. These technologies include, in particular, laser interference lithography (LIL), which allow a periodic structure formation in the interference light field. Most of research is based on the use of various types of photoresists including a photoresist SU-8 [1-3]. The main problem of the use of this material when the thickness of the layer is large is the problem of residual solvent removing. This paper discusses the processes of three-dimensional structures formation in photopolymerized acrylate compositions and nanocomposites [4, 5] excluding this problem. Along with the possibility of the relief structures formation by interference lithography, processes based on holographic methods are also considered.

When implementing the method LIL structure are formed when exposed in interference field, and followed washing in izopropanole to remove the uncured phase. Depending of the quality of the structures on the exposure parameters established the possibility of relief structures obtaining only for small exposures. Increase in the exposure leads to the formation of bridges (Fig.1b), and then the processes of mass transfer take place and the structures formed are based on the nanoparticles distribution. Possibility of nanoparticles transport in the nanocomposite materials was considered in a number of papers [6-8]. On the photo (Fig.1g) in a cut of structure can be seen nanoparticles (zinc oxide) in the gaps between the polymeric elements.

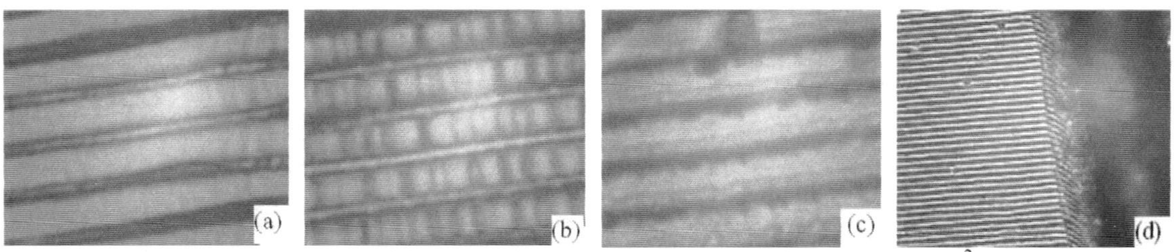

Figure 1. Kinetics of formation of the periodic structure. Power density of 8 mW/cm^2 . Wavelength of the recording light is 325 nm. Period is 6 microns. Exposure Duration 2 (a), 4 (b), 8 (c), 20 (d) s.

Also, process for structures fabrication with a refractive index modulation formed as a result of mass transfer in interference field was investigated. The process for obtaining the structure consists of two stages - the exposure in interference field and the post-exposure UV illumination (without chemical treatment). The dependence of the diffraction properties of periodic structures in the nanocomposite with silica nanoparticles (with a size of 14 nm and 7 nm) on the frequency of the interference field, layer thickness, exposure parameters was investigated. The possibility of obtaining high refractive structures with a diffraction efficiency of 80% at a layer with thickness of 100 microns and a frequency of 300-400 mm^{-1} was found. Increasing in the frequency leads to a decrease in diffraction efficiency. By increasing the layer thickness an increase in light scattering is observed. Thus the scattering elongated in a direction perpendicular to the diffraction plane as a result the selective properties of the Bragg grating. Examination of cross-sections of structures obtained on a microtome showed that the structures are formed in the all thickness of the layer and the structures are volume (Fig.2a). Fig. 2b shows the results of measurements of the angular selectivity of the structures with different periods.

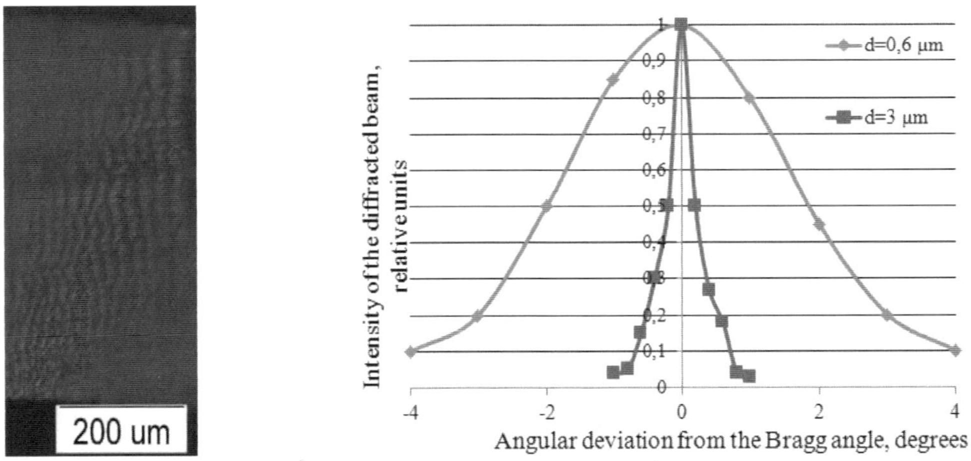

a b

Figure 2. The cut of structure (a), angular selectivity of volume periodic structures (b).

In addition to the one-dimensional structures structure at multi beams interference (Fig.3c).were obtained. The structures were formed in the layers of composition with a thickness of 100 microns deposited on a glass or film substrates.

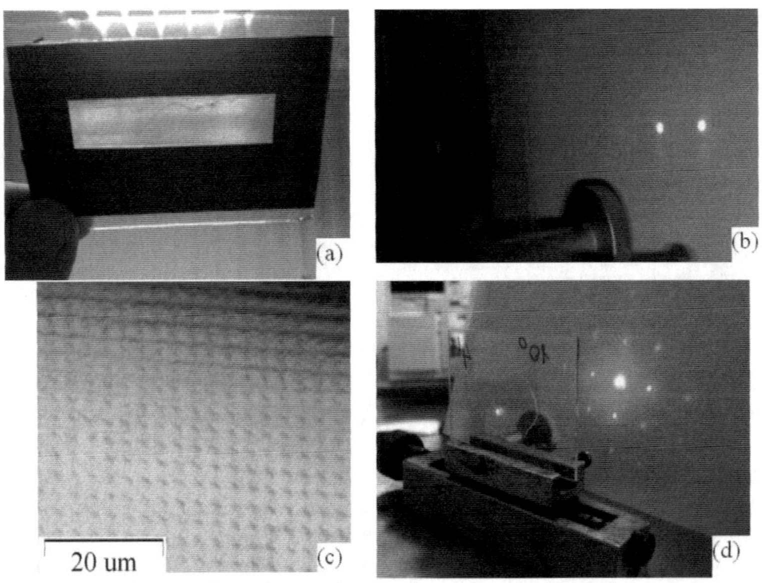

Figure 3. The diffractive element based on volume structure (a) view of the structure obtained by four-beam interference (c) and the diffraction patterns (b, d).

Results given above were obtained when recording with continuous radiation at a wavelength of 442 nm. We investigated also the possibility of pulsed radiation using with a wavelength of 532 nm and the dependence of the diffraction properties of the structures of the pulse frequency and pulse energy. The structures were formed with a period up 3 microns. The highest diffraction efficiency - 74% is obtained at a frequency of 10 Hz and a pulse energy 2mJ and is comparable with that obtained at a given wavelength when recording with continuous radiation.

It was researched the possibility of a three-dimensional polymeric elements forming with the by projection to the volume of photopolymerized material of three-dimensional intensity of the laser radiation distribution formed by hologram. The main problem of the implementation of this approach is to display the configuration of the projected wavefront in the material volume. Solution was achieved as a result of oxygen inhibiting the polymerization process and the corresponding radiation absorption in layer. The holographic image of model objects with flat, spherical, sloped and stepped surfaces was projected in the volume of the photopolymerized photocurable composition. On Fig. 4 present a view of the three-dimensional polymeric elements that were obtained as a result displaying of the wave front configuration.

Figure 4. Three-dimensional polymeric elements formed by a holographic method.

The elements are formed as a result of one light action what determines the advantage of the proposed method relatively the widely used technologies of laser stereolithography or 3D printing [9, 10] which to obtain similar object require a significantly longer time.

The results can be used to obtain, diffractive elements, the elements for light beams forming, elements with arbitrary three-dimensional surface, elements of the photonics with high performance characteristics - optical transparency, high heat resistance and light moisture when simplicity and low cost of processes for their formation

References

[1] Y. Miklyayev, SV Karpeev, PN Dyachenko, VS Paul SD Poletaev 2008 (*Computer Optics* (32, № 4, c. 357-360)).

[2] Zhong Y., Zhou J., Wong K. S. 2012 *Two-photon fabrication of photonic crystals by single-beam laser holographic lithography* (Journal of Applied physics. – 2010. – V. 107, № 074311. – P.1-3.)

[3] Lai N.D., Lin J.H.,. Do D. B, Liang W.P., Huang Y.D., Zheng T.S., Huang Y.Y., Hsu C.C. 2011 *Fabrication of Two- and Three-Dimensional Photonic Crystals and Photonic Quasi-Crystals by Interference Technique* (Holography, Research and Technologies, Prof. Joseph Rosen (Ed.).– 2011. – P. 253-279)

[4] Burunkova Yu. E., Semjina S. A., Kaporsky L. N., Levichev V.V. 2008 *Nanomodified optical acrylate composites* (Optical journal, 2008, V. 75, No. 10, P. 54-58.)

[5] Denisyuk, I. Yu., Burunkova Yu. E., Semjina S. A. 2012 *Investigation of the influence of nanoparticles SiO_2 on the self-organization of acrylate composites cured by ultraviolet radiation* (Optical journal, 2012, V. 79, No. 2, P. 67-71.)

[6] Tomita Y. and Suzuki N. 2005 *Holographic manipulation of nanoparticle distribution morphology in nanoparticle-dispersed photopolymers* (Optics letters.— 2005.—V. 30, № 8.— P. 839–841.)

[7] Smirnova TN, Kohtich LM, OV Sakhno, Stump I. 2011 *Holographic recording nanocomposites periodic structure of the polymer - nanoparticles. II. The mechanism of formation of the periodic structure of the polymer volume - LF and formative influence of the parameters of the field on the effectiveness* (Optics and Spektroskopiya.- 2011.- t. 110, № 1.- C. 143-150.)

[8] I. Yu. Denisyuk, Ju. A. Burunkova, S. Kokenyesi, V. G. Bulgakova, M. I. Fokina. 2012 *Optical nanocomposites based on high nanoparticles concentration and its holographic application*, (Nanocrystals, Sudheer Neralla (Ed.), InTech Europe, Croatia, 2012.)

[9] S.Maruo, K.Jkuta. 2002 *Submicron stereo-lithography for the production of freely movable mechanisms by using single-photon polymerization.* (Sensorsand Actuators. 2002. V.100. P. 70-76)

[10] K. B.Shishkin, M. V.Samusev, M. F.Rybin, Yu.Limonov, S. Kivshar', A. Gaidukeviciute, R. V. Kiyan, B. N. Chichkov (JETP Letters. **95**, No. 9, p.457-461)

Glauber-Sudarshan P function for a single-emitter laser

Vikentyev A.V., Goryainov V.Y., Larionov N.V.

Department of Theoretical Physics, Peter the Great Saint Petersburg Polytechnic University, Polytechnicheskaya 29, Saint Petersburg, 195251, Russia

E-mail: `larionov.nickolay@gmail.com`

Abstract. We consider the model of a single-emitter laser: an incoherently pumped single two-level system interacting with a single cavity mode of finite finesse. In the stationary regime using normally-ordered diagonal representation of the density matrix over the coherent states we derive equation for the phase-averaged Glauber-Sudarshan P function. In the strong-coupling regime between atom and cavity mode the asymptotic solution of this equation is obtained. This solution is a linear combination of two basis solutions. We analyze one of them which was ignored in previous work without any discussions.

1. Introduction

The master equation written in terms of coherent states is often used by different authors for investigation of a single-emitter laser problem (see e.g. [1, 2, 3, 4]). In the stationary regime from this equation one can obtain the system of differential equations for the Glauber-Sudarshan P-function (hereinafter just P-function) and some additional quasi-probability or equivalent closed equation only for the P-function. The latter is the second order differential equation with the polynomial coefficients – one of the simplest equations which is nonintegrable in explicit form. This equation have been obtained in [4], but its derivation was not so clearly: it was made incorrect assumption, that two dimensional quasi-probability current density $\vec{J}(z, z^*)$ defined in this work is equal to zero when its divergence is vanished (see the text after Eqs.(36) in [4]). Also, without any discussions it was ignored the second basis solution in the asymptotic solution of the equation for the P-function in the case of strong-coupling regime.

Thus in our paper at first we shall perform accurate derivation of the stationary equation for the P function and show that latter equation can be done only for the phase-averaged P function. After we shall discuss the second asymptotic basis solution of this equation which arises in the strong coupling regime.

2. Stationary equation for the phase-averaged P function

The single-emitter laser consists of the incoherently pumped two-level atom which interacts with the single cavity mode of finite finesse. Just four constants characterize this laser: $\kappa/2$ – the decay rate of the cavity mode, $\gamma/2$ – the atomic polarization decay rate, Γ – the pumping rate from the low level $|1\rangle$ to the excited level $|2\rangle$ and g – the coupling constant between atom and cavity mode. Following the lead of [4] let us write the master equation for the normally-ordered diagonal representation of the density matrix of our laser $\hat{\rho}(z,z^*)$ over the coherent states $|z\rangle$ of the field and over the projections on the atomic states (Eqs.(36) in [4])

$$\begin{aligned}
\frac{\partial P}{\partial t} &= \frac{\partial}{\partial z}J(z,z^*) + \frac{\partial}{\partial z^*}J^*(z,z^*) = div\vec{J}(z,z^*),\\
\frac{\partial D}{\partial t} &= (\Gamma-\gamma)P - (\Gamma+\gamma)D + \frac{\partial}{\partial z}\left(\frac{\kappa}{2}zD + g\rho_{21}\right)\\
&+ \frac{\partial}{\partial z^*}\left(\frac{\kappa}{2}z^*D + g\rho_{12}\right) - 2g\left[z^*\rho_{21} + z\rho_{12}\right],\\
\frac{\partial \rho_{21}}{\partial t} &= -\frac{(\Gamma+\gamma)}{2}\rho_{21} + \frac{\partial}{\partial z}\left(\frac{\kappa}{2}z\rho_{21}\right) + \frac{\partial}{\partial z^*}\left(\frac{\kappa}{2}z^*\rho_{21}\right)\\
&+ g\left[zD - \frac{1}{2}\frac{\partial}{\partial z^*}(P+D)\right],
\end{aligned} \tag{1}$$

where was introduced the following quasi-probabilities: $\rho_{ik}(z,z^*) = \langle i|\hat{\rho}(z,z^*)|k\rangle$ for $i,k = 1,2$, the difference $D(z,z^*) = \rho_{22}(z,z^*) - \rho_{11}(z,z^*)$, and the P-function $P(z,z^*) = \rho_{11}(z,z^*) + \rho_{22}(z,z^*)$ was included in the vector of quasi-probability current density $\vec{J}(z,z^*) = (J(z,z^*), J^*(z,z^*))$ as $J(z,z^*) = \frac{\kappa}{2}zP(z,z^*) - g\rho_{21}(z,z^*)$.

The physical meaning of the additional quasi-probabilities $D(z,z^*)$ and $\rho_{21}(z,z^*)$ are expressed through the following integrals $\langle\hat{D}\rangle = \int D(z,z^*)\,d^2z$ and $\langle\hat{\sigma}\rangle = \int \rho_{21}(z,z^*)\,d^2z$, where $\langle\hat{D}\rangle$ and $\langle\hat{\sigma}\rangle$ are the mean values of the atomic inversion and polarization correspondingly.

Let us show that in the stationary regime we can equate to zero only the quantity $j(|z|)$ which is a result of averaging of $\vec{J}(z,z^*)$ all over phase. For that we rewrite the right-hand member of the first equation in (1) in terms of polar coordinates I and φ as $z = \sqrt{I}\exp(i\varphi)$ and average it all over φ. The result is

$$\begin{aligned}
\langle div\vec{J}(z,z^*)\rangle_\varphi &= \frac{1}{2\pi}\int_0^{2\pi}[e^{-i\varphi}\left(\sqrt{I}\frac{\partial}{\partial I} - \frac{i}{2\sqrt{I}}\frac{\partial}{\partial\varphi}\right)J(I,\varphi)\\
+e^{i\varphi}\left(\sqrt{I}\frac{\partial}{\partial I} + \frac{i}{2\sqrt{I}}\frac{\partial}{\partial\varphi}\right)J^*(I,\varphi)]d\varphi &= \sqrt{I}\frac{\partial j(I)}{\partial I} + \frac{1}{2\sqrt{I}}j(I),
\end{aligned} \tag{2}$$

where $j(I) = \kappa\sqrt{I}P(I) - g[\rho_{21}(I) + \rho_{12}(I)]$ and $P(I) = (1/2\pi)\int_0^{2\pi}P(I,\varphi)d\varphi$, $\rho_{21}(I) = (1/2\pi)\int_0^{2\pi}e^{-i\varphi}\rho_{21}(I,\varphi)d\varphi$, $\rho_{12}(I) = (1/2\pi)\int_0^{2\pi}e^{i\varphi}\rho_{12}(I,\varphi)d\varphi$. In the stationary regime the obtained expression for $\langle div\vec{J}(z,z^*)\rangle_\varphi$ should be equated to zero, what yields

$$\sqrt{I}\frac{\partial j(I)}{\partial I} + \frac{1}{2\sqrt{I}}j(I) = 0. \tag{3}$$

The solution of this equation is $j(I) = j_0/\sqrt{I}$, where j_0 is a constant. To find the latter let us integrate the product $\pi\sqrt{I}\times j(I)$ all over I, the result is

$$\kappa\langle n\rangle - g(\langle\hat{a}^\dagger\hat{\sigma}\rangle + \langle\hat{\sigma}^\dagger\hat{a}\rangle) = \pi J_0\int_0^\infty dI, \tag{4}$$

where $\langle n\rangle = \int|z|^2P(z,z^*)d^2z = \pi\int_0^\infty IP(I)dI$ is the photon mean value and $\langle\hat{a}^\dagger\hat{\sigma}\rangle = \langle\hat{\sigma}^\dagger\hat{a}\rangle^* = \int z^*\rho_{21}(z,z^*)d^2z = \pi\int_0^\infty\sqrt{I}\rho_{21}(I)dI$. From Eq. (4) it immediately follows that $j_0 = 0$, what is easy to see obtaining the same equation from the Eqs. (1). Thus we accurately demonstrated that in the stationary regime the phase-averaged current density of the quasi-probability for a single-emitter laser is equal to zero, i.e. $j(I) = 0$.

Using the obtained equality $j(I) = 0$ together with the two last phase-averaged equations from Eqs. (1) one can derive the following stationary equation for the phase-averaged P-function (Eqs.(41)-(48) in [4])

$$P''(I) + p(I)P'(I) + q(I)P(I) = 0, \tag{5}$$

where prime indicates the derivative with respect to I and

$$p(I) = a_{12}(I - I_{-1})(I - I_{+1})/I^2(I - I_{00}), \tag{6}$$
$$q(I) = a_{22}(I - I_{-2})(I - I_{+2})/I^2(I - I_{00}). \tag{7}$$

All constants a_{ik} and $I_{\pm1}, I_{\pm2}, I_{00}$ depend on system parameters (see Appendix in [4]).

3. Asymptotic solution in the strong coupling regime

Let us return to Eq. (5). In the strong coupling regime between atom and cavity mode $g/\kappa >> 1$ in Eq. (5) the small parameter arises $\varepsilon = \kappa^2/4g^2$ in front of the higher derivatives. If we apply WKB method [5, 6] to this equation we obtain the following zeroth-order approximation result which consists of two basis solutions

$$P_0(I) = P_1(I) + P_2(I) = \mathbf{c}_1 (I_{-1} - I)^{f_1} (I_{+1} - I)^{f_2} \exp\left(\left|\frac{a_{22}}{a_{12}}\right| I\right) +$$
$$\mathbf{c}_2 (I_{-1} - I)^{-(f_1+1)} (I_{+1} - I)^{-(f_2+1)} (I_{00} - I)^2 I^{1-a_{12}} \times \exp\left(-\left|\frac{a_{22}}{a_{12}}\right| I\right) \exp\left(\frac{cI_s^2 r}{2}\frac{1}{I}\right). \tag{8}$$

In [4] the first solution $P_1(I)$ was investigated in detail. One of the result of this investigation reads that the function $P_1(I)$ is determined on the restricted domain $0 \le I < I_{-1}$. Now let us analyze the second basis solution $P_2(I)$. And the first conclusion about the latter is that the $P_2(I)$ is not bounded at $I = 0$. In the same time the solution $P_1(I)$ gives a clear physical result for some limiting cases when $I \to 0$ (see Eq. (52) in [4]). Thus it is natural to drop the second solution at least on the domain where $P_1(I)$ manifests physical behaviour, i.e. on the domain $0 \le I < I_{-1}$. Note that $P_2(I)$ can probably shows some physical behaviour after the point $I = I_{-1}$ and for the large values of the variable I, because of the exponential decreasing. But for considered in [4] range of the laser parameters such behaviour failed to reveal.

4. Conclusion

In conclusion, for the single-emitter laser we have considered the derivation of the stationary equation for the phase-averaged Glauber-Sudarshan P-function $P(I)$ and showed that the quasi-probability current density $\vec{J}(z, z^*)$ can be equated to zero only after averaging all over the phase. In the strong coupling regime we have analyzed the asymptotic solution of the equation for $P(I)$. This solution consists two basis solutions (8) and we have explained the neglecting of one of them.

Acknowledgments

Acknowledgments This work was supported by Russian Foundation for Basic Research (RFBR), project n° 13-02-00944.

References

[1] Karlovich T B and Kilin S Ya 2001 *Opt. Spectr.* **91** 343
[2] Karlovich T B and Kilin S Ya 2002 *J. Exp. Theor. Phys.* **95** 805
[3] Karlovich T B and Kilin S Ya 2008 *Laser Physics* **18** 783-789
[4] Larionov N V and Kolobov M I 2013 *Phys. Rev.* A **88** 013843
[5] Nayfeh A H 1973 *Perturbation Methods* (New York: Wiley)
[6] Wazov W 1965 *Asymptotic Expansions for Ordinary Differential Equations* (New York: J. Wiley and Sons Inc.)

International Conference of Young Scientists and Specialists "Optics-2015" IOP Publishing
Journal of Physics: Conference Series **735** (2016) 012049 doi:10.1088/1742-6596/735/1/012049

Measurement of coating thickness using laser heating

S A Martsinukov[1], D K Kostrin[1], V V Chernigovskiy[1] and A A Lisenkov[2]
[1] Saint-Petersburg State Electrotechnical University «LETI»
n.a. V.I. Ulyanov (Lenin), 197376, prof. Popova str. 5, Saint-Petersburg, Russia
[2] Institute of Problems of Mechanical Engineering, Russian Academy of Sciences, 199178, Bolshoy pr. V.O. 61, Saint-Petersburg, Russia

E-mail: sergm2006@mail.ru

Abstract. The analysis of thermal processes during the measurement of coating thickness with the use of heating with laser radiation is conducted. The obtained curves of the heating process allow determining thickness of the formed coatings.

1. Introduction

At present time to protect products from corrosion in aggressive environments and to enhance their heat resistance various types of coatings on metallic and dielectric surfaces are actively employed. For coatings of high thickness (100...500 μm) high-performance vacuum-arc deposition installations are usually used [1]. Drop phase in the products of sputtering may cause non-uniformity of film thickness. The uneven thickness of the coating reduces the lifespan of the products, because coating during service life is gradually destroyed and in the case of different thickness quickly enough on the product «clean» substrate areas appear, erosion of which will occur at high speed.

2. Analysis

Today there are a number of nondestructive methods of coating thickness control based on sensing using ultrasound or X-ray radiation, as well as measuring basic capacity of individual sections of a coating. Typically thickness of a substrate is several orders of magnitude greater than thickness of a coating and therefore use of X-ray diagnostics will be ineffective [2]. Measurement of thickness by elementary capacity is possible for dielectric coatings and it is not applicable for metal and semiconductor layers. Application of ultrasonic method is limited because of the strong attenuation of ultrasonic waves in thick coating layers, which hinders fixing of a weak reflected signal. In this regard development of nondestructive methods of coating thickness control, capable of exploring a significant surface of a large batch of products and determining status of the applied coatings, is very relevant. One way to solve this problem can be probing the surface by laser radiation with subsequent investigation of the dependence of sample temperature on time.

Lasers based on gas discharge, due to high density of the energy flow and high spatial uniformity, provide unique technological capabilities in this direction. The most widely applied in this case are molecular CO_2 lasers with a wavelength of 10.6 μm.

Let us review conditions of heating for measuring coatings thickness [3]. First of all we will consider the principles of heat conduction in one-dimensional approximation for unrestricted environment. It distributes heat flux in the direction of x axis and through the lateral surface of the object heat is not supplied. If environment settings are changed in the direction x and time t, then the heat flux density q is dependent on coordinates and time $q(x, t)$. In an infinite environment let us imagine a cylinder and consider an infinitely small section with length dx and cross section area S.

The amount of heat that comes in the direction x through the cross section of the cylinder S during time dt is $q(x)Sdt$. The amount of heat radiated during time dt through the opposite base of the cylinder will be $q(x + dx)Sdt$. Then the total amount of heat passing during time dt through an area of cylinder dx is equal to

$$[q(x)-q(x+dx)]Sdt=(\partial q/\partial x)Sdxdt.$$

On the other hand this heat is determined by expression $cdTdm$, where c is specific heat capacity; $dm=\rho Sdx$ – increment of the cylinder mass; ρ – specific density. Equating these two formulas and making conversion we will receive the following:

$$\rho c\partial T/\partial t=\partial q/\partial x. \quad (1)$$

Let us consider distribution of heat in infinite homogeneous plate with thickness d. On one side of the plate temperature T_1 is supported and on the other side – T_2 ($T_1>T_2$). The heat flow will be proportional to the temperature difference and inversely proportional to the plate thickness d. Then in this case heat flux can be written in the following form:

$$q=\lambda\Delta T/d, \quad (2)$$

where λ is coefficient of thermal conductivity.

Value of heat transfer by conduction depends on the temperature gradient, i.e. temperature difference at the ends of the cylinder to the distance between them $\Delta T/\Delta x$. It also depends on the cross section area of the cylinder and coefficient of thermal conductivity of the material, which characterizes the ability of a substance to conduct heat, and in general it also depends on temperature, structure, moisture, density and pressure.

Assuming that the plate is infinitely thin and x axis is pointing in the direction of decreasing temperature: $d=dx$; $T_1=T(x)$; $T_2=T(x+dx)$, thus

$$\frac{T_2-T_1}{d}=\frac{T(x+dx)-T(x)}{dx}=\frac{\partial T(x)}{\partial x}.$$

Then formula (2) transforms into the following equation:

$$q=-\lambda\Delta T/d. \quad (3)$$

Relation (3) is derived from the assumption that x axis is directed in the direction of decreasing temperature. This ratio is the Fourier law of heat conduction; the minus sign indicates that heat is propagated in the direction opposite the temperature gradient.

Taking into account ρ, c and λ equation (1) takes the following form

$$\rho c\frac{\partial T}{\partial t}=\lambda\frac{\partial^2 q}{\partial x^2},$$

if coefficient of thermal diffusivity $a=\lambda/(\rho c)$, then

$$\frac{\partial T}{\partial t}=a\frac{\partial^2 q}{\partial x^2}.$$

Solving problems on propagation of heat in environment is linked to the solution of a parabolic equation. To determine uniqueness of the parabolic equation solution in a finite region D bounded by a surface S, it is necessary to define initial and boundary conditions taking into account:

– geometrical parameters characterizing size and shape of the body, in which process flows;

– physical parameters characterizing physical properties of a body (thermal conductivity, density, heat capacity, power of internal heat sources, etc.).

Initial and boundary conditions are set by physical formulation of the problem. The initial conditions are defined as follows: $T(r,t)_{t=0}=T(r)$. Boundary conditions depending on the temperature regime at the borders may be different.

The heat source affects the surface of the single layer sample creating a constant heat flux, which gradually heats the surface (figure 1a).

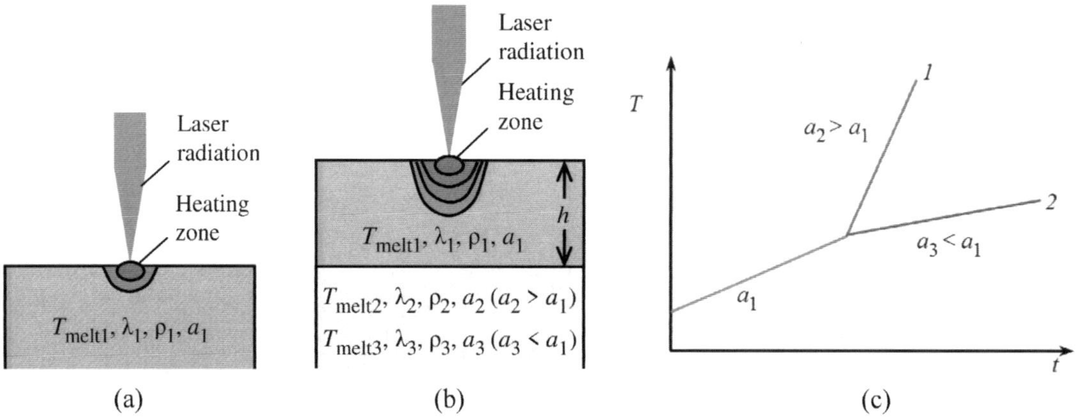

Figure 1. Heating of a sample surface by laser radiation: a – monolithic sample; b – two-layer sample; c – curves of the heating process.

If the heated surface is a two-layer sample with a finite thickness and is in metallic contact with the basis (figure 1b), the heat flows into a material with a thermal wave that travels a distance x in time t. Upon receiving heat from the heat source wave reaches the boundary between two media. Accounting that the thermal contact of a film to the substrate is ideal (adhesion of a film and substrate is ideal) this implies the equality of the temperatures and heat flows at their interface. Power dissipation in a film is uniformly distributed over the cross section, constant along the film thickness and constant throughout the duration of exposure.

If this surface is thermally insulated, then the rate of heat propagation at the boundary is reduced and temperature in the heating zone is raised (figure 1c, line *1*). If not, the heat will be dissipated from the coating faster and the temperature rise will be slowed down (figure 1c, line *2*). In this case, at the interface of two-layer material with an arbitrary initial distribution, boundary conditions of the fourth kind and heterogeneous asymmetric boundary conditions of the first kind at the external borders, when on the border of the body surface temperature is defined, are considered. For each layer its own equation of thermal conductivity is written:

$$\frac{\partial T_1(x,y,z,t)}{\partial t} = a_1 \Delta T_1(x,y,z,t), \quad \frac{\partial T_2(x,y,z,t)}{\partial t} = a_2 \Delta T_2(x,y,z,t),$$

while the boundary conditions are:

$$-\lambda_1 \left.\frac{\partial T_1(x,y,z,t)}{\partial z}\right|_{z=h} = -\lambda_2 \left.\frac{\partial T_2(x,y,z,t)}{\partial z}\right|_{z=h}, \quad T_1(x,y,z,t)_{z=h} = T_2(x,y,z,t)_{z=h},$$

where h is the boundary of a section (coating thickness).

Let us considered the analysis of thermal processes in the heated heat-conductive coating on the insulating substrate. Assuming substrate has thermal diffusivity so small compared to the coating that it can be neglected, at short times of heating the depth of coating heating $x = \sqrt{at}$ will be much less than the thickness of the coating. Thus, at the beginning of a process surface heating can be considered as heating of the monolithic sample with unlimited thickness. If heat is produced by a laser beam having a Gaussian distribution with radius r the following formula to determine the temperature in a center of a beam can be used [4]:

$$T_{11}(0,\ 0,\ t) = T1 = \frac{\alpha I_s r}{\lambda\sqrt{\pi}} \operatorname{arctg}\left[\left(\frac{4at}{r^2}\right)^{1/2}\right], \tag{4}$$

where $T_{11}(0,\ 0,\ t)$ is a surface temperature at the center of a beam (z = 0; r = 0); α – absorption coefficient of the sample; I_s – density of incident power. As the depth of the coating heating becomes comparable with the thickness a different mathematical expression for the temperature is needed:

$$T_{12}(0,\ 0,\ t) = T2 = \frac{\alpha I_s r^2}{4\lambda h}\ln\left[1+\frac{4at}{r^2}\right], \tag{5}$$

Transition from the expression (4) to the expression (5) is smooth in nature (figure 2). For it determination it is more appropriate to consider not function $T = f(t)$ but the change of temperature increase speed.

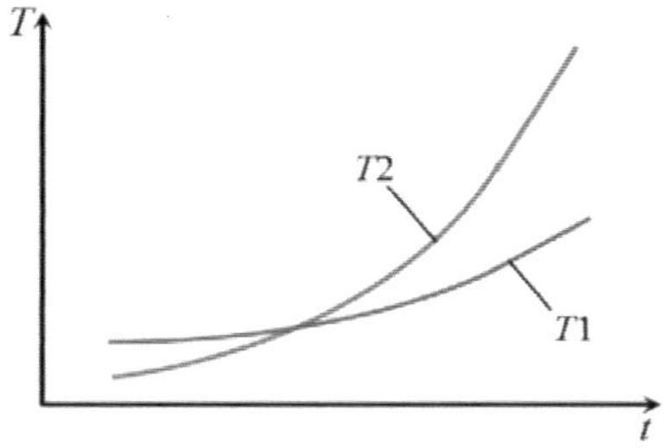

Figure 2. Time dependences of the laser heating.

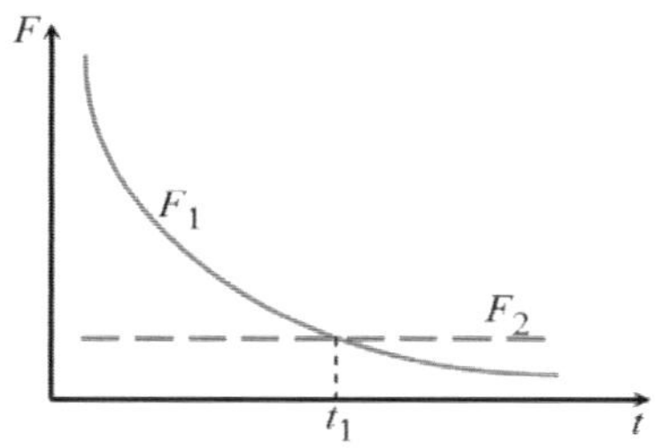

Figure 3. Function of the heating rate.

Let us differentiate equations (4) and (5) by time

$$\frac{dT_{11}}{dt} = \frac{\alpha I_s\sqrt{a}}{\lambda\sqrt{\pi}}\frac{1}{1+(4at/r^2)}\frac{1}{\sqrt{t}}, \quad \frac{dT_{12}}{dt} = \frac{\alpha I_s a}{\lambda h}\frac{1}{1+(4at/r^2)}. \tag{6}$$

Based on (6) auxiliary function F_i can be made:

$$F_i = \frac{\lambda}{\alpha I_s\sqrt{a}}\left[1+\frac{4at}{r^2}\right]\frac{dT_{1i}}{dt}, \tag{7}$$

where i = 1, 2. It is obvious that $F_1 = (\pi t)^{-1/2}$ and $F_2 = a^{1/2}/h$ = const. Hence there is a relatively simple method of finding h.

Simultaneously with the beginning of heating of the coating low-inertia radiation sensor detects the value of the continuously rising surface temperature, then using a PC in real time is taken the derivative dT/dt and the value of the auxiliary function F_i is calculated using (7). When the value of F_1 becomes equal to F_2 (time t_1 in figure 3) h is defined as $h = a^{1/2}/F_2$. In accordance with the physics of phenomena with increasing h the value of F_2 decreases and the time after which the heat flow reaches the insulating substrate grows. Naturally, after this time the rate of temperature rise becomes larger than it would be in the case of a monolithic sample.

Minimum thickness of the coating which can be evaluated by this technique is determined by the parameters of a coating and infrared receiver, as it has a certain response time $t_{\min}$: $h_{\min} = (a\pi t_{\min})^{1/2}$.

Let us consider coating of chromium carbide CrC (ρ = $6.68\cdot10^3$ kg/m^3; c = 0.54 J/(g·K); λ = 19.1 W/(m·K); a = $5.24\cdot10^{-6}$ m^2/s) deposited on the insulating ceramics. Then, taking $t_{\min}$ of the infrared receiver equal to $5\cdot10^{-4}$ s calculated $h_{\min}$ will be 90 μm.

Maximum thickness of the coating can be determined from the type of $F_1 = f(t)$ function. As can be seen from figure 3 with increasing time the steepness of this function falls that will hinder the fixation of the moment when F_1 becomes constant. If we assume that the minimum observed relative change of the function $\Delta F_1/F_1 = \gamma$, then the maximum measurement time is defined as $t_{\max} = \Delta t/2\gamma$.

Since $\Delta t = t_{min}$ then $h_{max} = (a\pi t_{max})^{1/2} = (a\pi t_{min} / 2\gamma)^{1/2}$. For the above example, assuming that $\gamma = 1$ % the maximum measured thickness of a coating will be equal to 640 μm.

Required laser power can be evaluated from the expression (4):

$$I_s = \frac{T_{max} \lambda \sqrt{\pi}}{\alpha r \operatorname{arctg}\left(\sqrt{4at / r^2}\right)}.$$

On the basis of this method was developed an experimental device for measurement of coatings thickness which consists of two main parts: test sample heating block and measurement unit. For the heating unit CO_2 laser with electro-magnetic modulator is used. For directing and focusing of the beam a system of mirrors and lenses is applied. The measuring device is a pyrometer, which is based on fixing temperature of an object by its own infrared radiation in the spectral range 2.0…3.1 μm. This pyrometer is adapted to operate in a wide temperature range with low inertia to continuously record the temperature. It includes a long-focus lens, allowing directing it to the treated area of a sample.

With the help of this complex experimental dependences of heating for samples of chromium carbide deposited on a ceramic substrate (figure 4a) and steel substrate (figure 4b) were obtained.

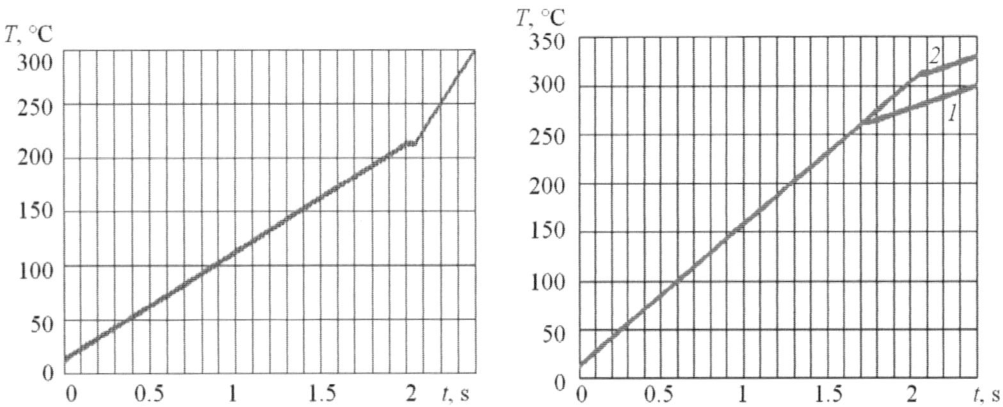

(a) (b)

Figure 4. Curves of the heating process allowing determination of the coatings thickness: a – chromium carbide on ceramic substrate; b – chromium carbide on steel substrate.

It is seen that while the thermal wave has not yet reached the boundary between two media, the temperature increases linearly. When thermal wave reaches boundary between materials a change in the course of schedule is observed. After that if the surface is insulated the rate of heat propagation at the boundary is slowed and temperature in the heating zone rises (figure 4a); otherwise temperature will be dissipated from the coating faster and the temperature rise will be slowed down (figure 4b). Also figure 4b shows curves for different coating thickness: *1* – 140 μm; *2* – 160 μm.

3. Conclusion

Considered method of determining coatings thickness extends the capabilities of the process control and diagnostics when receiving films and coatings of inorganic substances. When using a laser source with the controlled nature of radiation, this method allows determination of thickness of relatively thick coatings (100...500 μm) with high accuracy.

4. References

[1] Bystrov Yu A, Vetrov N Z, Lisenkov A A and Kostrin D K 2014 *Vakuum in Forschung und Praxis* **5** 19–23

[2] Potrakhov N N 2015 *Biomedical Engineering* **5** 237–40

[3] Tritt T M 2005 *Thermal Conductivity. Theory, properties and applications* (New York: Springer)

[4] Duley W W 1983 *Laser processing and analysis of materials* (New York: Plenum Press)

Journal of Physics: Conference Series **735** (2016) 012050 doi:10.1088/1742-6596/735/1/012050

Study on possibility for the improvement of corrosion resistance of metals using laser-formed oxide surface structure

J S Ruzankina, O S Vasiliev
ITMO University, 49 Kronverksky Pr., Saint-Petersburg, Russia

E-mail: ruzankinajulia@gmail.com

Abstract. The laser processes of oxidation are currently known and used extensively, in particular, to improve corrosion resistance of metals possessing certain properties and composition. In this regard, actuality is the methods of laser oxidation of metals and the determination of their modes of treatment in each specific case. Increase of corrosion resistance ST20 can carried out with the formation on the surface oxide films, as well as by reducing surface roughness. Studied various modes of processing of the steel surface. Corrosion resistance investigated for protecting a metal. Defocusing the beam to allow the surface treatment of a wide beam in the low temperature mode of processing. For further study of the irradiated surface on the corrosion resistance was conducted by chemical treatment in acid. Estimated phase composition of films formed under laser treatment simulated in the program astics. The study to increase the corrosion resistance of steel and titanium, have shown that under the chosen methods of processing of materials degradation observed.

1. Introduction

Every year about one third of steel produced is lost due to corrosion, and the tenth part of it was dissipating in the form of corrosion products. The damage caused by corrosion consists of direct and indirect losses. These are included: the cost of corroded equipment cost of replacement or repair machines and equipment, the cost of damaged of the reactants and products of chemical processes, payments to victims of accidents involving corrosion damage workshop equipment, etc. to take full account of economic and moral losses extremely difficult. In this regard, one of the urgent tasks of the modern world is the creation of corrosion-resistant materials for prolonged use in particularly harsh conditions. These materials were using in fuel and nuclear energy, space technology and oil and gas industry and in other industries.

Now there are works, for example, [1-5], in which they consider the protection of metals from corrosion due to the creation of, surface micro - or nanostructures when exposed to Femto - or picosecond lasers. However, the introduction of these lasers in industrial production is problematic due to the high complexity of their maintenance. More preferred are fiber lasers. Easy control of parameters such laser systems, pointing accuracy of the laser radiation to the treatment area and excellent technical and economic parameters, and in particular, reliability and long service life. However, for surface protection it is advisable to use the laser oxidation [6], which allows creating on the metal surface a dense oxide structure. Despite a certain amount of work on this subject [7, 8] significant challenges remain that need to be addressed. To them in the first place, is the quality of oxide surface structure, and mean treatment time (oxidation).

Thus, the study of physicochemical processes occurring on the surface of metals under laser irradiation, and introduction of technologies for the protection of metal from corrosion in the industry are actual

Journal of Physics: Conference Series **735** (2016) 012050 doi:10.1088/1742-6596/735/1/012050

scientific and practical task. Based on all of the above, the purpose of the investigational work is to improve the corrosion resistance of steel and titanium by forming a protective oxide film by laser radiation near infrared (1.07 μm). [9]

2. Research and results

In the process there were determined the optimal regimes of laser processing the surface of the metal during formation of the oxide film by radiation of a fiber laser. The study was conductING using two materials, namely Ti and ST20. To conduct the experimental work was used ytterbium fiber laser with wavelength 1064nm. General view of the laser setup is presented in Fig.1, where PC - personal computer; SW - software; BEADS - the control unit deflectors; RL is a beam expander; OS - deflecting system; BKL the control unit of the laser; BPL - power pack laser; FS - focusing system. This wavelength allows processing of most metals.

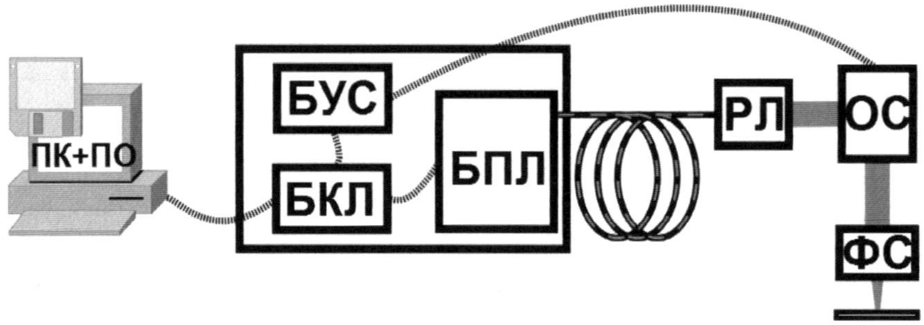

Figure 1. The block diagram of the laser system.

Increase of corrosion resistance ST20 can be carrying out with the formation on the surface oxide films, as well as by reducing surface roughness. This was investigating in various regimes and surface treatment of steel. Since these processes depend on T surface, the irradiation process was controlling by using an infrared camera, presented in Fig.2.

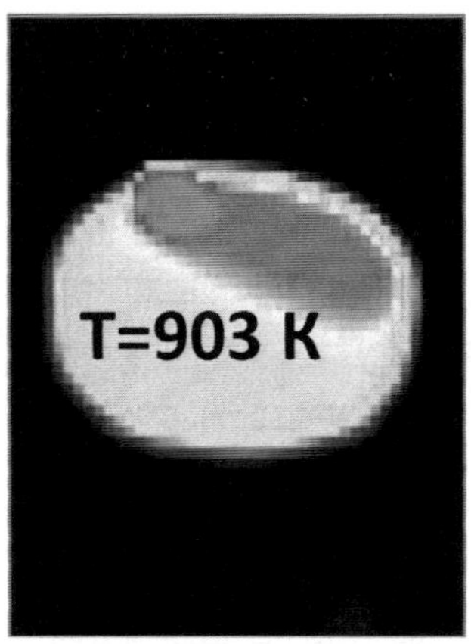

Figure 2. Termediary ST20

After the surface treatment of the steel examined using optical microscope (Fig.3). Thus T in the area of impact (1) reaches the optical 629, 85 S. the photos shown the change in appearance of the structure depending on the temperature. The field of laser radiation exposure are different from those that were establishing outside the zone of laser radiation, but also subjected to heat treatment.

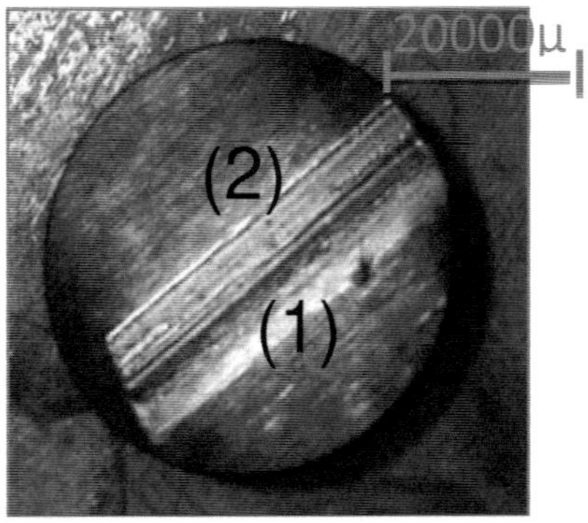

Figure 3. Optical photograph of the surface stali 1) the heat-affected zone; 2) zone of impact of laserradiation

The surface of the steel were determined using a profilometer. The surface roughness stale decreased, which follows from the data presented in table 1.

Table 1. The change of surface roughness of steel 20.

	Raw material	**The presence of the oxide layer**
Ra	0,53	0,48
Rz	3,66	2,66

For the study of the irradiated surface on the corrosion, resistance was conducting by chemical treatment in acid. As a chemical reagent is a mixture of copper sulfate and hydrochloric acid. In Fig.4A presents, the area was not treating with acid (1) and chemical treatment: (2A) t=10 s and (2B) t>10 s. In the course of the study, it was founding that the surface was covering with a special protective layer (Fig. 4).

Figure 4. a) the acid treated surface

Figure 4.b) the area of interaction of the acid with theuntreated surface

Because of the research revealed the feature of laser oxidation of steel, comprising forming the modified surface structure in the field of thermal impacts that have not been exposing to laser irradiation. In Fig.5 these patterns are halos surrounding impact area. The temperature in these areas was less than half that in the field of radiation exposure.

Figure 5. The haloing in the area of temperatures below TPL

For possible research, areas were presenting for protecting metal were investigating on the corrosion resistance. Before chemical processing of this region was creating during laser irradiation in the mode of defocusing. Defocusing the beam to allow the surface treatment of a wide beam in the low temperature regime treatment (Fig.6). According to the results of chemical treatment, you may notice that in the field of action of laser radiation, degradation of material was observing. Thus, the selected method of treatment of steel can improve its corrosion resistance.

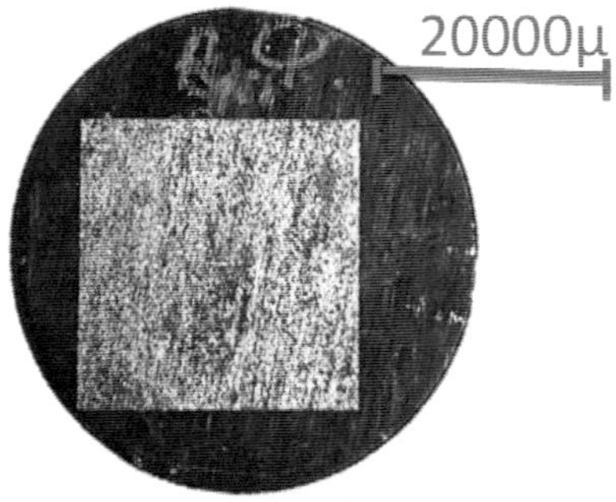

Figure 6. Processing stale in the low temperature regime

As for steel, was carrying out search of laser processing parameters, with the aim of improving the corrosion resistance of titanium. For further study of the irradiated surface on the corrosion, resistance was conducting by chemical treatment in acid. In the course of the study the following results were obtained:in Fig.7a shows the appearance of the surface of Titan to interact with a chemical reagent. In Fig.7b clearly shows that the appearance of the titanium sites, which were exposing to laser radiation, virtually unchanged, in contrast to the plots that were not subjecting to irradiation. Thus, the difference between irradiated and non-irradiated area amounted to 0.17 μm.

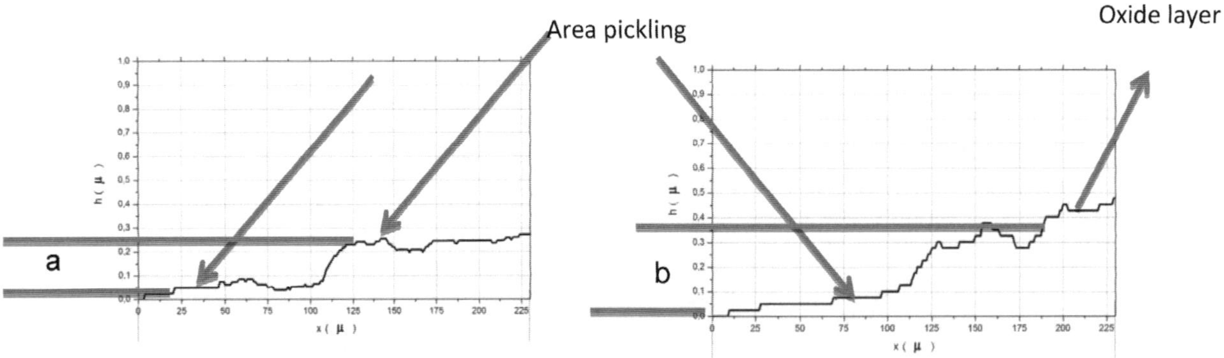

Figure 7. The titanium surface a) before

Figure 7. The titanium surface b) after interaction with a chemical reagent

Journal of Physics: Conference Series **735** (2016) 012050 doi:10.1088/1742-6596/735/1/012050

3. Conclusion

In the course of the study solved the problem by definition of optimum modes of laser processing the surface of the metal during formation of the oxide film by radiation of a fiber laser.

Was investigated the corrosion resistance of the treated surface material. To improve the corrosion resistance of steel and titanium were choosing, certain modes of surface treatment of metals. The desired protective layers were obtaining because of established conditions.

The proposed methods and approaches for the protection of metal from corrosion may be applicable to virtually all metals.

4. References

[1] M.F. Montemor Functional and smart coatings for corrosion protection: A review of recent advances. // Surface and Coatings Technology.2014.-Vol. 258.-p. 17-3

[2] D. Höche, J. Kaspar, P. Schaaf Laser nitriding and carburization of materials.// Laser Surface Engineering.2015.- p.33-38

[3] Alexandre M. Emelyanenko, Farida M. Shagieva, Alexandr G. Domantovsky, Ludmila B. Boinovich Nanosecond laser micro- and nanotexturing for the design of a superhydrophobic coating robust against long-term contact with water, cavitation, and abrasion.// Applied Surface Science.2015. Vol.332. – p. 513-517

[4] A.Y.Vorobyev, Chunlei Guo Multifunctional surfaces produced by femtosecond laser pulses. //Journal of Applied Physics. 2015. Vol. 117

[5] Liana Maria Muresan Corrosion Protective Coatings for Ti and Ti Alloys Used for Biomedical Implants.// Intelligent Coatings for Corrosion Control.2015. p.585-602

[6] GV Odintsova "Research and development of technology for color laser marking metals by local oxidation of" the Abstract. St. Petersburg, 2014

[7] Scripcenco AI, VL Popov, Kondratiev SY Features of laser surface modification machine parts. // Magazine "Rhythm". -2010

[8] I Kistyakovsky. Applied Physical Chemistry. Part 1. L. 1926

[9] AB Kolesnik, IV Gladyshev "Chemical resistance of materials and corrosion protection" - Tambov Univ Thumb. state. Tech. University Press, 2008. - 80c. -100 Copies. -ISBN 978-5-8265-0710-0.

A numerical solution algorithm and its application to studies of pulsed light fields propagation

V A Banakh, L O Gerasimova, I N Smalikho and A V Falits
Laboratory of wave propagation, V.E. Zuev Institute of atmospheric optics SB RAS, Tomsk, Russian Federation

E-mail: lilyan@iao.ru

Abstract. A new method for studies of pulsed laser beams propagation in a turbulent atmosphere was proposed. The algorithm of numerical simulation is based on the solution of wave parabolic equation for complex spectral amplitude of wave field using method of splitting into physical factors. Examples of the use of the algorithm in the case the propagation pulsed Laguerre-Gaussian beams of femtosecond duration in the turbulence atmosphere has been shown.

1. Introdaction

The development of femtosecond lasers and its applications in atmospheric problems determines the urgency of investigation of the propagation of short pulse laser radiation in the turbulent atmosphere [1, 2]. The problems of propagation such beams in the turbulent atmosphere are solved on the basis of equations for statistical moments of the complex amplitude of the wave field at different frequencies obtained in the Markov approximation from the stationary parabolic wave equation [3, 4]. However, in this case, even the equation for the second statistical moment under arbitrary turbulent conditions of propagation has no rigorous solution, which do not allow to the full research of the propagation of short pulse laser beams without use of numerical methods [3]. Solution of the problems of propagation of pulsed optical radiation in the turbulent atmosphere is possible based on the algorithm of numerical solution of parabolic wave equation for the complex spectral amplitude of the wave field by method of splitting into physical factors [5]. In this study, the algorithm are applied to calculation of statistical characteristics of the energy density of the broadband pulsed optical radiation depending on the turbulent conditions in the propagation path.

2. Theoretical background

The parabolic wave equation [3] for complex spectral amplitude of the field $U(x,\boldsymbol{\rho},f)$ which describes the propagation of pulsed radiation in a turbulent atmosphere has the form

$$\frac{\mathrm{i}4\pi f}{c}\cdot\frac{\partial U(x,\boldsymbol{\rho},f)}{\partial x}+\Delta_{\perp}U(x,\boldsymbol{\rho},f)+2\left(\frac{2\pi f}{c}\right)^{2}n'(x,\boldsymbol{\rho})U(x,\boldsymbol{\rho},f)=0\,, \qquad (1)$$

where $\boldsymbol{\rho} = \{y, z\}$ is the radius vector in the plane perpendicular to the optical axis x, $\mathrm{i} = \sqrt{-1}$, f is the linear frequency, c is the speed of light in vacuum, $\Delta_\perp = \frac{\partial^2}{\partial y^2} + \frac{\partial^2}{\partial z^2}$ is the transverse Laplace operator, $n'(x, \boldsymbol{\rho})$ are fluctuations of the refractive index caused by turbulent variations of the air temperature. In equation (1) is taken into account that in the atmosphere $|n'(x, \boldsymbol{\rho}, f)| << 1$ and in the dispersive medium the dependence of turbulent fluctuations of the refractive index on the frequency f can be neglected.

We assume that pulsed laser radiation propagation in the turbulent atmosphere along the axis $x \geq 0$ and in plane $x = 0$ is given by equation:

$$E(0, \boldsymbol{\rho}, t) = E^0_{nm}(\boldsymbol{\rho}) \exp\left(-\frac{t^2}{2\tau_0^2} - 2\pi \mathrm{i} f_0 t \right), \tag{2}$$

where the complex strength of the electric field of the wave E^0_{nm} at the point $(x = 0, \boldsymbol{\rho})$ is representated in the form of Laguerre-Gaussian beam [6]

$$E^0_{nm}(\boldsymbol{\rho}) = (-\mathrm{i})^m \frac{1}{2}\left(-\frac{\boldsymbol{\rho}}{a} \right)^m \exp\left[-\frac{\boldsymbol{\rho}^2}{2a^2} + j\psi_0 + jm\theta \right] L_n^m\left(-\frac{\boldsymbol{\rho}^2}{a^2} \right), \tag{3}$$

where parameter a is the radius of the Gaussian beam, $\boldsymbol{\rho}, \theta$ are polar coordinates, $L_n^m = x^{-1} \exp(x) \frac{d^n}{dx^n}\left[x^{n+1} \exp(-x) \right]$ is the Laguerre polynomial, integer numbers m and n are for the Laguerre-Gauss modes E^0_{mn}, excited in a confocal cavity with circular mirrors, ψ_0 is the wave phase independent of $\boldsymbol{\rho}$ and t, t is a time, f_0 is the frequency at the point of maximum of the radiation spectrum, τ_0 is a pulse duration determined from decrease of $|E(0,0,t)|^2$ to the level e^{-1}. The pulse duration $\tau_p(x)$ determined from decrease of the radiation power $\int_{-\infty}^{+\infty} \mathrm{d}^2 \rho \,|E(x, \boldsymbol{\rho}, t)|^2$ to the level ½ to the right and to the left from the point of maximum $t = t_{\max}$, in the plane $x = 0$ is related with τ_0 by the equation: $\tau_p(0) = 2\sqrt{\ln 2}\tau_0$. The values m=n=0 in equation (3) correspond to the lowest mode – Gaussian beam. In this case, parameter a describes the radius of the Gaussian beam at the level e^{-1} of the beam intensity normalized to the maximal value.

In the absence of radiation absorption by air and aerosol particles and nonlinear effects, the complex spectral amplitude of the wave field

$$\tilde{E}(x, \boldsymbol{\rho}, f) = U(x, \boldsymbol{\rho}, f) \exp\left(\frac{2\pi \mathrm{i} f < n(f) > x}{c} \right) = \int_{-\infty}^{+\infty} \mathrm{d}t\, E(x, \boldsymbol{\rho}, t) \exp(2\pi \mathrm{i}\, ft), \tag{4}$$

described by the equation (1) with the boundary conditions, according to (2) и (4),

$$\tilde{E}(0, \boldsymbol{\rho}, f) = \sqrt{2\pi}\tau_0 E^0_{mn}(\boldsymbol{\rho}, f) \exp\left(-\frac{(f - f_0)^2 (2\pi\tau_0)^2}{2} \right). \tag{5}$$

In equation (4) $< n(f) >$ is the average value of the refractive index.

Journal of Physics: Conference Series **735** (2016) 012051 doi:10.1088/1742-6596/735/1/012051

3. Procedure of numerical simulation

The statistical characteristics of the energy density of the radiation pulse of short duration was studied through simulation of the laser radiation propagation based on numerical solution of the wave equation in the parabolic approximation (1) by the splitting method. This approach is based on the procedure of replacement of the continuous medium, the radiation propagates through, with a set of thin phase screens imitating turbulent distortions acquiring by the optical wave in the process of propagation, between which the propagating wave experiences only diffraction. The essence of the method as applied to our case consists in the following. The entire propagation path of the length L is divided into N layers every with thickness $\Delta x = L/N$. But in contrast to the continuous-wave (cw) in the spectrum short pulse beam will attend more than two frequency components as a decrease in the pulse width increases the width of its spectrum. Therefore it is necessary to calculate the wave equation for the complex spectral amplitude $U(x_j, \mathbf{\rho}, f)$ for each component of the spectrum f where $x_j = j\Delta x$ and $j = 0, 1, ..., N$ consecutively at transition from one layer to another. Then at the end of the propagation path sum $U(x_j, \mathbf{\rho}, f)$ over all components of the frequency spectrum.

From simulated values of the complex spectral amplitude $U(x, \mathbf{\rho}, f)$ calculated the spectral intensity

$$S_I(L, \mathbf{\rho}, f) = |U(L, \mathbf{\rho}, f)|^2 \tag{6}$$

and radiation energy density, which in terms of frequencies has the form [7],

$$W(L, \mathbf{\rho}) = \int_{-\infty}^{+\infty} \mathrm{d} f S_I(L, \mathbf{\rho}, f) . \tag{7}$$

4. Results of numerical simulation

Figures 1-4 shows examples of simulated random realizations of normalized energy densities $W(L,\mathbf{\rho})/W(0,0)$ pulsed radiation with $\tau_P(0)$= 3 fc(τ_0= 1,8 fc) in comparison with distribution of normalized intensity $I(L,\mathbf{\rho})/I(0,0)$ cw radiation. Note that for $\tau_0 \to \infty$ transition to cw radiation, which intensity with the frequency f_0 does not change, and, consequently, $W(L,\mathbf{\rho})\big|_{\tau_0 \to \infty} = I(L,\mathbf{\rho})$.

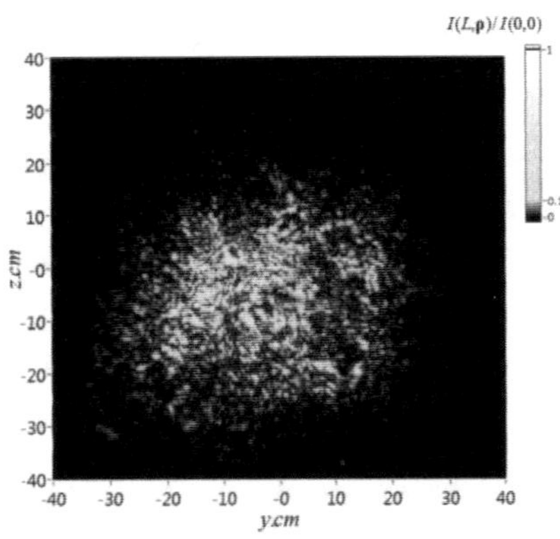

Figure 1. Two-dimensional random realization of the normalized energy density of cw radiation at $C_n^2 = 10^{-13}$ m$^{-2/3}$.

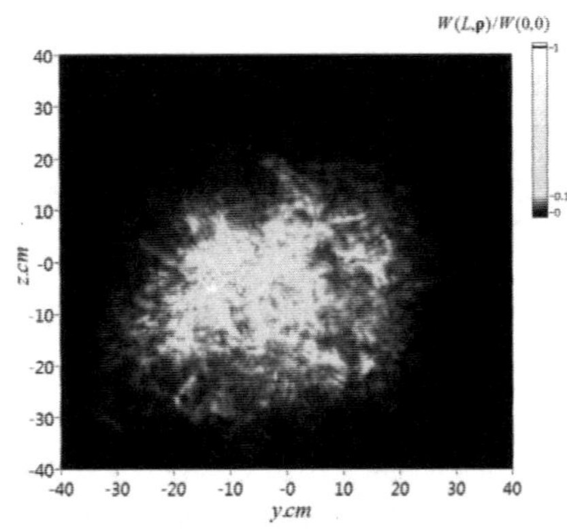

Figure 2. Two-dimensional random realization of the normalized energy density of pulsed radiation at $C_n^2 = 10^{-13}$ m$^{-2/3}$.

Journal of Physics: Conference Series **735** (2016) 012051 doi:10.1088/1742-6596/735/1/012051

The comparison of statistical characteristics of the energy density of pulsed radiation $W(L,\mathbf{\rho})/W(0,0)$ and the intensity $I(L,\mathbf{\rho})/I(0,0)$ of continuous-wave radiation from the same realizations of arrays of random phase screens allows us to perform the comparative analysis of turbulent fluctuations of the pulsed and cw radiation.

For numerical simulation of random realizations $W(L,\mathbf{\rho})$ we set the following parameters: $a = 5$ cm, $L = 3$ km $\lambda_0 = 1\mu$m for mode E^0_{02} Laguerre-Gaussian beam propagations in turbulence atmosphere with $C_n^2 = 10^{-13}$ m$^{-2/3}$ (figures 1, 2) and $C_n^2 = 10^{-12}$ m$^{-2/3}$ (figures 3, 4). At every node of the computational grid $M\times M$=1024×1024 with the step $h = 3$ mm the values of the complex spectral amplitude were calculated for 41 spectral channels with the width $\Delta f = 10$ THz, that is, in the discrete representation the frequency takes values $f = f_1 + k\Delta f$ where $f_1 = 100$ THz, $k = 0, 1, ..., K$ and $K = 40$. Twenty five phase screens with the structure characteristics of the refractive index were set along the propagation path.

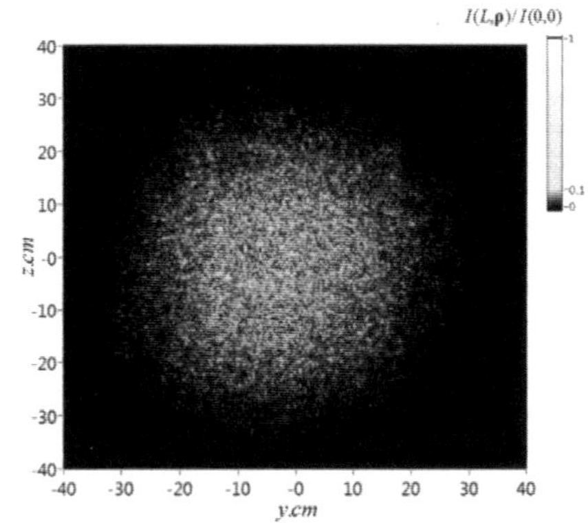

Figure 1. Two-dimensional random realization of the normalized energy density of cw radiation at $C_n^2 = 10^{-12}$ m$^{-2/3}$.

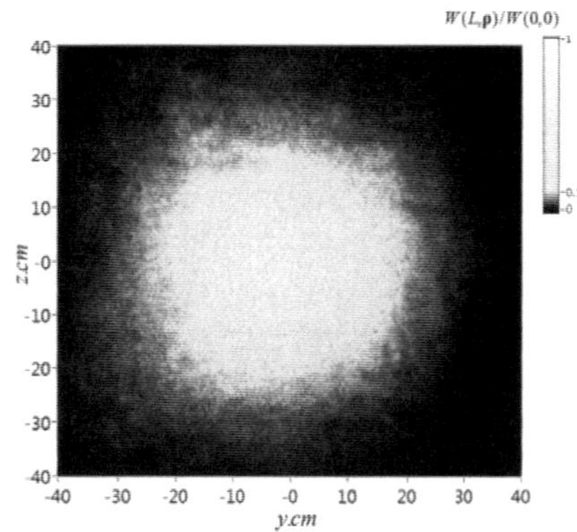

Figure 2. Two-dimensional random realization of the normalized energy density of pulsed radiation at $C_n^2 = 10^{-12}$ m$^{-2/3}$.

Comparison of the simulation results showed that with an increase of optical turbulence the diffraction structure of the intensity distribution in the beam cross section is fully blurred as for cw (figures 1, 3) so for pulsed (figures 2, 4) beams and the energy density distribution becomes close to the structure of Gaussian beam. From figures 1 and 2 one can see that the distribution $W(L,\mathbf{\rho})/W(0,0)$ (figure 2) is more smoother than that of $I(L,\mathbf{\rho})/I(0,0)$ (figure 1). In the conditions of strong optical turbulence with $C_n^2 = 10^{-12}$ m$^{-2/3}$ is a difference in the results of the simulation becomes more pronounced (figures 3, 4).

5. Conclusions

The algorithm of the numerical modeling of propagation of short pulse laser beams in a turbulent atmosphere on the basis of solution of the parabolic wave equation for the complex spectral amplitude of the wave field by method of splitting into physical factors was presented. The described algorithm allows to study of the statistical characteristics of energy density fluctuations of pulsed radiation with femtosecond pulse duration depending on diffraction and turbulent conditions in the propagation path.

6. Acknowledgements
This work was supported by the Russian Scientific Foundation for Maintenance and Development (Project 14-17-00386).

References
[1] Bagaev S N and Matvienko G G eds. *Femtosecond Atmospheric Optics* 2010 (Novosibirsk: Siberian Branch of the Russian Academy of Sciences) p 238
[2] Kandidov V P and Shlyonov S A 2012 *Atmos. Oceanic Opt.* **25**(3) pp 192–198
[3] Zuev V E , Banakh V A and Pokasov V V 1988 *Optics of the Turbulent Atmosphere* (Leningrad, Gidrometeoizdat) p 270
[4] Tjin D E, Kelly T S and Andrews L C 1999 *Waves Random Media* **9**(3) pp 307–325
[5] Banakh V A, Gerasimova L O and Smalikho I N 2015 *Quantum Electronics* **45**(3) pp 258–264
[6] Banakh V A and Falits A V 2014 *Opt. Spectrosc.* **117**(6) pp 969–975
[7] Vinogradova M B, Rudenko O V and Sukhorukov A P 1979 *Theory of Waves* (Moscow: Nauka) p 383

Calculation of energies of the isoelectronic series of atoms ($Z \leq 29$) using the Hartree–Fock method

I N Eremkin[1], **Yu B Malykhanov**[2] **and M V Gorshunov**[2]
[1]ANO Center for Professional Training, Saransk 430003, Russia
[2]M.E. Evsev'ev Mordovian State Pedagogical Institute, Saransk 430007, Russia

E-mail: eryomkin@mail.ru

Abstract. Energy of atoms from hydrogen to cuprum and all their ions up to the one-electron atom that were calculated in the algebraic approximation of the Hartree–Fock method were compared with the experimental values obtained from ionization potentials. It was established that the calculated energies of atoms with $Z \leq 29$ were only 0,3-1,0% greater than the experimental values. This indicated that the approximations underlying the Hartree–Fock method were correct.

1. Introduction

Knowledge of the energies of ionized states in addition to the ground-state energy is necessary in several areas of physics. In particular, highly ionized atoms play an important role in plasma physics. The ground-state energy of atoms and also of their ions can be found experimentally if ionization potentials for all of them up to the one-electron atom are measured. In turn, measurement of the last from atomic spectra is a complicated problem in atomic spectroscopy. Ionization potentials of all atoms in the Periodic Table were found through multi-year atomic spectroscopy studies [1, 2]. However, only first ionization potentials were measured for heavy atoms ($Z \geq 30$). Therefore, theoretical calculation of the energies of atoms and ions based on one quantum-mechanical approximation method or another is critical. Energies of light atoms are currently calculated with sufficiently high accuracy using variational methods and methods that take into account correlation [3]. However, the calculation of energies of heavy atoms in such an approximation is a complicated problem so that calculations of such atoms can be performed in practice using, for example, the Hartree–Fock (HF) method [4, 5]. The HF method was created in order to calculate the energies of atoms and ions [4]. However, it is approximate in nature because it is based on a one-electron approximation. An additional simplification, an approximation of the central field that enables radial and angular parts to be separated in one-electron functions, is used for atoms in the framework of this method. The HF equations for any atomic configurations can be solved by a numerical method [4] and in the framework of an algebraic approach [5-9] owing to these simplifications. Herein energies of the first 29 atoms of the Periodic Table and all their ions up to the one-electron atom inclusive are calculated using the algebraic approximation of the HF method.

All ionization potentials of 29 atoms (up to Cu inclusive) were found experimentally from spectra [1, 2]. Highly reliable data were obtained only for the first 20 atoms. Only approximate estimates of several ionization potentials were obtained for the other atoms. Energies of atoms and ions that were

[1] To whom any correspondence should be addressed.

calculated by us using the HF approximation were compared with the experimental values using the first 29 atoms as examples. This enabled the accuracy of the energies of atoms and ions given by the HF method to be evaluated.

Table. Energies (in at. units) of ground states of neutral atoms from H to Cu and their first two ions calculated in the algebraic approximation of the Hartree–Fock method compared with the experimental values

n	$-E_{RHF}$	Δ%	n	$-E_{RHF}$	Δ%	n	$-E_{RHF}$	Δ%
	H			**He**			**Li**	
1	0,500000	–0,05	2	2,861679	0,49	3	7,432726	0,60
			3	2,000000	–0,01	4	7,236415	0,59
						5	4,500000	–0,002
	Be			**B**			**C**	
4	14,57302	0,65	5	24,52906	0,52	6	37,68861	0,44
5	14,27739	0,34	6	24,23757	0,47	7	37,29222	0,39
6	13,61129	0,33	7	23,37599	0,22	8	36,40849	0,37
	N			**O**			**F**	
7	54,40093	0,38	8	74,80939	0,40	9	99,40934	0,39
8	53,88800	0,35	9	74,37260	0,31	10	98,83172	0,33
9	52,81579	0,32	10	73,10019	0,29	11	97,60897	0,27
	Ne			**Na**			**Mg**	
10	128,5470	0,21	11	161,8589	0,35	12	199,6146	0,34
11	127,8178	0,17	12	161,6769	0,34	13	199,3718	0,33
12	126,3721	0,30	13	159,9973	0,31	14	198,8308	0,32
	Al			**Si**			**P**	
13	241,8767	0,34	14	288,8543	0,34	15	340,7187	0,35
14	241,6746	0,33	15	288,5731	0,34	16	340,3497	0,35
15	241,0307	0,32	16	287,9958	0,33	17	339,6449	0,34
	S			**Cl**			**Ar**	
16	397,5048	0,38	17	459,4820	0,41	18	526,8175	0,43
17	397,1731	0,37	18	459,0485	0,40	19	526,2745	0,42
18	396,3327	0,37	19	458,2264	0,39	20	525,3043	0,42
	K			**Ca**			**Sc**	
19	599,1647	0,46	20	676,7581	0,49	21	759,7357	0,41
20	599,0175	0,46	21	676,5700	0,48	20	759,5391	0,40
21	597,8916	0,45	22	676,1543	0,48	19	758,9808	0,42
	Cr			**Co**			**Ti**	
24	1043,356	0,62	27	1381,414	0,82	22	848,4059	0,50
23	1043,139	0,62	26	1381,128	0,82	21	848,2033	0,52
22	1042,577	0,62	25	1380,423	0,83	20	847,5563	0,54
	Mn			**Ni**			**V**	
25	1149,866	0,68	28	1506,870	0,86	23	942,8843	0,56
24	1149,649	0,67	27	1506,568	0,87	22	942,6707	0,55
23	1149,109	0,67	26	1506,030	0,86	21	941,9816	0,57
	Fe			**Cu**				
26	1262,443	0,67	29	1638,963	0,92			
25	1262,213	0,67	28	1638,694	0,92			
24	1261,564	0,66	27	1638,084	0,91			

n – the number of electrons in an atom or ion.
E_{RHF} is the energy calculated by the Roothaan–Hartree–Fock method (our calculation).
Δ% – $(E_{RHF} - E_{EXP})/E_{EXP}$, in %.

2. Calculation of energies in the algebraic approximation on the HF method

Energies of neutral atoms from He to Xe that were calculated with high accuracy in the framework of Roothaan–Hartree–Fock atomic theory (Roothaan–Bagus method [5]) [7] were reported. This method is based on the formula for the energy of an atom in the central field approximation in which integration over angular variables is carried out. The radial parts of orbitals are represented as expansions in basis functions [atomic orbitals (AO)]. Solving the HF equations in the algebraic approximation resulted in the solution of two interrelated problems, i.e., solution self-consistent field equations in generalized form into eigen values and calculation of optimal values of AO. In the program for solving SCF equations, the cyclic Roothaan method is used; then, if the process of reaching self-consistency diverges, the steepest descent method that is based on gradient and has a wider domain of convergence is employed. To optimize the orbital exponents AO, the direct optimization methods (the zero-order methods), for example, the Powell conjugate direction method [9], are commonly used. However, the convergence of these methods is slow and they provide limited accuracy of the optimized parameters. It is more promising to apply the second-order minimization methods that were described before [7]. Necessary condition for an energy minimum determines the rigorous criterion for the completion of calculation in such algorithms. A multistep scheme for optimizing the exponents that allows us to perform optimization with a high accuracy was used in the program. The resulting calculations exceeded all known calculations in the HF algebraic approximation with respect to the accuracy of the virial theorem (10^{-15}-10^{-17}). It was shown that energies corresponding to the HF-limit could be obtained in the framework of the algebraic method [7]. The radial parts of AO calculated in analytic form can be multiplied by the corresponding spherical functions, and the orbitals obtained can be used to solve the equations of perturbation theory in the HF approximation. It is precisely this approach that was implemented in the calculations of polarizability [6].

3. Results

Due to the comparison of the theoretically calculated by the HF method energies of atoms from H to Cu and of all their ions the authors got the results characterizing the accuracy of the calculated values. For atoms ($1 \leq Z \leq 20$) the error of theoretically obtained values of energy by the HF method is only 0.3%-0.5%. For the isoelectronic series of Sc, Ti, V error of experimental calculations is around 0.6%. However, it starts to grow, and for Ni and Cu atoms is up to 1%. At the same for such atoms the accuracy of experimental data, used for comparison, falls.

4. Conclusions

The results presented in this paper show that the use of the HF method in the calculation of atom and ion energy is quite effective. The error of the theoretical calculations, not exceeding 1%, taking into account the uncertainty of the experimental data, allows to confirm the correctness of the approximations, that compile the base of the Hartree–Fock method.

References

[1] Lide D R 2004 *Handbook of Chemistry and Physics* (New York: CRC Press) p 2475
[2] Moore C E 1970 *Natl. Stand. Ref. Data* **34** 22
[3] Yutsis A P 1978 *Theory of Polyelectronic Atoms. Selected Works* (Vilnius: Mokslas)
[4] Hartree D R 1957 *The Calculation of Atomic Structure* (New York: J. Wiley)
[5] Roothaan C C J and Bagus P S 1963 *Methods in Computational Physics* (New York: Academic Press) **2** 47-94
[6] Eremkin I N and Malykhanov Yu B 2011 *Zh. Prikl. Spektrosk.* **74** (3) 325-332
[7] Malykhanov Yu B, Gorshunov M V, Evseev S V, Eremkin I N and Chadin R M 2013 *Opt. Spektrosk.* **114** (3) 355-362
[8] Malykhanov Yu B and Gorshunov M V 2013 *Zh. Prikl. Spektrosk.* **80** (5) 649-654
[9] Koga T, Watanabe S, Kanayama K, Yasuda R and Thakkar A J 1995 *J. Chem. Phys.* **103** (8) 3000-3005

Journal of Physics: Conference Series **735** (2016) 012053 doi:10.1088/1742-6596/735/1/012053

The analysis of optical wave beams propagation in lens systems

I Kazakov, S Mosentsov, O Moskaletz
St.Petersburg University of Aerospace Instrumentation, 67, B. Morskaya, 190000, St. Petersburg, RUSSIA

E-mail: vasilykazakov@mail.ru

Abstract. In this paper some aspects of the formation and propagation of optical wave beams in lens systems were considered. As an example, the two-lens optical information processing system was considered. Analysis of the two-lens optical circuit has been made with a systems approach perspective. As part of the radio-optical analogies had been applied certain provisions of the theory of dynamical systems to the spatial optical system. The lens system is represented as a simple series-connected optical elements with known spatial impulse response. General impulse response of such a system has been received, as well as consider some special cases of the impulse response. The question of the relationship between the parameters and the size of the input aperture lenses for undistorted transmission of the optical signal has been considered. Analysis of the energy loss resulting from the finite aperture of the lens. It's based on an assessment of the fraction of radiation that propagates beyond the lens. Analysis showed that the energy losses depend explicitly on the following parameters: radiation wavelength, distance between input aperture and lens, and ratio of the input aperture and lens aperture. With the computer help simulation the dependence of losses was shown on the above parameters

1. Introduction

In system of optical information processing (optical signals), a central issue is the formation of wave beams with the desired properties. Device forming such beams represent a plurality of layers are free space and lenses. Analysis of the propagation of optical signals in a lens system based mainly on the principles of geometric optics. This approach greatly simplifies the task of calculating the losses and distortions introduced by the system, but however, has several drawbacks. Firstly, the principles of geometrical optics ignore diffraction divergence of the optical beams, since the path of rays in the free space is rectilinear. Secondly, according to the principles of geometrical optics, an optical beam focused by a lens into a point, which corresponds to an infinitely large energy density and inappropriate physical picture of the phenomenon. In some cases this description is not acceptable. The above-mentioned disadvantages of the principles of geometrical optics do not allow to describe transformation spatial optical signal by optical system. This gives rise to the need to apply the methods of wave propagation analysis of optical beams in optical systems. However, the analysis of wave beams in lens system insufficient attention is given. This is due primarily to the complexity of the calculations required to obtain the fields at the system output.

Any optical information system is characterized by the size of the input aperture, lengths of layers of space and lens systems apertures[1,2]. In terms of structure, an important issue is the relationship between the size of the input aperture of the optical information processing system, and lens aperture systems.

For the analysis of energy losses in the two-lens optical system is sufficient to consider a simple optical system comprising:
- input aperture of finite size, on which supply the optical signal;
- a layer of free space, the impact of which is determined by Fresnel diffraction;
- finite lens aperture, the amount of which exceeds the size of the input aperture.

This simple optical system is a part of complex optical information systems. Known mathematical analysis of such optical systems supposes the lens system of infinite size. However, the real structure of any optical system involves the installation of a lens system with an aperture of finite size. This leads to losses of radiation energy affecting further processing of the diffracted optical field. Loss estimate is based on the part of optical radiation, which partly extends outside of the lens system.

2. Evaluation of energy losses two-lens system

Simple optical system is a link in a complex optical information systems (figure 1).

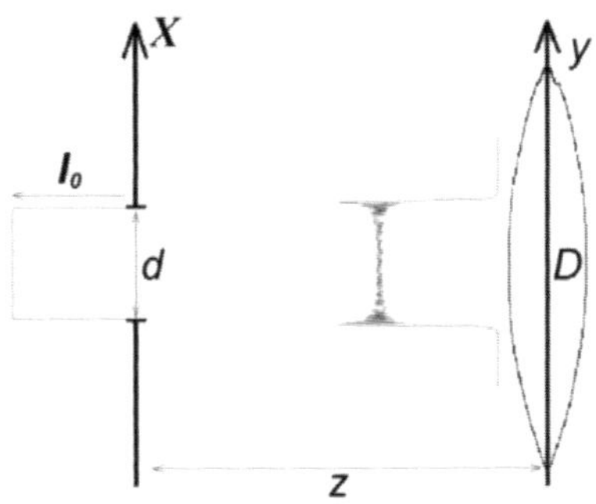

Figure 1. Simple optical system consisting of the input aperture size *d*, space length layer *z* and the lens diameter *D*

Consider the process of converting an input field with intensity I0 in this system. It is obvious that the lens diameter D should be larger than the input aperture d, otherwise it will lead to loss of the diffracted field propagating outside the lens aperture. However, the question of the relationship between the parameters *d*, *D* and *z*, in which the loss of fields are permissible is small, so far was not raised. Finite apertures lenses cause error of the optical system signal. In this case, the criterion of error for evaluating is the relative proportion of the energy which is not involved in the further conversion of the optical signal. It refers to that portion of the optical radiation, which partially extends outside the lens system.

The losses of optical radiation, arising due to the limited aperture of the lens, can be written as:

$$\Delta = 1 - \frac{\left(\int_{-\frac{D}{2}}^{\frac{D}{2}} I(y)\mathrm{d}y\right)}{\left(\int_{-\frac{d}{2}}^{\frac{d}{2}} I_0 \mathrm{d}x\right)}, \tag{1}$$

where I – optical field intensity.

The distribution of the field intensity $I(y)$ can be written as[3]:

$$I(y) = \frac{I_0}{2}\left[(C(\xi_2) - C(\xi_1))^2 + (S(\xi_2) - S(\xi_1))^2\right], \tag{2}$$

where $C(\xi)=\int_0^{\xi}\cos(\frac{\pi t^2}{2})dt$, $S(\xi)=\int_0^{\xi}\sin(\frac{\pi t^2}{2})dt$ – Fresnel integrals, $\xi_1(x)=-\sqrt{\frac{k}{\pi z}}(\frac{d}{2}+x)$, $\xi_2(x)=\sqrt{\frac{k}{\pi z}}(\frac{d}{2}-x)$ – integration variables, $k=\frac{2\pi}{\lambda}$ – wavenumber.

From the equation (2) implies that the losses of the diffracted optical field explicitly dependent on the following parameters: wavelength, distance between the input aperture and lens, lens aperture, which allows to evaluate the distortion of the transmitted signal depending on these parameters. With the help of Matlab the program was created that allows calculating the losses of optical radiation in the considered system. The calculation results are shown in figure 2 and figure 3.

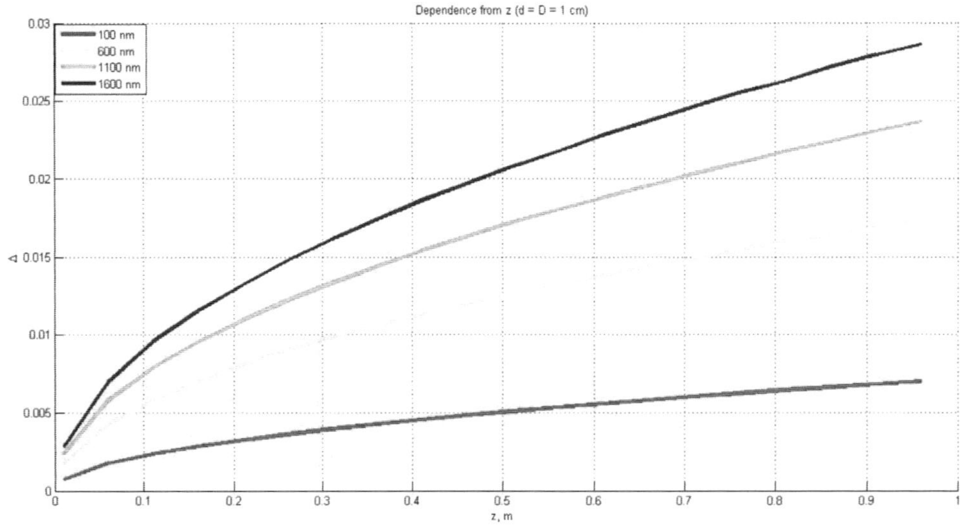

Figure 2. The calculation results of dependence of losses from z

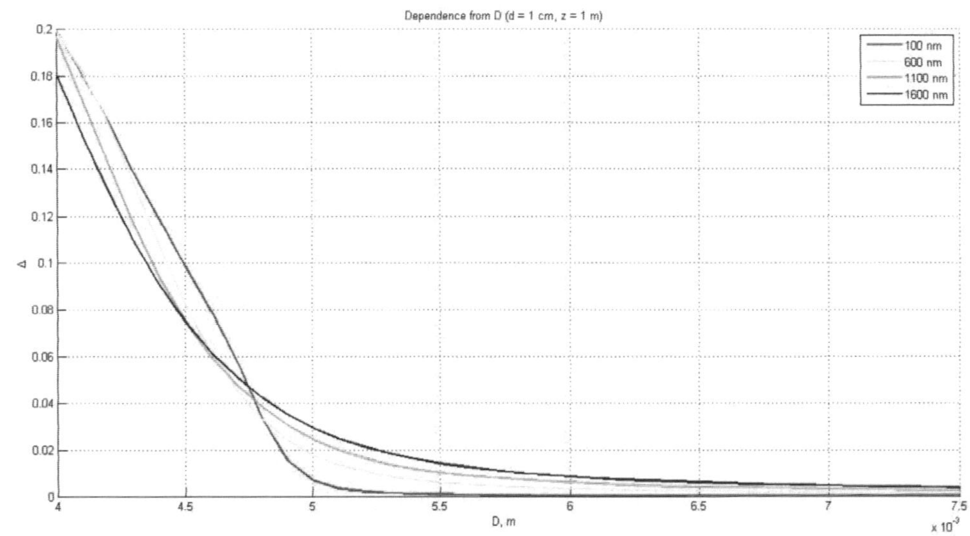

Figure 3. The calculation results of dependence of losses from D

Method of errors estimation can be applied to more complex optical systems that consist of several layers and lenses. A graphical representation of the results enables the selection of parameters for the selected lens in advance margin of error that is acceptable for further calculations.

As a more complicated optical system can be considered two-lens optical circuit (figure 4). Two-lens optical scheme (figure 1) is sequentially located on the one optical axis of the collimating (1) and focusing (2) lenses with the focal length *f*. A feature of such schemes, in contrast to dual scheme spatial Fourier transform is that the distance between the lenses is chosen arbitrarily.

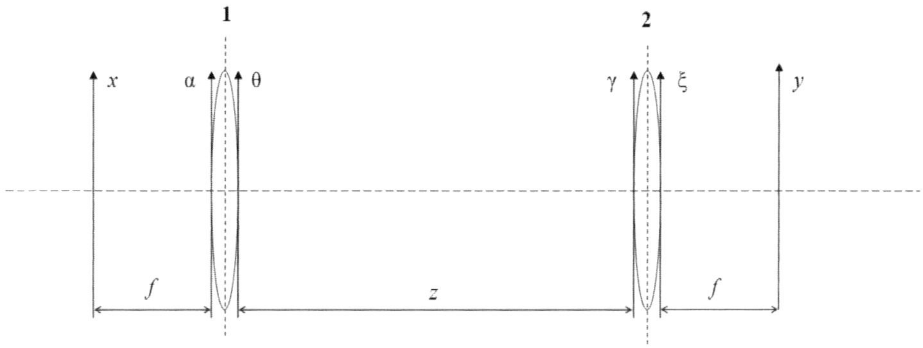

Figure 4. The two-lens optical scheme (*f* - focal length lens, *z* - the distance between the lenses)

In the paper we used some of the theory of a systematic approach to the analysis of the spatial optical lens systems. Known ratio, allowing establishing the input-output linear system in the form of:

$$U_{out}(y) = \int_{-\infty}^{\infty} U_{in}(x) \cdot h(x, y) \mathrm{d}x, \tag{3}$$

where $h(x, y)$ – impulse response of the system.

The lens system is represented as a simple series-connected optical elements with known spatial impulse response. In the case of such a two-lens system are: a layer of free space in front of the lens 1, the lens 1, a layer of free space between the lenses 1 and 2, the lens 2, a layer of free space after the lens 2. For further analysis, the basis of the known relationship was taken, establishing a total impulse response of two cells in series A and B[4]:

$$h_{BA}(t, \xi) = \int_{-\infty}^{\infty} h_B(t, \eta) \cdot h_A(\eta, \xi) \mathrm{d}\eta. \tag{4}$$

where $h_A(t, \eta)$ and $h_B(\eta, \xi)$ – impulse response elements B and A, respectively.

It should be noted that this ratio is known for dynamic systems. For spaces systems, such a transition is based on the use of radio-optical analogy[5], for a two-lens optical system, such an analysis is made for the first time.

The main result of this research is to establish a ratio between input and output of two-lens system which allows for a given input field $u(x)$ by convolution with a total impulse response of the system to obtain the distribution function of the field at the output of $u(y)$. For the two-lens optical system impulse response can be written as:

$$h(x, y) = \exp(i\left[y^2(\beta - \frac{\beta^2}{\mu}) \right]) \cdot \exp(i\beta x^2) \cdot \delta(x + y). \tag{5}$$

where $\beta = \dfrac{k}{2f}$, $\mu = \dfrac{k}{2z}$, $k = \dfrac{2\pi}{\lambda}$, f – lens focal length, z - distance between lenses.

Thus, the ratio is obtained, which depends directly on two variables: the focal length lens (f) and the distance between the focusing and collimating lenses (z), which can be used to calculate the form of the optical field at the output of the system.

To illustrate the use of the expression (5), consider a special case of a two-lens optical system - the double Fourier transform scheme (the distance between lenses is $2f$).The result of a double spatial Fourier transform is a function that is equal to the original. But the sign of the argument is converted[5].

Apply the equation (5) to calculate the field at the output of the scheme, which perform a double Fourier transform. Taking into account that $z=2f$, get $\mu=\beta/2$. Thus, obtained:

$$
\begin{aligned}
U(y) &= \int_{-\infty}^{\infty} u(x)\cdot \exp(i\left[y^2(\beta - \frac{\beta^2}{\mu}) \right])\cdot \exp(i\beta x^2)\cdot \delta(x+y)dx = \\
&= \int_{-\infty}^{\infty} u(x)\cdot \exp(i\left[y^2(\beta - \frac{2\beta^2}{\beta}) \right])\cdot \exp(i\beta x^2)\cdot \delta(x+y)dx = \\
&= \int_{-\infty}^{\infty} u(x)\cdot \exp(-i\beta y^2)\cdot \exp(i\beta x^2)\cdot \delta(x+y)dx = \\
&= u(-y)\cdot \exp(i\beta y^2 - i\beta y^2) = u(-y).
\end{aligned}
\tag{6}
$$

Thus, the expression (6) shows that the field at the output is a mirror image of the field at the input scheme of the double spatial Fourier transform.

Conclusion

In this work it was discussed in general terms two-lens optical circuit of formation of optical beams. As result of analysis based on the use of radio-optical analogies, was obtained spatial impulse response of the system at its various representations. These representations of impulse responses allowed to establish a connection between input and output of system at any distance between the lenses. This representation of impulse response allows the simple, but at the same time clearly and fully describe the process of transformation of the field in lens system. Detailed analysis of expression is shows that the field at the output of the system is equivalent to a mirror image of the field at the entrance, but with additional phase distortions. Since in optics traditionally accepted to work with the intensity of the optical field, i.e., with the square of the modulus of the optical field, phase distortions can be neglected.

This work was funded by the RFBR - The Russian Foundation for Basic Research: project No. 15-37-20446\15

References

[1] Tarasov K I 1968, Spectral devices (Leningrad.: Mechanical engineering).
[2] Belyakov Y M and Pavlycheva N K 2007, Spectral devices (Kazan: publishing state tech. Un-ty)
[3] Gudmen G 1970, Introduction to Fourier optics (Moscow: MIR)
[4] Zade L and Desoer C 1970, The theory of linear systems (Method of state space) (Moscow: Science, Home edition of Physical and Mathematical Literature)
[5] Papulis A 1971, Systems theory and transformation in optics (Moscow: MIR)

Journal of Physics: Conference Series **735** (2016) 012054 doi:10.1088/1742-6596/735/1/012054

Linearization of acousto-optic modulator transmission function

G Korol[1], D Moskaletz[2], O Moskaletz[1]

[1]St. Petersburg State University of Aerospace Instrumentation. 67, B. Morskaya, 190000, St. Petersburg, RUSSIA
[2]St. Petersburg Electrotechnical University. 5, Professora Popova, 197376, St. Petersburg, RUSSIA

E-mail: g-king7@yandex.ru

Abstract. The procedure of linearization of nonlinear transmission function of the optical transparency in the form of an acousto-optic modulator by the methods of nonlinear functional analysis is described. The transmission function of a pair of acousto-optic modulators is linearized in the context of generalized superposition principle.

1. Introduction

Acousto-optic modulator is an input device of dynamic signals in optical information processing techniques by acousto-optics methods. Here the basic operations of the processed radio signals are spectrum analysis and calculation of correlation functions and convolutions. The spectrum analysis assumes a linear regime of radio signal processing, and the calculation of correlation functions and convolution is non-linear operation performed with two radio signals [1], that is fully corresponds to the concepts of functional analysis [2]. In both cases it is assumed linear mapping of input to an optic computing device being processed signals.

The most important characteristic of acousto-optic modulator is its transmission function, which is non-linear mapping of input radio signals in optical systems of signal processing. Salvation requires transmission function acousto-optic modulator linearization. Usually this problem is solved in the context of acousto-optic modulator operation in the Raman – Nath regime at weak sound approach. Really this acousto-optic modulator operation is the exception rather than the rule. This requires problem solving of linearization of acousto-optic modulator transmission function at the most common form. This solution can be obtained on the basis of theoretical optics position [3] and nonlinear operators theory [2], and this determines the linear processing of radio signal by acousto-optical spectrum analyzer [4], which contains one acousto-optic modulator. In this case we are talking about the possibility of superposition principal applying.

Acousto-optic device of correlation processing of radio signals and calculations of their convolution consist in the pair of acousto-optic modulator, in this case the most problem is the solving of problem of establish of transmission function of the pair of acousto-optic modulators at the condition of the linearization of transmission function of each of them. In this case we are talking about the processing of radio signals appearing in the product. This problem can be solved in the context of generalized superposition principle [5].

2. Transmission function of acousto-optic modulator

In general, the transmission function of a transparent or translucent object is expressed in the form [3]

$$F(\xi,\eta)=\frac{V(\xi,\eta)}{V_0(\xi,\eta)} \tag{1}$$

where $V(\xi,\eta)$ - light field passing through the object; $V_0(\xi,\eta)=A\exp(ik(l_0\xi+m_0\eta))\exp(-i\omega t)$, here l_0 and m_0 - the direction cosines.

Transmission function $F(\cdot)$ is the basis for describing of the action of the diffraction grating structure and lattice like structure shaped acousto-optic modulator [3]. In the latter case, the transmission function is described by spatial - temporal dependence

$$F(\xi,\eta,t)=\frac{V(\xi,\eta,t)}{V_0(\xi,\eta)} \tag{2}$$

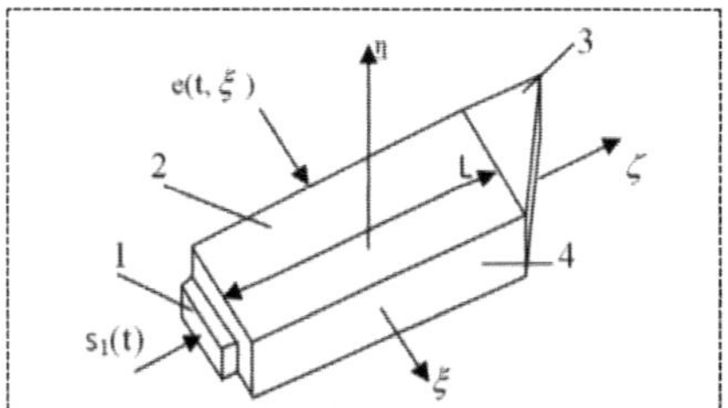

Figure 1. Acousto-optic modulator.

From formula (1) it is follows that the optical field at the output side of acousto-optic modulator, i.e. at the corresponding input stage to the optical information processing system, expressed in the form

$$e(\xi,\eta,t,\zeta)= F(\xi,\eta,\zeta,t)A\exp(-i\omega t)\cdot\exp\left(ik\left(l_0\xi+m_0\eta\right)\right) \tag{3}$$

where further relies $l_0=1$, $m_0=1$.

Acousto-optic modulator transmission function $F(\xi,\eta,\zeta,t)$ is non-linear mapping of acoustic wave $u(\xi,\eta,\zeta,t)$, propagating in the acousto-optic interaction medium. This wave is initiated by electric oscillations $s(t)$, which is input to the piezoelectric transducer. Acousto-optic modulator transmission function can generally be expressed by

$$F(\xi,\eta,\zeta,t)=\hat{\mathrm{N}}u(\xi,\eta,\zeta,t) \tag{4}$$

where $\hat{\mathrm{N}}$ – the nonlinear operator defined by diffraction regime.
The nonlinear operator $\hat{\mathrm{N}}$ is represented by the formula of finite increments i.e. in the form of sum of linear term and a term of order higher than the first concerning $\|h\|$ [2]:

$$\hat{N}(x+h) = \hat{N}(x) + \hat{L}(x)h + \omega(h) \tag{5}$$

where $\hat{N}' = \hat{L}$ - the first Frechet derivative, i.e, linear operator.

3. The linearization function passing acousto-optic modulator

With regard to the problem of determining the transmission function acousto-optic modulator naturally to assume $x = 0$, there is no signal at the input of the piezoelectric transducer and, $h = u(\xi,\eta,\zeta,t)$ then transmission function acousto-optic modulator written as:

$$F(\xi,\eta,\zeta,t) = \hat{N}(0) + \hat{L}(0)\beta u(\xi,\eta,\zeta,t) + \omega(u(\xi,\eta,\zeta,t)) \tag{6}$$

where $\omega(u(\xi,\eta,\zeta,t))$ defines nonlinear distortion, β - dimensional coefficient, $\hat{N}(0) = F_0 = \text{const}$.

Equation (6) is the basis of linear description of the transmission function of acousto-optic modulator at the solving of various problems of information processing by acousto-optics methods.
To use the formula of finite increments (6), it is necessary to find the first derivative of Frechet. This is done as follows. Initially, the Gateaux derivative is established [6]

$$\hat{N}'(x+th) = \lim \frac{\hat{N}(x+th) - \hat{N}(x)}{t}, t \to 0 \tag{7}$$

and its continuity near x is established. If Gateaux derivative is continuous, then Frechet derivative exists and coincides with the Gateaux derivative [6].
At Raman - Nath regime at approximation of weak sound transmission function of acousto-optic modulator is given by the well-known expression

$$F(\xi,\eta,\zeta,t) = \hat{N}u(\xi,\eta,\zeta,t) = A\exp(i\alpha u(\xi,\eta,\zeta,t)) \tag{8}$$

where A, α - dimensional coefficients.

Substituting into expression (7) $h = A\exp(i\alpha u(\xi,\eta,\zeta,t))$ and $x = 0$ gives the relation

$$\hat{N}'(t\exp(i\alpha u(\xi,\eta,\zeta,t))) = \lim \frac{A\hat{N}(t\exp(i\alpha u(\xi,\eta,\zeta,t)) - \hat{N}(0)}{t} = Ai\alpha u(\xi,\eta,\zeta,t), t \to 0 \tag{9}$$

where Gateaux derivative is continuous with respect to $i\beta u(\xi,\eta,\zeta,t)$ and therefore it coincide with the first Frechet derivative.

In expression (8) nonlinear operator $\hat{N}$ is analytical [6], it is represented in the form:

$$\hat{N}(x+h) = \sum_{k-0}^{\infty} \hat{N}^{(k)}(0)h^k \tag{10}$$

which implies the ratio (9).

In the context of accepted idealized description of the acousto-optic modulator

$$F(\xi,\eta,\zeta,t) = F_0 + \hat{\mathrm{I}}\,\beta u(\xi,\eta,\zeta,t) = F_0 + \beta u(\xi,\eta,\zeta,t) \quad (11)$$

where $\hat{\mathrm{I}}-$ the unit operator.

Using the general relation (11) different kinds of acousto-optic modulator transmission function determined by various acoustic beams $u(\xi,\eta,\zeta,t)$ with a base in the form of a rectangular piezoelectric transducer can be expressed. For example, in paper [7], acoustic beam is based at attenuation of the acoustic wave in the acousto-optic interaction medium, and in paper [8] the effect of the diffraction divergence of the acoustic beam is considered.

At the decision of principle, the general problems of the theory of optical information processing systems by acousto-optics methods acoustic beam in the shape of a cuboid is assumed, i.e. at distortion-free acousto-optic interaction medium

$$u(\xi,\eta,\zeta,t) = \chi(\xi)\chi(\eta)\chi(\zeta)u(\zeta - vt) = B\chi(\xi)\chi(\eta)\chi(\zeta)s(\zeta - vt) \quad (12)$$

where B - the dimensional coefficient, $\chi(\cdot)-$ characteristic functions of corresponding intervals determined by the size of the rectangular piezoelectric transducer and an aperture diaphragm; v - the velocity of acoustic waves.

Equation (12) defines acousto-optic modulator transmission function

$$F(\xi,\eta,\zeta,t) = F_0 + \chi(\xi)\chi(\eta)\chi(\zeta)\beta u(\zeta - vt) = F_0 + B\chi(\xi)\chi(\eta)\chi(\zeta)s(\zeta - vt) \quad (13)$$

Relation (13) is the most special case of the general relation (6), that form acousto-optic modulator transmission function usually is used for solving variety common, fundamental problems of optical information processing by acousto-optics methods.

Equations (9) and (13) allow us to write acousto-optic modulator transmission function at Raman - Nath regime at weak sound relation

$$F(\xi,\eta,\zeta,t) = \chi(\xi)\chi(\eta)\chi(\zeta)(F_0 + i\gamma u(\zeta - vt)) = F_0\chi(\xi)\chi(\eta)\chi(\zeta)(1 + i\kappa s(\zeta - vt)) \quad (14)$$

where $\kappa = \dfrac{\mu}{F_0} << 1$

Equations (13) and (14) set fulfilment of superposition principle at input of the electrical signal acousto-optic information processing system, in particular in acousto-optic spectrum analyzer.

4. The function of sequentially arranged pair of acousto-optic modulators

Figure 2 shows a pair of acousto-optic modulators, placed close. It is assumed that the transmission function of each of them is given by the form (8) .With regard relations (2) their joint transmission function defined as

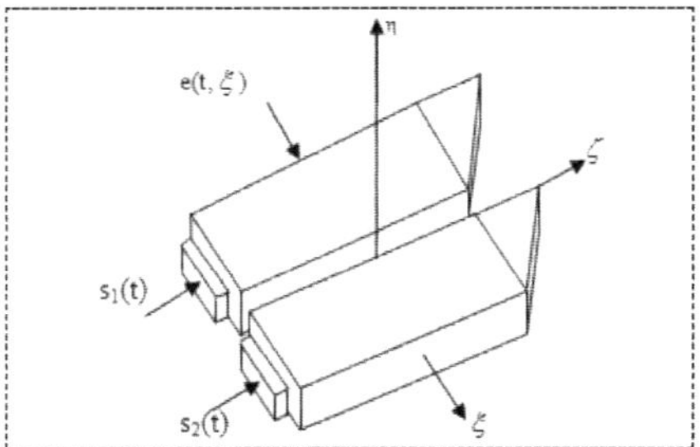

Figure 2. Acousto-optic modulator.

$$F_{1,2}(\xi,\eta,\zeta,t) = A_1 \exp(i\alpha_1 u(\xi,\eta,\zeta,t)) \cdot A_2 \exp(i\alpha_2 u(\xi,\eta,\zeta,t)) = \\ = A_1 A_2 \exp(i(\alpha_1 u(\xi,\eta,\zeta,t) + \alpha_2 u(\xi,\eta,\zeta,t)). \quad (15)$$

The Raman-Nath in the weak sound relation (15) takes the form:

$$F_{1,2}(\xi,\eta,\zeta,t) = A_1 A_2 (2 + (i(\alpha_1 u(\xi,\eta,\zeta,t) + \alpha_2 u(\xi,\eta,\zeta,t)) \quad (16)$$

Relation (16) can be regarded as a form of generalized superposition principle [5].

5. Conclusion

The aim set out in this paper researches is the application general methodology of theory of nonlinear operators to the problem of linear approximation of nonlinear operator describing acousto-optic modulator transmission function. Namely, this approach allows us to solve the problem of acousto-optic signal processing in the context of linear theory. Since these researches are based on common methodology of theory of non-linear operators, the proposed approach is suitable for solving of problems of linearization of any dynamic systems.

References:

[1] Petrovich N T, Razmahnin M K 1969 *Systems due to noise-like signals* (Moscow: Soviet radio)
[2] Kolmogorov A N, Fomin S V 1981 *Elements of the theory of functions and functional analysis* (Ed. the second. Moscow: Nauka) p 554
[3] Born M, Wolf E 1970 *Principles of Optics* (Trans. from English. Moscow: Science) p 856
[4] Lambert L B 1962 *Wide-band instantaneous spectrum analyzers employing delay-line light modulators* (IRE Int. Conv. Rec. 6) p 69-78
[5] Gonorovsky I S 1986 *Radio Circuits and Signals* (Moscow: Radio and communication)
[6] Lyusternik L A, Sobolev V I 1982 *A short course of functional analysis* (Moscow: Higher School) p 271
[7] Molotok V V Razzhivin B P *Impact attenuation of acoustic waves on the characteristics of acousto-optic spectrum analyzers* (Acousto-optical methods and techniques of information processing. Coll. scientific. articles. Leningrad: LIAP-LETI) p 10-15
[8] Egorov Y V, Ushakov V N *Accounting Fresnel diffraction of the acoustic beam in single-channel acousto-optical correlator* (Acousto-optical methods and techniques of information processing. Coll. scientific. articles. Leningrad: LIAP-LETI) p 33-41

Journal of Physics: Conference Series **735** (2016) 012055 doi:10.1088/1742-6596/735/1/012055

Spectrometric control of coatings deposition process

D K Kostrin[1], **A A Lisenkov**[2] **and A A Uhov**[1]
[1] Saint-Petersburg State Electrotechnical University «LETI»
n.a. V.I. Ulyanov (Lenin), 197376, prof. Popova str. 5, Saint-Petersburg, Russia
[2] Institute of Problems of Mechanical Engineering, Russian Academy of Sciences, 199178, Bolshoy pr. V.O. 61, Saint-Petersburg, Russia

E-mail: dkkostrin@mail.ru

Abstract. It is shown that one of the most important technological tasks for the process of reactive magnetron sputtering is to control the composition of a gas mix. Possibility of determining the composition of a gas mix using spectrometric equipment on example of argon–oxygen mix during deposition of tantalum pentoxide films is demonstrated. Results of experiments show that it is possible to control concentration of oxygen not only by change of intensity of oxygen lines, but also by relative change of intensity of argon lines.

1. Introduction

Magnetron sputtering is one of the most common technologies for deposition of films and coatings in opto-, micro- and nanoelectronics [1–3]. Compared to other methods of thin film deposition magnetron sputtering has several advantages, the main of which are:

– low temperature of the substrate;
– good adhesion of film to the substrate;
– high deposition rate;
– good uniformity of thickness.

In the discharge gap of magnetron systems simultaneously with the electric field a magnetic field that allows increasing the concentration of electrons to raise the density of the plasma is applied. In this case use of a magnetic field is equivalent to an increase of gas pressure.

In the magnetron sputtering system abnormal glow discharge in crossed electric and magnetic field is implemented. Discharge exists between cathode *1* and anode *2* (figure 1). Energy of the moving electrons is spent on ionization of the working gas atoms and maintaining of plasma discharge.

2. Analysis

When using a magnetron sputtering system electrons *3* beaten out from the cathode are accelerated in the field of cathode dark space with width *d* and with massive amounts of energy enter the plasma region *4*, where they drift parallel to the surface of the cathode in closed cycloid trajectories *h*. The transition to a path more remote from the cathode surface is possible only after a collision in a field of plasma *4* of electron *3* with atom of plasma gas *6*.

Fast free electrons are locked in the system in a kind of trap that is created, on one hand, by the magnetic field that returns electrons to the cathode, and on the other hand, by a surface of the target that repulses them. In this trap electrons move in complex trajectories and are located until and unless there are some ionizing or exciting collisions with atoms of a working gas, in which they lose energy received from the electric field. Prolonged circulation of electrons increases concentration of positive ions *5* at the surface of a target. This, in turn, leads to an intense bombardment of the target and allows obtaining the current density by two orders of magnitude greater than for cathode sputtering. Sprayed flux of material *7* is directed to the substrate and is deposited on it in a form of a film.

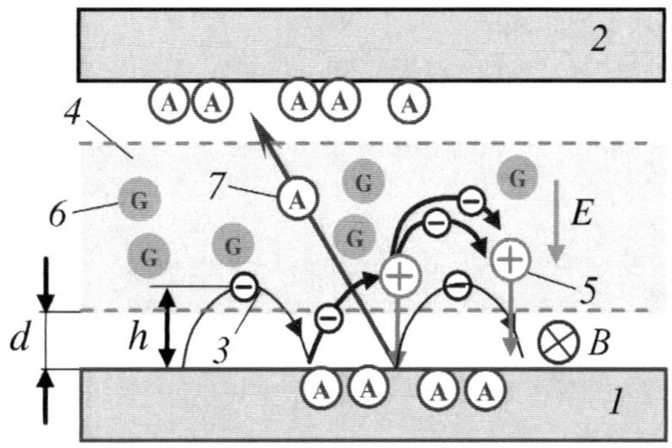

Figure 1. Discharge gap of a magnetron sputtering system.

Magnetron sputtering enables creation of thin films with a desirable composition free of impurities and defects with a high degree of repeatability [4–6]. Feature of reactive magnetron sputtering is the addition into the working chamber of a reactive gas (nitrogen, oxygen, methane, etc.) in supplement to the inert gas for directional changes in the composition of a coating. In reactive sputtering film is formed by interaction of reactive gas with atoms of the magnetron cathode material. Method of reactive magnetron sputtering is much more sensitive to the parameters of the technological process compared with conventional magnetron sputtering.

One of the main parameters requiring control is the partial pressure of reactive gas in the technological chamber. Minimum value of reactive gas pressure in the mix with the working gas is determined by the ratio of stoichiometric coefficients in the chemical formula of the coating material. The excess reactive gas in the technological chamber is undesirable, as in the case when oxygen is used as reactive gas it can lead to oxidation of the target and to changes in the dispersion parameters, and also to rapid failure of diffusion and preevacuation pumps. Since composition of the gas mix has a significant influence on the process of reactive sputtering and properties of the resulting films, maintaining a desired composition of the gas mix is an important task. Method of controlling the composition of the gas mix in the technological chamber carried out with the use of spectrometric equipment is perspective and finds greater application.

Spectrometric control is widely used in ion-plasma processes of etching and deposition of materials. Advantage of using spectrometric equipment for monitoring ion-plasma processes is not only simple integration of such systems into the technological process, but the lack of contact with the vacuum volume of the technological chamber, as well as possibility of diagnosing processes in real time. Depending on a task output of radiation can be made either integrally from the total volume of the technological chamber or selectively from a particular zone of interest for adjustment of the process. Optical fiber is most commonly used for light input to the spectrometer.

To conduct such research at the department of electronic devices and instruments of Saint-Petersburg State Electrotechnical University «LETI» was developed a universal spectrometric complex ISM3600 [7, 8], allowing control of ion-plasma processes, determination of coatings characteristics and properties of various substances, and to diagnose finished product of microelectronic production. Main characteristics of this complex are not inferior to similar device produced by leading companies such as «Avantes» and «Ocean Optics».

Most of spectrometric research methods need a panoramic observation of a spectrum. In mentioned device when using a diffraction grating with 300 lines per mm maximum observed wavelength range is limited to the region of 200...1000 nm, determined by the optical properties of the used silicon CCD

Journal of Physics: Conference Series **735** (2016) 012055 doi:10.1088/1742-6596/735/1/012055

photosensor. Optical resolution of the used device is determined by a diffraction grating and has a value about 1.0 nm. In terms of a tuned technological process it is not required to have a high optical resolution due to the presence of weak and strong lines in the spectrum of each substance; and if the focus is on the latter to monitor the presence of the most important technology components above mentioned optical resolution is sufficient.

Let us consider control process of reactive magnetron sputtering on example of depositing tantalum pentoxide (Ta_2O_5) coatings, applicable in various fields of technology, particularly in electronics when creating components operating in extreme conditions [9, 10]. This material has several properties that distinguish it from other compounds, namely high dielectric permittivity, density and melting point.

With the help of developed spectrometric complex, experiments were conducted on the study of plasma radiation of argon and oxygen mix in a reactive magnetron sputtering installation. Figure 2a shows a panoramic emission spectrum of the gas mix. This spectrum contains a number of lines of oxygen, argon and other substances involved in the process, which significantly complicates its analysis.

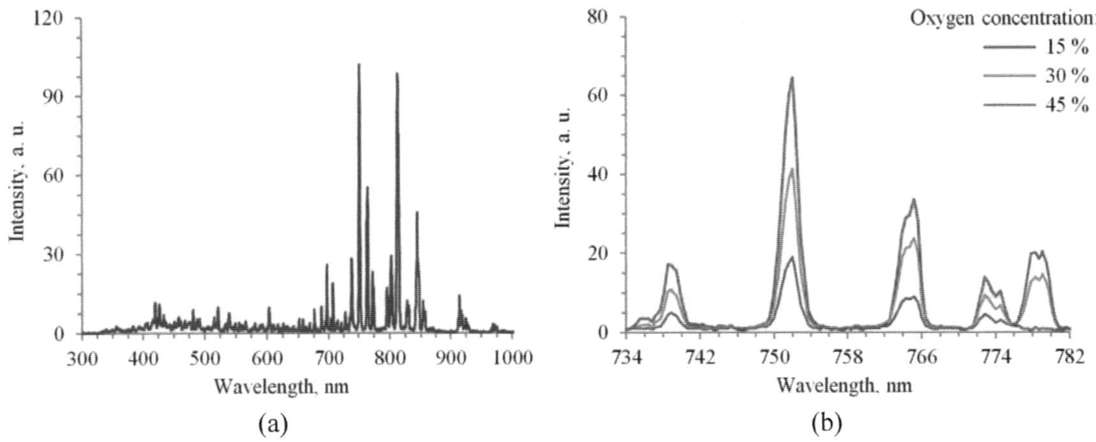

Figure 2. Emission spectrums of plasma during deposition of tantalum pentoxide films: a – panoramic spectrum; b – fragment of spectrum at different concentrations of oxygen.

To control the composition of the gas mix it is efficient to proceed to consideration of spectrum fragment that contains the most intense lines. Figure 2b shows a portion of plasma radiation spectrum for a fixed value of a discharge current during deposition of Ta_2O_5 films at different volumetric oxygen concentrations.

In the considered range of wavelengths argon is characterized by a number of intense spectral lines (738.4, 751.5, 763.5, 772.4 nm), and many of them are close in wavelength to the emission lines of oxygen. Taking into account low resolution of the spectrometer these lines will merge with each other. For atomic oxygen the greatest intensity has a triplet of lines, which is perceived as one line with a maximum at wavelength of about 777.5 nm. A distinctive feature of this oxygen line is its remoteness from the spectral lines of argon, which significantly increases its value in spectral analysis.

Experimental results show that emission line of oxygen 777.5 nm when the volumetric concentration of oxygen is less than 10–15 % is very weak compared with the lines of argon, and to judge the dependence of its intensity on the pressure of oxygen is not possible. With increasing concentration of oxygen there is an increase in the intensity of this line, also there is a significant decrease of intensity of argon lines, especially for the most intensive line of argon with wavelength 751.5 nm.

Volumetric concentration of oxygen during Ta_2O_5 film deposition in the majority of cases is low and is not more than 20 %. Experimental results show that control of oxygen concentrations should be performed not so much by the change of oxygen lines intensity, how many on the relative change of the intensity of argon lines (figure 3).

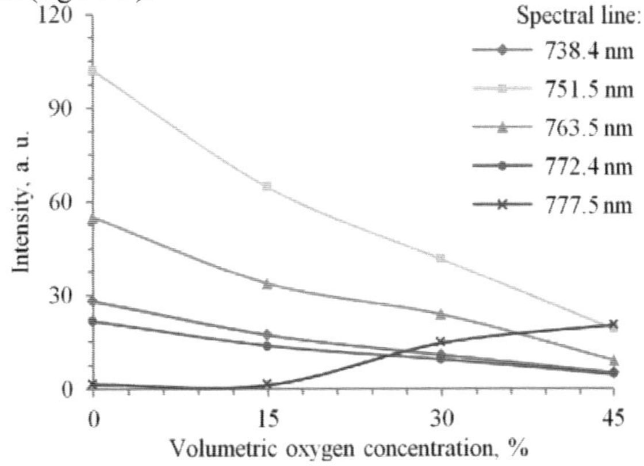

Figure 3. Dependence of the intensity of spectral lines of argon and oxygen from the volumetric concentration of oxygen.

Decrease in the intensity of emission lines of argon with a simultaneous increase in the intensity of the oxygen lines is explained by a transfer of excitation from the metastable energy levels of argon to oxygen according to the following scheme:

$$\mathrm{Ar} + \mathrm{O}(^3P) \rightarrow \mathrm{Ar} + \mathrm{O}^*(3p^5).$$

Energy of the first metastable level of argon is approximately 11.6 eV, excitation energy of the oxygen line 777.5 nm is 10.74 eV. In addition the creation of excited states of oxygen can occur via dissociation:

$$\mathrm{Ar}^* + \mathrm{O}_2 \rightarrow \mathrm{Ar} + \mathrm{O}(^3P) + \mathrm{O}^*(3p^5).$$

3. Conclusion

This study demonstrates the effectiveness of optical spectroscopy for monitoring composition of gas mix in the process of reactive magnetron sputtering, which allows controlling parameters of the obtained coatings. This method of control is not only simple in implementation, but also allows producing diagnostics in real time without any impact on the technological process.

4. References

[1] Wasa K 2012 *Handbook of sputter deposition technology: fundamentals and applications for functional thin films, nano-materials and MEMS* (Norwich: William Andrew)

[2] Wasa K, Kitabatake M and Adachi H 2004 *Thin film materials technology: sputtering of compound materials* (Norwich: William Andrew)

[3] Safi I 2000 *Surface and Coatings Technology* **2–3** 203–218

[4] Bystrov Yu A, Laska V L, Vol'pyas V A, Govako E A, Timofeev D E and Troshkov V V 2002 *Technical Physics Letters* **3** 173–5

[5] Barybin A A, Bystrov Yu A, Komlev A E, Mezenov A V and Shapovalov V I 2006 *Technical Physics Letters* **1** 79–81

[6] Soshnikov I P, Dubrovskii V G, Sibirev N V, Barchenko V T, Veretekha A V, Cirlin G E and Ustinov V M 2006 *Technical Physics Letters* **6** 520–2

[7] Vasilevskii A M, Konoplev G A, Svetlov D A, Gerasimov V A, Kostrin D K, Ukhov A A 2014 *Biomedical Engineering* **1** 13–6

[8] Uhov A A, Gerasimov V A, Kostrin D K and Selivanov L M 2014 *J. Phys.: Conf. Ser.* **567** 012039

[9] Bartic C, Jansen H, Campitelli A and Borghs S 2002 *Organic Electronics* **3** 65–72

[10] Chaneliere C, Autran J L, Devine R A B and Balland B 1998 *Materials Science and Engineering: R: Reports* **6** 269–322

The photoluminescence and phase composition of lead sulphide – cadmium sulphide layers obtained by chemical bath deposition

E.V. Maraeva[1],A.A. Shupta[1], A.A. Bobkov[1], V.S. Levitskii[1], A.I. Maximov[1], V.A. Moshnikov[1,2]

[1] Saint Petersburg Electrotechnical University "LETI", ul. Prof. Popova, 5, 197376, Saint Petersburg,Russia

[2]Peter the Great St.Petersburg Polytechnic University, 29, Polytechnicheskaya st., 195251, St.Petersburg, Russia

E-mail: *jenvmar@mail.ru*

Abstract. The study concerns optical properties and phase composition studies of the layers based on cadmium sulfide – lead sulfide synthesized by the hydro-chemical deposition. The presence of two peaks in the photoluminescence spectra presumably correspond to two different CdS modifications, i.e. the cubic structure of zinc blende and hexagonal structure (wurtzite type). X-ray phase analysis (XRPA) confirmes the presence of both types of crystallites.

1. Introduction

It is known that the layers based on lead chalcogenide and cadmium chalcogenide are traditionally used for creating infrared (IR) photosensitive devices and IR LEDs operating at room temperature [1]. Lead sulfide is a typical representative of $A^{IV}B^{VI}$ compounds class, it is a semiconductor with a band gap of 0.41 eV [2-4]. The crystal PbS has face-centered cubic lattice of the NaCl type. Cadmium sulfide is a representative of $A^{II}B^{VI}$ compounds class. It is a semiconductor with a band gap of 2.42 eV at 300 K. CdS is used in optoelectronics, in both photoreceivers, emitters and solar cells. Cadmium sulfide is crystallized in both the zinc blende (sphalerite) structure and wurtzite structure. Wherein each of the structural modifications can contribute to photoluminescence spectra and determine the optical properties of the created layers.

In this study the aspects of obtaining nanoparticles of lead chalcogenides and solid solutions formed by chemical bath deposition are researched, and its characterization with the use of such methods as X-ray diffraction, photoluminescence measurements is discussed.

1. Experiment

To obtain layers based on lead sulfide – cadmium sulfide the method of chemical deposition from aqueous solutions with the assistance of the following precursors: lead acetate $Pb(CH_3COO)_2$; sodium citrate Na_3Cit; cadmium chloride $CdCl_2$ and thiourea $(NH_2)_2CS$ was used. The pH level was adjusted by adding NH_4OH aqueous ammonia solution. The layers were obtained through the technology developed in the Ural Federal University named after the first President of Russia B. N. Yeltsin [5-8].

AFM experiments were performed using NTEGRA-Therma nanolaboratory (NT-MDT, Zelenograd, Russia). Commercial etched silicon tips NSG 01 with typical resonance frequency of 150 kHz were used as AFM probes. Photoluminescence measurements were carried out with spectrometer LabRam HR800 combined with a confocal microscope (manufactured by Jobin-Yvon Horiba). X-ray diffraction experiments were carried out on «DRN Farad» (Cr-Kα).

2. Results and discussion

The samples with different concentration proportions of $Pb(CH_3COO)_2$ and $CdCl_2$ in initial solution were obtained on vitroceramic-based substrates, and their properties were examined by analyzing X-ray diffraction data and photoluminescence spectra measurements. The concentration proportions of $Pb(CH_3COO)_2$ and $CdCl_2$ in initial solution was 2 (sample 1) and 2.7 (sample 2).

After deposition, the samples were subjected to heat treatment at a temperature T = 200 °C, T = 500 °C, the microstructure was subject to the atomic force microscope investigation. Figure 1 presents AFM images of lateral forces of one of obtained layers before and after annealing.

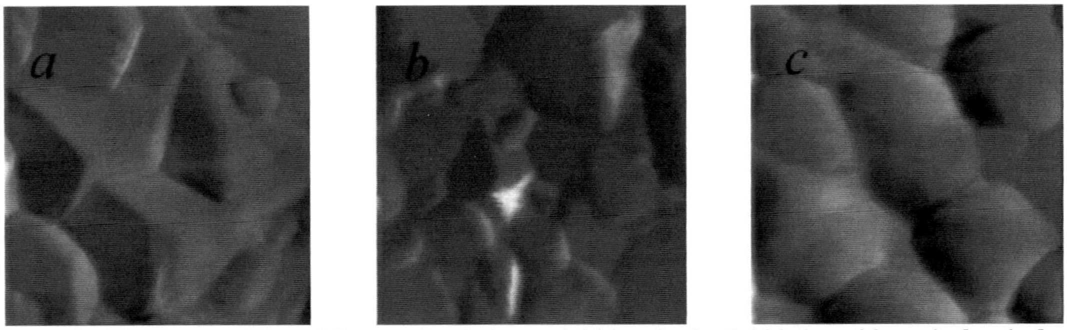

Figure 1. AFM image of lateral forces of cadmium sulphide – lead sulphide based layer before before heat treatment (a, b), after before heat treatment (c). Scan size area is 1 x 1 μm

As it can be seen from figure 1, in some cases crystallites with hexagonal structure were formed. Some crystallites had cubic or pyramidic form.

To determine the phase composition the X-ray phase analysis (XRPA) was used. Fig. 2. shows a bar chart of the layer of the composition 1 before heat treatment. The details of the methods of atomic force microscope investigation and XRPA used are considered in [9, 10].

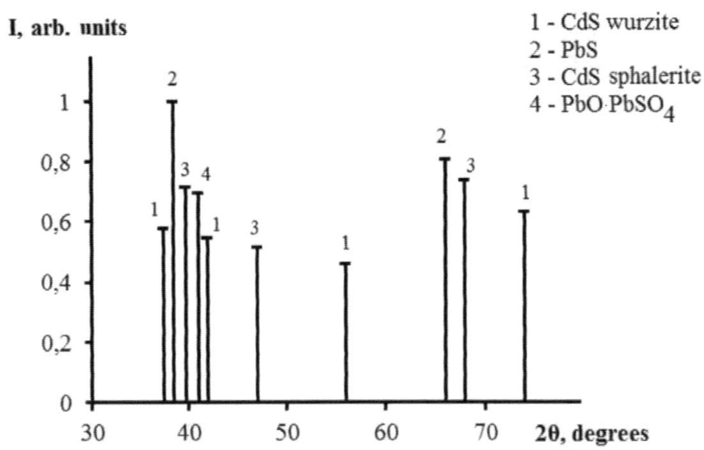

Figure 2. – Bar chart of cadmium sulphide – lead sulphide based layer before heat treatment

As Figure 2 shows, in the bar chart there are lines that correspond to the PbS and CdS phases. This shows that PbS and CdS layers having a crystalline structure were obtained through the hydro-chemical deposition method. It should be specially noted that cadmium sulfide can be seen in two modifications, CdS-wurtzite and CdS-sphalerite. There is also $PbO{\cdot}PbSO_4$ complex oxide phase.

It was found that after a low-temperature treatment (T = 200 °C) processing the intensity of the lines corresponding to $PbO{\cdot}PbSO_4$ oxysulfate lead increases. No active oxide phases of cadmium in such conditions were recorded.

Figure 3 shows a bar chart of the layer 1 annealed at 500 °C.

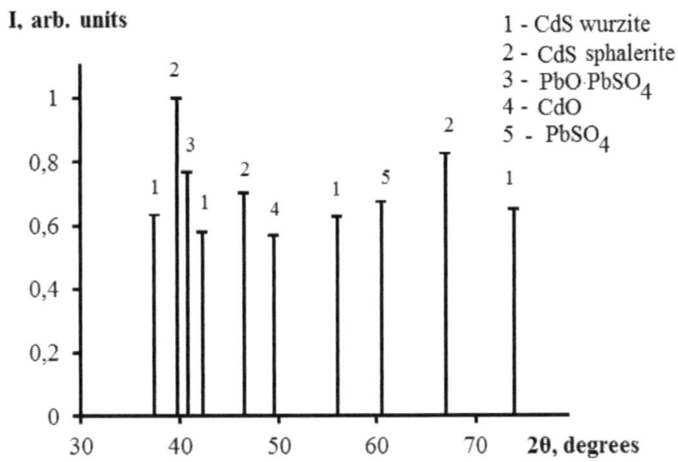

Figure 3. Bar chart of cadmium sulphide – lead sulphide based layer after heat treatment, T = 500 °C, t = 60 min

As Figure 3 shows, cadmium sulfide is still recorded in two modifications, CdS-wurtzite and CdS-sphalerite. There is CdO cadmium oxide phase under the strong temperature influence, and there is also oxysulfate lead. The lines corresponded to CdS – sphalerite are left-shifted in compare to those ones before annyaling. There will be no peak corresponding to PbS phase. Thus pointing to its transition to the oxide phase and also a part of lead sulphid goes to solid solution $Pb_xCd_{1-x}S$ forming.

It should be noted that earlier authors conducted a set of studies on the oxidation effect of polycrystalline layers of lead selenide (and the $Pb_{1-x}Cd_xSe$ solid solutions) on the spectral characteristics of luminescence [11-16]. Despite the variety of technologies for obtaining (thermal spraying and hydrochemical deposition) and replacing selenides with sulfides, the formation of the oxide phases has some common features. At the same time there is a major difference. In fact, the hydrochemical method concerns the coexistence of very limited solid solutions similar in composition to the PbS and CdS phases. Presumably, this is due to lower synthesis temperatures at which the formation of binary solid compounds occurs without the interaction between the PbS and CdS phases to form a solid solution. At higher temperatures, the solubility limit of Pb in CdS rises (especially along the PbS – CdS section).

The photoluminescence spectra were obtained using a spectrometer LabRamHR800 (manufactured by Jobin-YvonHoriba) combined with a confocal microscope. The second harmonic Nd: YAG-laser (wavelength 532 nm) was used as an excitation source. The laser beam is focused to a spot with a diameter of ~1-2 μm on the surface of the sample that allows the measurement of optical properties in different parts of the samples of the inhomogeneous composition. For example, Fig. 4 shows the spectra of the photoluminescence layer of composition 2 before the heat treatment. Lines 1 and 2 correspond to different positions of the laser spot on the sample (No. 1 is the homogenous portion of surface, No. 2 denotes the area corresponding to the exit of individual agglomerates on the surface).

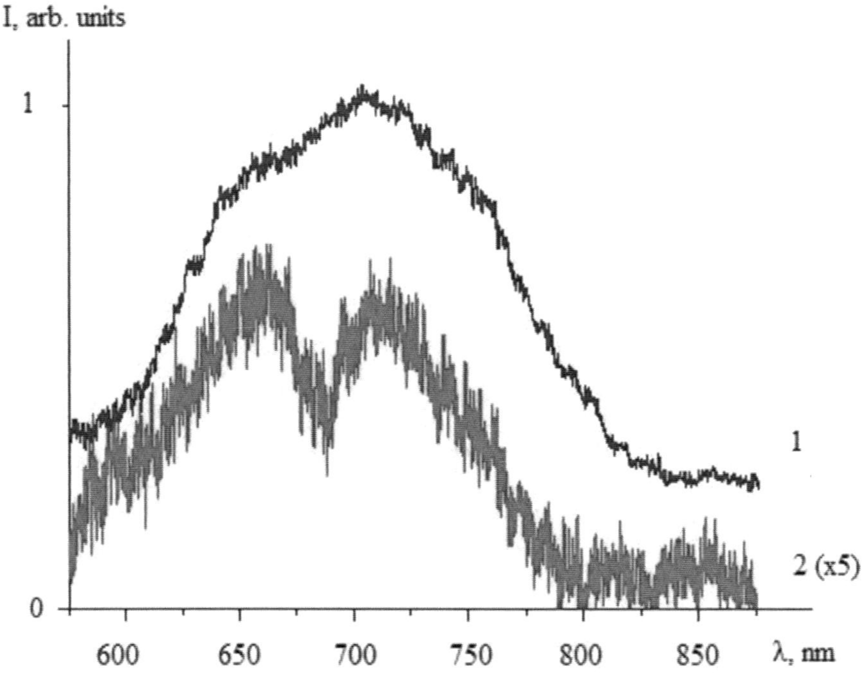

Figure4. PL spectra obtained with homogenous portion of surface of the layer of composition 2 (curve 1) and of the CdS nanocrystals (curve 2) on the surface

Figure 4 shows that the spectrum obtained with the free surface of the film has a wide loop in the range from 550 to 800 nm.

The spectrum obtained from the nanocrystals is in the same range but it has two distinct maxima near 670 nm and 725 nm. A similar pattern was observed by the authors of the paper [17]. They connect the appearance of two peaks of luminescence with the presence of nanocrystals of two structural modifications of CdS with different bandgap values. Each modification contributes to the optical properties of nanoparticles and leads to the appearance of two luminescence peaks, the first of which can be attributed to the wurtzite structure of CdS, and the second one to CdS nanocrystals of the zinc blende.

It should be noted that the intensity of the photoluminescence spectra obtained from CdS nanocrystals (line 2) is significantly less (by about 5 times) than from the free surface of the film which may be due to multiple reflections of light from the facets of the crystals onto the surface.

Conclusions

With the use of chemical bath deposition method layers based on lead sulfide and cadmium sulfide are obtained, their microstructure, phase composition and luminescence are studied. The atomic force microscopy method revealed that the surface layers before the heat treatment system represent a system of crystallites of the hexagonal and pyramidal shapes. Two peaks are observed on the spectral characteristics of luminescence presumably corresponding to two different CdS modifications, the cubic structure of the zinc blende and the hexagonal structure (wurtzite type). The luminescence spectra measurement results are well in line with X-ray phase analysis data confirming the creation of CdS crystallites in two different modifications.

Acknowledgments
The reported study was supported by RFBR (project №14-02-31680) mol_a.

References

[1] Butkevich V G, Bochkov V G, Globus E R 2001 *Prikladnaya fizika* **6** 66
[2] Alexandrova O A, Maximov A I, Moshnikov V A, Chesnokova D B 2008 *Chalcogenides and Oxides of IV Group Element* (Saint-Petersburg) http://www.twirpx.com/file/514792/
[3] Aven M, Prener D S 1967 *Physics and Chemistry of $A^{II}B^{VI}$ Compounds* (Amsterdam) http://www.twirpx.com/file/1366247/
[4] Zimin S P, Gorlachev E S 2011 *Nanostructured lead chalcogenids* (Yaroslavl) http://www.twirpx.com/file/1400782/
[5] MarkovV, MaskayevaL, Ivanov P 2006 *Hydrochemical deposition of metallic sulphide films: simulation and experiment* (Ekaterinburg) http://www.twirpx.com/file/1022619/
[6] Markov V F, Maskaeva L N, Kitaev G A 2000 *Inorganic Materials* **36** 1194
[7] Markov V F, Maskaeva L N 2010 *Russian Journal of Physical Chemistry* A. **84** 1288
[8] Katysheva A S, Markov V F, Maskaeva L N 2013 *Russian Journal of Inorganic Chemistry* **58** 833
[9] Spivak Yu M, Moshnikov V A 2010 *Journal of surface investigation: X-ray, synchrotron and neutron techniques* **4** 71
[10] Maraeva E V, Chesnokova D B, Moshnikov V A 2010 *The technical university of varna. Annual proceedings* **1** 131
[11] Chesnokova D B, Moshnikov V A, Gamarts A E et all V 2010 *Journal of Non-Crystalline Solids* **356** 2010
[12] Moshnikov V A, Gamarts A E, Chesnokova D B, Maraeva E V 2011 *Inorganic Materials* **47** 18
[13] Gamarts A E, Moshnikov V A, Chesnokova D B, Lebedev V M 2004 *Semiconductors.* **38** 1160
[14] Gamarts A E, Moshnikov V A, Chesnokova D B 2006 *Semiconductors.* **40** 662
[15] Maraeva E V, Moshnikov V A, Tairov Y M 2013 *Semiconductors* **47** 1422
[16] Golubchenko N V, Moshnikov VA, Chesnokova D B 2006 *Glass Physics And Chemistry* **3** 337
[17] Piven N G, Sherback L P, Feichuck P I et all 2006 *Condenced media and interphases boundaries* **8** 315

Investigation of optical pulse propagation in optical fiber

O Moskaletz, I Osmakov
St.Petersburg University of Aerospace Instrumentation, 67, B. Morskaya, 190000, St. Petersburg, RUSSIA

E-mail: ilya.osmakov@gmail.com

Abstract. The optical analog of dispersion-time method of measurement of radio spectrum is offered. The ratio between the duration of the input pulse and the differential delay group of dispersion system is established for satisfaction spectrum analyzes of optical pulses. Determining spectral measurements errors method is offered.

1. Introduction

Spectrum analysis is one of the most important field in optical measurements. In compare, with radiofrequency measurements it is impossible oscilloscopic observation of optical signals. Only one way exists to research dynamic signal optical frequency – it's spectrum analysis. Therefore, the most urgent goals at present time and future is improving of known methods and apparatus spectrum analysis. Offered researches allow building the basis of another new optical spectrum analyzer.

In this paper the possibility of new optical spectra measurement like as propagation of optical pulses in a dispersive system (optical fiber) is considered. Such spectrum analyser creating is based on a well-known in a radio range dispersion-time method of measuring radio spectrum [1].

2. Complex spectrum spread function of spectrum analyzer

The functional diagram of the optical spectral device is shown in figure 1.

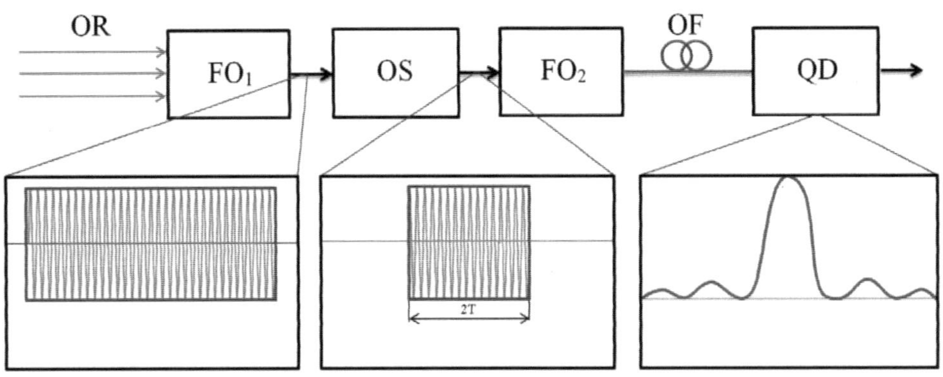

Figure 1. Functional diagram of definition complex spectrum spread functions dispersion-time spectrum analyser, where OR – optical radiation, FO_1,FO_2 - forming optics, OS - optical shutter, OF – optical fiber and QD - quadratic detector.

A well-known theory of dispersion-time analysis of radio spectrum is based on inequality ΔT >> 2T, where ΔT – is length of the analyzed radio pulse; 2T – is differential delay. In the frame theory of spectral measurements the problem of ratio ΔT/2T is not detected. The goal of this paper is researching of transformation optical pulse in dispersive medium, for example in an optical fiber for determining the minimum allowable ratio ΔT/2T.

The result of measuring the complex spectrum $S_a(\omega)$ dynamic signal s(t):

$$S_a(\omega) = \int_{-\infty}^{\infty} K(\omega,\omega')S_0(\omega')d\omega, \tag{1}$$

where $S_0(\omega)$ – mathematic spectrum of analyzed signal s(t), K(ω, ω') – complex spectrum spread function of spectrum device.

In the case of "real spectral device in perfect performance" witch making is Fourier transform of its complex spectrum spread function have form:

$$K(\omega,\omega') = \frac{\sin(\omega,\omega')T}{\omega-\omega'}, \tag{2}$$

where 2T – time of analyzing.

The complex spectrum spread function of optical spectral devices is defined as a reaction to the homogeneous plane monochromatic wave. When using optical fiber as resolving system of a spectral device, it is only required to study propagation in an optical fiber of a single optical pulse $e_0(t)$ shaped as monochromatic oscillation with duration 2T (figure 1).

3. Mathematic modeling of evolution optical pulse in dispersive media.

To determine the degree of closeness function describing the response of the optical fiber impulse $e_0(t)$ to the complex spectrum spread function (2) corresponding mathematical modeling was performed at the following approximation of the transfer function of the optical fiber:

$$\dot{K} = \begin{cases} K\exp\left\{-i\left[\dfrac{(\omega-\omega_0)^2}{2v} + (\omega-\omega_0)t_0\right]\right\} \text{ if } \omega_0 - \dfrac{\Delta\omega_0}{2} \le \omega \le \omega_0 + \dfrac{\Delta\omega_0}{2} \\ 0 \text{ if } \omega < \omega_0 - \dfrac{\Delta\omega_0}{2},\ \omega > \omega_0 + \dfrac{\Delta\omega_0}{2}, \end{cases} \tag{3}$$

where, ω_0 – average frequency bandwidth;
t_0 – constant delay;
υ – the slope of the linear group delay characteristics;
$\Delta\omega_0$ – bandwidth.

One of the most important characteristics of the dispersive medium which using the transfer function (3) is the differential delay ΔT.

Mathematical modeling of evolution $e_0(t)$ in optical fiber with different ΔT was performed with corresponding with formula:

$$I(y) = \frac{I_0}{2}\left[\left(C(\xi_2)-C(\xi_1)\right)^2 + \left(S(\xi_2)-S(\xi_1)\right)^2\right], \tag{4}$$

where, I(y) – oscillation intensity at the output of optical fiber; $C(\xi)=\int_0^{\xi}\cos(\frac{\pi t^2}{2})dt$, $S(\xi)=\int_0^{\xi}\sin(\frac{\pi t^2}{2})dt$ – Fresnel integrals.

The results if mathematical modeling of the optical pulse evolution in the form of truncated monochromatic oscillation are presented at the following diagrams:

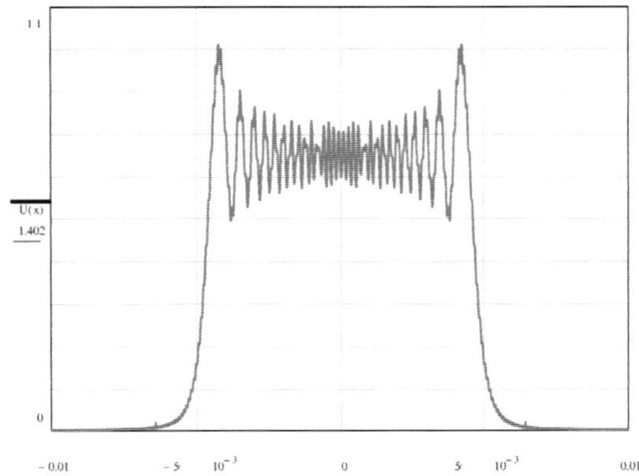

Figure 2.

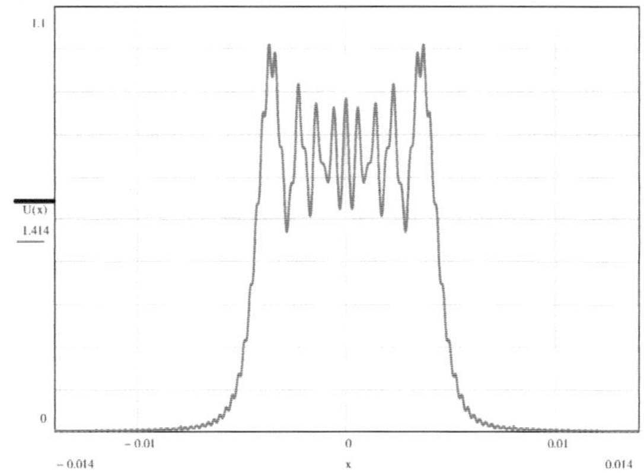

Figure 3.

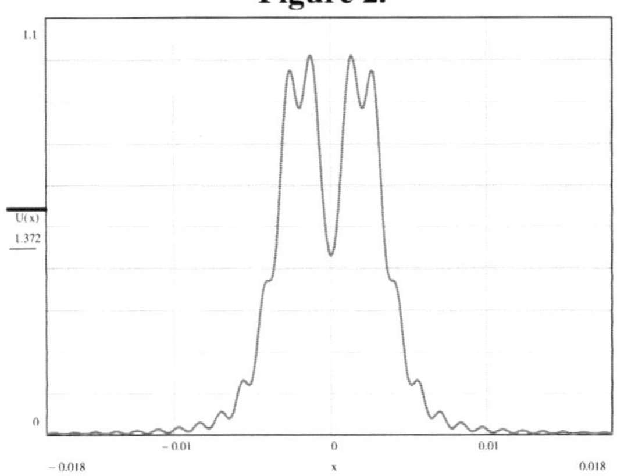

Figure 4.

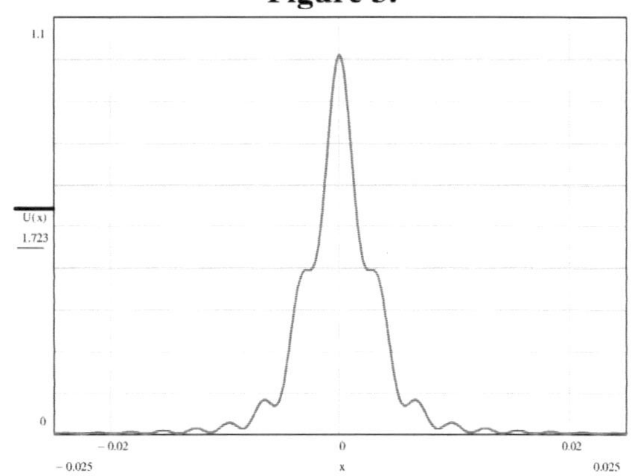

Figure 5.

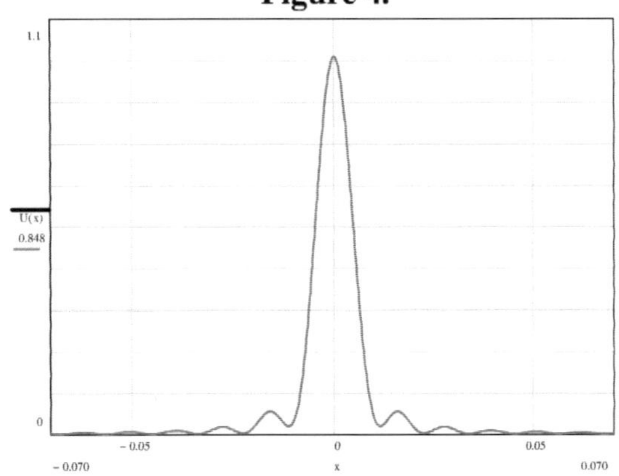

Figure 6.

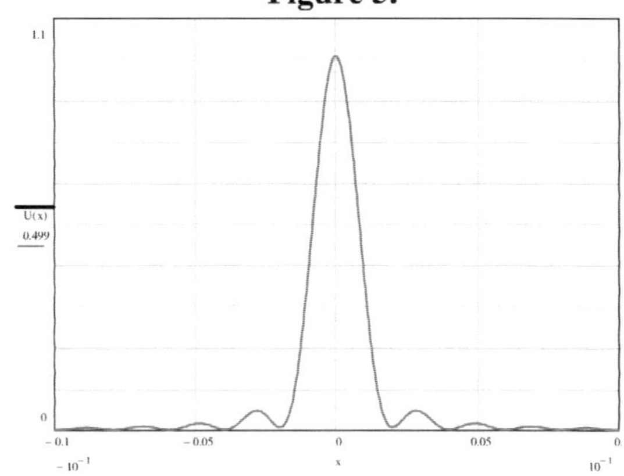

Figure 7.

From these figures, it follows that a good approximation to the required complex spectrum spread function (2) is obtained when ΔT/2T = 50 (Figure 7).

4. Errors of the spectral measurements.

In the ratio (1) the complex spectrum spread function is a reproducible kernel for spectral function which is described by entire functions of exponential type of degree T. For this reason, the kernel (1) is complex spectrum spread function of "real spectral device in perfect performance"

In real spectrum measurements with help of optical fiber the complex spectrum spread function as shown by figure 7 is described by relation which differs from (2). It leads to spectral measurement errors because the results of spectrum analyzing is described in the form (1), where $K(\omega, \omega') \neq \sin(\omega - \omega')/(\omega - \omega')$.

The error of spectral measurements ΔS have form:

$$\Delta S = \int_{-\infty}^{\infty} d\omega \left[\int_{-\infty}^{\infty} S_0(\omega') \frac{\sin(\omega - \omega')T}{\omega - \omega''} d\omega' - \int_{-\infty}^{\infty} S_0(\omega') K(\omega, \omega'') d\omega' \right]^2, \quad (5)$$

this function can be simplified to:

$$\Delta S = \int_{-\infty}^{\infty} d\omega \left[\int_{-\infty}^{\infty} S_0(\omega') \Delta K(\omega, \omega') d\omega' \right]^2, \text{ where } \Delta K(\omega, \omega') = \frac{\sin(\omega - \omega')}{\omega - \omega'} - K(\omega, \omega') \ . \quad (6)$$

The equation (6) allows to raise the question of such complex spectrum function $S_0(\omega)$ which is maximum distorted at given complex spectrum spread function $K(\omega, \omega')$. This problem can be solved by methods of calculus of variation.

5. Conclusion

In this paper method of measuring optical spectrum which based on dispersion properties of optical fiber is offered. The ratio between the length of the analyzed signal and differential delay dispersion system is detected. This ratio allows to say about the satisfactory performance of the analysis of the spectrum of the optical pulse signal. The method of determining the spectral measurement errors is offered. This method can be used for spectral measurements in any spectral instruments.

6. Reference

[1] V I Tverskoy 1974, *Dispersion-time measurement methods of radio spectrum* (Moscow: Soviet radio)

Journal of Physics: Conference Series **735** (2016) 012058 doi:10.1088/1742-6596/735/1/012058

Changes in spectral characteristics of thin molybdenum films with atmospheric annealing

V.D. Paranin
Samara State Aerospace University, 34, Moskovskoe shosse, Samara, 443086, Russian Federation

E-mail: vparanin@mail.ru

Abstract. In this work the spectral transmission, reflection and absorption of molybdenum films with 35 nanometers thick are investigated. It is established that atmospheric annealing at 600 degrees considerably changes optical characteristics of films. At annealing time of 1 min. the absorption maximum on the wavelength of 532 nanometers is observed. At annealing time more than 1 min. the monotonous increase of scattering caused by increase in a roughness of a sample surface is noted. The studied effects allow to increase resolution of laser record of microstructures.

1. Introduction

Thin molybdenum films are used in the technology of laser thermal chemical writing in producing diffractive optical elements and mask plates [1, 2]. Various physical effects, such as oxidation, melting, evaporation etc. can be observed when a focused laser beam interacts with a molybdenum film [3, 4]. Sub-100 nanometer writing resolution in thick molybdenum films 35-70 nm and improvement of the quality of diffractive elements can be achieved by changing the wavelength, intensity and exposure. A large number of the effects observed make it necessary to investigate physical properties of molybdenum films under different thermal and optical impacts. The aim of the study was to analyze the influence of thermal atmospheric annealing on the optical characteristics of thin molybdenum films.

2. Experimental investigation

To carry out the investigation specimens of molybdenum films 35 nm thick were deposited on a silica substrate 2.6 mm thick in vacuum. The surface of the substrate was polished to class 14b according to GOST (State Standard) 2789-59. Spectral transmission of the substrate in the range of 400-700 nm was close to uniform and amounted to 90±1%. After the deposition the molybdenum films were smooth, without visible pinholes or scratches. The films were oxidized at 600°C during 0-3 minutes with a pitch of 0.25-0.50 minutes in a muffle electrically-heated furnace. Optical absorption of the specimens in the visible range was analyzed on a Shimadzu UV-2450 spectrophotometer. The measurement pitch was 1 nm, the spectral width of the slit- 1 nm with a single scan mode and average scan speed. The device was heated during 2 hours prior to measuring. Measurement noises did not exceed ±0.2% in case of the setting specified.

Transmission $T(\lambda)$ was measured in the visible range in a convergent beam with the aperture angle of ±5° and ellipticity of polarization 3:1 - 4:1. Fig. 1 shows measured spectral transmission as the function of the annealing time.

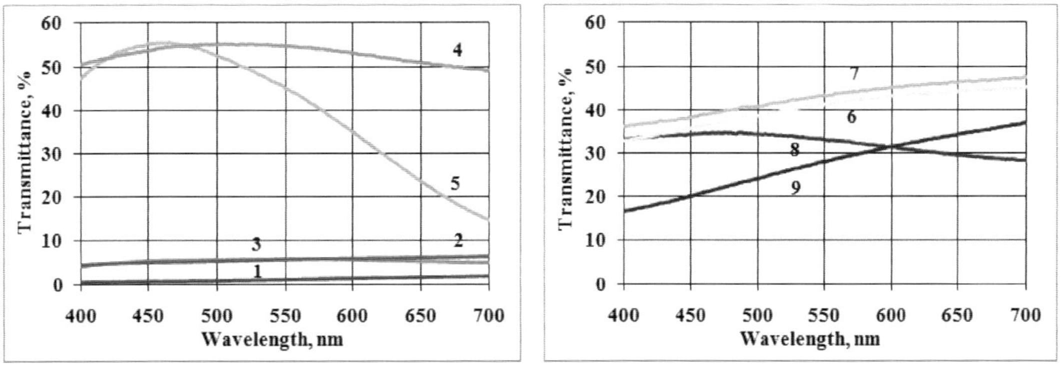

a) b)

Figure 1. Film reflection measured for different times of annealing: a) 1-0 min., 2-0.25 min., 3-0.5 min., 4- 0.75 min.,5-1.0 min., b) 6-1.5 min., 7-2.0 min., 8-2.5 min.,9-3.0 min.

Fig. 2 shows the reflection of specimens R (λ) measured at the striking angle of 5°. The reflection of the film corresponds to the description of its state analyzed by the method of spectral transmission. Scattering of the shortwave part for the specimens with the times of annealing equal to 2.5 min., 3.0 min. (relationships 8, 9 in fig.2, b) is less pronounced due to the influence of aperture properties of the spectrometer reflex attachment used for the analysis of the reflection.

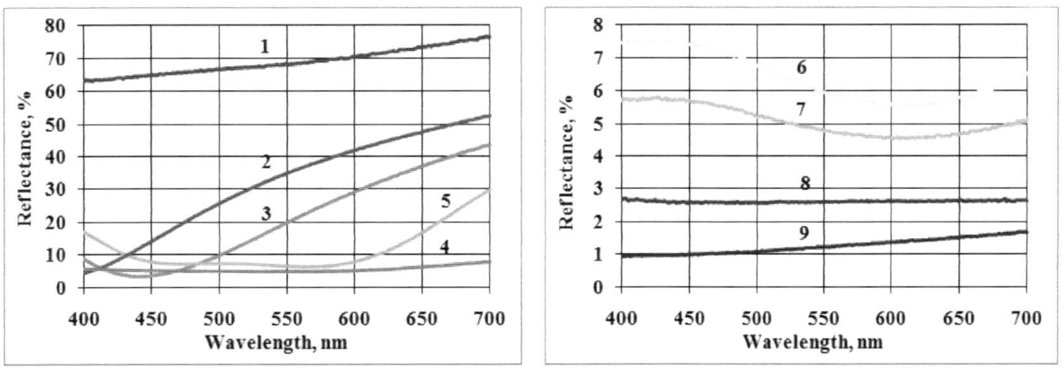

a) b)

Figure 2. Film reflection measured for different times of annealing: a) 1-0 min., 2-0.25 min., 3-0.5 min., 4- 0.75 min.,5-1.0 min., b) 6-1.5 min., 7-2.0 min., 8-2.5 min.,9-3.0 min.

Fig. 3 shows calculated ingestion A (λ) of specimens with the initial molybdenum film thickness of 35 nm. The formula A (λ) =1-R(λ)-T(λ) where R(λ), T(λ) are spectral reflection and transmission, respectively, was used in computations. In case of short time of annealing (0.25-0.75 min.) absorption is increased due to oxidation of the molybdenum film, while in case of a longer period of annealing (0.75-3.0 min.) it happens because the roughness of the oxide film increases in the process of its sublimation at 600 °C.

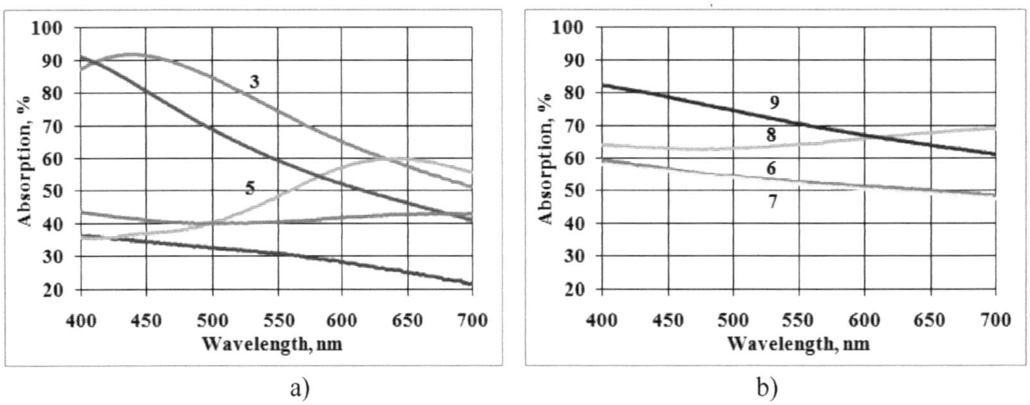

Figure 3. Film absorption computed for different times of annealing: : a) 1 – 0 min., 2 – 0.25 min., 3 – 0.5 min., 4 – 0.75 min., 5 – .1.0 min., b) 6 – 1.5 min., 7 – 2.0 min., 8 – 2.5 min., 9 – 3.0 min.

Figures 4, 5 show changes in transmission, reflection and absorption of the CLWS-200 photomask generator at the wavelength λ=532 nm.

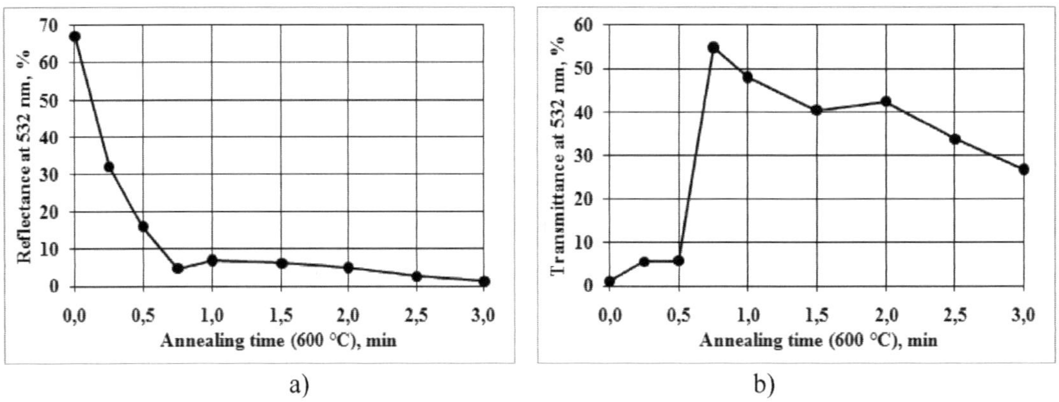

Figure 4. Measured dependence of the film reflection (a) and transmission (b) at λ=532 nm on the time of annealing

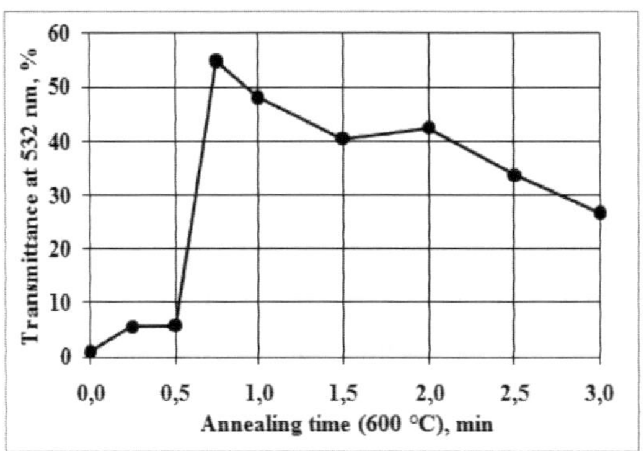

Figure 5. Computed dependence of film absorption at λ=532 nm on the time of annealing

The action of several effects can be seen from fig.4 The first effect is related to the oxidation of the molybdenum film and is accompanied by a sharp decrease of reflection (fig.4, a) and increase of transmission (fig.4,b) of the film in the annealing time range from 0 to 0.75 min. Partial removal of the molybdenum oxide film with the annealing time of 0.50-0.75 min. is also confirmed by the shift of the interference transmission maximum towards short wavelengths (relationships 3,4 in fig.1). The shift corresponds to a decrease of the film design thickness from 140 nm to 110 nm, which also leads to a decrease of absorption (relationships 3, 4 in fig.5 and point 0.50 min. in fig.5).

If the time of annealing exceeds 0.75 min simultaneous decrease in transmission and reflection is observed. This is caused by the scatter of radiation due to increased roughness of the film in its high-temperature sublimation at 600 °C. This is confirmed by the analysis of the annealed film roughness using the Zygo New View 7300 interferometer-profilometer (table 1). As film roughness reduces the transmission and reflection measured the absorption calculated by the expression $A(\lambda)=1-R(\lambda)-T(\lambda)$ increases (fig.5, time 0.75-3.0 min.).

Table 1. RMS of sample profile

Annealing time (min)	RMS (nm)
0	1
0.25	2
0.50	2
0.75	40
1.00	120
1.50	120
2.00	115
2.50	120
3.00	280

3. Conclusion

It follows from the data obtained that the oxidation of a molybdenum film at the initial stage of annealing results in a significant increase of the film's optical absorption, from 32% to 78 % (fig.5). This contributes to increased heat extraction and accelerated ablation of molybdenum-molybdenum oxide films at the centre of the laser Gaussian beam with λ=532 nm, where radiation intensity is the highest. Such coordinate-dependent increase of absorption is equivalent to sharper beam focusing and serves to increase the resolution in case of laser formation of microstructures using the CLWS-200 photomask generator. It should be noted that the absorption measured is typical mostly of molybdenum-molybdenum oxide films. Absorption of silica substrates annealed in the same temperature remained practically unchanged and did not exceed 0.5% in the range of 500 – 600 nm.

Acknowledgments

This work was financially supported by the Russian Ministry of Education and Science.

References

[1] Volkov A V, Moiseev O Yu, Poletaev S D 2013 *Computer Optics* **37** N2 220.
[2] Volkov A V, Moiseev O Yu, Poletaev S D, Chistyakov I V 2014 *Computer Optics* 2014 **38** N4 757.
[3] Volkov AV, Kazansky N L, Moiseev O Yu, Poletaev S D *Technical Physics* 2015 **85** N2 107.
[4] Surovoy E P, Borisova N V *Russian Journal of Physical Chemistry A* 2008 **82** N11 2120.

Optomechanical control of transforming Bessel beams in a c-cut of lithium niobate

V D Paranin[1], S V Karpeev[1,2], S N Khonina[1,2]
[1] Samara State Aerospace University, 34, Moskovskoe shosse, Samara, 443086, Russian Federation
[2] Russian Academy of Sciences Image Processing Systems Institute, 151, Molodgvardeyskaya Street, Samara, 443001, Russian Federation

E-mail: vparanin@mail.ru

Abstract. Transformation of zero-order Bessel beams into a second-order vortex beam in the process of propagation in a c-cut of lithium niobate $LiNbO_3$ crystal has been investigated experimentally. The possibility of controlling beam transformation by means of changing the curve radius of the illuminating beam is shown. The possibility of Bessel beam transforming by compact devices on the basis of thin c-cuts of uniaxial crystals with a diffraction mask formed on their surface is proved.

1. Introduction

High-order Bessel laser beams are used for optical trapping and displacement of atoms [1], microparticles [2], microbiological objects [3, 4]. The use of anisotropic crystals is one of the ways of producing devices and methods of forming Bessel beams with a vortex phase. Theoretical studies [5-8] have shown that periodic transformation of zero-order Bessel beams propagating off-axis into a second-order vortex beam takes place in uniaxial crystals. Higher efficiency of transformation and reduction of the spatial period of transformation in case of increased numerical aperture of the beam and the crystal birefringence is noted in experimental studies [9, 10]. As was shown earlier [8], the beam recurrence interval in the crystal depends on the crystal refraction and the axicon numerical aperture. For axicons with the numerical aperture of 0.15-0.5 the period is tens of microns. For a beam to be formed the thickness of a crystal should be maintained with micron tolerance, which is difficult because of the production complexity.

It is implied, as a rule, that the axicon is illuminated by a plane wave. As this takes place, the greater the diameter of the beam illuminating the axicon, the longer the illuminated section being formed. In most cases the plane wave is obtained with the help of a telescopic system that increases the size of the beam emitted by the laser.

If the distances between the components of the telescopic system vary the radius of the curve of the beam wavefront at the system's output will also vary. The illumination of the axicon by the spherical wavefront is equivalent to the addition of phase functions of the axicon and the lens (the element is known as "lensicon" [11, 12]). It can be easily shown that the numerical aperture of the formed Bessel beam determining the distance between the rings will vary depending on the radius of curvature of the wavefront of the beam that illuminates the axicon. Consequently, the period of beam transformation in the crystal will also vary. This phenomenon can be used for the adjustment of the optical system to the crystal parameters. It can be adjusted both toward the production of a diverging spherical wavefront (the period will increase) and toward a converging one (the period will decrease).

If, for some reason, it is not convenient to adjust the position of the collimator lens this can be accomplished with an additional lens mounted between the axicon and the crystal. In this case the converging wavefront can be obtained only under certain conditions of the formation of a real image $d>4f$, where d is the distance between the axicon and the crystal. The magnification coefficient can be calculated by the known formulas of the geometric-optical theory of image formation.

Variations in longitudinal magnification coefficient known from geometrical optics [13] is the main disadvantage of this approach (taking place as a result of using lenses as imaging systems). It leads to changes in the transformation period along the axis inside the crystal, while ring periods at the input and output of the crystal vary. An additional telescopic system is used to reduce the effect [14]. However, these phenomena are insignificant for low magnification coefficients and moderate numerical apertures. Comparing this optomechanical approach with other possible methods of optics adjustment to the crystal parameters, namely, heating the crystal or changing the wavelength we should note the undeniable advantages of the optomechanical approach, such as simplicity of implementation and a wide range of tuning.

2. Experimental investigation

An optical setup was assembled for the investigation of optomechanical transformation of a zero-order Bessel beam into a second-order vortex beam. The diagram and the external view of the setup are shown in figs. 1, 2. The setup comprised a source of radiation, a beam expander, a diffraction axicon, a c-cut of a crystal, a microobjective and a CCD array. The C-axis of the crystal was oriented parallel to the optical axis of the setup. Accurate bringing of the ordinary and extraordinary beams to a point was provided due to a two-dimensional angular mount.

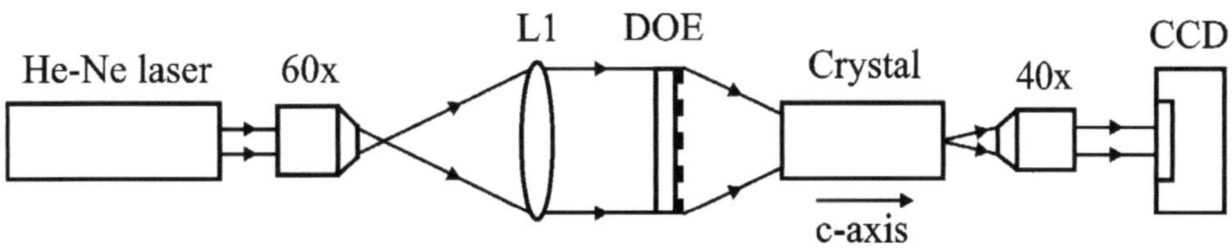

Figure 1. Scheme of experimental setup

Figure 2. Photo of the experimental setup

To form a zero-order Bessel beam a 40 mm-diameter amplitude diffraction axicon on a glass substrate with the ring period of 4 mkm was produced (for λ=0,6328 it corresponds to the numerical aperture α=0,1582). Mask exposure was performed on the CLWS-200 installation in the vector mode, which provided absence of topology defects and high quality of the Bessel beam formed. To detect the beam outside crystal its size was increased by a 40x objective with a NA=0.65 aperture in excess of the axicon numerical aperture. The image of the diffraction mask (chrome 100 nm thick) is presented in fig. 3. The width of the light and dark rings is the same and is equal to 2 microns. There are two rings of increased width, necessary for the alignment of the laser beam, in the central part of the axicon. Since the radius of the outer broad chrome ring does not exceed 50 mkm the distortion of the Bessel beam introduced by it will take place at a distance of up to 350 mkm with an numerical aperture α=0,1582. As the crystal was placed at a distance of 5 mm from the axicon and the diameter of the axicon's illuminated part was 15 mm the presence of such defects did not affect the results of the experiments.

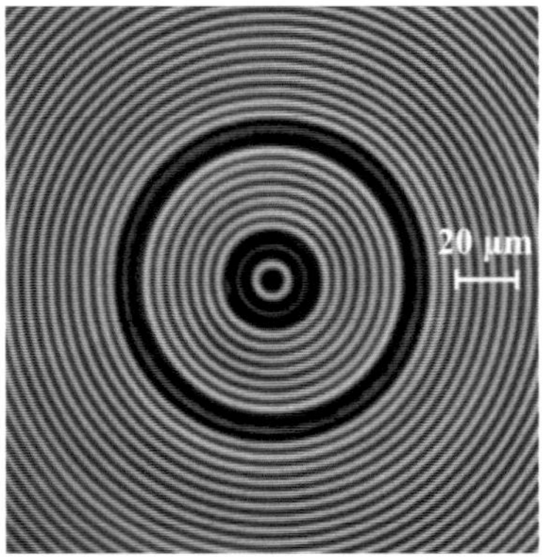

Figure 3. Electron microscope image of the axicon with 4 mkm period

The laser beam was expanded by a 60x objective and an L1plane-convex lens with a focusing distance of 20 cm. The L1 lens could be moved along the optical axis causing changes in the curvature of the vortex beam that illuminated the axicon. Fig. 4 shows the experimental results of Bessel beam transformation in a c-cut of a $LiNbO_3$ crystal 0.5 mm thick for different positions of the L1 lens.

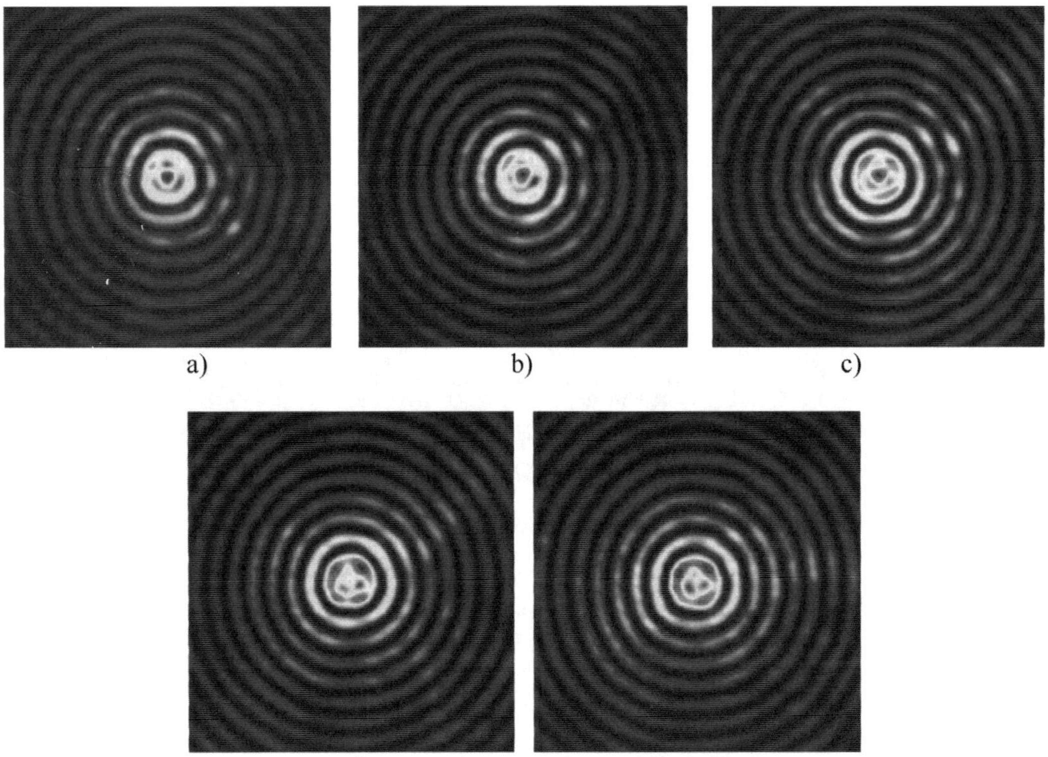

a) b) c)

d) e)

Figure 4. Transformation of Bessel beam in c-cut of lithium niobate with changing of distance from objective to lens L1: a) L=0 cm, b) L=5 cm, c) L=10 cm, d) L=15 cm, e) L=20 cm

From figure 4 it follows that the variation of the wavefront curvature by moving the illuminating lens makes it possible to transform a zero-order Bessel beam into a second-order beam. The magnitude of movement required for complete transformation is determined by the crystal length, the birefringence and the numerical aperture of the axicon.

3. Conclusions

In case of thin crystals considerable shifting of the lens is required, comparable with the focusing distance. This reduces the operation speed of adjustment. This drawback can be eliminated by using axicons with a greater numerical aperture formed directly on the crystal surface. The use of electrically operated lenses based on liquid crystals and electrooptical materials is a promising approach. The methods mentioned can be incorporated in a single optical element comprising control circular electrodes and a diffraction microrelief.

Acknowledgments

This work was financially supported by the Russian Ministry of Education and Science.

References

[1] Tao S H, Lee W M, Yuan X C 2003 *Opt. Lett.* **28** 1867.
[2] Garces-Chavez V. et al. 2014 *Phys. Rev. A* 2002 **66** 063402.
[3] Soifer V A, Kotlyar V V, Khonina S N 2004 *Physics of Particles and Nuclei* **35** N6 733.
[4] Skidanov R V et al. *Optical Memory & Neural Networks 2007* **16** N2 91.
[5] Khilo N A, Petrova E S, Ryzhevich A A *Quantum Electronics* 2001 **31** N1 85.
[6] Stepanov M A *Optics Communications* 2002 **212** 11.
[7] Ciattoni A, Cincotti G, Palma C 2003 *J. of the Opt. Soc. of Am. A.* 2003 **20** N1 163.
[8] Khonina S N, Kharitonov S I *Journal of Modern Optics*, 62(2), 125-134 (2015)
[9] Khonina S N, Morozov A A, Karpeev S V *Laser Physics* 2014 **24** N5 056101.
[10] Khonina S N, Paranin V D, Karpeev S V, Morozov A A 2014 *Computer Optics* **38** N4 598.
[11] Koronkevich V P et al. 1993 *Appl. Opt.* **34**, 5761.
[12] Khonina S N et al. 2011 *J. Opt. Technol.* **78** N11, 724.
[13] Khonina S N et al. 2004 *Proc. of SPIE: Photon Management* **5456** 244.
[14] Vaicaitis V, Paulikas S 2003 *Optical and Quantum Electronics* **35** 1065.

Journal of Physics: Conference Series **735** (2016) 012060 doi:10.1088/1742-6596/735/1/012060

Photo detection process and power spectrum estimation of optical radiation by the multichannel resonant spectrum analyzer

O D Moskaletz, A S Paraskun, M A Vaganov
St. Petersburg University of Aerospace Instrumentation, 67, B. Morskaya, 190000, St. Petersburg, RUSSIA

E-mail: artur.paraskun@icloud.com

Abstract The problem of receiving of an energy spectrum estimation of optical radiations in the new analyzer of optical signals is considered. It is the parallel resonant optical spectrum analyzer (SPECTRUM ANALYZER). Its resolving system is a set of narrow–band optical resonators in the form of interference filters. Each optical resonator is equivalent to a system with lumped parameters. This allows us to consider only oscillations of an optical field in the form of a scalar functions and adopt as a model of analyzed signal harmonized scalar random process. The photodetector operation and average of photocurrent using an integrator and integrating circuit is considered too. On the basis of the application prolate entire spheroidal wave function theory energy spectrum estimation by the integral of photocurrent is obtained. This energy spectrum estimation is consistent and asymptotically unbiased.

1. Introduction

The measurement of the optical signals spectrum is the one of the most important types of optical measurements and an energy spectrum estimation of a stationary random process is the one of the most important problem in the theory of statistical measurements.

There are several methods of receiving of energy spectrum estimation: using predefined the correlation function estimation; on the basis of processing of the instantaneous frequency spectrum; using the filtration method. This work is devoted to the implementation of the filtration method by the multichannel optical SPECTRUM ANALYZER which is considered as the multi-channel narrow-band system [1].

2. The multichannel optical specrum analyzer

The block diagram of the multichannel optical spectrum analyzer which allows to receive an energy spectrum estimation is shown in figure 1.

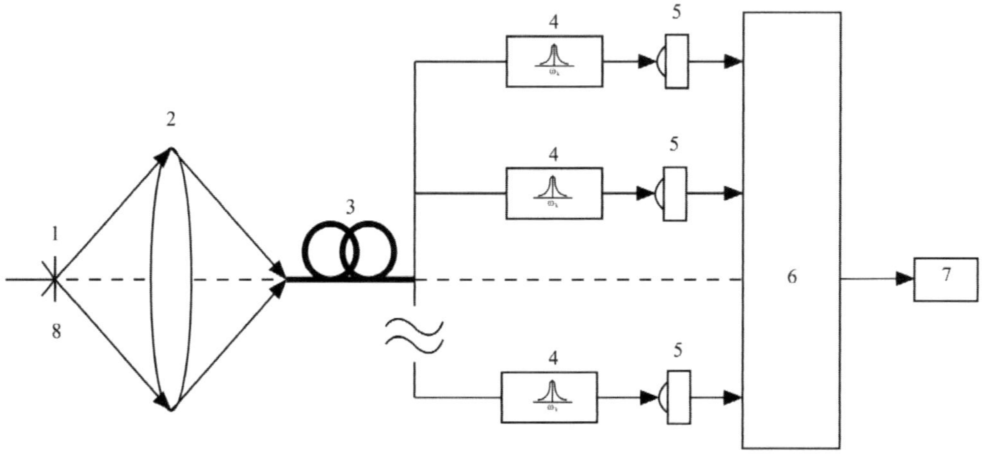

Figure 1. The block diagram of the multichannel optical spectrum analyzer

Where 1, the following notation is introduced: 1 – a source of optical radiation; 2 – forming optics; 3 – fiber optic harness; 4 – resonator units; 5 – photo detectors; 6 – spectroscopic information processing unit; 7 – recorder.

The process of optical filtering in one channel of spectrum analyzer is shown in figure 2.

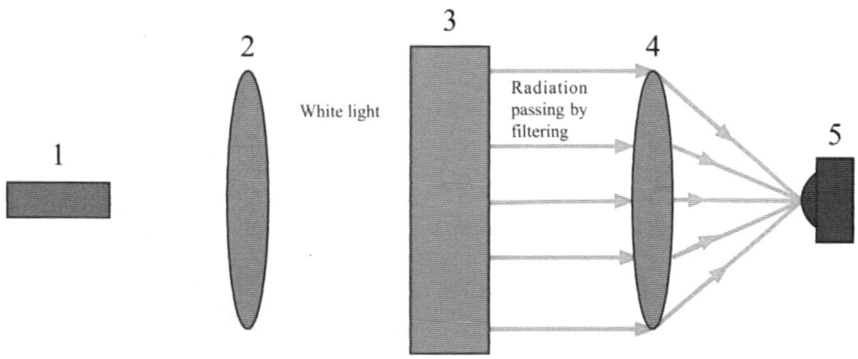

Figure 2. The process of optical filtering in one channel of spectrum analyzer

1 – is output of optical fiber, 2 - is first collimating lenses, 3 – is optical filter, 4 – is second collimating lens, 5 – is photodiode

The spectral characteristic of the optical filter is shown in figure 3.

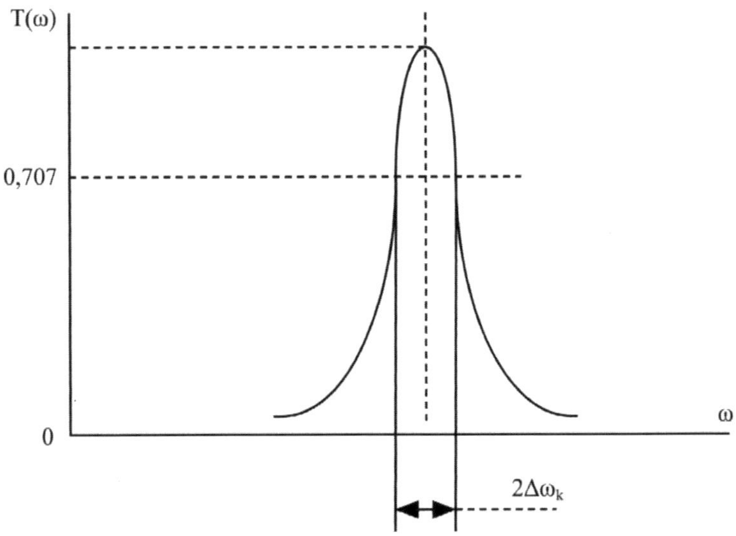

Figure 3. The spectral characteristic of the optical filter

The algorithm of receiving of an energy spectrum estimation of an optical signal is implemented by the following sequence of operations: a narrow-band filtration of an optical radiation – square-law photo detection – temporal integration. The functional diagram of receiving of an energy spectrum estimation of an optical signal is shown in figure 4.

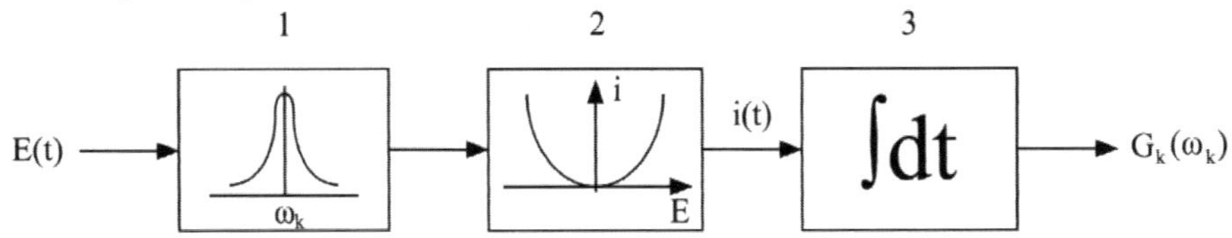

Figure 4. The functional diagram of receiving of an energy spectrum estimation of an optical signal

1 – resonator, 2 – quadratic detector, 3 – integrator, $G_k(\omega_k)$ – evaluation of the energy spectrum of the optical signal at the output k–th channel of multichannel spectrometer.

The output current of the photo detector

$$i = \gamma \cdot P = \gamma \cdot \iint_{\Delta S} \left|\vec{\Pi}\right| \cdot ds = \frac{c \cdot \gamma}{4\pi} \cdot \sqrt{\frac{\varepsilon}{\mu}} \cdot \iint_{\Delta S} E^2 ds. \quad (8)$$

Fluctuations of the electric field strength in the plane of the photo detector are defined as:

$$\overline{E_k} = \overline{E_k}(t) = \frac{\overline{e_l}}{2\pi} \cdot \left(\int_{-\omega_k - \Delta\omega}^{-\omega_k + \Delta\omega} S(\omega) \cdot K_k(\omega) \cdot B_k(\omega) \cdot e^{i\omega t} d\omega + \right.$$
$$\left. + \frac{\overline{e_l}}{2\pi} \int_{\omega_k - \Delta\omega}^{\omega_k + \Delta\omega} S(\omega) \cdot K_k(\omega) \cdot B_k(\omega) \cdot e^{i\omega t} d\omega \right). \quad (9)$$

Substitution of expression (11) in (6) gives the Pointing vector:

$$\overline{\Pi} = \frac{c}{4\pi^2}\cdot\sqrt{\frac{\varepsilon}{\mu}}\cdot\bar{e}\cdot$$
$$\cdot\int_{\omega_k-\Delta\omega}^{\omega_k+\Delta\omega} S(\omega)\cdot K_k(\omega)\cdot B_k(\omega)\cdot e^{i\omega t}d\omega\cdot \tag{10}$$
$$\cdot\int_{\omega_k-\Delta\omega}^{\omega_k+\Delta\omega} S^*(\omega')\cdot K_k^*(\omega')\cdot B_k^*(\omega')\cdot e^{-i\omega' t}d\omega'.$$

Using the integrator in scheme is very important because the result of estimation is defined by the accumulated energy. This means the necessity of integration of the Pointing vector in time:

$$\int_{-T}^{T}\overline{\Pi}(t)dt = \int_{-\mathrm{T}}^{\mathrm{T}}\mathrm{dt}\int_{\omega_k-\Delta\omega}^{\omega_k+\Delta\omega} S(\omega)\cdot K_k(\omega)\cdot B_k(\omega)\cdot e^{i\omega t}d\omega\cdot$$
$$\cdot\int_{\omega_k-\Delta\omega}^{\omega_k+\Delta\omega} S^*(\omega')\cdot K_k^*(\omega')\cdot B_k^*(\omega')\cdot e^{-i\omega' t}d\omega'. \tag{11}$$

Application of the theory of prolate spheroidal wave functions to the expression (14) gives[2]:

$$G_k(\omega)=(\Delta\omega_k)^2\cdot P_k(\omega_k)\cdot$$
$$\cdot\int_{-\Delta\omega_k}^{\Delta\omega_k}\left|K_{kk}(\omega_k,\omega')\right|^2\cdot\left|B_{kk}(\omega')\right|^2\cdot\left|S_0(\omega')\right|^2 d\omega'. \tag{12}$$
$$\cdot\int_{-\Delta\omega_k}^{\Delta\omega_k}\left|K_{kk}(\omega_k,\omega')\right|^2\cdot\left|B_{kk}(\omega')\right|^2\cdot\left|S_0(\omega')\right|^2 d\omega'.$$

This expression determines the energy density accumulated in the estimating process of the energy spectrum of optical radiation.

The accumulated energy during this process is expressed by:

$$W_k = \iint_{\Delta S} G(\omega)ds \tag{13}$$

This expression determines an energy spectrum estimation of optical radiation as a stationary random process.

Previously it was assumed that the process of integration is performed perfectly. However it is necesspectrum analyzerry to consider using integrating circuit instead of the integrator. It is described by the expression.

If the time constant of the integrating circuit is more than the processing time, then the process is close to a perfect integration. If this condition doesn't perform, the energy spectrum estimation is distorted.

3. Conclusion

The performed analysis has demonstrated that the properties of energy spectrum estimation depend on the characteristics of the transfer function of the narrow-band filter and the time of perfect integration. Distortions of energy spectrum estimation can be defined on basis of the transfer function.

Using a simple integrator as an integrating circuit allows to implement perfect integrating and defining distortions during long period of observation.

Journal of Physics: Conference Series **735** (2016) 012060 doi:10.1088/1742-6596/735/1/012060

Given energy spectrum estimation of optical radiation is a consistent and asymptotically nondisplacement [3].

4. Reference

[1] Moskaletz, O.D., Vaganov, M.A., Kulakov, S.V., Preslenev, L.N., Arkhipov, I.N., Kotlikov, E.N. and Prokashev, V.N. Parallel spectrum analyzer of signals of optical range. Patent of Russian Federation № 86734 (2009).

[2] Vaganov, M.A., Moskaletz, O.D., Parallel spectrum analysis of dynamic signals. // Informacionno-upravljajushhie sistemy (Information and control systems), Vol. 54, Iss 5, 15-22 (2011).

[3] Bendat J.S., Piersol A.G. Measurement and analysis of random data. John Wiley & Sons, Inc. New York-London-Sydney. 1967.

This work was funded by the RFBR- The Russian Foundation for Basic Research: project No. 15-37-20446\15.

Effect of angular intensity distribution of radiation on a conoscopic pattern of crystal

O Y Pikoul, L L Kovalenko
Far Eastern State University of Transportation, Khabarovsk, Russia

Abstract. Angular intensity distribution of radiation in incident converging beam effects on the appearance of a conoscopic pattern of optical crystal. The peculiarities of a «Maltese cross» formation in the conoscopic patterns of optical crystals are considered.

1. Introduction

Conoscopic pattern of the crystal is determined by the spatial distribution of differences of course and direction of the vectors E at the output of the crystal in different areas of the cone and becomes visible to the analyzer.

2. Analysis

The intensity of radiation in the incident beam Gaussian distribution, that is,

$$I = I_0 e^{-k\beta^2}, \tag{1}$$

where I_0 - intensity of radiation along the axis of the beam; k - coefficient characterizing the radiation intensity distribution in the beam; β - the angle of incidence on the plate.

Fig. 1 shows the intensity distribution for the three cases differing in the value of the coefficient k. In the case of k = 0, the intensity of radiation in the beam is distributed uniformly.

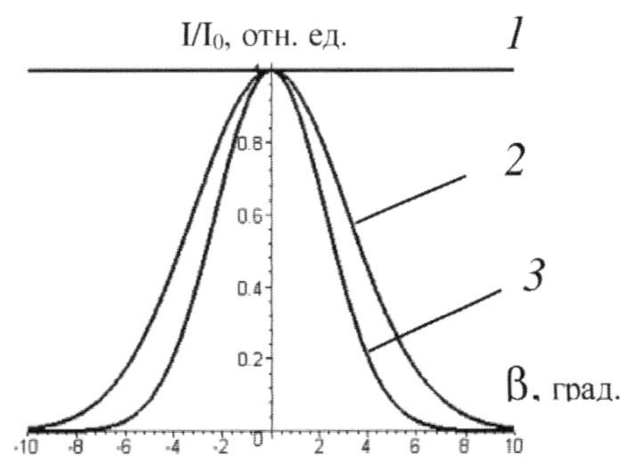

Figure 1. The intensity distribution of the radiation in a converging beam. The values of the coefficient k: 1 - 0.0; 2 - 0.05; 3 - 0.1

After passing through the beam of rays converging system consisting of a polarizer, crystal plate and the analyzer, the radiation intensity is given by

$$I_1 = I_0 e^{-k\beta^2} \frac{1}{2}(1-\cos\Delta)\sin^2 2\varphi, \tag{2}$$

where Δ - the phase difference between the ordinary and extraordinary rays, which depends on the angle β; φ - the azimuth angle measured, for example, the transmission axis of polarizer (from one of the branches of the "Maltese Cross" in the conoscopic pattern).
Equation (2) allows you to build a conoscopic pattern in polar coordinates, putting the angle of incidence β of the radius vector. The angle φ - the azimuth angle (the angle of rotation of the radius vector).

Figure 2 shows the conoscopic pattern under certain simplifying assumptions. The black circle correspond to minima conoscopic pattern (Δ = 2πn, where n = 0, 1, 2, ...). If you increase the thickness of the crystal plate diameter of these circles is reduced.

When building a "Maltese cross" (figure 2), it was assumed that the boundary of visibility of the picture. That is, the intensity values in the conoscopic picture shown in black; values are shown in white.

It is clearly seen that a uniform distribution of intensity on the corner angular width "Maltese cross" is the smallest and does not depend on the values of β (figure 2a).

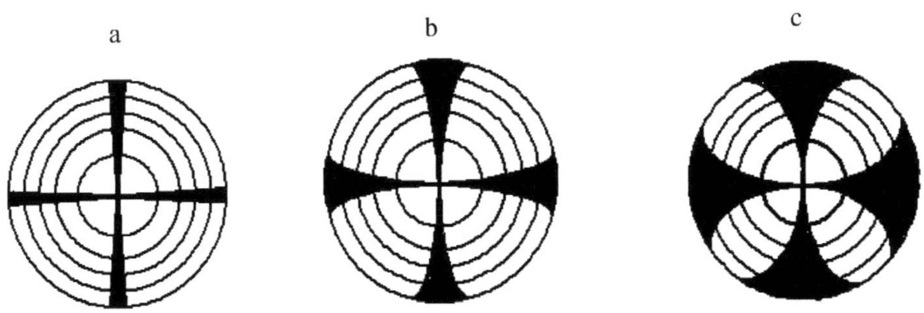

Figure 2. Conoscopic pattern, designed for radiation beams with different angular distribution of intensity. The coefficient k: a - 0; b - 0.05; c - 0.1

The faster decreases the intensity of radiation in a converging angle of the beam, the greater the "Maltese cross" (figure 2b, c). The width and the type of "Maltese cross" does not depend on the thickness of the crystal plate.

Calculation of the borders "Maltese cross" is implemented on the basis of the expression (2), converted to the following form:

$$\sin 2\varphi = (I_1 / I_0 e^{-k\beta^2})^{1/2}. \quad (3)$$

If necessary, by the form "Maltese cross" in conoscopic pattern can be used to evaluate the structure of the convergent beam.

The literature indicated the existence of black "Maltese cross" in the conoscopic pattern and light "Maltese cross" [2]. However, the description of the kind of "Maltese cross" at arbitrary angles between the axes of the polarizer and analyzer transmission is omitted.

Study changes in the intensity of radiation on the rings-isochromen conoscopic pattern at different positions of the transmission axes of the polarizer and analyzer showed that all the charts there are corners φ, in which all isochromen with different phase shifts have the same intensity.
When the polarizer and analyzer are parallel (χ = 0°), there are four such radial directions and are located through 90° (figure 3,a). The intensity of izohrom crossing with these directions is the same, that is formed by the light "Maltese cross."

When the polarizer and analyzer are crossed, then there are also four such radial direction and they are arranged as through 90°, but the intensity izohrom in these areas is equal to zero, that is, formed black "Maltese cross" (figure 3, e).

When the angle χ between the axes of the polarizer and analyzer transmission is arbitrary, then there are 8 of these radial directions in which the intensity izohrom same (figure 3, b, c, d).

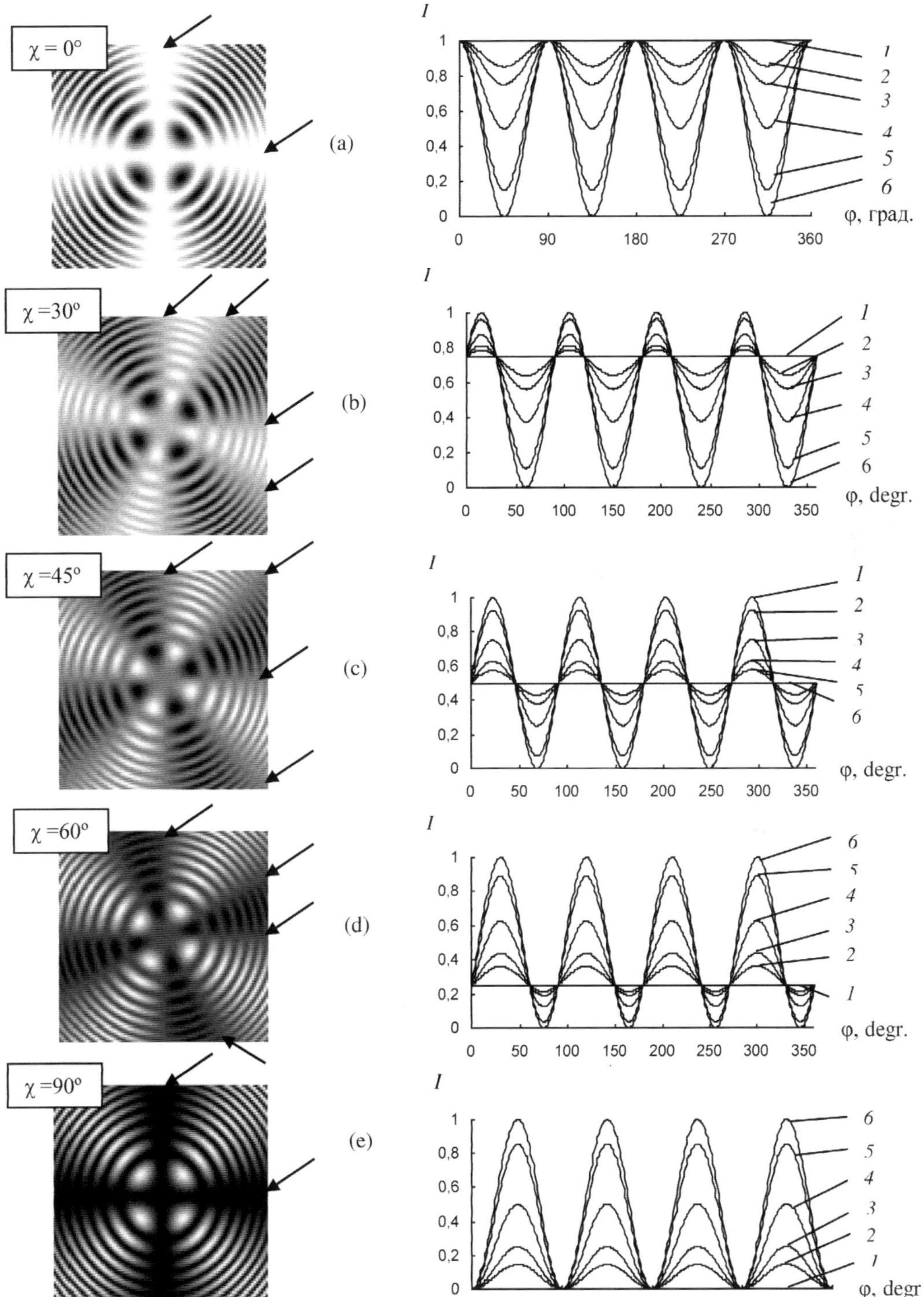

Figure 3. Conoscopic pattern of lithium iobate ($LiNbO_3$), designed to Maple at respective angles between the axes χ transmission polarizer and analyzer. The arrows indicate the "Maltese cross." On the right side - the intensity of the radiation transmitted polarizer - crystal plate - the analyzer angle φ. The values of the phase difference Δ, degrees: 1 - 0; 2 - 45; 3 - 60; 4 - 90; 5 - 135; 6 - 180.

With further increase of the angle between the transmission axes of the polarizer and analyzer highs and lows on the chart are reversed. In all cases, there are four fixed radial direction: 0°, 90°, 180° and 270°, forming one fixed "Maltese cross" with a gradually varying intensity.

The remaining four radial direction form the second "Maltese cross", which rotates around a fixed cross at an angle corresponding to the angle of rotation of the analyzer.

The change in the angle between the polarizer and analyzer from 0° to 90° leads to the same change in the intensity of both crosses in the range of zero. The intensity in the center of the conoscopic pattern corresponds to the intensity of "Maltese cross."

Similar changes, only in reverse order, occur during rotation of the analyzer from a position in which the transmission axis of the polarizer and analyzer are parallel. Bright "Maltese cross" is divided into two cross, the intensity of which gradually decreases, reaching a minimum value at the time of the merger of the two crosses in the dark "Maltese cross" with the axes of the orthogonal transmittance of the polarizer and analyzer.

The results of experiment on the "Maltese cross" shown in figure 4.

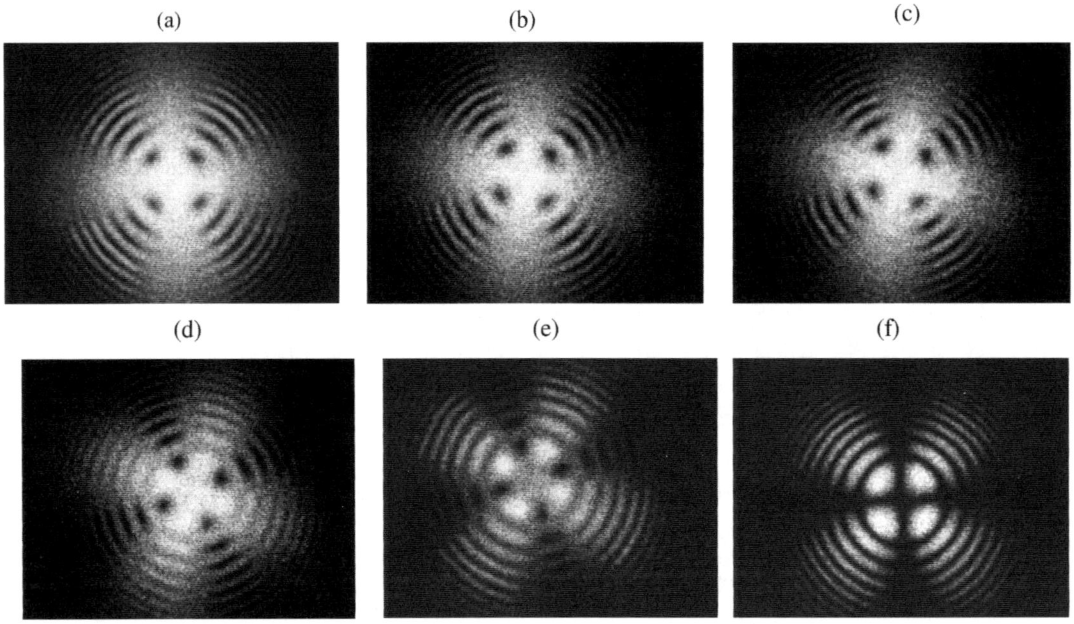

Figure 4. Photos "Maltese cross" on the conoscopic pattern of lithium niobate. The angle between the transmission axes of the polarizer and analyzer, degrees: a - 0; b - 20; c - 30; d - 45; e - 60; f - 90.

4. Conclusion

For crystal plate TeO_2 (uniaxial gyrotropic crystal) such a change conoscopic pattern while rotating analyzer much more worse. The reason for this is the lack of a "Maltese cross" in the center of the picture due to the optical activity [2, 3]. The peripheral part of the picture, where the branches of "Maltese Cross" there, quickly obscured, making it difficult for observation.

5. References

[1] Pikoul O, Alekseyeva L, Povkh I et al. 2004 *IVUZ. Priborostr.* **12** 53

[2] Stoiber R and Morse S 1972 *Microscopic identification of crystals* (New York: The Ronald Press Company)

[3] Pikoul O 2010 *Appl. Cryst.* **43** 955

Periodic chain of resonators: gap control and geometry of the system

I.V.Blinova, A.S.Melikhova, I.Yu.Popov[1]
ITMO University, Kronverkskiy, 49, St. Petersburg, 197101, Russia

E-mail: popov1955@gmail.com

Abstract. A periodic system of coupled resonators is considered. It has a form of zigzag line. A band structure of the spectrum is studied. Solvable model based on the theory of self-adjoint extensions of symmetric operators is used. The position of gaps depends on the zigzag line angle. It opens a way to control the gap position by geometrical parameters. Particularly, it is possible to construct a system with predetermined gap position

1. Introduction

Chain structures constructed from micro- or nano-resonators are often used in optical systems. Particularly, it allows one to modify optical waveguides system by adding such a chain playing the role of a SCISSOR device [1]. In other application it ensures a photonic crystal property to optical waveguide system [2] (an analogous effect can be reached by using of periodically coupled waveguides [3, 4]). This system is also used for constructing optical delay line [5]. The property of the operator spectrum for the system allows one to use it for creation of biosensors having many advantages in comparison with another ones [6].

One meets such systems in nanoelectronics too. An example of a chain of coupled conglobate quantum resonators (not unique but very interesting) is so-called nano peapod. It is a nanotube filled by a chain of fullerene molecules. There is a number of experimental works describing properties of such systems and the processes of its creation [7-12].

To describe the system, we use solvable model based on the operator extensions theory [13, 14]. It allows one to obtain the dispersion equation for the model operator in an explicit form. Particularly, our consideration is based on a variant of this approach known as a model of point-like window (zero-width slit) [15 - 18]. We consider two-dimensional zigzag-like chain of coupled resonator (see Fig. 1) and describe its spectrum. Bands and gaps positions depend on zigzag angle. It gives an opportunity to control the gap position by geometrical parameters.

---

[1] I.Yu.Popov is the corresponding author

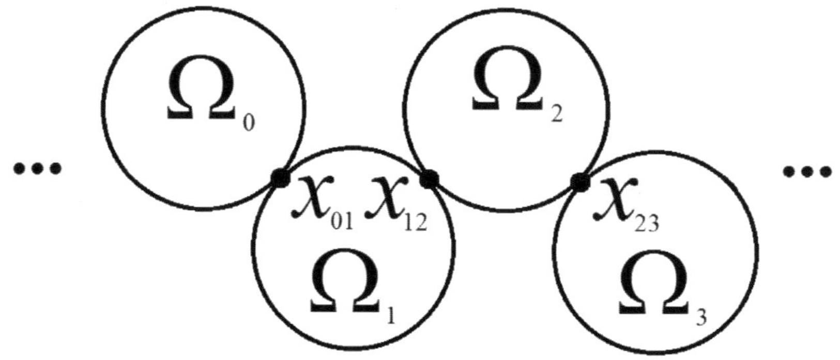

Fig. 1. Geometry of the system

2. Model construction and spectral analysis

First, consider two coupled two-dimensional identical resonators Ω_0, Ω_1 (disks) having common boundary point x_{01}. Let $-\Delta$ be the orthogonal sum of the Laplace operators with the Neumann boundary conditions in $L_2(\Omega_0)$ and $L_2(\Omega_1)$. Restrict the operator on the set of smooth functions vanishing at point x_{01}. We obtain symmetric non-self-adjoint operator with deficiency indices $(2,2)$. Its self-adjoint extension gives us the model of resonators coupled through point-like opening. The self-adjoint extension is a restriction of the adjoint operator. The domain of the adjoint operator consists of elements (u_0, u_1) such that

$$u_{0,1}(x) = a_{0,1} G(x, x_{01}, k_0) + \breve{u}_{0,1}(x), \quad x \in \Omega_{0,1},$$

where $G(x, x_{01}, k_0)$ is the Green function for the resonator, k_0 is some imaginary number, $\breve{u}_{0,1}(x)$ is an element from the domain of the Fridrichs extension, $\breve{u}_{0,1}(x_{01}) = \mathrm{b}_{0,1}$. To construct the self-adjoint extension we should restrict the adjoint operator, i.e. to establish a relation between $a_o, a_1, \mathrm{b}_o, \mathrm{b}_1$. We choose the following (natural, see [17]) condition

$$a_o = -a_1, \; b_o = b_1 \tag{1}$$

As for the periodic system in question, we use the transfer-matrix approach (see, e.g., [19]) To find the transfer-matrix, let us consider one of the resonators (they are identical), say, Ω_1. A non-trivial solution corresponding to a value of the spectral parameter k^2 has the form

$$\alpha_{01}^{+} G(x, x_{01}, k) + \alpha_{11}^{-} G(x, x_{11}, k).$$

Here plus and minus mark the right and left coefficients for the common boundary point. Correspondingly,

$$\begin{cases} a_{01}^{+} = \alpha_{01}^{+}, \\ b_{01}^{+} = \alpha_{01}^{+} \left(G(x, x_{01}, k) - G(x, x_{01}, k_0) \right)\Big|_{x = x_{01}} + \alpha_{11}^{-} G(x_{01} x_{11}, k). \end{cases}$$

For the next point, one has (we take into account conditions (1)):

$$\begin{cases} a_{11}^{+} = \alpha_{11}^{+} = -\alpha_{11}^{-} = \left(\alpha_{01}^{+} g(k_1 k_0) - b_{01}^{+} \right) \dfrac{g(k_1 k_0)}{G(x_{01}, x_{11}, k)}, \\ b_{11}^{+} = b_{11}^{-} = \alpha_{01}^{+} G(x_{01}, x_{11}, k) - \alpha_{11}^{-} \left(G(x, x_{12}, k) - G(x_1, x_{12}, k_0) \right)\Big|_{x = x_{12}}, \end{cases}$$

where

$$g(k, k_0) = \left(G(x, x_{01}, k) - G(x, x_{01}, k_0) \right)\Big|_{x = x_{01}}.$$

Hence,

$$b_{11}^{+}=\alpha_{01}^{+}G(x_{01},x_{11},k)-\left(b_{01}^{+}-\alpha_{01}^{+}g(k,k_0)\right)\frac{g(k,k_0)}{G(x_{01},x_{12},k)}.$$

Thus, one gets

$$\begin{pmatrix}\alpha_{11}^{+}\\ b_{11}^{+}\end{pmatrix}=\begin{pmatrix}\dfrac{g^2(k,k_0)}{G(x_{01},x_{12},k)} & -\dfrac{g(k,k_0)}{G(x_{01},x_{12},k)}\\ G(x_{01},x_{12},k)+\dfrac{g(k,k_0)}{G(x_{01},x_{12},k)} & -\dfrac{g(k,k_0)}{G(x_{01},x_{12},k)}\end{pmatrix}\begin{pmatrix}\alpha_{01}^{+}\\ b_{01}^{+}\end{pmatrix}=M\begin{pmatrix}\alpha_{01}^{+}\\ b_{01}^{+}\end{pmatrix}.$$

Here M is the transfer-matrix. Let us find its eigenvalues by solving the equation

$$\det(M-\lambda I)=\lambda^2-\lambda\left(g(k_1k_0)-1\right)\frac{g(k,k_0)}{G(x_{01},x_{12},k)}-\frac{g^3(k,k_0)}{G^2(x_{01},x_{12},k)}+g(k,k_0)=0$$

The eigenvalues of the transfer-matrix are as follows.

$$\lambda_{1,2}=\frac{\dfrac{(g-1)g}{G}\pm\left(\dfrac{(g-1)^2g^2}{G^2}+\dfrac{4g^3}{G^2}-4g\right)^{1/2}}{2},\quad G=G(x_{01},x_{12},k)$$

It is known (see, e.g., [20], [21]) that k^2 belongs to the continuous spectrum (band) of the model operator if

$$|\lambda_{1,2}|=1.$$

In our case, it means

$$(g-1)^2+4g-\frac{4G^2}{g}<0,$$

or

$$(g+1)^2g<4G^2. \qquad (2)$$

One can see that the left hand side of (2) does not depend on the form of the zigzag line. In contrast, the right hand side depends on the position of the disks touching points, i.e. on the form of the zigzag line (zigzag line angle).

Acknowledgments

This work was partially _nancially supported by the Government of the Russian Federation (grant 074-U01), by Ministry of Science and Education of the Russian Federation (GOSZADANIE 2014/190, Projects No 14.Z50.31.0031 and No. 1.754.2014/K)), by grant MK-5001.2015.1 of the President of the Russian Federation.

References

[1] Heebner J E, Boyd R W, and Park Q-H, 2002 SCISSOR solitons and other novel propagation effects in microresonator-modified waveguides, *J. Opt. Soc. Am.* **19**, 722–73

[2] Popov I Yu, Trifanov A I and Trifanova E S 2010 Coupled dielectric waveguides with photonic crystal properties. *Comput. Math. Math. Phys.* **50** (11) 1830-1836

[3] Melnichuk O P and Popov I Yu, 2005 Coupled dielectric waveguides: Variational estimations, *J. Math. Phys.* **46**, 073501

[4] Popov IYu 2002 Asymptotics of bound states and bands for laterally coupled waveguides and layers, *J. Math. Phys.* **43** (1), 215–234

[5] Lenz G, Eggleton B J, Madsen C K, and Slusher L E, 2001 Optical Delay Lines Based on Optical Filters, *IEEE J. Quantum Electron.* **37**, 525–532

[6] Boyd R W and Heebner J E, 2001 Sensitive disk-resonator photonic biosensor, *Appl. Opt.* **40**, 5742–5747

[7] Rotkin S V, Subramoney Sh 2006. *Applied Physics of Carbon Nanotubes: Fundamentals of Theory, Optics and Transport Devices.* (Berlin: Springer)

[8] Wang Y, Zhang H J, Lu L, Stubbs L P, Wong C C and Lin J. 2010 Designed Functional Systems from Peapod-like Co@Carbon to Co3O4@Carbon Nanocomposites. *ACS Nano*, **4** (8), 4753-4761

[9] Yang Y, Li L and Li W 2013. Plasmon Absorption of Au-in-CoAl2O4 Linear Nanopeapod Chains. *J. Phys. Chem.* **C 117** (27), 14142-14148

[10] Enyashin A N,. Ivanovskii A L 2010 Nanotubular composites: modeling of capillary filling of nanotubes of disulfide of molybdenum by molecules of TiCl4. *Nanosystems: Phys. Chem. Math.* **1** (1) 63-71

[11] Krivanek O L, Dellby N,.Murfitt M F Chisholm M F, Pennycook T J, Suenaga K and Nicolosi V. Gentle STEM:ADF imaging and EELS at low primary energies. *Ultramicroscopy.* **110,** 935-945

[12] Abou-Hamad E. Kim Yo, Talyzin A.V., Goze-Bac Ch., Luzzi D.E., Rubio A. and Wagberg T. 2009. Hydrogenation of C60 in Peapods: Physical Chemistry in Nano Vessels. *J. Phys. Chem.* **C, 113**, 8583-8587

[13] Albeverio, S., Gesztesy, F., Hoegh-Krohn, R. and Holden, H. 2005 *Solvable Models in Quantum Mechanics.* 2nd ed., with an appendix by P. Exner. (Providence (RI): AMS Chelsea)

[14] Pavlov, B. S. 1987 Extension theory and explicitly solvable models. *Uspekhi Mat.Nauk.* **42** (6), 99 - 131

[15] Popov I.Yu. 1992 The resonator with narrow slit and the model based on the operator extensions theory. *J. Math. Phys.* **33** (11) 3794-3801

[16] Popov I.Yu.. 2012 Operator extensions theory model for electromagnetic field-electron interaction. *J. Math. Phys.* **53** (6) 063505

[17] Popov, I. Yu. 1990 Extensions theory and localization of resonances for domains of trap type. *Mat. Sbornik.* **181** (10), 1366-1390

[18] Popov I.Yu 2013 Model of point-like window for electromagnetic Helmholtz resonator. *Zeitschrift Anal. Anwend.* **32**, 155 - 162

[19] Anikevich A.S.. 2013 Negative eigenvalues of the Y-type chain of weakly coupled ball resonators. *Nanosystems:Phys. Chem. Math.* **4** (4) 454-549.

[20] Duclos P., Exner P., and Turek O. 2008 On the spectrum of a bent chain graph, *J. Phys. A: Math. Theor.* **41**, 415206.

[21] Popov I. Yu. and Smirnov P. I., 2013 Spectral problem for branching chain quantum graph, *Phys. Lett. A* **377**, 439–442

Journal of Physics: Conference Series **735** (2016) 012063 doi:10.1088/1742-6596/735/1/012063

Spatial-temporal dynamics of broadband terahertz Bessel beam propagation

V A Semenova[1], M S Kulya and V G Bespalov
ITMO University, Department of Photonics and Optoinformatics, Kadetskaya line, Saint Petersburg 3/2B, Russia

E-mail: varvara.semenova@niuitmo.ru

Abstract. The unique properties of narrowband and broadband terahertz Bessel beams have led to a number of their applications in different fields, for example, for the depth of focusing and resolution enhancement in terahertz imaging. However, broadband terahertz Bessel beams can probably be also used for the diffraction minimization in the short-range broadband terahertz communications. For this purpose, the study of spatial-temporal dynamics of the broadband terahertz Bessel beams is needed. Here we present a simulation-based study of the propagating in non-dispersive medium broadband Bessel beams generated by a conical axicon lens. The algorithm based on scalar diffraction theory was used to obtain the spatial amplitude and phase distributions of the Bessel beam in the frequency range from 0.1 to 3 THz at the distances 10-200 mm from the axicon. Bessel beam field is studied for the different spectral components of the initial pulse. The simulation results show that for the given parameters of the axicon lens one can obtain the Gauss-Bessel beam generation in the spectral range from 0.1 to 3 THz. The length of non-diffraction propagation for a different spectral components was measured, and it was shown that for all spectral components of the initial pulse this length is about 130 mm.

1. Introduction

The unique properties of non-diffractive beams and, in particular, Bessel beams, have led to the various applications and also a great variety of new optical and physical studies. The most common application of the Bessel beams in a terahertz diapason is terahertz imaging, where using Bessel beams one can significantly enhance the depth of focusing and resolution [1].

Another possible application for the Bessel beams is optical wireless communications [2] where Bessel beams can be used to increase a beam propagation distance and to remain the beam intensity and its transverse distribution unchanged for a longer distance. The potential of Bessel beams for the wireless communications was studied, for example, in [3]. It was shown that Bessel beams possess several important features for wireless communications, such as beam robustness against small scale obstacles and dimensional scaling.

For the terahertz communications it is crucial to minimize the diffraction in the propagation channel due to features rising from the frequency shift to the terahertz range, such as higher attenuation and diffraction [4]. In the case of broadband near-field terahertz communications it is even more important to minimize diffraction and increase the propagation distance due to the wide range of propagating

[1] To whom any correspondence should be addressed.

Journal of Physics: Conference Series **735** (2016) 012063 doi:10.1088/1742-6596/735/1/012063

wavelengths. Therefore, the study of spatial-temporal dynamics of the broadband terahertz Bessel beams is needed.

Here we present a numerical simulation of propagating Bessel beam generated by interaction of the broadband terahertz pulse with conventional conical axicon. The aim of the work is to study spatial-temporal dynamics of the broadband terahertz Bessel beam and to investigate the potential of the conventional axicon for the wireless broadband terahertz communications.

2. Methods

In this work, we have simulated the conversion of the input Gaussian beam into Gaussian-Bessel beam using conventional axicon model. It was show in [3] that conventional axicon is one of the most effective ways for Bessel beam generation. Pseudo-nondiffracting beam generated by an axicon has high quality and the energy conversion is more efficient in contrast to the Fourier filtering, for example.

The initial beam represents 2 ps terahertz Gaussian pulse with frequency range from 0.1 to 3 THz and with central frequency of 0.6 THz (500 mkm). The spatial amplitude distribution of the initial pulse is shown in figure 1(a) and the simulated axicon is shown in figure 1(b). The simulated axicon material is Teflon with refraction index 1.46 for the terahertz range. It should be noted that in this work we are not considering the attenuation and dispersion effects of the terahertz radiation in Teflon and in the air but these effects will be considered in the further work. The maximum thickness of the axicon is 9.1 mm, the diameter is 50 mm and the base angle is 20°.

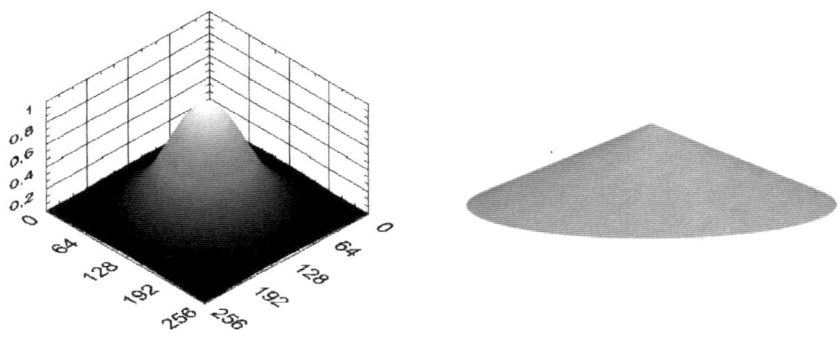

a b

Figure 1. Spatial amplitude distribution of the initial pulse (a) and simulated axicon model (b).

For the field distribution calculations of the propagated Bessel beam we used an algorithm based on the scalar diffraction theory, which is described in details in [5]. We considered the incident terahertz beam as a sum of monochromatic components and therefore, the propagation of each frequency component was calculated independently taking into account the source spectral shape, which is shown in figure 2b. The field distributions were calculated using angular spectrum method at the distances [5]

$$l \leq N\frac{\Delta x^2}{\lambda}, \qquad (1)$$

where N – number of elements in a calculation grid, λ – radiation wavelength, Δx – sampling step in the calculation grid. We exploited the convolution method for the field distribution calculation of each monochromatic component of the broadband source spectrum at the distances [5]

$$l \geq N\frac{\Delta x^2}{\lambda}. \qquad (2)$$

Journal of Physics: Conference Series **735** (2016) 012063 doi:10.1088/1742-6596/735/1/012063

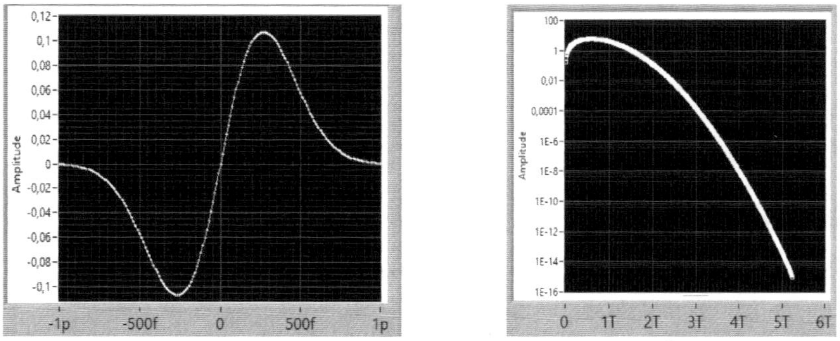

a b

Figure 2. The initial pulse shape (a) and its spectrum (b).

3. Results and discussion

Using methods described above, we simulated the formation of an optical vortex in the axicon illuminated by a broadband terahertz pulse with a Gaussian amplitude distribution. The results presented here were obtained at transverse dimensions Dx × Dy = 50 × 50 mm of the calculation domain and a grid of 256 × 256 elements. The field structure is calculated at the distances 10-200 mm.

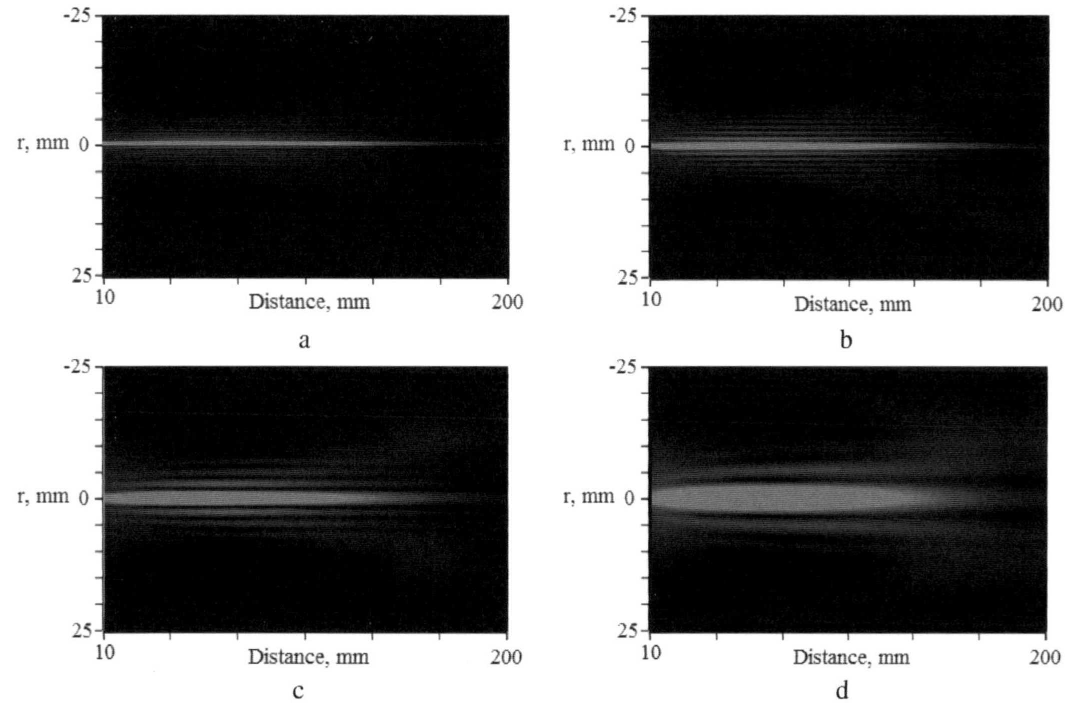

Figure 3. Transverse distribution of the Gauss-Bessel beam at the distances 10-200 mm from the axicon for the frequencies a) 2.4 THz; b) 1.2 THz; c) 0.6 THz; d) 0.3 THz.

The results of numerical calculations of the transverse amplitude distribution of the Gauss-Bessel beams at the frequencies 2.4 THz, 1.2 THz, 0.6 THz, and 0.3 THz for different distances from the axicon plane are presented in figure 3. The evolution of the Gauss-Bessel beam amplitude distribution at the frequency 0.6 THz (500 mkm) with the propagation distance increasing is shown in a figure 4.

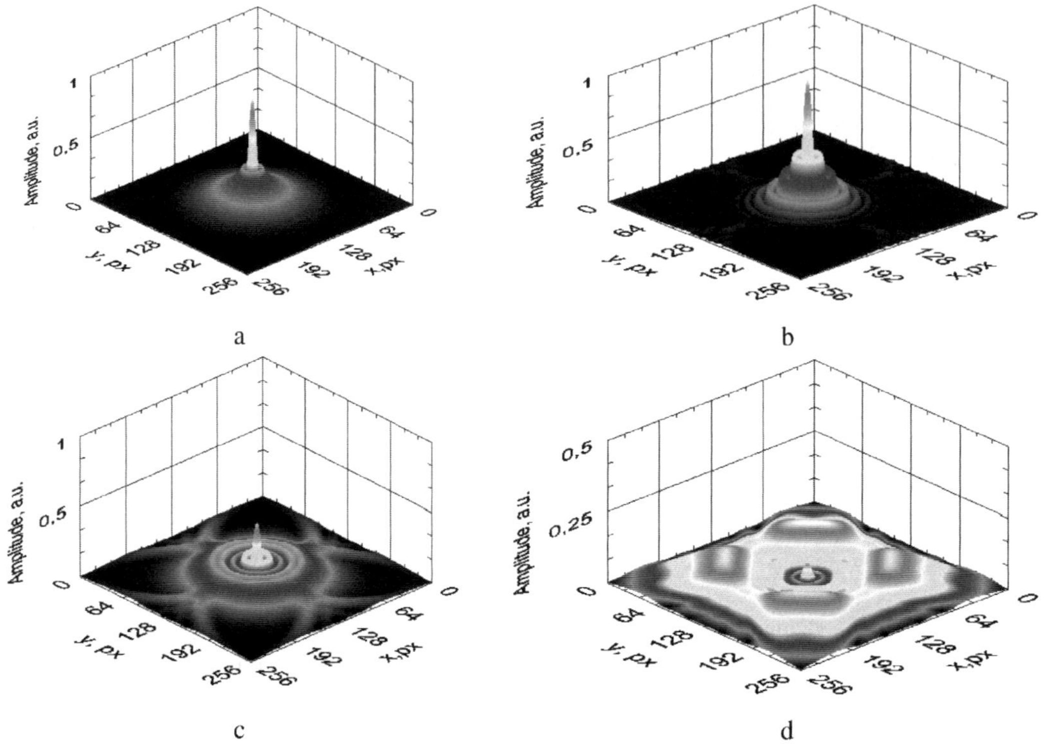

Figure 4. The evolution of the Gauss-Bessel beam amplitude distribution at the frequency 0.6 THz (500 mkm) at the distances from the axicon a) 20 mm; b) 80 mm; c) 140 mm; c) 190 mm.

The length of non-diffraction propagation for a different spectral components was also measured, and it is shown in the figure 5 that for all spectral components this length is about 130 mm. It is also shown in figure 5 that the relative central kern energy decreases with the distance from the axicon plane for all spectral components after 130 mm because of the energy transfer from the central maximum to the collateral maximums.

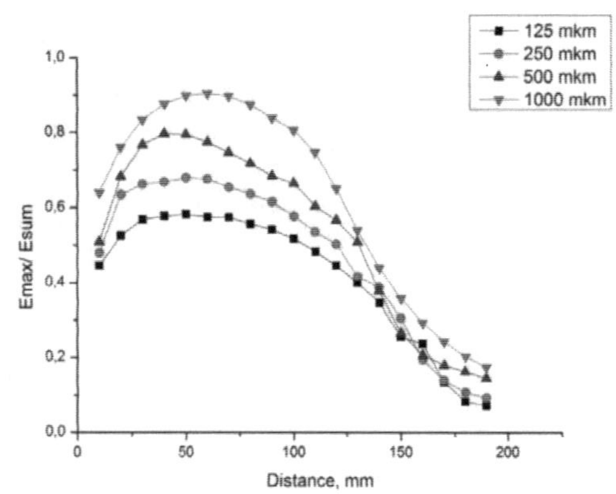

Figure 5. The dependence of the relative central kern energy from the propagation distance for the frequencies 2.4 THz (125 mkm), 1.2 THz (250 mkm), 0.6 THz (500 mkm), 0.3 THz (1000 mkm).

In conclusion, the simulation-based study of the propagating in non-dispersive medium broadband terahertz Gauss-Bessel beams generated by the axicon is presented. The numerical simulation showed that for the given parameters of the axicon lens one can obtain the Gauss-Bessel beam generation in the spectral range from 0.1 to 3 THz. The length of non-diffraction propagation for a different spectral components was measured, and it was shown that for all spectral components of the initial pulse this length is about 130 mm. Therefore, the axicon lens can significantly increase the propagation distance for all spectral components of the initial broadband pulse.

Acknowledgments
This work was supported by Ministry of Education and Science of Russian Federation under grant № 3.1675.2014/K.

References

[1] Bitman A, Moshe I and Zalevsky Z 2012 *Opt. Lett.* **37**(19)
[2] Semenova V and Bespalov V 2015 *Photonics Russia* **51**(3) 126-141
[3] Kollarova V, Medrik T, Celechovsky R, Bouchal Z, Wilfert O and Kolka Z 2007 *Proc. SPIE, Unmanned/Unattended Sensors and Sensor Networks IV (Italy, Florence, 17 September 2007) (* SPIE Vol. 6736*)* ed E Carapezza 67361C p 1-9
[4] Mandehgar M, Yan Y and Grischkowsky D 2013 *Opt. Lett.* **38** (17) 3437-40
[5] Petrov N V, Pavlov P V and Malov A N 2013 *Quantum Electron.* **43** 582–7

Journal of Physics: Conference Series **735** (2016) 012064 doi:10.1088/1742-6596/735/1/012064

Numerical simulation of broadband vortex terahertz beams propagation

V A Semenova[1], **M S Kulya and V G Bespalov**
ITMO University, Department of Photonics and Optoinformatics, Kadetskaya line, Saint Petersburg 3/2B, Russia

E-mail: varvara.semenova@niuitmo.ru

Abstract. Orbital angular momentum (OAM) represents new informational degree of freedom for data encoding and multiplexing in fiber and free-space communications. OAM-carrying beams (also called vortex beams) were successfully used to increase the capacity of optical, millimetre-wave and radio frequency communication systems. However, the investigation of the OAM potential for the new generation high-speed terahertz communications is also of interest due to the unlimited demand of higher capacity in telecommunications. Here we present a simulation-based study of the propagating in non-dispersive medium broadband terahertz vortex beams generated by a spiral phase plate (SPP). The algorithm based on scalar diffraction theory was used to obtain the spatial amplitude and phase distributions of the vortex beam in the frequency range from 0.1 to 3 THz at the distances 20-80 mm from the SPP. The simulation results show that the amplitude and phase distributions without unwanted modulation are presented in the wavelengths ranges with centres on the wavelengths which are multiple to the SPP optical thickness. This fact may allow to create the high-capacity near-field communication link which combines OAM and wavelength-division multiplexing.

Introduction

In optical communications different physical properties of an optical wave such as amplitude, phase, polarization and wavelength are used for data encoding and multiplexing. However, it was discovered in 1992 that particular light beams also possess an angular orbital moment (OAM), which represents a new informational degree of freedom. Contrary to spin angular momentum (SAM), which is associated with circular polarization of light and thus can only assume two possible states; orbital angular momentum is related to the spatial distribution of the intensity and phase of an optical field and has an infinite number of eigenstates [1].

The Poynting vector of light beams carrying OAM (also called vortex or spiral beams) has an azimuthal component along the beam, thus, such beams have an optical vortex along the axis. The well-known example of such beam is the one with an azimuthal phase dependence of $\exp(il\varphi)$, where φ is an azimuthal coordinate and l is topological charge (the number of twist in a wave front per unit wavelength [2]). Such beams have helical phase fronts with the number of intertwined helices and the handedness depending on the magnitude and the sign of topological charge, respectively [1]. The topological charge indicates the orbital angular momentum state number and can take any integer number. The intensity

[1] To whom any correspondence should be addressed.

Journal of Physics: Conference Series **735** (2016) 012064 doi:10.1088/1742-6596/735/1/012064

distribution of the vortex beam in transverse direction has a 'doughnut' shape with the dark area in the center.

Due to the unique properties of the vortex beams, they can be used in a number of different fields, for example, for the micro-manipulation in biology; for the contrast enhancement and resolution improvement in phase contrast microscopy; for the direct detection of extrasolar planets in astronomy; for the faster data manipulation in quantum information processing [3] .
Since OAM has the infinite number of eigenstates it has the potential to exceedingly increase the capacity, security and spectral efficiency of communication systems [4].
Utilization of OAM for communications is based on the fact that each OAM beam with a different ℓ are mutually orthogonal [5]. That means that the group of beams this different OAM states presents the set of orthogonal modes. Thus, coaxially propagating light beams with different OAM states can be efficiently separated, enabling efficient multiplexing at the transmitter and demultiplexing at the receiver with low crosstalk [6].

High capacity data transmission exploiting quasi-monochromatic OAM beams have been reported for free-space [6] and fiber [7] optical communications. Recent reports have shown that OAM beams can be used for data transmission in radio [8] and millimeter-wave [9] communication links as well as in optical links. However, there is great interest to also show advanced multiplexing approaches at terahertz frequencies due to an increasing demand for much higher speed wireless communications, which can be possibly overcomed by the operating frequency shift to the terahertz diapason. Wei *at al.* demonstrated basic functionalities for terahertz (THz) orbital angular momentum (OAM) communications, including the generation, detection, conversion, multicasting and manipulation of OAM at 0.1 THz [10] and 0.3 THz [11] using fabricated 3D printed spiral phase plates (SPPs). However, there is also a possibility to use OAM multiplexing for broadband terahertz communications, therefore, it is necessary to provide the detailed study of the broadband terahertz vortex beam generation and propagation.

Here we present a numerical simulation of propagating vortex beam generated by interaction of the broadband terahertz pulse with spiral phase plate. The aim of the work is to investigate the potential of the conventional SPP for the broadband terahertz communications.

2. Methods

OAM-carrying beams can be generated by modulating the phase delay distribution with phase modulating devices, such as q-plates, fork diffraction gratings, spiral phase plates (SPPs) and diffractive hologram displayed on spatial light modulator. In our work, we have simulated the conversion of the input Gaussian beam into OAM beam using SPP model. The SPP is a refractive optical element that can impose azimuth dependent phase retardation with an azimuthally varying thickness. The optical thickness of our SPP model increases in proportion to the azimuthal angle around the central point of the plate, and keeps constant in the radial direction. Thus the SPP introduce a linear phase delay equal to

$$\varphi(x, y) = k\,(n - 1)h(x, y), \tag{1}$$

where $k=2\pi/\lambda$, λ – radiation wavelength, n – refraction index, a $h(x,y)$ – function describing relief of the plate.
The initial beam represents 2 ps terahertz Gaussian pulse with frequency range from 0.1 to 3 THz and with central frequency of 0.6 THz (500 mkm). The spatial amplitude distribution of the initial pulse is shown in figure 1(a) and spatial thickness distribution of the SPP is shown in figure 1(b). The simulated SPP material is Teflon with refraction index 1.46 for the terahertz range. It should be noted that in this work we are not considering the attenuation and dispersion effects of the terahertz radiation in Teflon and in the air. The maximum thickness of the SPP is 1.162 mm thus SPP produce the vortex beam with topological charge one at the frequency 0.56 THz.

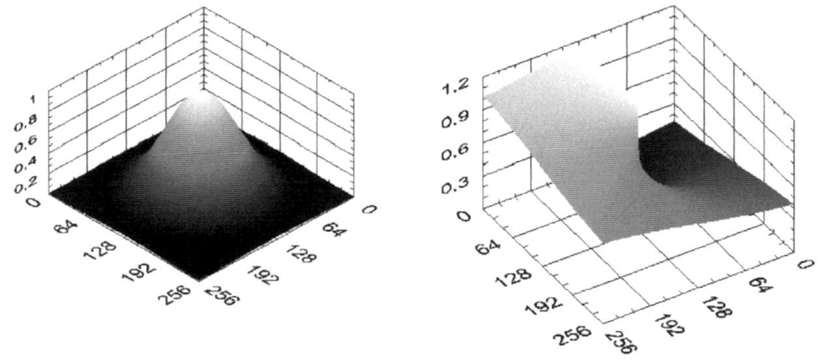

a b

Figure 1. Spatial amplitude distribution of the initial pulse (a) and simulated spatial thickness distribution of the spiral phase plate (b).

For the field distribution calculations of the propagated helical beam we used an algorithm based on the scalar diffraction theory, which is described in details in [12]. We considered the incident terahertz beam as a sum of monochromatic components and therefore, the propagation of each frequency component was calculated independently.

3. Results and discussion

Using methods described above, we simulated the formation of an optical vortex in the SPP illuminated by a broadband terahertz pulse with a Gaussian amplitude distribution. The results presented here were obtained at transverse dimensions Dx × Dy = 50 × 50 mm of the calculation domain and a grid of 256 × 256 elements. The field structure is calculated at the distances 20-80 mm.

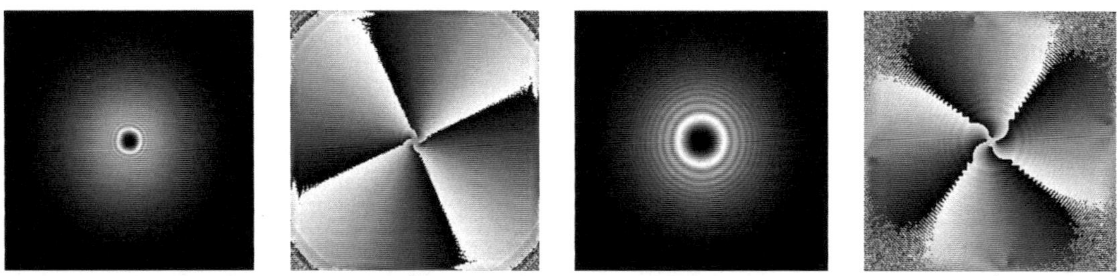

a b c d

Figure 2. The amplitude (a, c) and phase (b, d) distribution of optical vortex at 2.2 THz (134 mkm) for the distances of 20 mm (a, b) and 80 mm (c, d). The vortex has topological charge four.

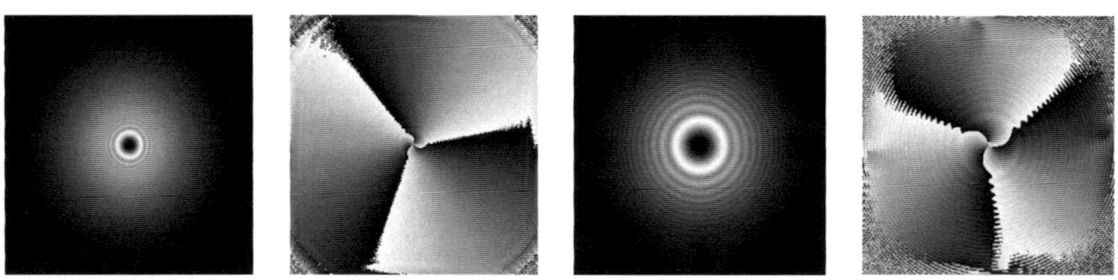

a b c d

Figure 3. The amplitude (a, c) and phase (b, d) distribution of optical vortex at 1.7 THz (178 mkm) for the distances of 20 mm (a, b) and 80 mm (c, d). The vortex has topological charge three.

As can be seen from equation (1), there are only a few wavelengths for which a total phase shift is an integer value multiple to 2π and, thus, at which the clear vortex structure can be presented. One can see that due to our SPP parameters the clear vortex structure can be presented at 2.2 THz, 1.7 THz, 1.1 THz, 0.56 THz. The results of numerical calculations of the amplitude and phase distribution of optical vortices at these wavelengths for the distances 20 and 80 mm are presented in figures 2-5. The vortices produced by SPP on the listed wavelengths have different topological charges due to the difference in the total phase shifts for each wavelengths.

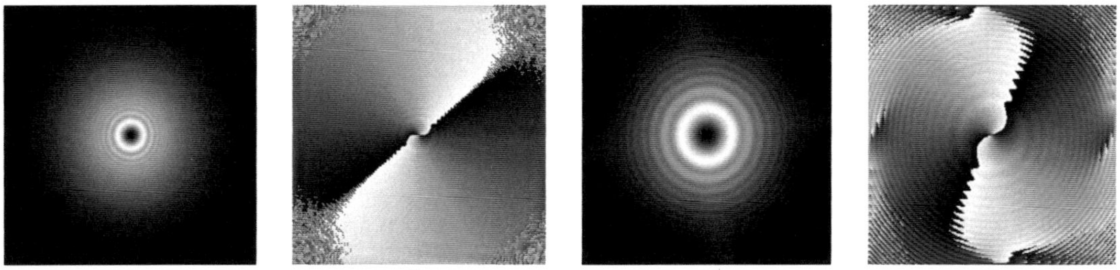

a b c d

Figure 4. The amplitude (a, c) and phase (b, d) distribution of optical vortex at 1.1 THz (267 mkm) for the distances of 20 mm (a, b) and 80 mm (c, d). The vortex has topological charge two.

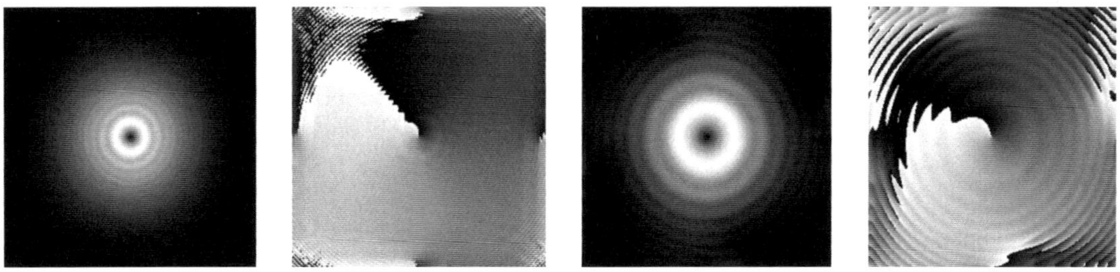

a b c d

Figure 5. The amplitude (a, c) and phase (b, d) distribution of optical vortex at 0.56 THz (535 mkm) for the distances of 20 mm (a, b) and 80 mm (c, d). The vortex has topological charge one.

Amplitude and phase distributions on the other wavelengths are expected to have the unwonted modulation in the amplitude and phase distributions as shown in figure 6. However, as figure 7 illustrated, this modulation is not presented or is very slight in the frequency ranges around frequencies at which pure vortex structure is presented, for example, for ranges 2.2 - 2.3 THz; 1.66 - 1.7 THz; 1.1 - 1.15 THz and 0.54 - 0.58 THz. Therefore, these spectral ranges are channels with different OAM states and it is possible to use wavelengths-division multiplexing for each channel.

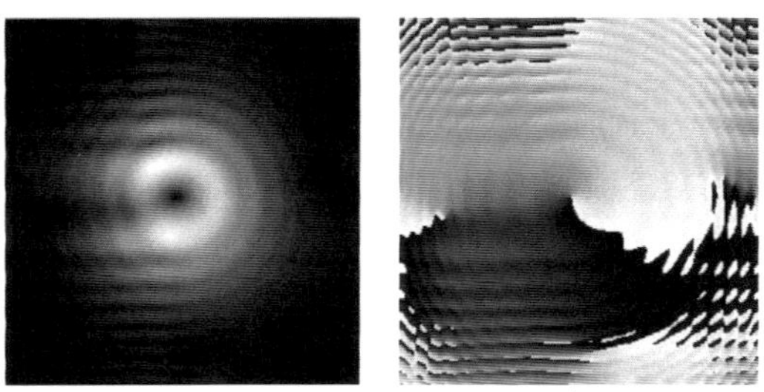

a b

Figure 6. The example of amplitude (a) and phase (b) distribution of optical vortex with unwanted modulation at 0.42 THz (714 mkm) at the distance 50 mm.

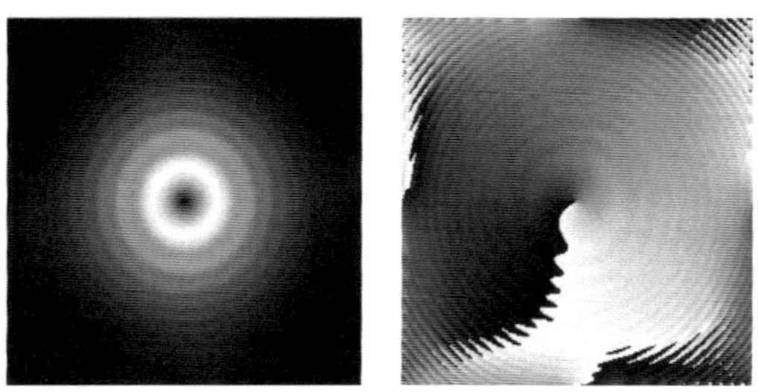

a b

Figure 7. The example of amplitude (a) and phase (b) distribution of optical vortex without unwanted modulation at 0.55 THz (545 mkm) at the distance 50 mm.

In conclusion, the simulation-based study of the propagating in non-dispersive medium broadband terahertz vortex beams generated by a spiral phase plate showed that the amplitude and phase distributions without unwanted modulation are presented in the wavelengths ranges with central wavelengths which are multiple to the SPP optical thickness. This fact may allow to create the high-capacity near-field communication link which combines OAM and wavelength-division multiplexing.

Acknowledgments
This work was supported by Ministry of Education and Science of Russian Federation under grant № 3.1675.2014/K.

References

[1] Yao A M and Padgett M J 2011 *Adv. Opt. Photon.* **204** 161–204
[2] Vaity P and Rusch L 2015 *Opt. Lett.* **40** 597–600
[3] Massari M, Ruffato G, Gintoli M, Ricci F and Romanato F 2015 *Appl. Opt.* **54** L74 4077–83
[4] Gibson G, Courtial J, Padgett M, Vasnetsov M, Pas'ko V, Barnett S and Franke-Arnold S 2004 *Opt. Express.* **12** 5448–56
[5] Willner A E, Huang H, Yan Y, Ren Y, Ahmed N, Xie G, Bao C, Li L, Cao Y, Zhao Z, Wang J, Lavery M P J, Tur M, Ramachandran S, Molisch A F, Ashrafi N and Ashrafi S 2015 *Adv. Opt. Photonics* **7** 66–106
[6] Ren Y, Wang Z, Liao P, Li L, Xie G, Huang H, Zhao Z, Yan Y, Ahmed N, Lavery M P J, Ashrai N, Ashrafi S, Linquist R D, Tur M, Djordjevic I B, Neifeld M A and Willner A E 2015 *OSA Technical Digest (online)*: *Optical Fiber Communication Conf. (California United States, Los Angeles, 22–26 March 2015)* M2F.1(Optical Society of America, 2014) pp 1–3
[7] Bozinovic N, Yue Y, Ren Y, Tur M, Kristensen P, Huang H, Willner A E and Ramachandran S 2013 *Science* **340** 1545-8
[8] Tamburini F, Mari E, Sponselli A, Thide B, Bianchini A and Romanato F 2012 *New J. Phys.* **14** 033001 1–17
[9] Yan Y, Xie G, Lavery M P J, Huang H, Ahmed N, Bao C, Ren Y, Cao Y, Li L, Zhao Z, Molisch A F, Tur M, Padgett M J and Willner A E 2014 *Nat. Commun.* **5** 4876
[10] Zhu L, Wei X, Wang J, Zhang Z, Li Z, Zhang H, Li S, Wang K, Liu J and National W 2014 *OSA Technical Digest (online)*: *Optical Fiber Communication Conf. (California United States, San Francisco, 9–13 March 2014)* M2F.1 (Optical Society of America, 2014) pp 1–3
[12] Wei X, Liu C, Zhang Z, Zhu L, Wang J, Wang K, Yang Z and Liu J 2014 *Proc. SPIE, Infrared, Millimeter-Wave, and Terahertz Technologies III* (China, Beijing, October 09, 2014) *(*SPIE vol 9275) ed C Zhang et al. p 92751P-1
[13] Petrov N V, Pavlov P V and Malov A N 2013 *Quantum Electron.* **43** 582–7

Journal of Physics: Conference Series **735** (2016) 012065 doi:10.1088/1742-6596/735/1/012065

Spatial-temporal dynamics of the terahertz field generated by femtosecond filament

S V Smirnov, Ya V Grachev, A N Tsypkin, M S Kulya, S E Putilin, V G Bespalov
ITMO University, Saint-Petersburg, Russia

E-mail: onlywww@list.ru

Abstract. We present the study on spatial distribution of the maximum of terahertz field amplitude in time domain when generated by a femtosecond filament. It is shown that as a result of the propagation of the terahertz field forms a spherical wave front, on the edge of which the maximum of amplitude has a temporary delay in contrary to its central part.

1. Introduction

In recent years there has been a significant shift in the field of terahertz studies due to the development of femtosecond lasers and microelectronics. The difficulty of obtaining electromagnetic radiation in the terahertz frequency range is that in microwave it is practically impossible to produce radiation at frequencies much greater than several hundreds of GHz due to modification of the emitters, although such radiation can be harmonically multiplied to terahertz range. Currently, three methods of producing extremely short THz pulses with femtosecond laser sources are the most developed: generation in the surface layer of semiconductors (photoconductive antenna) [1, 5-11], nonlinear optical difference frequency generation and optical rectification [4, 12-14] and generation using a femtosecond filament in gases [2, 3, 19, 20]. This paper considers the dynamics of the spatial-temporal distribution of the terahertz field generated by a femtosecond filament.

Ultrashort THz pulses have the potential for internal and external communications in integrated circuits, for time-resolved spectroscopy in far-infrared and chemical determination of the composition of complex compounds for the detection and localization of explosives and other non-metallic substances which are nondifferentiable in other ranges of electromagnetic radiation, and also for creating THz radars for THz time-resolved imaging.

2. Experimental setup

The experimental setup is shown in figure 1. The radiation from a femtosecond laser system Regulas 31F (FL) is divided by the beam splitter at a ratio of 5% (probe beam) and 95% (pump beam), respectively. The pump beam passes through the optical modulator (ch) and lens F = 10 mm (L), and in the caustic a femtosecond filament with the length of 1 cm is formed. Diverging from the central point of the filament, the terahertz radiation is collimated by a parabolic mirror (PM). The collimated beam passes through the scanning diaphragm (D). After that, radiation is focused by a lens (L) to the electrooptical crystal (ZnTe).

The probe beam after the beam splitter passes phase plate L / 2, Glan prism (GP) and a delay line, and then reflected from the dichroic mirror (DM) and is sent to the path of electrooptical detection system consisting of electrooptical crystal (ZnTe), the phase-shifting plate $\lambda/4$, Wollaston prism (WP) and the balanced detector (BD). The signal from the balanced detector was also detected at the operation frequency of optical-mechanical modulator (ch) using the lock-in amplifier (lock-in amp), then was digitized (ADC) and recorded on a computer (PC).

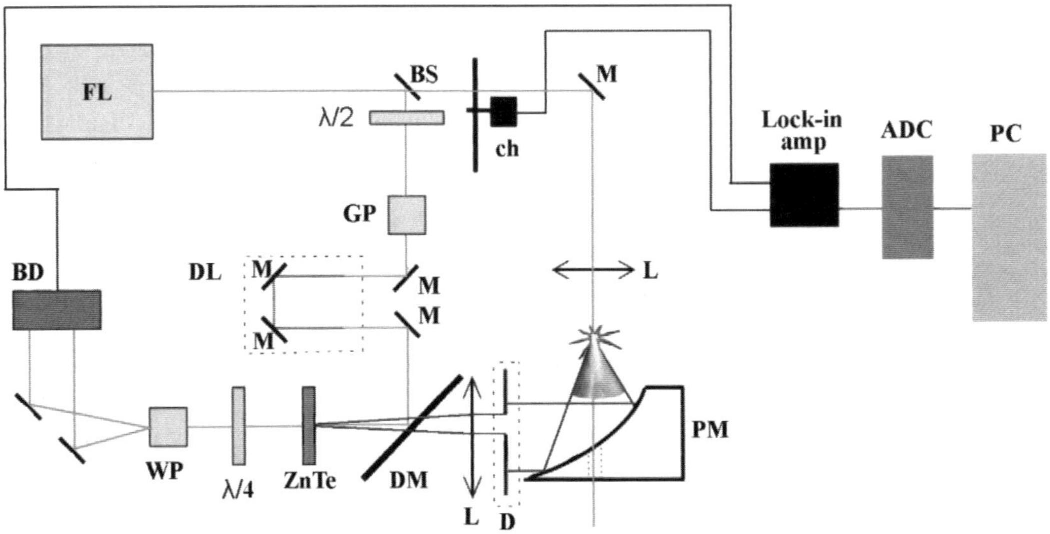

Figure 1. Schematic representation of the experimental setup

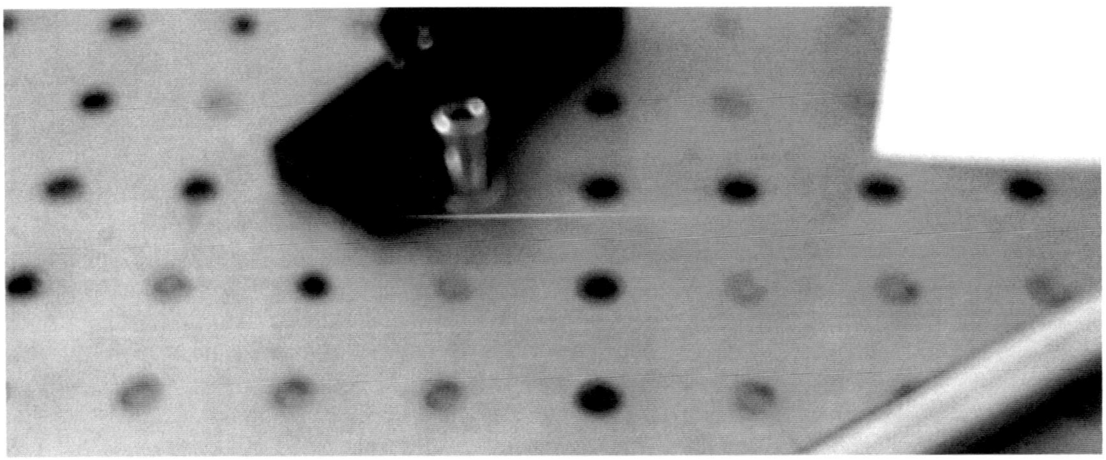

Figure 2. Photography Femtosecond filament

In the collimated terahertz beam between the parabolic mirror and the detection system path is placed scanning aperture with 5 mm diameter. THz field measurements were performed by raster scanning with moving increment of 2.5 mm, using a program written in the program environment of LabVIEW. Measurements were performed when taking into account the optimal parameters of raster scanning of THz field to optimize the experiment time without losing quality signal recovery [15].

3. Results

Figure 3 shows the experimental results of the spatial distribution of the maximum of terahertz field in time domain. This time-of-flight images show that there is a time delay between the maximum of amplitude of THz field and it has a spherical form. Probably, the temporary shift occurs due to the interference of the terahertz field generated by the various points of a femtosecond filament and it requires further research.

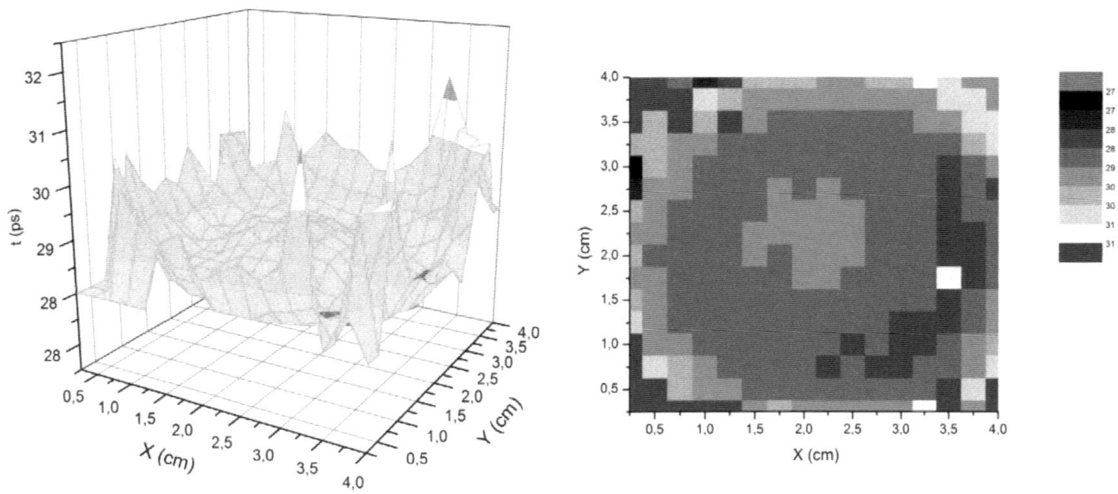

a) б)

Figure 3. Spatial and temporal distribution of the maximum amplitude of the THz field in time domain (time-of-flight pictures). a) side view, b) top view

4. Conclusion

As a result, the THz pulse time-of-flight structure in dependence of the spatial location of the generated THz field at filamentation was demonstrated.

Acknowledgments

This work was supported by Ministry of Education and Science of Russian Federation under grant № 3.1675.2014/K.

References

[1] S Greene B I *et al.* 1992 *Quantum Electron* **28** 2302–2312.
[2] Cook D J *et al.* 2000 *Opt. Lett* **25** 1210–1212.
[3] Loffler T *et al.* 2000 *Appl. Phys. Lett.* **77** 453.
[4] Xu L *et al.* 1992 *Appl. Phys. Lett.* **61** 1784.
[5] McIntosh K A *et al.* 1995 *Appl. Phys. Lett.* **67** 3844–3846.
[6] Sarukura N *et al.* 1998 *J Appl. Phys* **84** 654–656.
[7] Kondo T *et al.* 1999 *Jpn. J. Appl. Phys.* **38** 1035–1037.
[8] Heyman J N *et al.* 2001 *Phys. Rev. B.* **64** 085202.
[9] Davies A *et al.* 2002 *Phys. Med. Biol.* **47** 3679–3689.
[10] Carey J *et al.* 2002 *Appl. Phys. Lett.* **81** 4335.
[11] Schneider A *et al.* 2003 *Opt. Comm.* **224** 337–341.
[12] Hoffmann M C 2007 *Opt. Exp.* **15** 11706–11713.
[13] Kress M *et al.* 2004 *Opt. Lett.* **29** 1120–1122.
[14] Zhong H *et al.* 2006 *Appl. Phys. Lett.* **88** 1103.
[15] Kulya M S *et al.* 2014 *JPCS* **536** 012010.

Influence of resonance radiation transfer on ionization balance in a positive column plasma

Yu Golubovskii[1] and A Syasko
Faculty of Physics, St Petersburg State University, Ulyanovskaya 3, 198504 St Petersburg, Russia

E-mail: yu_golubovski@yahoo.com

Abstract. A method of self-consistent solution of charged particles balance equation, which is described by a differential equation of ambipolar diffusion, and an equation of resonance atom balance, which is described by an integral equation of radiation transfer, is proposed. The method is related to a replacement of an integral operator and a differential operator by a system of linear algebraic equations. The difference between a precise solution and a solution in the approximation of the effective resonance transition probability is shown. The influence of highest diffusion and radiation modes becomes apparent during transition to a contracted state.

1. Introduction

This article describes a method of precise self-consistent solution of an ambipolar diffusion equation and a resonance radiation transfer equation. A role of resonance radiation trapping on positive column plasma parameters discussed as well. It is significantly important to take into consideration the processes of transfer precisely when the excitation sources radically differ from the fundamental modes of diffusion and radiation problems (contracted discharge, skin effect, localization of excitation sources in a limited volume etc.). At present time there are only a few works [1,2] where diffusion and resonance transfer equation were solved along. Current state of problem is described in article [3]. A precise solution of a problem when resonance transfer is taken into consideration and a solution of an approximation of the effective resonance transition probability according to Biberman [4], Holstein [5] are compared. The difference between the solutions obtained for simple model concentration dependences of ionization and recombination terms are discussed. Symmetric cylindrical positive column and large absorption coefficients according to the Lorentz contour of a spectral line considered during the calculations.

2. Self-consistent solution of an ambipolar diffusion equation and a resonance radiation transfer equation considering a three-level ionization balance model

The difference between the diffusion of particles and photons is evident when a simple three-level scheme of ionization balance, which takes into account an electron excitation of a resonant state from the ground state at rate Z, a stepwise ionization with a probability W_i,a two-particle recombination on

[1] To whom any correspondence should be addressed.

a resonance state Γ, an ambipolar diffusion of charged particles $D_a \Delta n$ and an output of resonance radiation with the effect of trapping, is considered.

This scheme can be described by two initial balance equations for resonance atoms (1) and charged particles (2).

$$\begin{cases} Z + \Gamma = N_r(\vec{r})W_i + G(\vec{r}) & (1) \\ N_r(\vec{r})W_i = \Gamma - D_a \Delta n & (2) \end{cases},$$

where an integral operator of resonance radiation transfer $G(\vec{r})$ equals to

$$G(\vec{r}) = AN_r(\vec{r}) - \int_V AN_r(\vec{r}')K(|\vec{r} - \vec{r}'|)d^3r', \quad (3)$$

$N_r(\vec{r})$ - resonance atom concentration at a position $\vec{r}$,
A - spontaneous decay probability of resonance atoms,
$K(|\vec{r} - \vec{r}'|)$ - kernel of an integral operator.

Thus, initial system requires a self-consistent solution of an integral equation of resonance radiation transport (1) and a differential equation of an ambipolar diffusion (2). It should be also noticed that both of them have non-linear dependences on electron concentration. An approximate solution of this problem can be obtained in a common approximation of the effective transition probability according to Biberman [4] which assumes that the kernel $K(|\vec{r} - \vec{r}'|)$, as a function of the distance $|\vec{r} - \vec{r}'|$, decreases much faster than the atom density $N_r(\vec{r})$. This allows to take the density $N_r(\vec{r})$ at $\vec{r}$ outside the integral in equation (3) and replace the integral operator $G(\vec{r})$ by $N_r(\vec{r})A_{eff}$, where A_{eff} is an effective resonance transition probability.

But if the resonance atom density strongly changes in space on the scale of a decrease of the kernel $K(|\vec{r} - \vec{r}'|)$, the effective transition probability approximation is not appropriate. A promising and precise method for the solution of initial equation is a replacement of the integral operator by a system of linear algebraic equations. The direct method for the solution of equation (1) uses discretization of the entire plasma volume V into small cells ΔV_i inside which the resonance atom density $N_r(r_i)$ can be considered as a constant. Then, it is possible to remove it from the integral from equation (3) by substitution of the corresponding value in the cell center r_i. In case of large absorption coefficients the calculation of matrix elements can be carried out using asymptotes of the spectral line wings [3]. After the described substitution equation (3) has the form:

$$G(r) = \sum_j N_r(r_j)A\left[\delta_{jm} - \int_{\Delta v_j} K(|r_k - r'|)d^3r'\right],$$

where δ_{jm} - is the Kronecker symbol. The term in square brackets is taken as a matrix element $b_{jm} = \delta_{jm} - \int_{\Delta v_j} K(|r_k - r'|)d^3r'$. In matrix B positive diagonal elements b_{jj} describe a decay of resonance atoms and the negative non-diagonal elements b_{jm} describe a redistribution of the resonance atoms over the plasma volume.

The diffusion operator, describing the ambipolar diffusion, is reduced to a tridiagonal matrix A by finite differences. A dimension N of both matrices is determined by a number of fragmentations of an interval $[0; R]$, where R is a tube radius.

After a replace of the operators by the described matrices A, B in the initial equations (1), (2) and introducing new variables: $x = r/R$, $y = \frac{n(x)}{n_0}$, $z = \frac{N_r(x)}{n_0}$, $\tilde{I}(y) = \frac{I(n)}{I_0}$, $a = \frac{I_0}{\Gamma_0}$, where index zero signifies the value at the axis of a discharge, the initial system has a form which is convenient for solving:

$$\begin{cases} \sum_j (b_{jm} z_j) + z_m \left(\frac{W_{i_m}}{A} \right) = \frac{\Gamma_0}{A n_0} (a \tilde{I}_m - \tilde{\Gamma}_m) \\ \sum_j (a_{jm} y_j) + z_m \left(\frac{W_{i_m}}{D_a} \right) = \frac{\Gamma_0}{D_a n_0} \tilde{\Gamma}_m \end{cases}.$$

In this system a relative electron concentration $y_j = n_j / n_0$ and a relative resonance atom concentration $z_j = N_{r_j} / n_0$ are unknown variables. Matrix A contains boundary conditions for an electron concentration. The resonance atoms integral equation does not require any of boundary conditions. To apply a method of iterations, the matrix of diffusion and the matrix of resonance radiation transfer should be combined into a single matrix

$$M = \begin{pmatrix} A & 0 \\ 0 & B \end{pmatrix}.$$

Model rates of ionization and recombination, which provide the switch from diffuse state to a contracted one with an increase of current, were introduced ($\tilde{I} = y^5$, $\tilde{\Gamma} = y^2$).

Considered approach allows to present dependences of plasma discharge parameters on current

$$i = 2\pi R^2 \int_0^R j(r) r dr,$$

where $j(r)$ - current density, $j(r) \sim n$. In relative variables this equation has a form

$$i = 2\pi R^2 j_0 \int_0^1 y(x) x dx.$$

It gives an opportunity to determine a dimensionless effective cross section of current flow

$$S_{eff} = \int_0^1 y(x) x dx,$$

where j_0 - current density at the axis. S_{eff}=0,215 for the Bessel radial concentration distribution and S_{eff}=0,5 for the square-shaped profile.

3. Comparison of model problem solutions

Solutions of model problem, which were obtained using a common approximation of the effective transition probability by Biberman [4] and the matrix method described above, are compared in figure 1. It can be seen that the resonance radiation trapping leads to a radial broadening of the electron and resonance atoms densities with a growth of an electron concentration value at the axis in case of precise solution with taking into account the transfer of resonance radiation transfer. Oppositely, in case of Biberman approximation those profiles are shrinking.

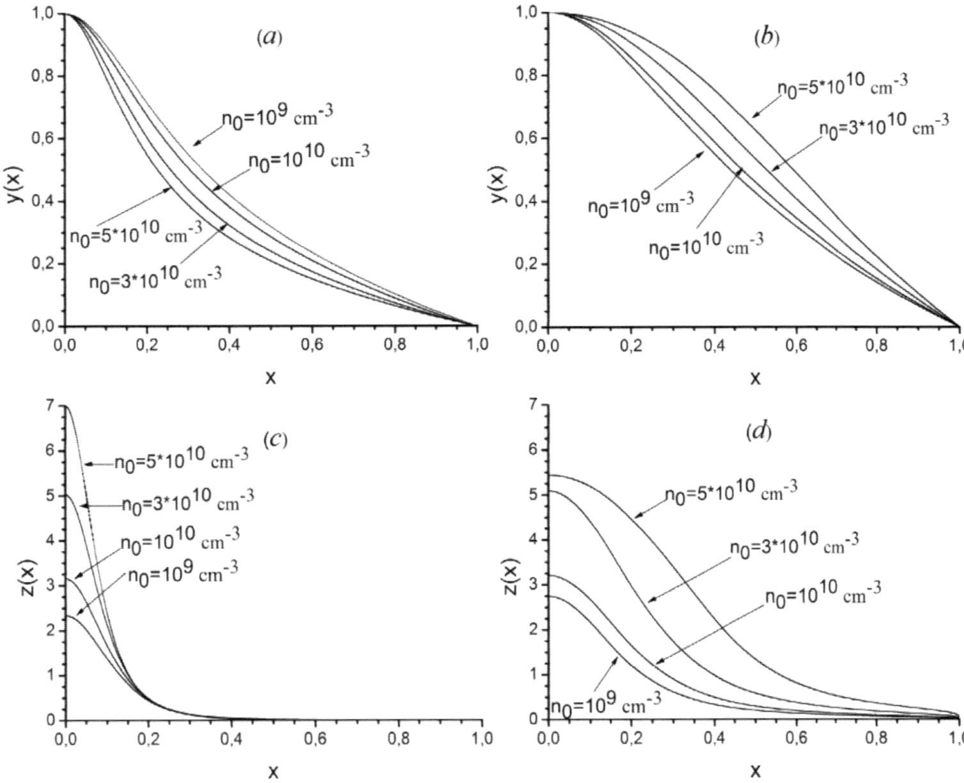

Figure 1. Radial relative electron concentration distribution (above) in the approximation of the effective transition probability (*a*) and with taking into account the transfer of resonance radiation (b). Radial relative resonance atom concentration distribution (below) in the approximation of the effective transition probability (*c*) and with taking into account the transfer of resonance radiation (d).

Dependences of an effective cross section of current flow S_{eff}, electron and resonance atom concentrations at the axis on current are presented in figure 2. It illustrates that in case of the local approximation an increase of current leads to a contraction, whereas the effect of trapping causes spreading of the current cord (figure 2 (a)).

The influence of resonance radiation trapping on electron and resonance atom concentration formation can be seen in figure 2 (b), (c). While current is relatively small (when radial profiles of electron concentration decrease smoothly), two solutions give a good agreement with each other. With a growth of current the difference in radial decreases starts to appear and leads to a significant distinguish in the values of plasma parameters. This fact demonstrates an influence of highest modes in a formation of plasma properties. At low current radial distributions are close to fundamental modes of diffusion and radiation problems and a solution, obtained in the approximation of the effective transition probability, is precise enough. But when the radial distributions differ from fundamental modes, the disadvantages of local approximation appear. The extent of accuracy of common approximation can be seen on presented results.

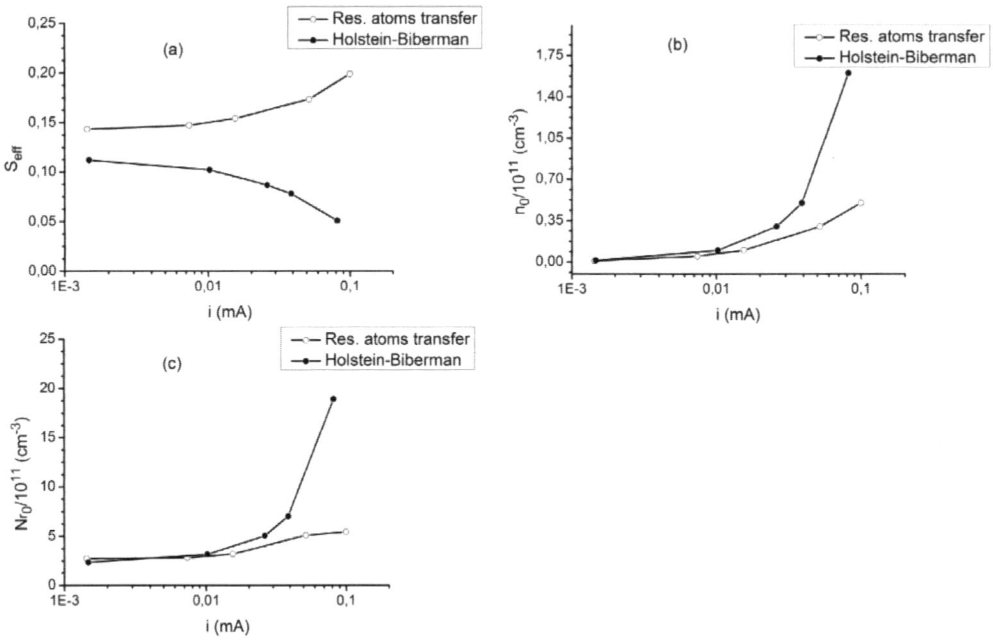

Figure 2. Dependences of an effective cross section current flow (a), an electron concentration (b) and a resonance atom concentration (c) on current.

4. Conclusion

The main result of this work is a development of a new approach for solving the balance equations, which is related to an accurate consideration of a resonance radiation transfer, since at present time almost all works are made in the approximation of the effective transition probability by Biberman [4]. Proposed technique of obtaining a solution of the integral radiation transfer equation has number of advantages. It is possible to obtain accurate numerical solutions by dividing the interval of integration into a greater number of steps and then subsequently compare a new solution with the previous. Thus one can obtain the spatial distribution of the resonance atoms at the same level of accuracy as the distribution of the electrons, which can be calculated by precise methods of numerical integration of the differential equation of an ambipolar diffusion. Eigenvalues and eigenmodes of the radiation transport operator can be easily calculated by using the discussed matrix method. It makes it possible to solve various time-dependent problems accurately by approximating the excitation sources and the expected solution by a system of the radiation modes alike in case of the diffusion modes. In addition it is possible to calculate the Green functions as an inverse matrix of the radiation problem.

Acknowledgement

Financial support by Saint-Petersburg State University (grant 11.38.203.2014) is gratefully acknowledged.

References

[1] Giuliani J L, Petrov G M, Apruzese J P and Davis J 2005 *Plasma Sources Sci. Technol.* **14** 236
[2] Golubovskii Yu, Gorchakov S, Loffhagen D and Uhrlandt D 2007 *Eur. Phys. J. Appl. Phys.* **37** 101
[3] Golubovskii Yu B, Gorchakov S and Uhrlandt D 2013 *Plasma Sources Sci. Technol.* **22** 023001
[4] Biberman L M 1947 *Piz. Zh. Eksp. Teor. Fiz.* **17** 416
[5] Holstein T 1947 *Phys. Rev.* **72** 1212

THE CdSe/CdS QUANTUM DOTS LUMINESCENCE ENHANCEMENT NEAR SILICA LAYER WITH THE ION-SYNTHESIZED SILVER NANOPARTICLES

R R Shamilov[1], Yu G Galyametdinov[1,2], A A Nugaeva[1], V I Nuzhdin[2], V F Valeev[2] and A L Stepanov[1,2]
[1] Kazan National Research Technological University, Kazan, Russia,
[2] Kazan E. K. Zavoisky Physical-Technical Institute, Kazan Scientific Center, Russian Academy of Sciences, Kazan, Russia

E-mail: reedish@mail.ru

Abstract Photoluminescence characteristics of hybrid quantum dots CdSe/CdS deposited on the surface of silica containing the layer of ion-synthesized silver nanoparticles were studied. The quenching or enhancement of the luminescence depending on distance between silver nanoparticles and quantum dots layers was detected. The optimal spacer layer and excitation waveleghth for the highest intensity of their photoluminescence in the plasmon field of metal nanoparticles was defined.

1. Introduction

Recent years some interest was significantly increased in different luminescent materials due to their wide use in various light-emitting and optoelectronic devices, as well as in bio- and chemosensors. Considerable attention was attracted to quantum dots (QDs) based on metal chalcogenides, resulting in their photostability and size-dependent luminescence in the wide visible region [1]. The optical characteristics of the QDs depend on the presence in their vicinity of noble metal nanoparticles (MNP). Strong local plasmon field near MNP as result of conduction electron oscillations in external electromagnetic field could significantly increase the intensity of the QD luminescence [2]. The feature of MNP plasmon influence on the luminescence of QDs is determined by a combination of various factors such as the distance between MNP and QDs, their sizes, shape, type of chemical composition, the excitation wavelength etc [3–5].

Among different approaches for obtaining of MNP, such as chemical vapor deposition, vacuum deposition et al., the ion implantation is one of most interesting technology. This method allows to fabricate composite materials based on different dielectric matrices (silicate glass, polymers, crystals) containing in the surface layer ion-synthesized MNP of certain sizes [6]. Structures with MNP formed by ion implantation can be used to improve luminescence efficiency of different materials. For example, recently it was demonstrated the luminescence enhancement of thin ZnO films, deposited on the surface of silica containing ion-synthesized silver nanoparticles [7].

In present paper, we report on the designing a new system organized by a combination of a layer of silver nanoparticles produced by ion implantation in the surface region of the silica covered by thin film of chemically synthesized hybrid core-shell CdSe/CdS QDs, and a study an influence MNP plasmon field on the photoluminescence of QDs in such system.

2. Experimental Procedure

CdSe and CdSe/CdS QDs were obtained by colloidal method in an aqueous-organic media [8]. A solution of 0.72 g NaOH in 2 ml of bidistilled water was added 30 ml of glycerol and 10 ml of oleic acid. Further, the mixture was introduced 266 mg (1 mmol) of cadmium acetate dihydrate in 2 ml of water, after which the reaction mixture was heated up to 190 °C. At this temperature 1 mmol selenosulfate was added to a solution, and the mixture was stirred for 15 min. After cooling the reaction mixture obtained CdSe QDs were extracted with n-hexane.

Hybrid CdSe/CdS core/shell QDs were obtained in water-ethanol medium containing 1 mmol of cadmium acetate dihydrate and CdSe QDs, which was introduced into a solution of thioacetamide (1 mmol) at a temperature of 50 °C. These nanoparticles CdSe/CdS were purified by reprecipitation.

Plasmon structures on the base of silica (SiO_2) with the silver nanoparticles were prepared by low energy implantation by Ag^+ ions with an energy of 30 keV, a dose of $4{\cdot}10^{16}$ ion cm^{-2} at current density 8 $\mu A\ cm^{-2}$ with ILU-3 accelerator by the technique described in detail in [6].

The photoluminescence and photoluminescence excitation spectra of QDs were detected by the Cary Eclipse (Varian) fluorescence spectrophotometer. Optical absorption spectra were measured with a two-beam spectrometer Perkin Elmer Instrumental LAMBDA 35. QDs size were evaluated by dynamic light scattering with Malvern Zetasizer Nano which was equipped with a narrow filter at 632.8 nm. The average size of the QDs was estimated to be of ~ 9.3 nm.

Thin films of QDs with a thickness about 35 nm were deposited by spincoating equipment from solution in n-hexane on the surface of blank silica and silica with ion-synthesized silver nanoparticles.

For same samples, between a glass surface with MNP and film with QD separating polymethylmethacrylate (PMMA) layer from a polymer solution in chloroform of various thicknesses was deposited.

3. Results and Discussion

In present paper the influence of the plasmon field of silver nanoparticles on the optical properties of CdSe/CdS QDs was studied in comparison with control samples with QD layers of the same thickness (about 35 nm) deposited on the surface of the blank silica. Figure 1 shows the absorption spectra of a control sample (SiO_2 + QD) and a test sample (Ag:SiO_2 + QD) consisting of the superposition of the contributions from absorption layers of silver nanoparticles and QDs in transparent region of silica from 340 to 800 nm. The wide selective band with a maximum at 400 nm appears as result of the surface plasmon resonance in the silver nanoparticles [9].

The effect of the plasmon electromagnetic field of MNP on the QD luminescence in the samples was studied when the luminescence was exited at the wavelength of 400 nm, corresponding to maximum plasmon absorption (figure 2). It is seen that the test sample with silver nanoparticles has the luminescence intensity of QD layer approximately twice lower compared with the control sample. The observed luminescence quenching is obviously explained due to nonradiative electron transitions or energy transfer from the QD to the MNP which happens in the case of a relative close space position to each other MNP and QDs [4].

In order to eliminate QD luminescence quenching in systems with MNP it is proposed in the literature to use some organic or inorganic spacer layers for increasing the distance between MNP and QDs [9, 10]. Therefore, we formed special samples, in which between the QD thin film and the surface of the silica with silver nanoparticles a thin layer of PMMA was introduced.

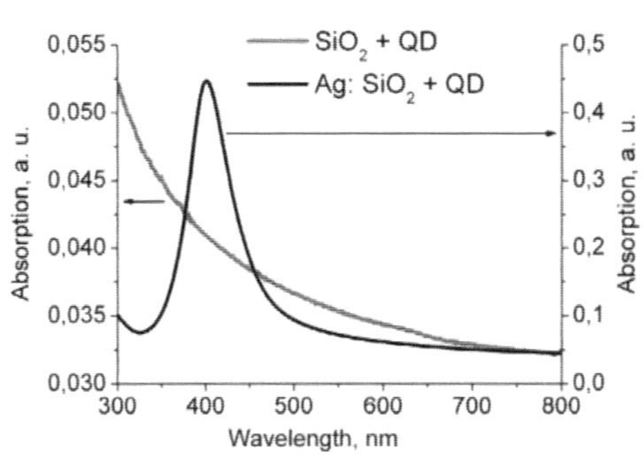

Figure 1. The absorption spectra of the samples.

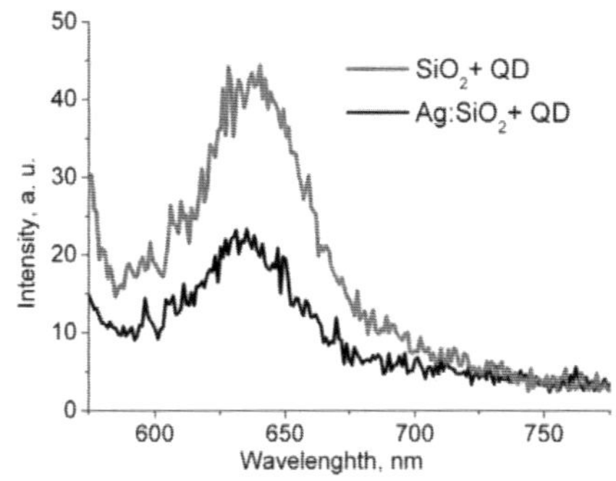

Figure 2. Photoluminescence spectra of the samples at an excitation wavelength of 400 nm.

The different thicknesses of the spacer layer were selected to estimate the impact of surface plasmon resonance of MNP on the QD luminescence intensity. Introduction a thin PMMA layer (~ 35 nm) in the structure of the sample lead to a decreasing of luminescence quenching. With increasing thickness of the spacer layer to 100 nm a luminescence quenching is replaced by a luminescence enhancement. The maximum increasing of the QD emission was defined for a layer with thickness of about 150 nm. Further increase in the thickness of the spacer layer results in a monotonic reduction of enhancement factor to 1.0 and the presence of silver nanoparticles in the such sample did not change the luminescence intensity of QDs.

As example, in figure 3 shows luminescence spectra of a test sample Ag:SiO_2 + PMMA + QD with a thickness of separation layer of 150 nm in comparison with a control sample SiO_2 + QD which were excited at a wavelength of 400 nm, as in case of figure 1. It is clear seen that the intensity of the QD photoluminescence at wavelength of 640 nm was increased by ~ 50 % at present case.

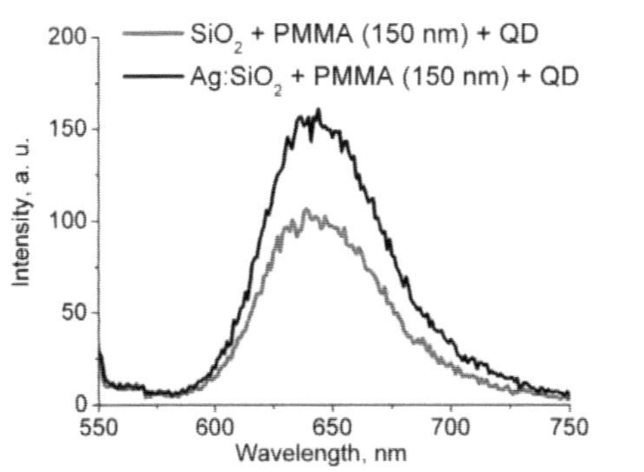

Figure 3. Photoluminescence spectra of the samples with the PMMA layer of 150 nm at an excitation wavelength of 400 nm.

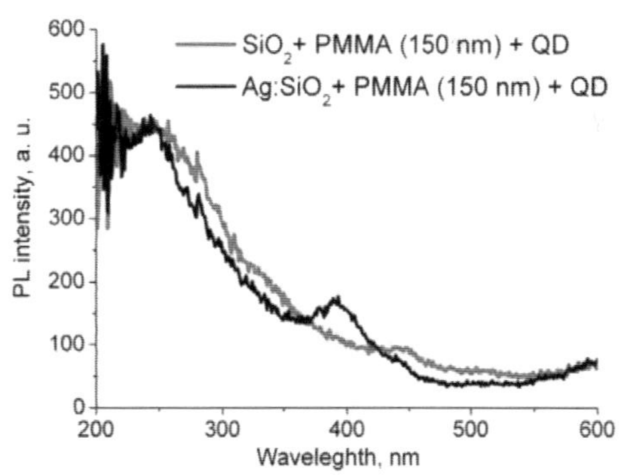

Figure 4. PLE spectra of the samples with the PMMA layer of 150 nm at a photoluminescence wavelength of 640 nm.

Thus in dependence on the distance between QD and MNP two competing processes could be occurred. First, it is nonradiative energy transfer from the QDs to the silver nanoparticles, which is possible only at very small distances between them. The second opposite effect is the luminescence enhancement of QDs located in a field near MNP, which decreases exponentially with a distance from the surface of MNP.

It is should be mention that the intensity increasing of QD luminescence is observed only when it is excited in the spectral region of the plasmon absorption of silver nanoparticles, as it is confirmed by

Journal of Physics: Conference Series **735** (2016) 012067 doi:10.1088/1742-6596/735/1/012067

the spectral dependence of the photoluminescence excitation (PLE) for a test sample Ag:SiO_2 + PMMA + QD (figure 4). The PLE spectrum of this sample presents a selective band, which is matching in spectral position with the surface plasmon resonance band of silver nanoparticles (figure 1).

4. Conclusion

The multilayer systems consisting of CdSe/CdS quantum dots deposited on the surface of the silica containing ion-synthesized silver nanoparticles exhibiting surface plasmon resonance were formed. The effect of plasmon field of metal nanoparticles on photoemission characteristics of quantum dots layer depending on their distance was studied. The optimal spacer layer and excitation waveleghth for the photoluminescence enhancement of the quantum dots were determined.

Acknowledgement

This work was supported by grant Russian Science Foundation Nr. 14-13-00758.

5. References

[1] Zhang F, Yi D, Sun H, and Zhang H 2014 *J. Nanosci. Nanotechnol.* **14** 1409–24
[2] Soganci I M, Nizamoglu S, Mutlugun E, Akin O and Demir H V 2007 *Optics Express* **15** 14289–98
[3] Haldar K K, Sen T, Mandal S and Patra A 2012 *ChemPhysChem* **13** 3989–96
[4] Zhang X, Marocico C A, Lunz M, Gerard V A Gun'ko Y K, Lesnyak V, Gaponik N, Susha A S, Rogach A L and Bradley A L 2012 *ACS Nano* **6** 9283–90
[5] Ito Y, Matsuda K and Kanemitsu Y 2007 *Phys. Rev.* B. **75** 033309
[6] Stepanov A L and Popok V N 2004 *Surf. Sci.* **566-568** 1250–54
[7] Xiao X H, Ren F, Zhou X D, Peng T C, Wu W, Peng X N, Yu X F and Jiang C Z 2010 *App. Phys. Lett.* **97** 071909
[8] Shukla R K, Galyametdinov Yu G, Shamilov R R and Haase W 2014 *Liquid Crystals* **41** 1889–96
[9] Seo J, Fudala R, Kim W J, Rich R, Tabibi B, Cho H, Gryczynski Z, Gryczynski I and Yu W 2012 *Optic. Mater. Express* **2** 1026–39
[10] Liu B T, Liao T H, Tseng S and Lee M H 2014 *Appl. Surface Sci.* **292** 615–19

Dependence of optical properties of calcium bismuthates on synthesis conditions

D S Shtarev[1], **A V Shtareva**[1,2]
[1] Department of the Natural Science, Far Eastern State Transport University, Khabarovsk, Russia
[2] Laboratory of Physical and Chemical Research Methods, Institute of tectonics and geophysics named after Yu. A. Kosygin of the FEBRAS, Khabarovsk, Russia

E-mail: shtarev@mail.ru

Abstract. The article studies optical properties of calcium bismuthate nanoparticles of different composition. For the first time the synthesis of these compounds was produced by the pyrolysis of organic precursors using an organic solvent. Characterization of particles was made by scanning electron microscopy and X-ray analysis. The optical properties were investigated by diffuse reflectance spectroscopy (DRS). It is shown that the type of crystal lattice of the particles of calcium bismuthate determines the possibility to control the optical properties of nanoparticles by varying their composition. The conclusions about the production process and the composition of calcium bismuthate, the most promising for use as a photocatalyst of visible light and solar cells, were made.

1. Introduction

The usage of wide-bandgap semiconductors for the water and air purification, solar energetics and sensors is a very promising and environmentally friendly technology. This offers the prospects of the use of clean, safe and renewable solar energy solutions to pressing environmental issues. Recently among all the materials with photocatalytic properties the most intensively studied, highlighted is the titanium dioxide TiO_2. It is considered one of the best photocatalysts for the oxidative degradation of organic compounds, due to their unique characteristics, chemical stability and high catalytic activity. However, the TiO_2 has a large band gap, so it has an excitation spectral range in (or near) the UV wavelength. The part of this spectral range in the whole sunlight irradiation is only 4%. In order to effectively use the visible spectrum of solar light (with wavelengths greater than 420 nm), we should develop the industry of photocatalysts with lower activation energy.

Up until now, for the displacement of spectral activity of titanium dioxide in the visible region one of the few technologies was used. The first is the introduction into the crystal lattice of titanium dioxide of transition metal ions and anions (such as the carbon, nitrogen, sulfur, and others) [1–4]. The second is to create compositions of TiO_2 - organic dye [5–6], or combination of several different semiconductors [7]. However, the use of such modified catalysts (obtained as the first and second method) was significantly limited due to significant reductions in the quantum yield for this type of sensitizations.

One of the known ways to enhance the catalytic activity is the doping different oxides - such as RuO_2, V_2O_5 and Co_3O_4, - by the metals [8–15]. The effect of catalytic activity increasing in this case can be explained by the effective charge transfer between the metal atoms and the oxide lattice, which increases the recombination time of photoexcited electron-hole pairs.

In recent years an interest [16–20] to the photocatalytic systems based on the bismuthates of the alkaline earth metals (in particular a composition based on the calcium bismuthates [17–18] and stroncium bismuthates [19–20]) has been growing. This is due to the fact that in the electronic structure of bismuth the $d^{10}s^2$ orbital exists, which is able to absorb the light in the visible spectral range.

Thus, in the Reference [17–18] it was shown, that the $CaBi_6O_{10}/Bi_2O_3$ photocatalytic composition initiates the destruction of methylene blue by the visible light irradiation. From this it is concluded that such class of photocatalysts can be used to purify water from various real pollutants. However, in these articles there is no experimental data on the photocatalytic degradation of any real water pollutants. In the Reference [21] at the first time the possibility of the photocatalytic degradation of diesel fuel by visible light in the presence of $CaBi_6O_{10}/Bi_2O_3$ photocatalytic composition was demonstrated. Also in this article on the base of biological testing was shown that neither the $CaBi_6O_{10}/Bi_2O_3$ catalyst composition, nor products of photocatalytic degradation of diesel fuel have more toxic impact than the initial solution of the diesel fuel.

However, no studies of the optical properties of the bismuthates of the alkaline earth metals, depending on their composition and synthesis conditions have been done. Otherwise it is impossible to identify the most promising compounds from the standpoint of their practical use as photocatalysts or solar cells. In this study, the optical properties of calcium bismuthate nanoparticles depending on its composition and production technology were analyzed.

2. Materials and Methods

The present work is studying the dependence between composition and preparation conditions of the calcium bismuthates nanoparticles and their optical properties. To solve this task the series samples of the calcium bismuthate nanoparticles with different relative concentration of bismuth and calcium and the annealing temperature were prepared (Table 1). For all samples, a well-known method was used [22]. The basis of this method is the ability of the bismuth nitrate to form at its grinding with the powders of the polyhydric alcohols (e.g., sorbitol) the transparent homogeneous viscous solution only by its own crystal water.

For the preparation of the calcium bismuthate nanoparticles in a ratio [Ca]:[Bi] = 1:5.25 the 10.0 g of the bismuth (III) nitrate pentahydrate $Bi(NO_3)_3{*}5H_2O$ and 0.9712 g of the calcium (II) nitrate tetrahydrate $Ca(NO_3)_2{*}4H_2O$ was rubbed in an agate mortar with 5.0 g of sorbitol to form a soluble mixture of organic complexes. Then, the resulting solution was allowed for 24 hours in air at room temperature until the spontaneous homogeneous mass of foam was formed. After that the organic matrix with homogeneously distributed calcium and bismuth atoms was annealed at 350 ^{0}C for 9 hours. As a result, in the entire volume of the mixture the amorphous nanoparticles of the calcium bismuthate were formed. At the next stage the amorphous calcium bismuthate nanoparticles annealed in a porcelain dish at 700 ^{0}C (for samples № 1-3) or 770 ^{0}C (for samples № 4-6) during 4 hours to form a crystalline lattice in the nanoparticles. Samples with different ratio [Ca]:[Bi] in the same manner with the appropriate amount of precursor recalculation (bismuth and calcium nitrates) were obtained. For comparison of the optical properties of the samples the bismuth oxide Bi_2O_3 were used. We used chemicals ACROS, purity of 99.5%.

It is known [23] that the calcium bismuthates could form the different types of the crystalline lattice, which depends on the annealing temperature (Fig. 1). Annealing at the lower temperature (700 ^{0}C) leads to the forming of the orthorhombic lattice with parameters a=6.81 Å, b=3.95 Å, c=27.83 Å. Increasing the annealing temperature up to 770 ^{0}C leads to the formation of a rhombohedral lattice with parameters a=b=3.95 Å, c=27.83 Å.

The X-ray analysis of the synthesized nanoparticles was performed by using desktop X-ray diffractometer MiniFlex II – Rigaku with the cooper cathode and CuKα-radiation. The scanning electron microscopy researches were performed on an EVO-40 SEM (Carl Zeiss) equipped by the

EDX (Oxford Instruments). The DRS research was carried out with using the "SDL-41" spectrophotometer (Russia).

Table 1. The compound and preparation terms of the calcium bismuthates nanoparticles.

Sample №	The atomic ratio		Annealing temperature, ^{0}C
	Bi	Ca	
1	6.14	1	700
2	5.25	1	700
3	4.26	1	700
4	6.14	1	770
5	5.25	1	770
6	4.26	1	770

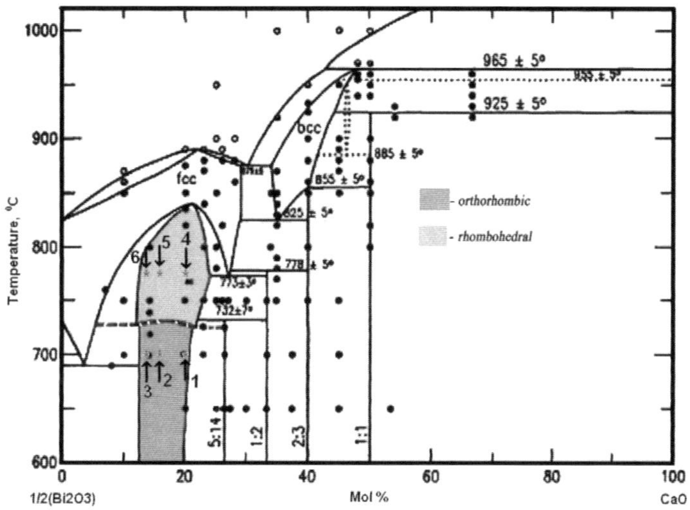

Figure 1. The calcium bismuthates phase diagram [23]. Select the areas of formation of orthorhombic and rhombohedral crystal lattices. The arrows with numbers mark synthesized nanoparticles of the calcium bismuthate. The numbering follows the Table 1.

3. Results and Discussion

3.1. SEM and X-ray characterizations

To determine the morphological characteristics the synthesized samples were studied by scanning electron microscopy. Fig. 2 shows the SEM picture of the powder of calcium bismuthate nanoparticles (sample № 2 in Table 1). It is seen from the figure that the resulting samples of the calcium bismuthates are powder composed of nanoparticles with size of 50-450 nm.

In the Reference [23] a method to determine the type of crystal lattice of the synthesized calcium bismuthates is proposed. The method is based on an analysis of the relative intensities of lines of the doublet peaks on the X-ray diffraction patterns near $2\theta=70°–71°$. Orthorhombic crystal lattice is characterized by the equal intensity of the lines $\alpha 1$ and $\alpha 2$, while for the rhombohedral crystal lattice it is characteristically that intensity of the α1-line is significantly higher of the α2-line. Fig. 3 highlighted areas near diffraction patterns characteristic Bragg angles (insets a, b). All synthesized samples correspond to the calcium bismuthate $Ca_{0.4}Bi_{1.6}O_{2.8}$ (ICDD №41-309), which is consistent with the known data [23].

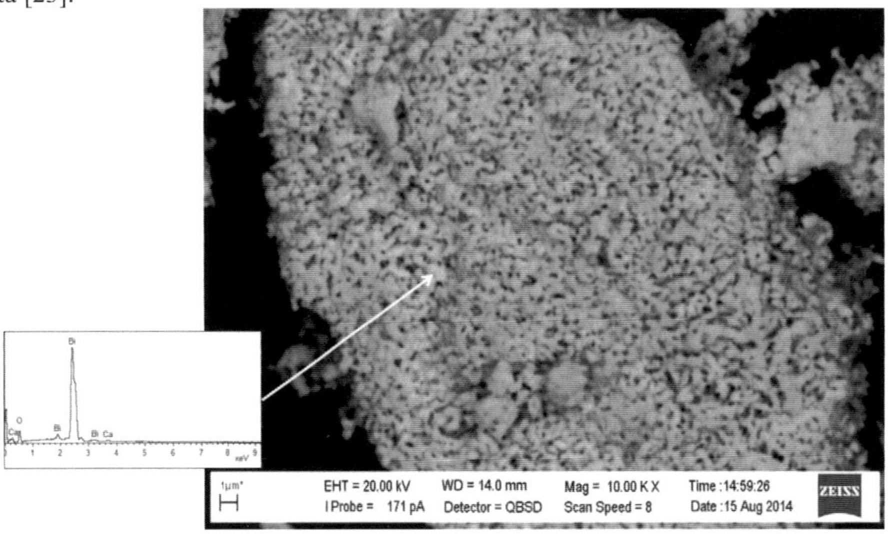

Figure 2. The SEM image of the powder of the calcium bismuthate nanoparticles (sample № 2 in Table 1). The inset - the elemental composition of a sample obtained by energy dispersive analysis.

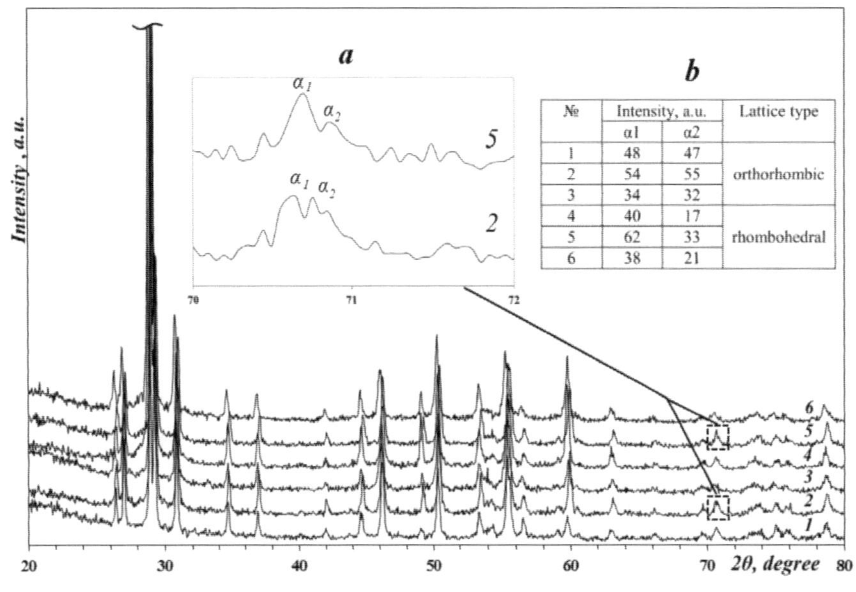

№	Intensity, a.u.		Lattice type
	α1	α2	
1	48	47	orthorhombic
2	54	55	
3	34	32	
4	40	17	rhombohedral
5	62	33	
6	38	21	

Figure 3. The X-ray diffraction patterns of the synthesized nanoparticles of the calcium bismuthate. Numbering corresponds to Table 1. a - the region of the characteristic Bragg angles, b - the intensity values of lines α1 and α2.

3.2. DRS characterization

Fig. 4 shows the diffuse reflectance spectrum of nanoparticles of calcium bismuthate of various compositions. It is seen that the annealing temperature at which formation of the crystalline lattice occurs has a significant influence on the optical properties of the calcium bismuthates nanoparticles. In general, samples with orthorhombic crystalline lattice are characterized by a strong absorption of radiation in the range 420-600 nm.

Also, when comparing all six DRS (Fig. 4) it is clear that they are clearly divided into two groups. The first group of samples can be distinguished by low temperature samples 1-3. Feature of the DRS of these calcium bismuthates is the proportion of calcium in the composition with respect to the proportion of bismuth (3 → 2 → 1) there is a proportional increase in the absorption of radiation in the visible range. The second group includes high-temperature samples 4-6, in which a similar shift of DRS is virtually nonexistent.

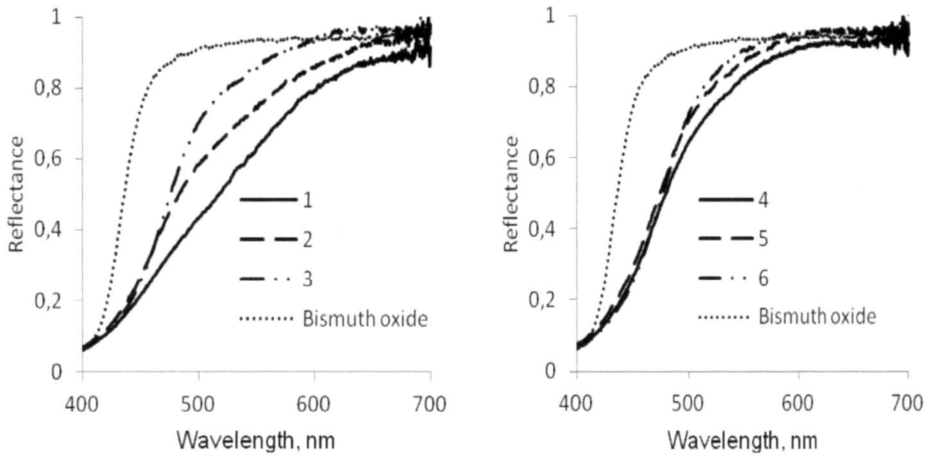

Figure 4. The diffuse reflectance spectrum of nanoparticles of calcium bismuthate of various compositions.

4. Conclusions

Apparently, the explanation of the shown differences in the diffuse reflectance spectrums between high- and low-temperature samples is associated with a denser packing of calcium bismuthate atoms inside the unit cell of the rhombohedral type which is formed at higher annealing temperature.

The amount of absorbed light in the visible range of the semiconductor determines the prospects of its use as optical sensors. Therefore, the studies show that the most promising visible light optical sensors can be obtained based on calcium bismuthate having an orthorhombic crystal lattice and the highest relative concentration of calcium.

5. References

[1] Choi W, Termin A and Hoffmann M R 1994 The Role of Metal Ion Dopants in Quantum-Sized TiO_2: Correlation between Photoreactivity and Charge Carrier Recombination Dynamics J. Phys. Chem. **98(51)** 13669

[2] Ashi R, Morikawa T, Ohwaki T, Aoki K and Taga Y 2001Visible-light photocatalysis in nitrogen-doped titanium oxides Science **293(5528)** 269

[3] Sakthivel S and Kisch H 2003 Daylight photocatalysis by carbon-modified titanium dioxide Angew. Chem., Int. Ed. **42(40)** 4908

[4] Chunjing H, Jing L, Zailei Zh, Yongjun J, Hanhui Zh, Fangxing X, Dan W, Bin L and Fabing S 2015 Enhancement of photocatalytic properties of TiO_2 nanoparticles doped with CeO_2 and supported on SiO_2 for phenol degradation Applied Surface Science **331** 17

[5] Hirano K, Suzuki E, Ishikawa A, Moroi T, Shiroishi H and Kaneko M 2000 Sensitization of TiO_2 particles by dyes to achieve H_2 evolution by visible light J. Photochem. Photobiol. A: Chem. **136(3)** 157

[6] Qin G, Sun Zh, Wu Q, Lin L, Liang M and Xue S 2011 Dye-sensitized TiO_2 film with bifunctionalized zones for photocatalytic degradation of 4-cholophenol J. Hazard. Mater. **192(2)** 599

[7] Akurati K K, Vital A, Dellemann J-P, Michalow K, Graule T, Ferri D and Baiker A 2008 Flame-made WO_3/TiO_2 nanoparticles: Relation between surface acidity, structure and photocatalytic activity Appl. Catal. B: Environmental **79(1)** 53

[8] Tang J W, Zou Z G and Ye J H 2004 Efficient removal of organic contaminants by a visible light driven photocatalyst $Sr_6Bi_2O_9$ Angew. Chem., Int. Ed. **43** 4463

[9] Luan J F, Pan B C, Paz Y, Li Y M, Wu X S and Zou Z G 2009 Structural, photophysical and photocatalytic properties of new Bi_2SbVO_7 photocatalyst under visible light irradiation Physical Chemistry Chemical Physics **11** 6289

[10] Ikarashi K, Sato J, Kobayashi H, Saito N, Nishiyama H and Inoue Y 2002 Photocatalysis for Water Decomposition by RuO_2-Dispersed $ZnGa_2O_4$ with d^{10} Configuration J. Phys. Chem. B **106(35)** 9048

[11] Kato H, Kobayashi H and Kudo A 2002 Role of Ag^+ in the Band Structures and Photocatalytic Properties of $AgMO_3$ (M: Ta and Nb) with the Perovskite Structure J. Phys. Chem. B **106(48)** 12441

[12] Tokunaga S, Kato H and Kudo A 2001 Selective Preparation of Monoclinic and Tetragonal $BiVO_4$ with Scheelite Structure and Their Photocatalytic Properties Chem. Mater. **13(12)** 4624

[13] Dan-Lian H, Cong W, Piao X, Guang-Ming Z, Bing-An L, Ning-Jie L, Chao H, Cui L, Mei-Hua Zh, Juan-Juan X and Xiang-Ying L 2015 A coupled photocatalytic-biological process for phenol degradation in the Phanerochaete chrysosporium-oxalate-Fe_3O_4 system International Biodeterioration & Biodegradation **97** 115

[14] Aslam M, Iqbal M Ismail I, Salah N, Chandrasekaran S, Tariq Qamar M and Hameed A 2015 Evaluation of sunlight induced structural changes and their effect on the photocatalytic activity of V_2O_5 for the degradation of phenols Journal of Hazardous Materials **286** 127

[15] Zhao-Jin W, Wei H, Ke-Ke C, Zhi-Fang G and Ping W 2014 Sustainable synthesis of metals-doped ZnO nanoparticles from zinc-bearing dust for photodegradation of phenol Journal of Hazardous Materials **278** 91

[16] Kunduz S, Selda G and Soylu P 2015 Highly active $BiVO_4$ nanoparticles: The enhanced photocatalytic properties under natural sunlight for removal of phenol from wastewater. Separation and Purification Technology Separation and Purification Technology **141** 221

[17] Yongjiao W, Yiming H, Tingting L, Jun C, Mengfei L and Leihong Zh 2012 Photocatalytic degradation of methylene blue on $CaBi_6O_{10}/Bi_2O_3$ composites under visible light Chemical Engineering Journal **189–190** 473

[18] Yongjiao W, Yiming H, Tingting L, Jun C, Mengfei L and Leihong Zh 2012 Novel $CaBi_6O_{10}$ photocatalyst for methylene blue degradation under visible light irradiation Catalysis Communications **18** 161

[19] Shan Z C, Xia Y J, Yang Y X, Ding H M and Huang F Q 2010 A one-pot method to grow pyrochlore $H_4Nb_2O_7$-octahedron-based photocatalyst Mater. Lett. **63(20)** 1942

[20] Hu C, Hu X, Guo J and Qu J H Efficient Destruction of Pathogenic Bacteria with $NiO/SrBi_2O_4$ under Visible Light Irradiation Environ. Sci. Technol. **40(17)** 5508

[21] Shtarev D S and Shtareva A V 2013 Photocatalytic Degradation of the Diesel Fuel by Using the Calcium Bismuthate - Bismuth Oxide Photocatalyst Composition Applied Mechanics and Materials **377** 204

[22] Karpovich N .F, Makarevich K S, Chigrin P G and Lebukhova N V 2010 Production of copper molybdates by the extraction-pyrolytic method Theoretical Foundations of Chemical Engineering **44(4)** 471
[23] Burton B P, Rawn C J and Roth R S 1993 Phase Equilibria and Crystal Chemistry in Portions of the System SrO-CaO-Bi_2O_3-CuO, Part IV — The System CaO'Bi_20_3-CuO J. Res. Natl. Inst. Stand. Technol. **98(4)** 469

A method for solid phase synthesis of phosphors under increased pressure; creation of remote phosphors

E N Galashov[1], **A A Yusuf**[1] **and E M Mandrik**[1]

[1] Department of Applied Physics, Scientific Research Department Novosibirsk State University, 2 Pirogov's Street, Novosibirsk 630090, Russia

E-mail: chessplanet86@yandex.ru

Abstract. A method for phosphors solid-phase synthesis under high pressure is proposed. Using the method, phosphors based on rare earth garnets were synthesized at a temperature considerably lower (1300 °C) than the temperature of the synthesis by conventional method (above 1650 °C). Using organic, inorganic ligaments and synthesized phosphors, remote phosphors were created. The chromaticity coordinates of created LEDs on the base of blue LEDs and remote phosphors fall into the bin of white and green colours. Luminous efficiency reaches 140 lm/W.

1. Introduction

Light is an integral part of human life. The last few years were crucial for the sciences studying light. Light technologies are rapidly evolving. In many ways, it is possible thanks to their rather wide application [1, 2], both in everyday life and in various industries.

The role of lighting technology is appreciated by the world community. So, in 2014, – Nobel Laureate in Physics [3] was awarded for the creation of energy-saving and environmentally friendly light sources - blue light-emitting diodes (LED). The efficiency of the LED (80%) is 10-fold higher than the efficiency of incandescent and twice the efficiency of fluorescent lamps, which contain highly toxic mercury vapors. However, the service life of the light-emitting heterostructures more than 100 000 hours, it is many times longer than incandescent and fluorescent lamps.

To raise global awareness of how light-based technologies provide solutions to global problems in the areas of health, energy, education and agriculture, United Nations General Assembly declared 2015 - the International Year of the light and light technologies (IYL 2015).

Currently yellow phosphors based on rare-earth garnet ($Y_3Al_5O_{12}$:Ce) are used in most commercial LEDs. The use of such phosphors for white LEDs is primarily due to high conversion efficiency and optimum light performance. However, the commercial synthesis of these phosphors is quite complex and energy-intensive because it is being done at high temperatures (above 1650 °C). Phosphor powders of different fractions (typically 2 to 50 microns) for different applications are required and such high temperatures of the synthesis lead to the formation of agglomerated microparticles. This implies involvement of additional expensive processing methods to obtain powders of the desired gravimetric composition.

A method for the synthesis of phosphors based on rare earth garnets in a controlled high pressure is proposed in this paper. A prerequisite to the establishment of the proposed synthesis method is the fact that at higher pressures the cubic yttrium oxide phase becomes more disordered monoclinic yttrium oxide phase [4]. It is assumed that in a solid phase form reaction of $Y_3Al_5O_{12}$, the diffusion of aluminum atoms in the disordered structure of yttrium oxide is faster, as a result of the reaction rate increases, intermediate phase $YAlO_3$ disappears, and as a consequence, the total synthesis phase $Y_3Al_5O_{12}$ occurs at a lower temperature and a shorter time.

Modern commercial LEDs are coated optical silicone mixed up with the phosphor powder. The disadvantage of this system is that the silicone has a low thermal conductivity. In operation, the LED phosphor is heated, this leads to lower conversion properties and loss luminous efficiency. Furthermore, since the optical silicone is an organic compound, the degradation of the silicone occurs under the action of radiation emission of a blue LED, and this also leads to a reduction in light transmittance. It should be noted that the optical silicon has a high price - the order of $ 1200 / kg. In order to avoid light and conversion losses associated with the heating of the phosphor, one possible way is to use a remote phosphor.

2. Fabrication

For the synthesis of the samples, the following original reagents are used:

Table 1. Original reagents.

Reagent	Production	Purity, %	Grain size D50, μm
Y_2O_3	South Korea	99.99	2
Al_2O_3	China	99.99	2
CeO_2	Russia	99.95	2
Lu_2O_3	China	99.99	2

The starting reagents are hung in the ratios equal gross formulas $(Y_{0.98}Ce_{0.02})_3Al_5O_{12}$, $(Y_{0.78}Gd_{0.2}Ce_{0.02})_3Al_5O_{12}$ and $(Lu_{0.98}Ce_{0.02})_3Al_5O_{12}$ on laboratory scale (VC-3000) with a precision weighing 50 mg. Further starting reactants are placed into a mixer (Frich firm) of the type "drunken barrel" and mixed in ethanol for 12 hours. The mixture is placed in an oven desiccator and dried for 12 hours at a temperature of 120 °C. Then the dried blend is placed into an alumina crucible and covered with a lid made of the same material. The crucibles were placed in a high-pressure synthesis unit. Working chamber installation was evacuated to 10^{-2} Pa, and then a graphite heater heats up to a temperature of 1300 °C. When the temperature reached 1300 °C installation is filled with reducing argon - hydrogen gas mixture (95% Ar, 5% H_2) until the pressure reached 10^7 Pa. Under these conditions the samples were incubated for 3 hours. Thereafter, the heater power setting is disabled and installation is cooled to room temperature for 12 hours.

In the described method the following examples were prepared: $(Y_{0.98}Ce_{0.02})_3Al_5O_{12}$ and $(Y_{0.78}Gd_{0.2}Ce_{0.02})_3Al_5O_{12}$, $(Lu_{0.98}Ce_{0.02})_3Al_5O_{12}$

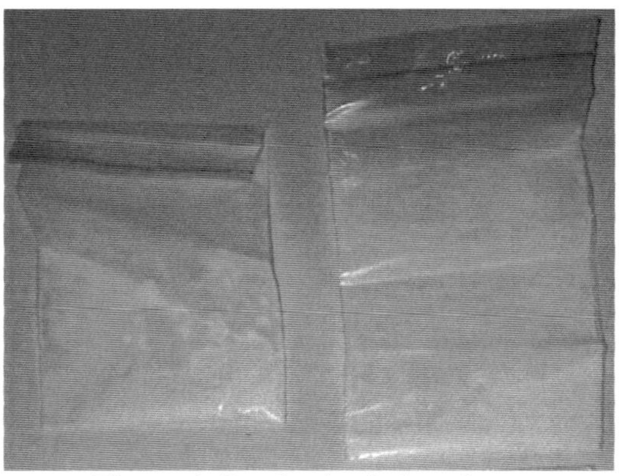

Figure 1. Synthesized phosphors: $(Lu_{0.98}Ce_{0.02})_3Al_5O_{12}$ and $(Y_{0.98}Ce_{0.02})_3Al_5O_{12}$.

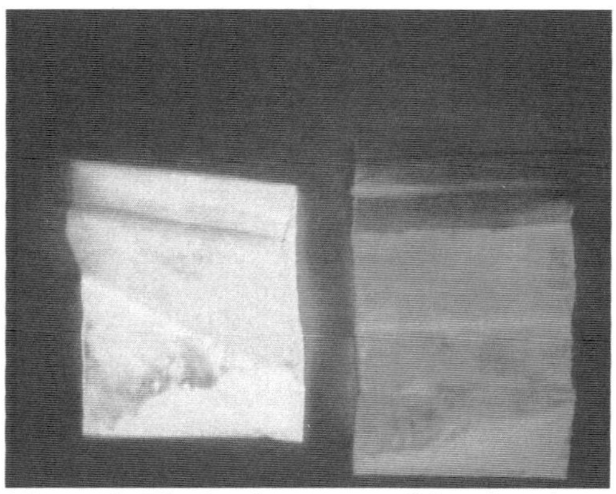

Figure 2. The luminescence of the phosphors due to irradiation of blue.

3. Results

X-ray diffraction analysis (copper radiation λ=1.5418 Å) of the samples showed that samples were synthesized at a temperature of 1300 °C and a pressure of 10^5 Pa (Figure 3) in addition to garnet phase have extraneous phases of reaction products of synthesis. At a temperature of 1650 °C and a pressure of 10^5 Pa (Figure 4) the amount of foreign phases is reduced, but not completely. Electron microscopy showed that the particles are agglomerated and have a non-uniform surface shape (Figure 5).

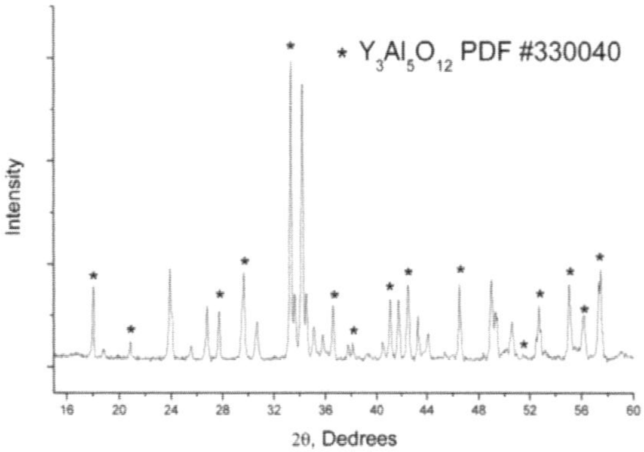

Figure 3. Synthesis of $(Y_{0.98}Ce_{0.02})_3Al_5O_{12}$ at a conditions T=1300 °C and P=10^5 Pa.

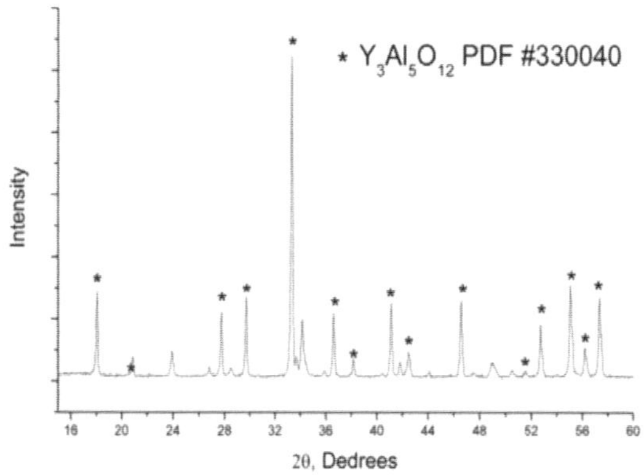

Figure 4. Synthesis of $(Y_{0.98}Ce_{0.02})_3Al_5O_{12}$ at a conditions T=1650 °C and P=10^5 Pa.

Figure 5. Electron microscopy of $(Y_{0.98}Ce_{0.02})_3Al_5O_{12}$ at a conditions of synthesis T=1650 °C and P=10^5 Pa.

X-ray diffraction analysis of the samples showed that samples were synthesized at a temperature of 1300 °C and a pressure of 10^7 Pa (Figure 6) have only garnet phase. Electron microscopy showed that the particles are not agglomerated and have faceted surface (Figure 7). According to the data of laser particle size analyser it was found that all the samples synthesized at high pressures have particle sizes from 10 to 20 μm.

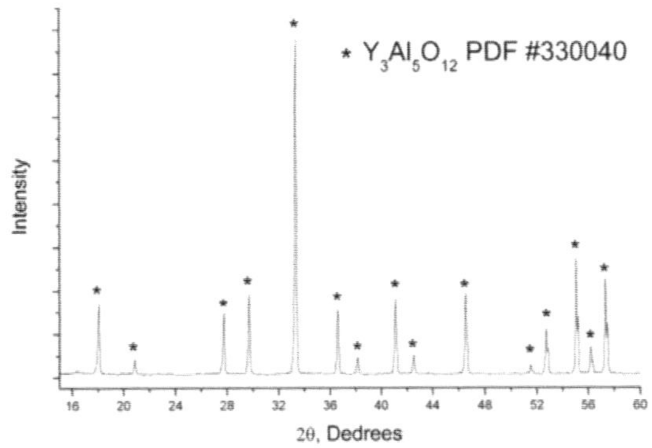

Figure 6. Synthesis of $(Y_{0.98}Ce_{0.02})_3Al_5O_{12}$ at a conditions T=1300 °C and P=10^7 Pa.

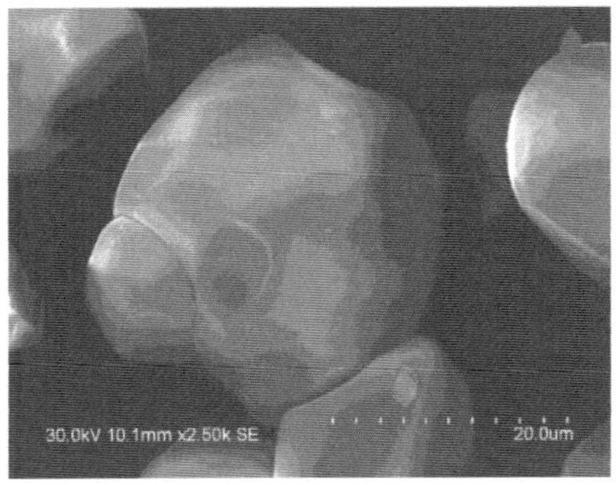

Figure 7. Electron microscopy of $(Y_{0.98}Ce_{0.02})_3Al_5O_{12}$ at a conditions of synthesis T=1300 °C and P=10^7 Pa.

Using organic, inorganic ligaments and synthesized phosphors remote phosphors (Figure 8) were created. The chromaticity coordinates of created LEDs on the base of blue LEDs and remote phosphors fall into the bin of white and green colors. Luminous efficiency was 140, 130, and 120 lm, respectively for $Y_3Al_5O_{12}$:Ce, $Gd_3Al_5O_{12}$:Ce, $Lu_3Al_5O_{12}$:Ce remote phosphor.

Figure 8. Remote phosphors on base of organic and inorganic ligaments and synthesized phosphors.

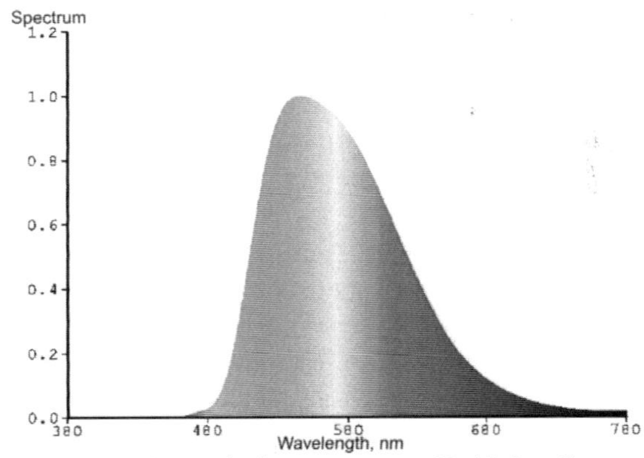

Figure 9. The emission spectrum $Y_3Al_5O_{12}$:Ce, as a result of irradiation with blue LED.

References

[1] S Muthu, F J P Schuurmans, M D Pashley, Red, green, and blue LEDs for white light illumination, *IEEE Journal on Selected Topics in Quantum Electronics*, **8** – 2, pp. 333–338, 2002

[2] Y Shimizu, Development of White LED light source, *Rare Earth*, **40**, pp. 150–151, 2002

[3] H Yusa, T Tsuchiya, N Sata, Y Ohishi, *InorgChem.*, **49**, pp.4478-4485, 2010

Multispectral high-resolution hologram generation using orthographic projection images

I Muniraj, C Guo, J T Sheridan[1]
School of Electrical, Electronic and Communication Engineering, University College Dublin, Belfield, Dublin 4, Ireland.

E-mail: john.sheridan@ucd.ie

Abstract. We present a new method of synthesizing a digital hologram of three-dimensional (3D) real-world objects from multiple orthographic projection images (OPI). A high-resolution multiple perspectives of 3D objects (i.e., two dimensional elemental image array) are captured under incoherent white light using synthetic aperture integral imaging (SAII) technique and their OPIs are obtained respectively. The reference beam is then multiplied with the corresponding OPI and integrated to form a Fourier hologram. Eventually, a modified phase retrieval algorithm (GS/HIO) is applied to reconstruct the hologram. The principle is validated experimentally and the results support the feasibility of the proposed method.

1. Introduction

The applications of digital holography have been applied widely as it captures and displays whole three-dimensional (3D) information of the real world objects [1, 3]. Nevertheless, practically, the holography imaging is a tedious process, as it requires coherent light source and demands scrupulous stability of the optical system that can work under darker environment [3]. In order to alleviate, approches that includes hybrid interfaces between Integral imaging (II) and holography has been suggested in the past decade [4-7].

The principle of II is based on Integral Photography (IP) that uses a circular or square lens array to capture and display 3D images. In this process, a series of multiple two-dimensional (2D) images, Elemental Image Array (EIA), are captured under white-light illumination and that can be used to reconstruct the 3D image. II makes the 3D image visualization and 3D depth extraction process easier and also it removes the necessity for camera calibration [3, 5].

In this paper, we propose a new method for the hologram generation using white-light (i.e., incoherent source) based orthographic images. The 3D objects are digitally recorded and then the orthographic projection images (OPI) are obtained from the captured EIA. Each OPI is numerically modulated by the plane wave (i.e., reference beam) propagating in the direction of the corresponding view angle and combined into a single complex value that refers to one pixel in the synthesized digital hologram [7]. By repating this process for all OPIs, we can generate the Digital Fourier Hologram (DFH) of the 3D objects. We note that the process of generating DFH is more compact and that does not requires a coherent illumination system [5].

This paper is organized as follows: In section 2 we describe the priniciple of OPI obtaining process. In section 3 we discuss about the generation and the reconstruction of an integral Fourier hologram. Experimental results are presented in section 4. Finally, in section 5 we conclude our discussion.

---

[1] To whom any correspondence should be addressed.

Journal of Physics: Conference Series **735** (2016) 012070 doi:10.1088/1742-6596/735/1/012070

2. Orthographic Projection Image (OPI)

Orthographic projection geometry describes the geometrical relationship between the 3D objects and the captured view image at the image plane (i.e., CCD or CMOS imaging sensor). As it can be seen from Fig.1, in this procedure, the projection lines are considered to be parallel with each other [7].

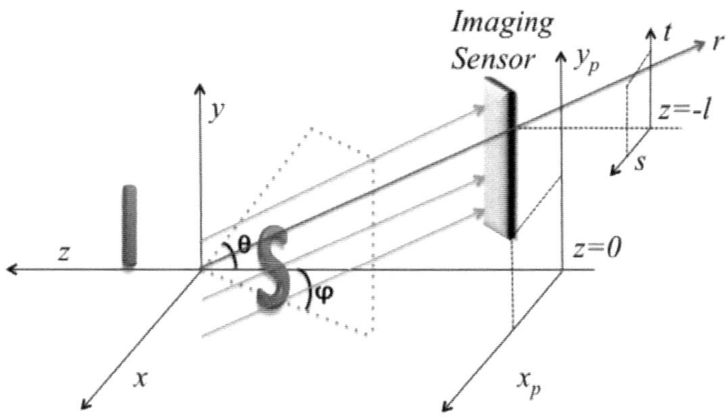

Figure 1. Orthographic Projection Geometry.

If we consider r as one of the projection lines (see Fig. 1 green color line), the angle φ is the angle between the projection line r and the $x-z$ plane. Similarly, the angle θ is the angle that the projection of r onto the $y-z$ plane. Thus, the projection image coordinates (x_p, y_p) is given as follows:

$$x_p = x + z\tan\varphi = x + {}^{zs}\!/_{l}, \qquad y_p = y + z\tan\theta = y + {}^{zt}\!/_{l}, \qquad (1)$$

where s,t and l are used to explain projection direction, conveniently. It is said that in [7], orthographic projection leads to the exact Fourier and Fresnal based hologram generation.

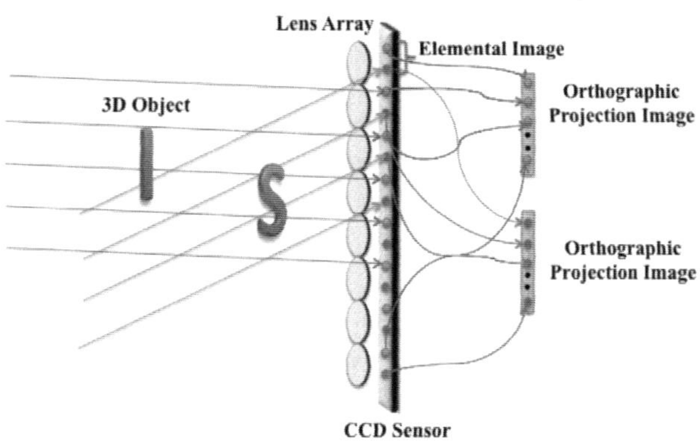

Figure 2. Process of generating orthographic projection images using lenslet array.

Figure 2 shows the process of generating OPI using lenslet array. As it can be noticed, the 3D object is imaged through the lens array and thus the captured elemental images represent a different perspective of the 3D object. It also can be seen from Fig. 2 that the orthographic projection image is a collection of pixels at the same local position from all the captured elemental images. For example, in Fig. 2, a pixel at the position of red dot is collected from all elemental images to form one orthographic projection image. Analogously, the pixels from blue dot are gathered to form another OPI. Since, one pixel is extracted from each EI the number of pixels in the synthesized OPI is same as captured EI. Therefore, the number of pixels in one EI gives the total number of the synthesized OPI [7, 8].

3. Digital Fourier Hologram (DFH)

Figure 3 shows the process of Fourier hologram generation using the orthographic projection images. At first, the OPIs are generated from the captured EIA. Then the each synthesized OPI is multiplied by the phase factor of the slanted plane wave (i.e., reference beam). This product is then combined into a single complex value (i.e., one pixel value in the synthesized DFH). Finally, by repeating these process for all OPIs, an entire Fourier hologram can be generated [7].

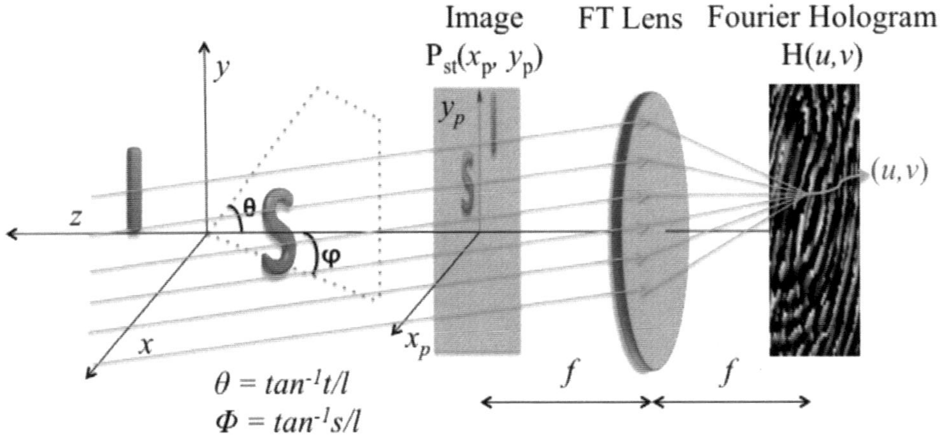

Figure 3. Fourier Hologram generation using the orthographic image.

Mathematically, the generation of Fourier hologram from OPI is derived as follows:

$$H(u,v) = \iint P_{st}(x_p, y_p)\exp\left[-j2\pi k(x_p s + y_p t)\right]dx_p dy_p, \tag{2}$$

where $P_{st}(x_p, y_p)$ denotes OPI and it directional information can be derived using s, t . k is a constant value used in [7] and we note that an exact Fourier hologram can be obtained using this process.

4. Simulation Results

Experiments and simulation results are presented in this section. We used two different 3D objects (see Fig. 4) longitudinally separated and illuminated using incoherent white light. We considered an elemental image in the size of 1024(H) x 1024(V). From these EIs, the orthographic projection images are generated as explained in section 2. Later, by using these OPIs the Fourier hologram can be generated using Eq. 2. In our experiments, we used GS/HIO phase retrieval algorithm [9] to numerically reconstruct the Fourier hologram.

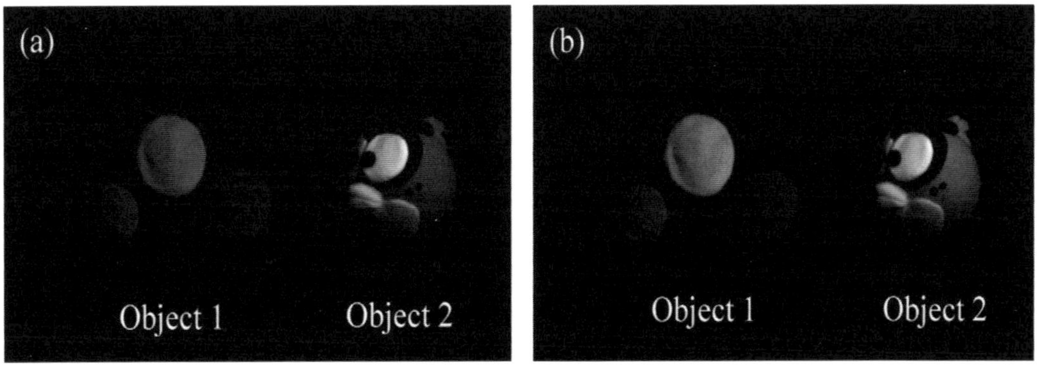

Figure 4. Objects used in our experiment. (a) Raw Bayer image; (b) Interpolated multispectral image.

Figure 5 shows the simulation results of the reconstruced 3D sectional image. It is clearly evident from Fig 5. that in the reconstructed plane, only one of the objects is focused.

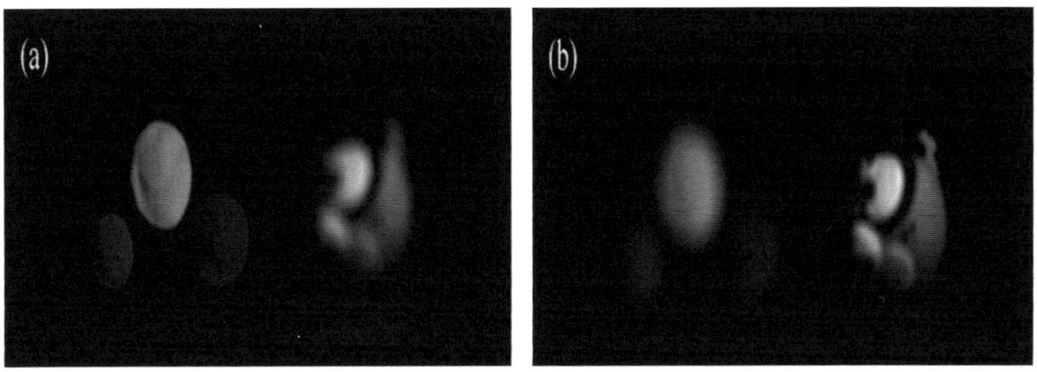

Figure 5. Reconstructed objects. (a) 1st object is in focus; (b)2nd object is focused.

Figure 5(a) shows the reconstruction in the best focus plane of the first 3D object while the second object is out of focus and Fig 5(b) shows the reconstruced focused second 3D object. By these results, it is evident that the volumetric information is indeed encrypted inside the hologram synthesized.

5. Conclusion

In this paper, we have presented a compact method of synthesizing digital Fourier hologram using orthographic projection images (OPI). OPIs are generated from computationally recorded Bayer patterened elemental images that captured under spatially incoherent illumination. Later, a slanted plane wave (i.e., reference beam) is multiplied with OPI and combined to generate Fourier hologram. We note that using computational elemental images improves the reconstructed hologram image quality. We verified this technique by experiments and simulations also the results proves the feasibility.

Acknowledgement

IM acknowledges the support of Irish Research Council (IRC). CG is supported under the UCD-China Scholarship Council (UCD-CSC) program. JTS thanks Science Foundation Ireland (SFI), and Enterprise Ireland (EI) under the National Development Plan (NDP).

References:

[1] Lohmann A W and Paris D P 1967 *Appl. Opt.* **6**, 1739
[2] Goodman J W and Lawrence R W 1967 *Appl. Phys. Lett.* **11**, 77
[3] Muniraj I, Guo C, Lee B G, and Sheridan J T 2015 *Opt. Exp.* **23**, 15907
[4] Abookasis D and Rosen J 2006 *Appl. Opt.* **45**, 6533
[5] Shaked N T, Rosen J, and Stern A 2007 *Opt. Exp.* **15**, 5754
[6] Kim M S, Baasantseren G, Kim N, and Park J H 2008 *J. Opt. Soc. Korea.* **12**, 269
[7] Park J H, Kim M S, Baasantseren G, and Kim N 2009 *Opt. Exp.* **17**, 6320
[8] Park J H, Kim J, and Lee B 2005 Opt. Exp. **13**, 5116
[9] Guo C, Liu S, and Sheridan J T 2015 *Appl. Opt.* **54**, 4698

Performance Study of optical Modulator based on electro-optic effect

V Palodiya, S K Raghuwanshi
Department of Electronics Engineering,
Indian School of Mines
Dhanbad, India
E-mail: vikrampalodiya@gmail.com

Abstract. In this paper, we have studied and derive performance parameter of highly integrated Lithium Niobate optical modulator. This is a chirp free modulator having low switching voltage and large bandwidth. For an external modulator in which travelling-wave electrodes length L imposed the modulating switching voltage, the product of V_π and L is fixed for a given electro optic material Lithium Niobate. We investigate to achieve a low V_π by both magnitude of the electro-optic coefficient for a wide variety of electro-optic materials. A Sellmeier equation for the extraordinary index of congruent lithium niobate is derived. For phase-matching, predictions are accurate for temperature between room temperature 250^0C and wavelength ranging from 0.4 to $5\mu m$. The Sellmeier equations predict more accurately refractive indices at long wavelengths. Theoretical result is confirmed by simulated results. We have analysed the various parameters such as switching voltage, device performance index, time constant, transmittance, cut-off frequency, 3-dB bandwidth, power absorption coefficient and transmission bit rate of Lithium Niobate optical Modulator based on electro-optic effect.

1. Introduction

$LiNbO_3$ Optical modulators (LNOM) widely used in optical networks like metropolitan area networks (MANs) [1]. LNOM are based on the modulation of the material refractive index of the electric field, induced by an applied input voltage. LNOM achieve high performance and high reliability. LNOM also look promising for next-generation systems, because these next-generation systems require that optical modulators have almost zero chirp, X-cut LN modulators can be used with differential output drivers, by adjusting phases and amplitude of electrical signals precisely [2,3]. Due to the easier technology and the need of piezoelectrically induced static charge problems, X-cut modulators have been the initial choice for high speed broadband communication [4, 5]. In Fig 1 shows a few solutions to the interaction region, implemented through coplanar RF distributed electrodes. For an unbuffered X-cut modulator which, due to large velocity mismatch between the RF and optical modes, has a better modulation bandwidth than the lumped modulator but cannot achieve 10 Gbps operation. Reduction of the velocity mismatch is possible through the use of a low-permittivity dielectric buffer (silicon oxide, but also polymers has been exploited to this end) and of metal electrodes thickened through electroplating. X-cut modulators can very well cover 10 Gbps applications and are able to reach the 40 Gbps goal with modulation voltages of the order of 2 V in X-cut configurations. In LN-based modulator the use of junctions allows enhancement of the applied field in the active region. Reducing the switching voltage is important in high-speed operation, since it makes the design of the driver stage less critical. A commonly cited Sellmeier equation for congruent lithium niobate is based on refractive-index measurements at room temperature for wavelengths ranging from 404.6 nm to 3.39 mm [6,7] Since electric-field-poled LN was first demonstrated [8], interest in using periodically poled lithium niobate (PPLN) for optical applications has been growing steadily. This material offers a wide range of phase-matching possibilities [9] and high nonlinearities. This is not surprising since the data used to derive the Sellmeier equation do not extend beyond 3.39 mm. As discussed the multiphonon absorption in oxide materials leads to an exponential increase in absorption at wavelengths near the IR cutoff of the transparency range. For LN, the absorption coefficient at the temperature increases, the absorption edge moves toward shorter wavelengths and therefore has an even greater effect on the refractive index

Journal of Physics: Conference Series **735** (2016) 012071 doi:10.1088/1742-6596/735/1/012071

in the mid-IR spectral region [10]. The potential applications for trace-gas sensing and countermeasures using mid-IR light generation in PPLN justify a revision of the Sellmeier equations to predict more accurately refractive indices at long wavelengths.

2. Theory of Lithium Niobate

$LiNbO_3$ modulator is used to get the highest electro-optic coefficient. Where (E_z), the externally applied electric field should be parallel to the X-axis shown in Fig.1. Therefore, in order to optimize the interaction between the electric field and the optical mode, in an X-cut wafer the electrodes should be located on both sides of the branches.

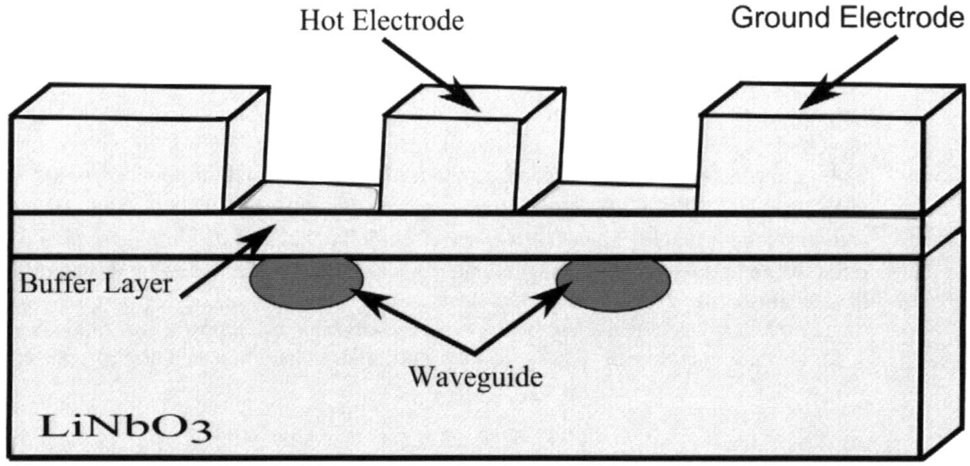

(a)

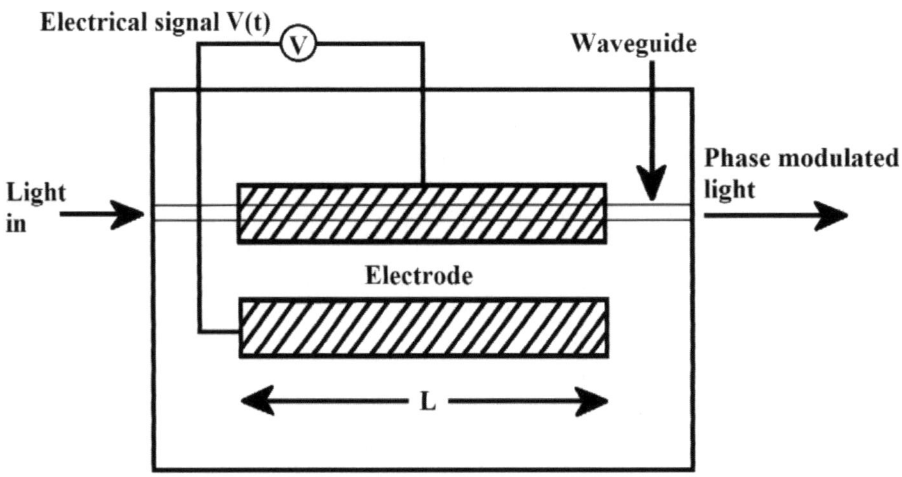

(b)

Figure 1(a) shows X-cut $LiNbO_3$ Phase Modulator and fig 1(b) Internal structure of $LiNbO_3$ Phase Modulator.

In these conditions, the extraordinary refractive-index change by electro-optic effect in $LiNbO_3$. In LN crystal symmetry, most of the electro-optic coefficients (r) in the matrix are zero, some of the electro-optic coefficients acquire equal values. However, the non-zero elements can be represented using the eq. (2.1)

$$\begin{cases} r_{13} = r_{23} = 9 \text{ pm/V} \\ r_{33} = r_{51} = r_{42} = 30\text{pm/V} \\ r_{22} = -r_{12} = -r_{61} = 6.6 \text{ pm/V} \end{cases} \tag{2.1}$$

However, the externally applied field will change all the components of the permittivity matrix. We confine the analysis for the case relevant to an application, and assumption, that the applied field is directed along the optical axis, i.e., that $\varepsilon_z = \varepsilon_3 \neq 0$ while both $\varepsilon_x = \varepsilon_1 = 0$ and $\varepsilon_y = \varepsilon_2 = 0$. In this case, considering only the nonzero element of r yield (for clarity, we partly revert to the expanded index notation in rectangular coordinates):

$$\Delta n_{xx} = \Delta n_{yy} = \Delta n_0 = -\frac{n_0{}^3 r_{13} \varepsilon_z}{2} \tag{2.2}$$

$$\Delta n_{zz} = \Delta n_e = -\frac{n_e{}^3 r_{33} \varepsilon_z}{2} \tag{2.3}$$

$$\Delta n_{yz} = \Delta n_{xz} = \Delta n_{xy} = 0 \tag{2.4}$$

Based on the proposed choice of the applied electric field, the off-diagonal elements of the permittivity matrix are always zero, and the principal axes remains unchanged with respect to the zero-field case. Moreover, the largest element r_{33} is exploiting, leading to a variation of the extraordinary index Δn_e, but the ordinary index Δn_0 experiences smaller variation in the X-cut configuration [11,12]. It is clear that; optical axis is parallel to the crystal surface and orthogonal to the waveguide axis. The waveguide axis runs along the x or y, while the crystal surface is an orthogonal to$y$ or x, respectively. The optical mode polarization is transverse electric and it is apparent that, the electric field polarization is again along the optical axis, more specifically, it is horizontal. In such type of conditions, the modal RI is $n_{TE} = n_e$. Application of an electric field, along the z axis, which is parallel to the optical field and to the crystal surface leads to a variation of the modal RI:

$$\Delta n_{TE} = \Delta n_e = \frac{n_e{}^3 r_{33} E_Z}{2} \tag{2.5}$$

where n_e is the effective refractive index of the material based electro-optic modulator device, r_{33} is the corresponding electro-optic coefficient in pm/Volt, E_z is the integrated Mach-Zehnder intensity modulator. The theoretical value of the half-wave voltage of the integrated electro-optic modulator or switching voltage can be calculated using the expression

$$V_\pi = \frac{\lambda g}{L_m n_e^3 r_{33} \Gamma} \tag{2.6}$$

where λ is the operating signal wavelength in μm, g is the gap between electrodes in μm, L_m is the modulator length in cm, and Γ is confinement factor. The transmittance of an optical signal through the modulator using the following equation:

$$T_m = e^{-\alpha} L_m \tag{2.7}$$

where α is the power absorption coefficient in dB/cm. The resistance of modulator device can be measured by estimating the device dimensions as the following equation:

$$R = \rho \frac{L_m}{wd} \quad (2.8)$$

where ρ is the resistivity, w is the modulator width and d is the modulator thickness. The time constant of the modulator device τ.

$$\tau = RC \quad (2.9)$$

The cut-off frequency of the modulator device is calculated from the RC time constant as the following:

$$f_{cut-off} = \frac{1}{2\pi RC} \quad (2.10)$$

The relative refractive index change induces in an optical phase modulation is written as:

$$\Delta\emptyset = k_0 L_m \Delta n_e \quad (2.11)$$

where k_0=2π/λ is the wave number. The electrical 3-dB bandwidth f_{3-dB} for which modulation voltage V_m is reduced by $(1/\sqrt{2})$.

$$f_{3-dB} = \frac{1}{\pi R \varepsilon_{eff} C L_m} \quad (2.12)$$

where C is the capacitance in pF, ε_{eff} is the effective RF relative dielectric constant.

The electro-optic device reduces the bandwidth. Under the perfect velocity matching condition achievable modulation bandwidth f_m is:

$$f_m = \frac{6.84}{\alpha L_m} \quad (2.13)$$

For $LiNbO_3$ material, the investigation of both the thermal and spectral variations of the effective waveguide refractive-index (n_e) require empirical equation. The set of parameters required to completely characterize the temperature dependence of the refractive-index is given by, Sellmeier equation;

$$n_e = \sqrt{A_1 + A_2 H + \frac{A_3 + A_4 H}{\lambda^2 - (A_5 + A_6 H)^2} + \frac{A_7 + A_8 H}{\lambda^2 - A_9^2} - A_{10}\lambda} \quad (2.14)$$

where λ is the optical signal wavelength in μm and $H = T^2 - T_0^2$, T is the ambient temperature in K, and T_0 is the room temperature (300 K). The set of parameters of equation coefficients ($LiNbO_3$) are recast and dimensionally adjusted as: A_1= 5.35583, $A_2 = 4.629 \times 10^{-7}$, A_3= 0.100473,

$A_4 = 3.862 \times 10^{-8}$, A_5=0.20692, $A_6 = -0.89 \times 10^{-8}$, A_7= 100, $A_8 = 2.657 \times 10^{-5}$, A_9 = 11.34927, and A_{10}= 0.01533. Equation (1.10) can be simplified as the following:

$$n_e = \sqrt{A_{12} + \frac{A_{34}}{\lambda^2 - A_{56}^2} + \frac{A_{78}}{\lambda^2 - A_9^2} - A_{10}\lambda} \tag{2.15}$$

where: $A_{12} = A_1 + A_2 H$, $A_{34} = A_3 + A_4 H$, $A_{56} = A_5 + A_6 H$ and $A_{78} = A_7 + A_8 H$. Then the first and second differentiation of Equation (1.11) with respect to operating signals wavelength λ, which gives:

$$\frac{dn_e}{d\lambda} = \left(\frac{-\lambda}{n_e}\right)\left(\frac{A_{34}}{(\lambda^2 - A_{56}^2)^2} + \frac{A_{78}}{(\lambda^2 - A_9^2)^2} - A_{10}\right) \tag{2.16}$$

$$\frac{d^2n_e}{d\lambda^2} = \left(\frac{1}{n_e}\right)\left(\frac{A_{34}\left(2 - (\lambda^2 - A_{56}^2)\right)}{(\lambda^2 - A_{56}^2)^2} + \frac{A_{78}\left(2 - (\lambda^2 - A_9^2)\right)}{(\lambda^2 - A_9^2)^2} - A_{10}\right) \tag{2.17}$$

The material dispersion based electro-optic modulator, D_{mat} which is given by the following equation:

$$D_{mat} = -\left(\frac{L_m(\Delta\lambda)\lambda}{c}\right)\left(\frac{d^2n_e}{d\lambda^2}\right) \tag{2.18}$$

The modal-dispersion delay, D_{modal} for a multi-mode step-index electro-optic device with length L_m, is given by:

$$D_{modal} = \frac{L_m n_e \Delta n_e}{c} \tag{2.20}$$

Where the total dispersion coefficient, $D_t = D_{mat} + D_{modal}$. In addition to providing sufficient power to the receiver, the system must also satisfy the bandwidth requirements imposed by the rate at which data are transmitted. A convenient method of accounting for the bandwidth is to combine the rise times of the various system components and compare the result with the rise time needed for the given data rate and pulse coding scheme. The system rise time is given in terms of the data rate for non-return to zero pulse code by:

$$B_R(NRZ) = \frac{0.7}{D_t} \tag{2.21}$$

In the same way, the device performance index (DPI) can be expressed as the following expression:

$$DPI = \frac{f_m}{V_\pi} \tag{2.22}$$

3. Results and Discussion

We have investigated the recent progress of $LiNbO_3$ based electro-optic modulator device in high-speed photonic networks over a wide range of the affecting operating parameters. Based on the model equations analysis, set of the operating parameters as shown in Table 1.

Table 1. Proposed operating parameters for z–cut modulator device

Operating Parameter	Symbol	Value
Operating signal wavelength	λ	1.3 μm-1.55 μm
Spectral line width of the optical source	$\Delta\lambda$	0.1 nm
Ambient temperature	T	300 K$\leq$ T $\leq$ 320 K
Room temperature	T_0	300 K
Relative refractive-index change	Δn_e	0.05$\leq \Delta n \leq$0.09
Modulator length	L_m	5 cm$\leq L_m \leq$10 cm
Modulator width	W	2 cm$\leq w \leq$5 cm
Modulator thickness	D	0.2 cm$\leq d \leq$1 cm
Speed of light	C	3×10^{10}cm/sec
Gap between electrodes	G	0.1-0.35 μm
Electro-optic coefficient	r_{33}	30.8×10^{-10} cm/Volt
Applied electric field	E	2Volt/cm
Resistivity	ρ	$2.66\times 10^{-3}\Omega$cm
Confinement factor	Γ	0.8-0.95
Modulator device capacitance	C	0.2-0.4 pF
Effective RF relative dielectric constant	ε_{ff}	17.85
Power absorption coefficient	α	0.1-0.4 dB/cm

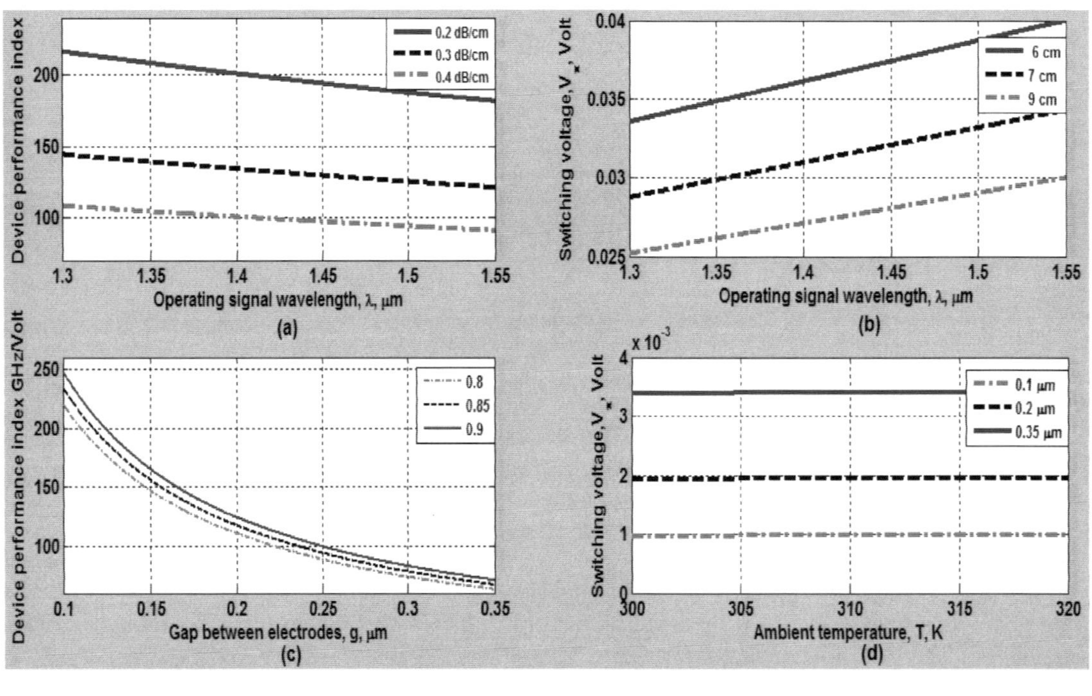

Figure 2: (a) Variations of the device performance index versus operating signal wavelength at the assumed set of parameters, (b) Variations of the switching voltage versus operating signal wavelength at the assumed set of parameters, (c) Variations of the device performance index against the gap between electrodes at the assumed set of parameters, (d) Variations of the switching voltage against ambient temperature at the assumed set of parameters.

Calculated results for λ as a function of DPI are plotted in Fig. 2 (a) for three values of power absorption coefficient α. This plot shows the operating optical signal wavelength increases, this leads to decrease in device performance index of the constant power absorption coefficient. As well, as power absorption coefficient increases, this results in decreasing of device performance at constant operating optical signal wavelength. In Fig. 2 (b) the operating optical signal wavelength increases, this leads to increase in switching voltage at constant modulator length. As well as modulator length increases, this results in decreasing of switching voltage at constant operating optical signal wavelength. Fig. 2 (c) has shown that as the gap between the electrodes increases, this leads to decrease in device performance index at constant confinement factor. As well as confinement factor increases, this result shows increasing of the device performance index at constant gap between electrodes. In Fig. 2 (d) has indicated that as ambient temperature increases, this leads slightly decrease in switching voltage at the gap between electrodes. As well as the gap between the electrodes increases, this result of increasing of switching voltage at constant ambient temperature.

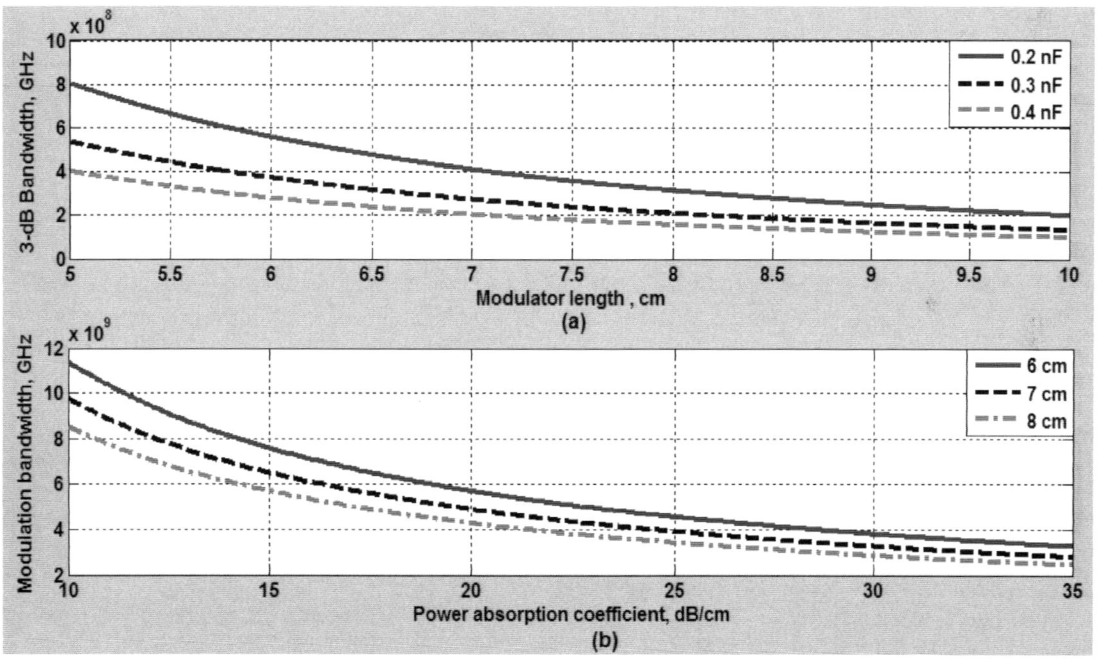

Figure 3: (a) Variations of 3-dB bandwidth against modulator length at the assumed set of parameters, (b) Variations of modulation bandwidth against power absorption coefficient at the assumed set of parameters.

As shown in Fig. 3 (a) has indicated that both modulator length and modulators capacitance increase, this results in decreasing of 3-dB bandwidth. In Fig. 3 (b) has shown that as both modulator length and power absorption coefficient increase, this leads to decrease of modulation bandwidth.

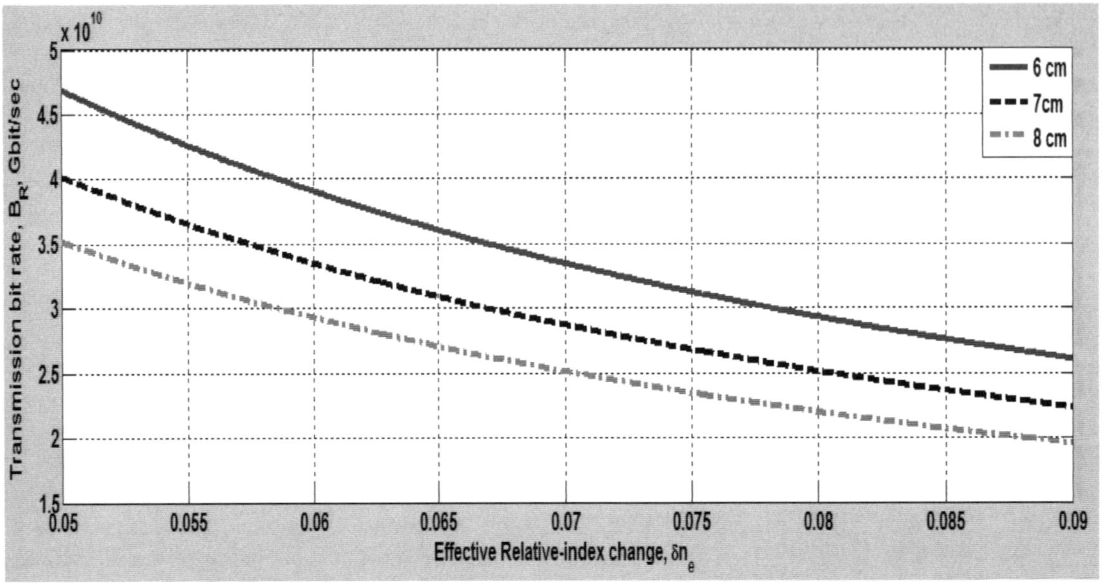

Figure 4: Variations of transmission bit rate versus effective refractive index of the assumed set of parameters.

In Fig. 4 shows that as effective relative refractive-index change and modulator length increases, these effects in decreasing in transmission bit rates of modulator devices.

4. Conclusion

We have investigated the control of light wave amplitude, phase and frequency using high speed lithium niobate optical modulation techniques. Optical MZ intensity modulators based on phase modulation can provide precise and agile control of light waves. In short, we deeply investigated the $LiNbO_3$ based electro-optic devices using a Sellmeier equation to handle the modulation bandwidth, switching voltage, cut-off frequency, transmission date rate and transmission bit rate length product. The optical modulator device performance parameters have been described and it is evident that the variation of these parameters leads to increase in the modulator device performance index.

5. References

[1]. Kawanishi T, Sakamoto T, and Izutsu M, 2007 *IEEE Journal of Selected Topics in Quantum Electronics* **13** 79–91.
[2]. Gheorma I, Savi P and Osgood Jr R, 2000 *IEEE Photon. Technol. Lett.* **12** 1618–1620.
[3]. Nagata H, 2000 *IEEE Photon. Technol. Lett.* **12** 1477–1479.
[4]. Shimotsu S, *et al.* 1995 *IEICE Trans. Commun.* 25–30.
[5]. Hu H, Ricken R, Sohler W and Wehrspohn R 2007 *IEEE Photonics Tech. Lett.* **19** 417-419.
[6]. Jundt H, Optics Lett. **22** 1553-1557.
[7]. Fieberg S, Streit L, Kiessling J, Becker P, 2012 Proc. SPIE 8240. doi: 10.1117/12.923571.
[8]. Wooten E L, Bloomfield CT, Kissa K. M, Yi-Yan A, Murphy, E. J, 2002 *Selected Topics in Quantum Electronics,* **6** 69 - 82.
[9]. Pham H.V and Okamura Y, *Journal of Advances in Optoelectronics.* **8** 1-8.
[10]. Lindsay, I. D, Lee C. J, Adhimoolam B and Boller K. J, 2006 Optics Express **14** 25-29.
[11]. Kondo J, Aoki K, kondo A, Eiiri T, 2005 *IEEE Photon. Tech. Lett.,* **17** 2077–2079.
[12]. Kawanishi T, Oikawa S, Higuma K, Matsuo Y and Izutsu M, 2001 *Electron. Lett.,* **37** 1244–1246.

Revealing of photon-number splitting attack on quantum key distribution system by photon-number resolving devices

A A Gaidash [1], V I Egorov [1], A V Gleim [1,2]

[1] ITMO University, Kronverkskiy pr. 49, St.Petersburg 197101, Russian Federation
[2] Kazan National Research Technical University, K.Marx st. 10, Kazan 420111, Russian Federation

E-mail: andrewdgk@gmail.com

Abstract. Quantum cryptography allows distributing secure keys between two users so that any performed eavesdropping attempt would be immediately discovered. However, in practice an eavesdropper can obtain key information from multi-photon states when attenuated laser radiation is used as a source of quantum states. In order to prevent actions of an eavesdropper, it is generally suggested to implement special cryptographic protocols, like decoy states or SARG04. In this paper, we describe an alternative method based on monitoring photon number statistics after detection. We provide a useful rule of thumb to estimate approximate order of difference of expected distribution and distribution in case of attack. Formula for calculating a minimum value of total pulses or time-gaps to resolve attack is shown. Also formulas for actual fraction of raw key known to Eve were derived. This method can therefore be used with any system and even combining with mentioned special protocols.

1. Introduction

To share a private message communicating sides must have a secure method to share a key first, what might be not obvious task. Quantum cryptography systems (QKD) [1] allow performing secure quantum key distribution between two or more users. The use of single photons in transmission technology provides the legitimate users (Alice and Bob) an ability to detect an eavesdropper (Eve) by monitoring quantum bit error level (QBER) on receiver side. QKD protocols are based on general principles of quantum physics, so unconditional security can be achieved. In this case protocols might guarantee security without any restrictions on the technological level of the eavesdropper.

Security of QKD imposes restriction on the source of light, which must be true single-photon. There are several types of heralded true single-photon sources [2-4], even some most promising that function at room temperatures [5-8]. However this technology is immature and does not provide high photon emission rates. An obvious and easy (and also low-cost) solution to this problem is to use attenuated laser light with average energy per pulse or time-gap (in case of continuing distribution) less than energy of one photon [9]. Probability of particular number n of photon per pulse or time-gap emitted by source of coherent states can be described by Poisson distribution:

$$P_{\mathrm{A}}(\bar{n}, n) = \frac{\bar{n}^n}{n!}\exp(-\bar{n}), \tag{1}$$

where $\bar{n}$ is average number of photons per pulse or time-gap and it can be chosen on purpose to make fraction of multi-photon states small. Since that attenuated coherent states source can be compatible with QKD system. Moreover several outstanding records were established by systems with weak laser source [10-12].

Nevertheless, security of these systems can in principle be compromised. Eavesdropper can exploit presence of multi-photon states and carry out photon-number splitting (PNS) attack [13]. She performs quantum non-demolition measurement (QND) on each of the quantum systems going from Alice to Bob; this technique claims to measure number of photons without disturbing quantum states [14 - 16]. If the number of photons is more than one, then Eve can divide photons (for example by using an adjustable beamsplitter) in two parts - one of them follows to Bob, and the rest Eve stores in her quantum memory. She can control the fraction of photons but set apart at least one photon. Since this attack introduces no errors, it cannot be detected by authorized partners if it is assumed that Bob has access only to the average detection rate, and not to the statistics of the photons he receives; this type of attacks is one of the most inconvenient.

PNS attack has been previously studied, and two counteractive methods were developed. The first is SARG04 protocol [17], which adds extra non-orthogonal bases and strong reference beam. The second commonly used technique is called decoy states [18] – PNS attack is detected with help of additional set of states with mean number of photons not equal to mean number $\bar{n}$ as for a signal state.

These methods do not rely on full analysis of photon number distribution although this might be another efficient way to detect PNS attack. The reason is that counter-PNS researches have been done before any of photon-number resolving (PNR) devices were demonstrated experimentally. There are two main methods to achieve photon-number resolving. One direct approach to reach photon-number-resolving capability is to simply break the detector active area into many distinct areas or pixels, so that each can register a photon independently of the others [19]. The second is to find proper material; there are devices whose output is inherently proportional to the number of photons, even if their detection efficiency might be low and their proportional response ultimately saturates at high input photons levels, for example [20]. Thus PNR devices provide information about number of photons in pulse or time-gap with high fidelity. One of the most promising devices is superconducting nanowire single-photon detector (SNSPD) [21]. It was shown that it can resolve up to 12 photons (depends on number of pixels). Detection efficiency for these devices is up to 60 % (using cavity) and repetition rate is up to 1 GHz [22]. We chose this type of detector for further simulations.

Goal of this paper is to consider possibilities of detecting PNS attack using photon-number resolving (PNR) detectors by Bob [23]. Compared to previous mentioned methods, using PNR detectors is efficient way to prevent eavesdropper actions with neither changes in protocol nor decreasing of key rate, which also can be implemented with any kind of QKD system, even combing with two existing methods.

2. Considered model

In this paper we expand previous research [24], where first attempts to describe advantages of PNR detectors were shown, by taking into account losses in fibers and detection efficiency of real devices. As a model for simulations we use ideal laser sources with $\bar{n} = 1$, 100 km of optical fiber SMF-28 with 0.2 dB/km losses and a detector with 60% quantum efficiency. All splicing and other small fractions of losses are supposed to be negligible. Eavesdropper is considered to operate at the beginning of quantum channel and to not know basis of each obtained photon; this kind of assumptions is to estimate upper bound of known-to-Eve part of key. Also we use approximation of infinitely long keys.

It is well-known that losses introduce changes in Poisson distribution as additional attenuation as follows:

$$P_{\mathrm{B}}(\bar{n},n)=\frac{(\gamma\cdot\bar{n})^{n}}{n!}\exp(-\gamma\cdot\bar{n}),\tag{2}$$

where γ is transmission efficiency of optical fiber line. However this way of representation is inconvenient because it provides not much information about interconnection of initial distribution $P_{A}(\bar{n},n)$ and final one $P_{B}(\bar{n},n)$, thus there is another one:

$$P_{\mathrm{B}}(\bar{n},n)=\sum_{k=n}^{\infty}\left[P_{\mathrm{A}}(\bar{n},k)\cdot\frac{k!}{n!\cdot(k-n)!}\cdot\eta^{k-n}\cdot\gamma^{n}\right],\tag{3}$$

where $\eta=1-\gamma$ is total loss. We should expect transform with the same kernel for detector as follows:

$$P_{\mathrm{D}}(\bar{n},n)=\sum_{k=n}^{\infty}\left[P_{\mathrm{B}}(\bar{n},k)\cdot\frac{k!}{n!\cdot(k-n)!}\cdot(1-QE)^{k-n}\cdot(QE)^{n}\right],\tag{4}$$

where QE is quantum efficiency of the detector.

In case of PNS attack, when Eve split one photon from each pulse or time-gap at the beginning of quantum channel where is no losses, distribution of multi-photon states can be shown as follows:

$$P_{\mathrm{E}}(\bar{n},n)=\begin{cases}P_{\mathrm{A}}(\bar{n},0),\text{for } n=0;\\ P_{\mathrm{A}}(\bar{n},1)+P_{A}(\bar{n},2),\text{for } n=1;\\ P_{\mathrm{A}}(\bar{n},n+1),\text{for } n\geq 2.\end{cases}\tag{5}$$

This distribution can be put into transformations (3) and (5) instead of $P_{A}(\bar{n},n)$ to obtain final distribution $P_{D-PNS}(\bar{n},n)$ in case of PNS attack, so the goal is to distinguish two final distributions.

3. Method description

If Bob has access to information of multi-photon distribution, he can discover PNS attack. In this case it modifies distribution of number of photons arrived to Bob; to estimate an order of the deviation one can apply rule of thumb as follows:

$$\frac{P_{D}(\bar{n},n)}{P_{D-PNS}(\bar{n},n)}\approx\frac{n+1}{\bar{n}},\text{for } n\geq 2.\tag{6}$$

However, difference in multi-photon distribution can be resolved within limits of appropriate confidence only for small number of photons. Actually there is a trade-off between increasing maximal number of photons which Bob should consider and increasing of total number of pulses or time-gaps per session as following:

$$N(n_{\max})>\left(z\cdot\frac{\sqrt{P_{D}(n_{\max})}+\sqrt{P_{D-PNS}(n_{\max})}}{P_{D}(n_{\max})-P_{D-PNS}(n_{\max})}\right)^{2},\tag{7}$$

where z is chosen according to "68–95–99.7" rule (also known as "3σ rule"): for z = 1 random number of pulses of time-gaps will be within limit of confidence with 68% probability, for z = 2 with 95% probability, for z = 3 with 99.7% probability and so on. Thus for chosen $n_{\max}$ there is minimum value of total pulses or time-gaps $N(n_{\max})$ to resolve PNS attack. For instance, $N(3)=3.1\cdot10^{9}$ for z = 5 and $\bar{n}=1$.

Using introduced notation it is easy to estimate potential known-to-Eve part of raw-key with no losses:

$$E_{\text{lossless}}=C_{\text{pr}}\cdot\frac{1-\exp(-\bar{n})-\bar{n}\cdot\exp(-\bar{n})}{1-\exp(-\bar{n})},\tag{8}$$

where $C_{\text{pr}} \leq 1$ is constant depending on particular protocol and information known by Eve about bases of photons. For estimations on upper bound here and later in text we assume $C_{\text{pr}} = 1$. Also we provide a general formula including total losses in fiber and QE of detector as follows:

$$E = \frac{1 - P_{SNSPD-PNS}(\bar{n}, 0) - \bar{n} \cdot QE \cdot \gamma \cdot \exp(-\bar{n})}{1 - P_{SNSPD-PNS}(\bar{n}, 0)}. \quad (9)$$

Thus potential known-to-Eve fraction of key is 50% for mentioned parameters of system.

However Eve can resend the taken photon. Let us consider this case, where eavesdropper may not know right basis of photon, so she prepares state in random basis. In part of events she will be right and keep proper amount of photons in pulse or time-gaps, but in other part of events resent photon will not recover distribution. Hence there will be mixture of expected distribution and distribution in case of pure PNS, so PNR device can be useful even in case of PNS with resend, however it requires higher amount of total pulses or time gaps to resolve attack.

4. Results and conclusions

As a result of research, a method of defending against PNS attacks on quantum key distribution system with coherent light source was analyzed, including brief review of the problem. The use of detectors which resolve multi-photon states allows revealing PNS attack by analyzing statistics on receiver's side. The ability of using these detectors to ensure security against PNS attack was studied. There is always a trade-off between increasing maximal number of photons which Bob should consider to reveal most actions of eavesdropper and increasing of total number of pulses, what might greatly increase the time of single session. In particular, expressions of the efficiency of PNS attack on quantum key distribution system with coherent light source were derived to provide one's general estimations. Moreover, the method does not exclude possibility of combining with other known methods and can be used with any quantum cryptography setup without any hardware additions.

References

[1] C Bennett and G Brassard 1984 *Proceedings IEEE International Conference on Computers, Systems and Signal Processing* **175**
[2] B Lounis and M Orrit 2005 *Rep. Prog. Phys.* **68** 1129
[3] M Eisaman, J Fan, A Migdall et al. 2011 *Rev. Sci. Instrum.* **82** 071101
[4] S Lukishova 2013 *Proceed. SPIE* **9056** 90650C
[5] S Lukishova, L Bissell, J Winkler, and C Stroud Jr 2012 *Optics Letters* **37** 7 1259
[6] I Aharonovich, A Greentree, S Prawler 2011 *Nature Photonics* **5** 397
[7] A Jechow, A Heuer, and R Menzel 2008 *Optics Express* **16** 17 13439
[8] M Fiorentino, S M Spillane, R G Beausoleil et al. 2007 *Opt. Express* **15** 7479
[9] ID3100/ID3110 Clavis Quantum Key Distribution System, (http://www.idquantique.com/photon-counting/clavis.html)
[10] H Takesue, Q Zhang, R H Hadfield et al. 2007 *Nature Photonics* **1** 343
[11] D Stucki, N Walenta, F Vannel et al. 2009 *New J. Phys.* **11** 075003
[12] B Korzh, C Ci Wen Lim, R Houlmann et al. 2015 *Nature Photonics* **9** 163-168
[13] M Dušek, O Haderka, M Hendrych 1999 *Optics Communications* **169** 103-108
[14] V Braginsky, Y Vorontsov, K Thorne 1980 *Science* **209** 4456
[15] B Johnson, M Reed, A Houck et al. 2010 *Nature Physics* **6** 663
[16] L Liang, G Lin, Y Hao 2014 *Physical Review* **A 90** 055801
[17] V Scarani, A Acin, N Gisin and G Ribordy 2004 *Phys. Rev. Lett.* **92** 057901

[18] H-K Lo, X Ma, K Chen 2005 *Physical Review Letters* **94** 230504
[19] D Kalashnikov, S Tan, M Chekhova 2011 *Opt. Express* **19(10)** 9352-63
[20] A Lita, A Miller, and S Nam 2008 *Opt. Express* **16** 3032
[21] Z Zhou, S Jahanmirinejad, F Mattioli et al. 2014 *Optics Express* **22** 3 3475
[22] D-K Liu, S-J Chen, L-X You et al. 2012 *Appl. Phys. Express* **5** 125202
[23] M Eisaman, J Fan, A Migdall et al. 2011 *Rev. Sci. Instrum.* **82** 071101
[24] A Gaidash, V Egorov, A Gleim 2014 *J. Phys.: Conf. Ser.* **541** 012062

The optical properties and spectral features of malignant skin melanocytes in the terahertz frequency range

A A Goryachuk[1], V A Begaeva[1], M K Khodzitsky[1] and A S Truloff[2]
[1] Photonics and Optical information technology Department, ITMO University, Saint Petersburg, Russia
[2] Saint-Petersburg State University, Saint Petersburg, Russia

E-mail: goryachuk-anna@yandex.ru

Abstract. The samples of cells of mice's melanocytes have been investigated. Their optical properties and spectral features were investigated by terahertz time-domain spectroscopy (TDS) in transmission mode. It was found that the optical properties of oncological melanocytes and normal cells are different and oncological cells have spectral features of absorption coefficient so it can be concluded that it is easy to discriminate mice's oncological skin melanocytes by using THz TDS.

1. Introduction

Present day cancer diseases are the most important problem of medicine which requires effective methods of diagnostic and treatment. During the last decade percentage of people who had undergone of different kind of cancer was increased by 15, 2 % in Russia [1]. Over the past three decades, more people all over the world have had skin cancer than all other cancers combined [2]. The most dangerous form of skin cancer is melanoma. Fortunately, melanoma is not the most often occurred of the skin cancer, because it caused the most deaths. Cells of melanoma, melanocytes, are multiply rapidly and form malignant tumors. If melanoma was diagnosed and treated early, it is almost always curable, but if it is not, the cancer can advance and spread to other parts of the body, where it becomes hard to treat and can be fatal.

Nowadays there are many techniques for investigation of skin tissues but the most traditional are in vitro. In case of melanoma if specialist would use in vitro technique it could cause most rapidly multiplying of melanocytes and sooner death accordingly. In vivo techniques include such methods as sonography, magnetic resonance tomography, magnetic resonance microscopy, Raman spectroscopy and so on. All these techniques were compared in case of frequency range, penetration depth and resolution and as the most relevance technique was chosen terahertz spectroscopy.

The main advantage of THz radiation over other techniques is that the characteristic energies of molecule's rotational and vibrational motions lay in THz frequency region, so many chemical and biological molecules can be identified by their characteristic resonant peaks. Also THz radiation is non-ionizing and might be applied in vivo diagnostics and therapy that is the big advantage for biological structures, because there are not appeared any destruction during the procedure [3-5].

The aims of this study was to investigate the optical properties of oncological skin melanocytes at THz frequency range, to research the spectral features of oncological skin melanocytes at THz frequency range and to provide invasive diagnostics of oncological skin melanocytes in terahertz frequency range from 0.1 to 1.5 THz.

2. Description of the sample preparation and experimental setup

Subject of study was a monolayer of cells of mice's melanocytes Clone M-3, which were taken from Bank of Cell Cultures of the Institute of Cytology of the RAS (Saint-Petersburg). The length of such cell varied from 20 μm to 30 μm while width is about 15 μm. The cell thickness was determined by using confocal scanning microscope Zeiss LSM-710 - 8 – 10 μm (figure 1).

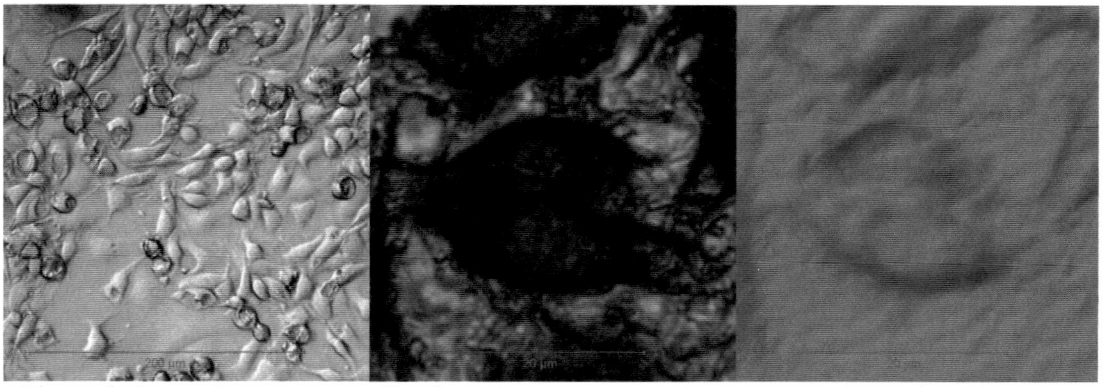

Figure 1. Photo of oncological cells of mice's melanocytes taken by confocal scanning microscope Zeiss LSM-710.

The research was performed by the THz time domain (TD) spectrometer in the transmission mode (figure 2). The characteristics of the spectrometer are following: the spectral resolution of 15GHz, the sensitivity of lock-in amplifier of 1mV, the lock-in amplifier time constant of 1s. The THz broadband pulsed radiation had the following parameters: the spectral range of 0.1–1.5THz, the pulse duration of 2.7ps, the average power of 30 mW. THz radiation was generated by femtosecond laser (Yb:KYW) irradiation of undoped indium arsenide crystal. The femtosecond laser parameters are following: the wavelength of 1040nm, the pulse duration of 120fs, the pulse repetition rate of 75MHz, the power of 1W. The sample is fixed between two THz lenses in there focuses so the THz radiation may transmit through it. The spectra in frequency domain were obtained by the Fourier transform of the measured time domain waveforms of the reference sample and the biological object.

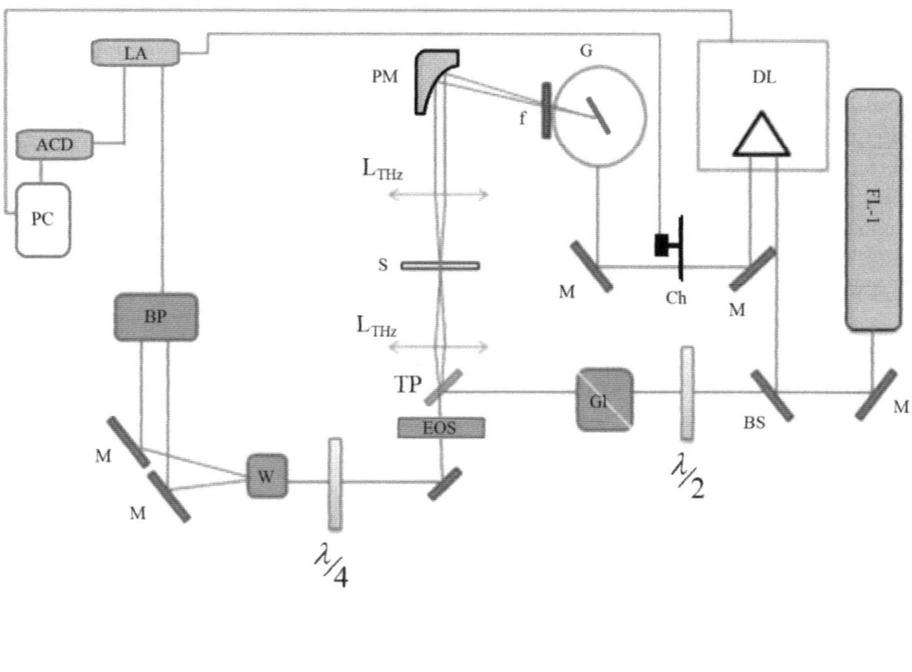

Figure 2. Scheme of the THz pulsed spectrometer in the transmission mode. FL-1 — femtosecond laser; M — mirrors; BS — beam splitter; DL — optical delay line; Ch — chopper; G — THz radiation generator: crystal of InAs in a magnetic system; f — IR cut filter; PM — parabolic mirror; S — biological object; L_{THz} — THz lens; TP — silicon beam splitter; EOC — electro optical crystal CdTe; λ/4 — achromatic quarter-wave plate; W — Wollaston prism; BP — balanced photodiodes; LA — lock-in amplifier; PC — personal computer; ADC — analog-digital converter.

2.1Method of optical properties extraction

In case of our study samples were appeared as a thin film, so it was impossible to calculate the optical properties by common equations (1 – 6).

$$n_{eff} = 1 + \frac{\left(\varphi_{eff}(f) - \varphi_{ref}(f)\right) \cdot c}{2 \cdot \pi \cdot f \cdot d} \quad (1)$$

$$\hat{n}_{eff} = n_{eff} + j \cdot k_{eff} \quad (2)$$

$$n_{eff} = \frac{n_s \cdot d_s + n_{sub} \cdot d_{sub}}{(d_{sub} + d_s)} \quad (3)$$

$$k_{eff} = -\frac{c}{2 \cdot \pi \cdot f \cdot d} \ln \frac{E_{eff}(\nu)}{E_{ref}(\nu)} \quad (4)$$

$$\hat{n}_s = n_s + j \cdot k_s \quad (5)$$

$$k_s = \frac{k_{eff}(d_{sub} + d_s) - k_{sub} \cdot d_{sub}}{d_s} \quad (6)$$

n_{eff}, n_s – real part of refractive indices of effective medium and sample accordingly, k_{eff}, k_s – imagine part of refractive indices of effective medium and sample accordingly, φ_{eff}, φ_{ref} – phases of object signal and reference signal accordingly, f – frequency, d – object thickness, E_{eff}, E_{ref} – amplitudes of object signal and reference signal accordingly.

For this research different methods of calculation of optical properties were compared and the most reasonable and efficient had been chosen. Five models of effective medium were considered - effective medium method [6], Maxwell – Garnett method [7], Bruggeman method [8], Landau-Lifshitz-Looyenga method [9], and method of thin films[10]. There are some limitations for each of these methods except the last one. The method of thin films is more accurate in obtaining the optical properties because it based not on correlation of dielectric permittivity of all components of the structure but on calculation of conductivity of thin film, which has to be investigated [11]. The samples usually are divided into an object under study and its substrate. The point of the method of thin films is that it allows obtaining the optical properties of object which cannot be investigated without substrate. For this paper a monolayer of mice's skin melanocytes on plastic substrate was investigated. As it was impossible to investigate cells itself, in experiments the effective structure of cells and plastic were investigated.

The calculation of complex film conductivity, in this case conductivity of monolayer of cells, was obtained accordingly to equation (7):

$$\hat{\sigma}(f) = \frac{\left[\frac{(\hat{n}_{sub}(f)+1)\cdot \hat{E}_0(f)}{\hat{E}(f)} - \hat{n}_{sub}(f) - 1\right]}{Z_0}, \quad (7)$$

where f — frequency, $\hat{n}_{sub}$ — complex refractive index of substrate, $\hat{E}_0(f)$ and $\hat{E}(f)$ — complex electric field amplitude of THz wave transmitted through the substrate and the effective medium accordingly, $Z_0 = 377\ \Omega$ — impedance of vacuum.

Then the complex permittivity of film was calculated as in equations (8):

$$\hat{\varepsilon}_f(f) = 1 + i\frac{\hat{\sigma}_f(f)}{2 \cdot \pi \cdot f \cdot \varepsilon_0 \cdot t_f}\ , \tag{8}$$

where ε_0 — dielectric constant, t_f— thickness of sample;

then refractive index of film as in equations (9):

$$n = \sqrt{\hat{\varepsilon}_f(f)}\ , \tag{9}$$

then absorption coefficient of film as in equations (10):

$$\alpha_s = \frac{4 \cdot \pi \cdot k_s}{\lambda}\ , \tag{10}$$

where k_s – imaginary part of refractive index of sample, λ – wavelength;

and finally optical penetration depth into film was calculated as in equations (11):

$$\delta = \frac{1}{\alpha_s}\ . \tag{11}$$

3. Results and Discussions

In our work the optical properties and spectral features of cells of mice's melanocytes were obtained. Averaging was made by seven measurements. Observed refractive index of oncological skin melanocytes varies from 3,0 to 3,25 at the frequency range 0,1 – 1,0 THz and has the maximum at the 0,34 THz. The real part of permittivity increases from 6,5 to 9,5 and then doesn't change at the frequency range from 0,35 to 0,75 THz and equals to 9 (figures 3 – 4).

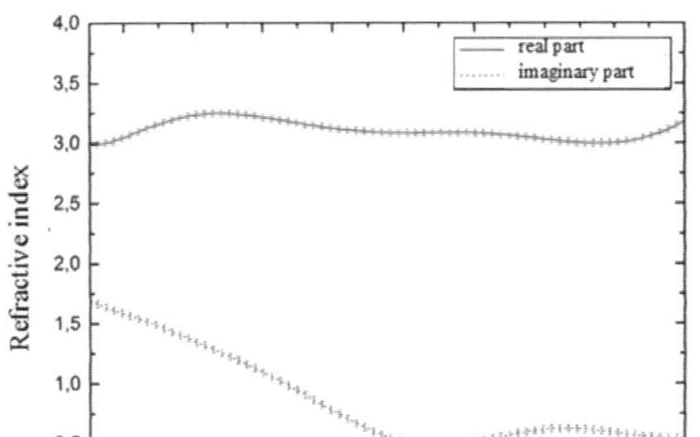

Figure 3. The dispersion of complex refractive index of oncological skin melanocytes in frequency range of 0,1 – 1,0 THz.

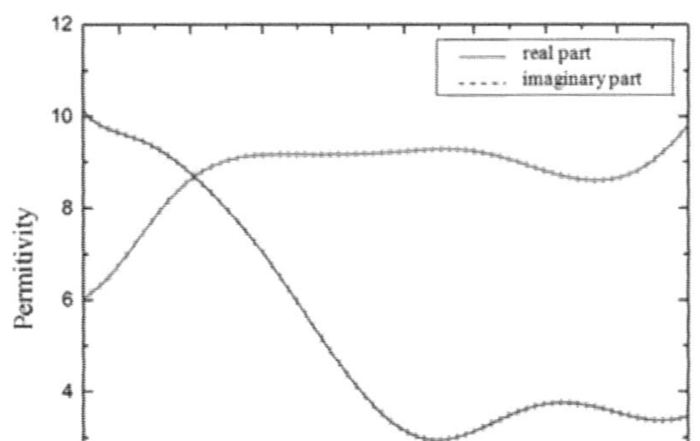

Figure 4. The dispersion of complex dielectric permittivity of oncological skin melanocytes in frequency range of 0,1 – 1,0 THz.

The optical penetration depth into sample has the maximum at frequency range from 0,5 to 0,75 THz with the central peak at the frequency of 0,63 THz and equals to 80 μm (figure 5).

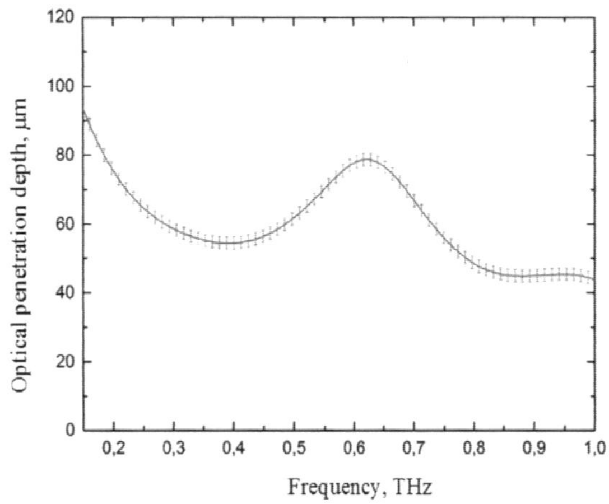

Figure 5. The dispersion of optical penetration depth into oncological skin melanocytes in frequency range of 0,1 – 1,0 THz.

The spectral features included the dispersion of absorption coefficient of oncological skin melanocytes was calculated, it changes from 125 to 225 cm^{-1} at the frequency range 0,1 – 1,0 THz. Also the maximum of absorption coefficient was revealed at the frequency of 0,4 THz, which may become a biomarker for determination of malignancy of skin melanocytes (figure 6). In the paper [11] normal and oncological tissues was observed (figure 7) and the comparison the results of dispersion of absorption coefficient in frequency range in paper [11] and the results of present work shows that the plots looks the same and only distinction between these two plots is intensity of the peaks. It may be explained by the thickness of investigated samples.

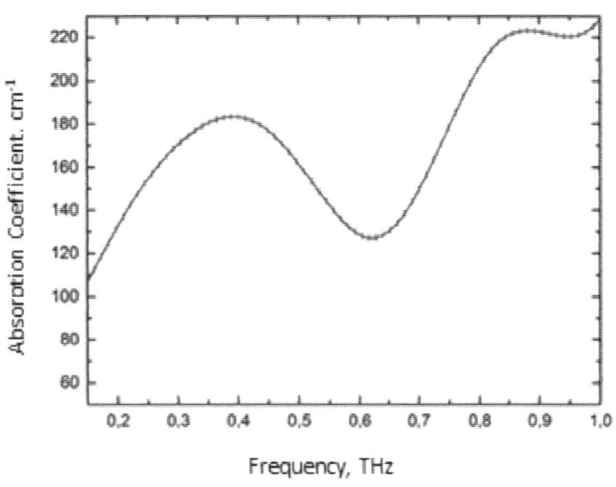

Figure 6. The dispersion of absorption coefficient of THz pulse of oncological skin melanocytes in frequency range of 0,1 – 1,0 THz.

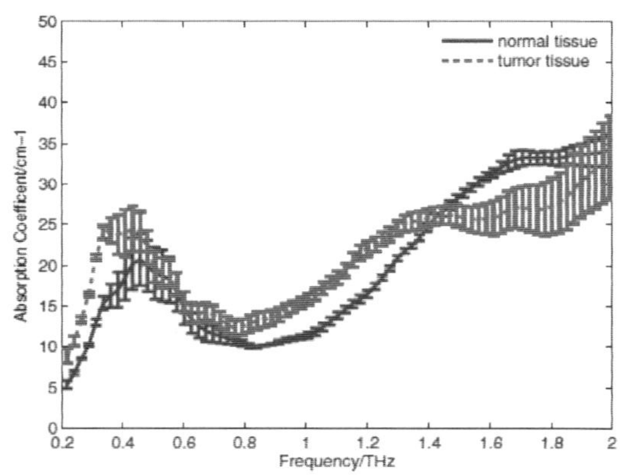

Figure 7. The dispersion of absorption coefficient of THz pulse of oncological and normal tissues in frequency range of 0,1 – 1,0 THz [11].

4. Conclusions

The unique optical and spectral features for oncological melanocytes was determined, so it can be concluded that it is easy to diagnose such objects in THz frequency range of 0.1–2.0THz by their dispersions of refractive index which is varies from 3,0 to 3,25 and their absorption coefficient which changes from 125 to 225 cm^{-1} at the frequency range 0,1 – 1,0 THz and have maximum at the frequency of 0,4 THz.

Acknowledgments
This work was financially supported by the Government of Russian Federation, Grant 074-U01.

References

[1] Chissov V I, Starinsky V V and Petrova G V 2012 *Malignant neoplasms in Russia in 2010 (Moscow: P. Herzen Moscow Oncology Research Institute Press)* p 216
[2] Stern R S 2010 *J. Arch. Dermatol.* **146** 279–282
[3] Gusev S. I., Borovkova M. A., Strepitov M. A., Khodzitsky M. K. 2015 *European Conferences on Biomedical Optics* **9537** 95372A
[4] Duka M. V., Dvoretskaya L. N., Babelkin N. S., Khodzitskii M. K., Chivilikhin S. A., Smolyanskaya O. A. 2014 *Quantum Electronics* **44(8)** 707
[5] Geyko I. A., Smolyanskaya O. A., Sulatsky M. I., Parakhuda S. E., Sedykh E. A., Odlyanitskiy E. L., Zabolotniy A. G. 2015 *European Conferences on Biomedical Optics* **9542** 95420E
[6] Golovan L A, Timoshenko V Yu and Kashkarov P K, Iwase N, Yamada T, Matsushita T, Kiyoku H and Sugimoto Y 2007 *Phys. Usp.* **50** 595–612
[7] Maxwell-Garnett J C 1904 *Philos. Trans. R. Soc. London ser. A* **203** 385–420
[8] Bruggeman D A G 1935 *J. Ann. Phys.* **24** 636–679
[9] Looyenga H 1965 *J. Physica* **31** 401–406
[10] Choon H G 2012 *J. Appl. Phys. Lett.* **101** 111609
[11] Hou D, Li X, Cai J 2014 *J. Phys. Med. Biol.* **59** 5423–5440

Journal of Physics: Conference Series **735** (2016) 012074 doi:10.1088/1742-6596/735/1/012074

Theoretical research of the distortion of quantum circuit in Grover's algorithm

K V Gubaidullina, S A Chivilikhin
ITMO University, Kronverkskiy, 49, St. Petersburg, 197101, Russia

E-mail: ksenya.gubaydullina@gmail.com

Abstract. Grover's algorithm is a quantum search algorithm among unsorted elements that can do several operations at the same time due to their wave like properties. In addition, it could solve problems of global optimization and graph coloring. This work presents the results of the simulation of Grover's algorithm and research of its resistance to the effects of perturbations of quantum logic circuit elements. These dependencies can be useful for creating optical circuit.

Introduction

Currently, the field of quantum information and quantum computing is a perspective direction of research. Quantum computers use quantum properties of matter, which gives to algorithms, working on the rules of quantum information, the ability to provide a significant increase in calculation speed compared to classical. For example, the problem of searching in unsorted list in classical algorithm actually reduces to sequential scan, and uses O (N) steps. Grover's algorithm is the fastest algorithm for solving such problems, giving quadratic speedup computations [1]. Using this algorithm to search, you can significantly speed up the well-known classical algorithms that require solving the problem of the search in an unsorted space [2].

Grover's algorithm applies for searching a specified element in unsorted list. The essence of the algorithm is to increase the amplitude of the state that satisfies the search condition, due to the decrease of the amplitudes of all other states. Therefore, amplitude of marked state increase, and increase the probability of detection correct answer.

Application of Grover's algorithm, to adapt it to accelerate the works of well-known classical algorithms, is promising. In [3, 4] proved the possibility of using Grover's algorithm for solving the problem of global optimization and shows the different ways of implementing the oracle.

In papers [5, 6] presented a quantum Dijkstra's algorithm, which improves the work of the classical algorithm. Classical Dijkstra's algorithm solves the problem of graph coloring and its quantum implementation can be used to finding shortest path in network routing.

Grover's algorithm is a perspective direction of research and has a number of applications that can be solved with its help. In this regard, there is a problems of creating an optical circuit that implements the algorithm and the calculation of the effect of the distortion circuit elements on the work of Grover's algorithm. In this research, we simulated Grover's algorithm and tested its resistance to the effects of perturbations of quantum logic circuit elements using the unitary distortion of quantum operators.

Journal of Physics: Conference Series **735** (2016) 012074 doi:10.1088/1742-6596/735/1/012074

Simulation of perturbations

Scheme of the Grover's algorithm is on figure 1-2. At the initial time, the system is in the zero state. Hadamard gate transforming the system into a state of superposition. Then follow iterative process consists from two functions - controlled phase rotation and diffusion function.

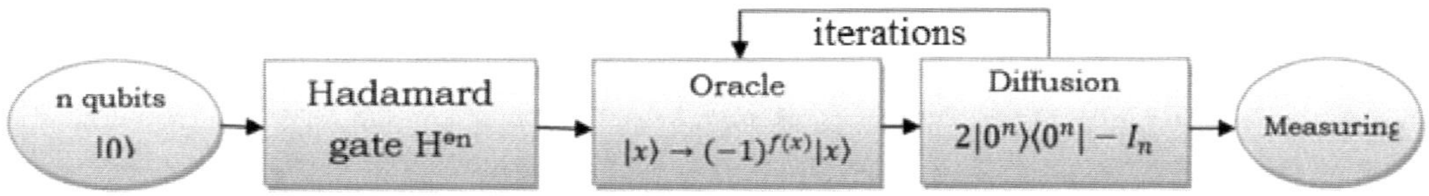

Figure 1. Flowchart of the Grover's algorithm.

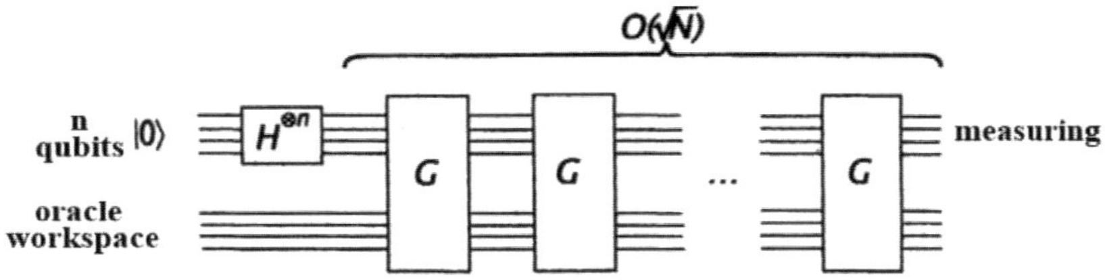

Figure 2. Logic scheme of the Grover's algorithm. G is a Grover's iteration.[7]

Phase rotation controlled by a search condition, which is enclosed in oracle. Oracle is a quantum operator that can recognize the decision and mark it by phase rotation.

Diffusion function makes a calculation of the amplitudes relatively its average. Therefore, amplitude of marked state increase, and increase the probability of its detection.

Histograms on figure 3 presented results of the simulation algorithm on 6 qubits. Such system has 64 different unsorted states, and if it is necessary to find 1 element from them, it should be better to use Grover's algorithm, which can find it for 6 iterations, instead of classically scan at least half of all elements.

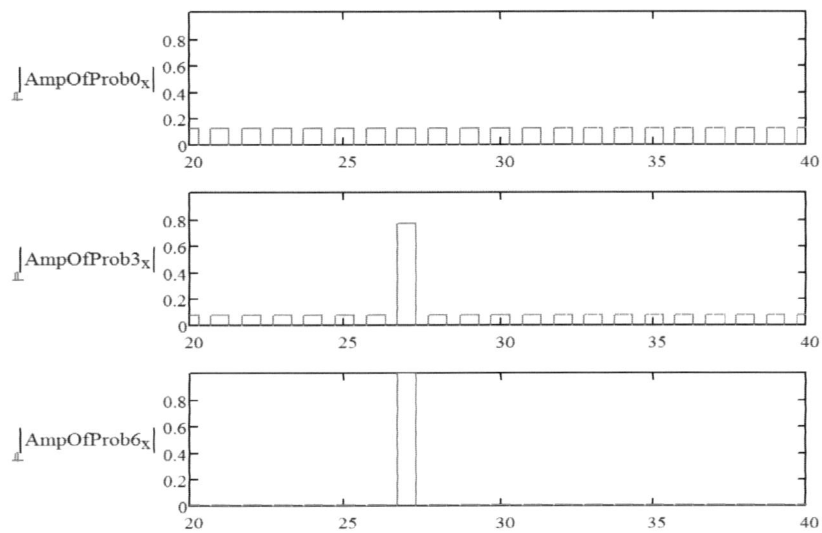

Figure 3. Results of the Grover's algorithm simulation on 6 qubits for superposition state, 3 and 6 iterations. Marked state equal 27.

Journal of Physics: Conference Series **735** (2016) 012074 doi:10.1088/1742-6596/735/1/012074

Obviously, that at the beginning probability of detection states is the same. After 3-d iteration probability of detection marked state increase by 59,1% and probability of detection other states equal 0,65%. After 6-th iteration marked state detects in 99,7% cases. Through formula (1) was calculated that six iterations is enough to reach highest probability.

$$I = \frac{\pi}{4}\sqrt{2^n} \tag{1}$$

To simulate the Grover's algorithm and research distortion system on 3 qubits was used. If n = 3, then we have $2^n = 8$ different states of qubits. In classical physics, each state has its own probability and sum of them equals one. In quantum physics, each state has its own amplitude, and the magnitude of the sum of all amplitudes squares is equal to one. All perturbations of this research is caused by unitary operators ensures the sum of probabilities of all possible outcomes of any event is always 1.

First step of the algorithm is the Hadamard operator that transform initial state of the system to the superposition state. In such system amplitudes of any states is the same, so probabilities of detection any states is the same too. For system on 3 qubits and after correct Hadamard transformation, amplitudes at the superposition state equal ≈ 0,354.

In this research have been explored different ways of setting the value of the perturbation. They could divided into 2 groups: light deviations and strong deviations (figure 4). For example if amplitude of marked state distorted for 10%, other states should distorted for 10% summary. In this case amplitude of other states distorted just a little, and amplitude of the marked state distorted more. This is light deviations (figure 4a). For strong deviations, amplitudes of other states distorted as well as amplitude of the marked state (figure 4b). Moreover, amplitude of marked state can increase on 10% or decrease on 10%. Therefore, we have 4 possible cases:

- Light deviations
  - amplitude of marked state increase (LI);
  - amplitude of marked state decrease (LD);
- Strong deviations
  - amplitude of marked state increase (SI);
  - amplitude of marked state decrease (SD).

a) Light deviations			b) Strong deviations		
$AmpFalse_i$ =		$AmpFalse_i$ =	$AmpFalse_i$ =		$AmpFalse_i$ =
0.351		0.356	0.371		0.335
0.351		0.356	0.371		0.335
0.351		0.356	0.371		0.335
<u>0.371</u>	or	<u>0.335</u>	<u>0.371</u>	or	<u>0.335</u>
0.351		0.356	0.335		0.371
0.351		0.356	0.335		0.371
0.351		0.356	0.335		0.371
0.351		0.356	0.335		0.371

Figure 4. Initial amplitudes of the states after 10% perturbations for light and strong deviations. Amplitude of marked state underlined and equal $|011\rangle$. Left values from a) and b) – amplitude of marked state increase, right values – decrease.

In the following histograms presented changes in the probability of detection states at different perturbations value. Zero percent means that algorithm working without any perturbations. It is ideal Grover's algorithm.

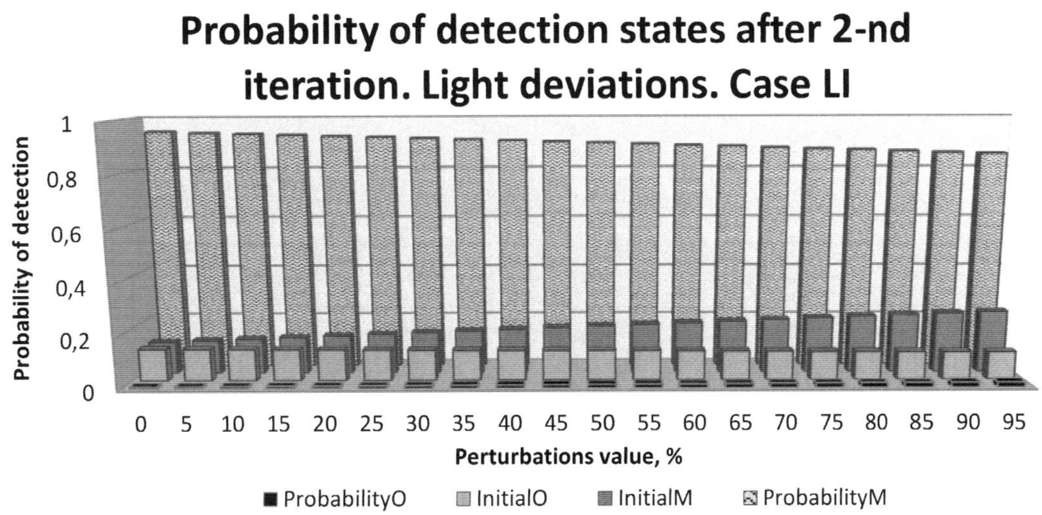

Figure 5. Dependencies of detection probability and value of perturbations for light deviations and case LI (when initial amplitude of marked state increase). ProbabilityM - probability of detection marked state, ProbabilityO – probability of detection other states, InitialM – initial probability of detection marked state, InitialO - initial probability of detection other states.

Histogram for light deviations and increases initial amplitude of marked state at figure 5 shows that probability of detection marked state decrease while perturbation value grows. Moreover, probability of detection other states increase, causing the growth of the frequency of registering wrong answer.

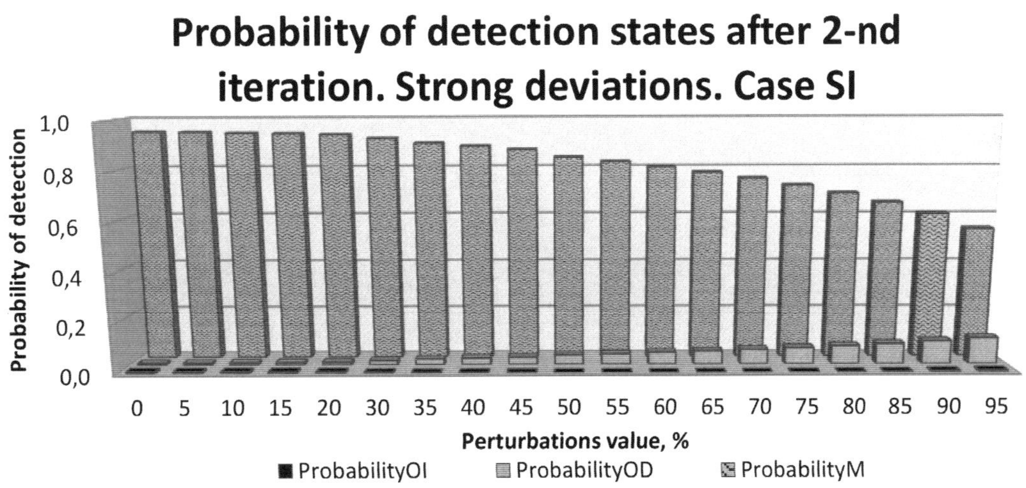

Figure 6. Dependencies of detection probability and value of perturbations for strong deviations and case SI (when initial amplitude of marked state increase). ProbabilityM - probability of detection marked state, ProbabilityOI – probability of detection other states, which initial amplitude increases, ProbabilityOD – probability of detection other states, which initial amplitude decreases.

Results for strong deviations and case SI is on figure 6. Simulation shows strong decrease of detection probability of marked state. Probability of detection other states increase strongly too.

Journal of Physics: Conference Series **735** (2016) 012074 doi:10.1088/1742-6596/735/1/012074

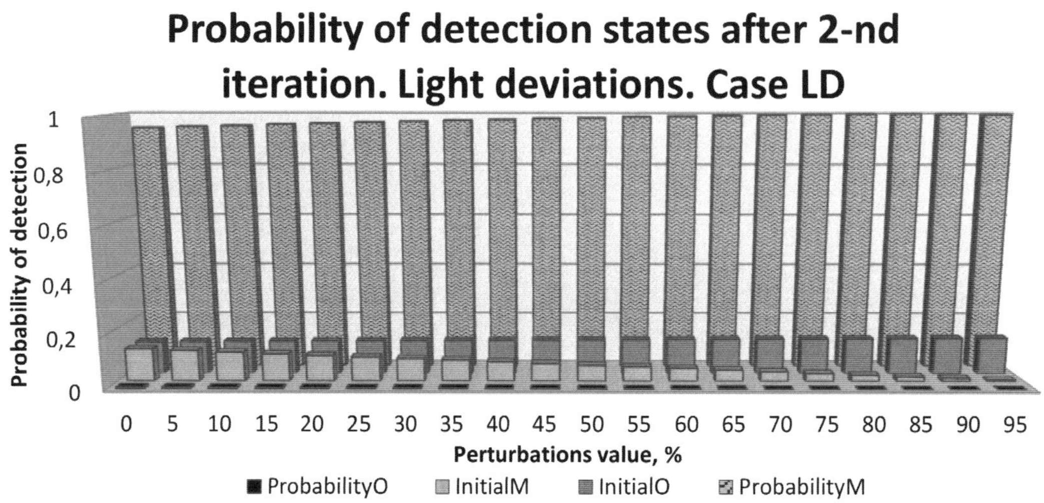

Figure 7. Dependencies of detection probability and value of perturbations for light deviations and case LD (when initial amplitude of marked state decrease). ProbabilityM - probability of detection marked state, ProbabilityO – probability of detection other states, InitialM – initial probability of detection marked state, InitialO - initial probability of detection other states.

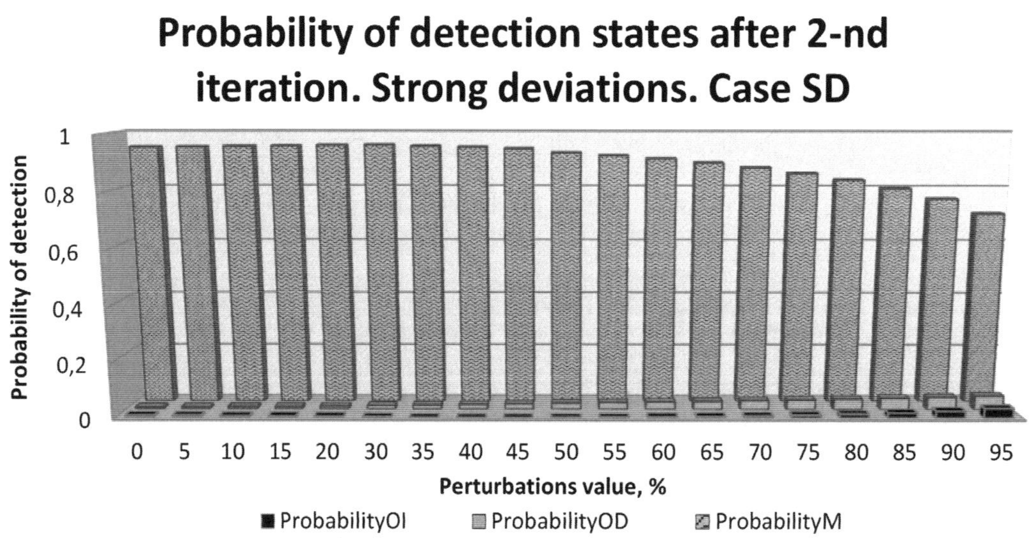

Figure 8. Dependencies of detection probability and value of perturbations for strong deviations and case SD (when initial amplitude of marked state decrease). ProbabilityM - probability of detection marked state, ProbabilityOI – probability of detection other states, which initial amplitude increases, ProbabilityOD – probability of detection other states, which initial amplitude decreases.

Light deviations and decreases initial amplitude shows effect of a little increase of detection probability of marked state (figure 7), but it's gone when deviation of states amplitude becomes much stronger (figure 8). On the figure 8 also shown, that probability of detection other states increase.

Results

In this research, we simulated Grover's algorithm and verified its resistance to the effects of perturbations of quantum logic circuit elements. The results is shown in table 1. As it shown, probability of detection marked state decreases. For highest perturbations values that researched in this work, probability decreases on 34%. Moreover, probability of detection other states increases, causing the growth of the frequency of registering wrong answer in 10 times for highest perturbation. On the other hand, probability of detection marked state is still higher others, so Grover's algorithm has good resistance to the effects of perturbations of quantum logic circuit elements.

Table 1. Results of Grover's algorithm simulation. Dependencies detection probability of states from perturbation value.

Perturbation value, %	Marked state, %	Other states, *100%
5	-0.06	1.21
20	-1	1.96
35	-3.1	2.9
50	-6.6	4.07
65	-11.9	5.52
80	-19.8	7.4
95	-34.1	10.3

Acknowledgments

This work was financially supported by the Government of Russian Federation, Grant 074-U01.

References

[1] Grover L K 1996 *Proc. 28th Ann. ACM Symp. Theory of Computing (STOC)* (New York: ACM Press) pp 212–221

[2] Dűrr C, and Høyer P 1996 A quantum algorithm for finding the minimum *Preprint* quant-ph/9607014

[3] Dűrr C, Heiligman M, Høyer P and Mhalla M 2004 Quantum query complexity of some graph problems *Lect. Notes in Comp. Sci.* **3142** (*Preprint* quant-ph/0401091)

[4] Baritompa W P, Bulger D W and Wood G R 2005 *SIAM J. Optimization* **15** 1170–1184

[5] Ray P 2014 *Int. J. Adv. Res. Comp. Sci. and Management Studies* **2** 30-43

[6] Aghaei M R S, Zukarnain Z A, Mamat A and Zainuddin H 2009 *J. Theoretical and Applied Inform. Technol.* 360-365

[7] Nielsen M A and Chuang I L 2006 *Quantum Computation and Quantum Information* (Moscow: Mir) p. 314

Journal of Physics: Conference Series **735** (2016) 012075 doi:10.1088/1742-6596/735/1/012075

The active-passive continuous-wave terahertz imaging system

Irina N Dolganova*, Kirill I Zaytsev, Anna A. Metelkina, and Stanislav O Yurchenko

Bauman Moscow State Technical University, 105005, 2nd Baumanskaya str. 5, Moscow, Russia

E-mail: *in.dolganova@gmail.com

Abstract. Active and passive terahertz (THz) imaging have recently become essential instruments of various THz applications. In this paper the active- and passive-mode THz imaging systems are studied and the hybrid active-passive THz imaging system is suggested. The concept of image contrast was used to compare the active and passive THz imaging results. In order to achieve better image quality the hybrid system is considered to be effective. The main advantage of the proposed system is the combination of the self-emitted radiation of the object with the back scattered source radiation. The experimental results demonstrate that the active-passive modality of THz imaging system allows retrieving maximum information about the object. An approach to synthesise the active-passive THz images was proposed using the false-color representation.

1. Introduction and Background

Terahertz (THz) imaging is widely used in such branches of fundamental and applied science as biomedical science and chemical physics [1–3], non-destructive evaluation of composite materials and constructions [4–5], quality-control in pharmaceutical industry [6], detection of concealed objects, drugs and explosives in security applications [7–9], and astrophysical science [10].

THz imaging systems can be classified as pulsed and continuous and the latter of these two types can be employed in either active or passive mode. While pulsed systems are efficient for two- and three-dimensional imaging [11–12], continuous systems using more powerful sources are applied for highly sensitive and remote imaging [13–16].

Active and passive imaging are different in technical realization and have different basic principles. Active imaging system [9,13–14,17–18] detects THz radiation reflected from the object, while passive imaging system [14,19–20] receive only thermal THz radiation of the object. Two modalities also have different imaging characteristics, such as contrast, resolution, dynamic range, and signal-to-noise ratio.

This paper compares two THz imaging modalities and describes a possible way of making hybrid active-passive system in order to increase imaging efficiency. To receive quantitative characteristic of THz imaging we used test-samples of different spatial frequencies and calculate Modulation Transfer Function (MTF). The obtained results show the potential advantages of hybrid THz imaging since it gives more information about the object.

2. Experimental Setup and Samples

To compare active and passive THz imaging modalities and to receive hybrid images we used the following experimental setup (figure 1(a)).

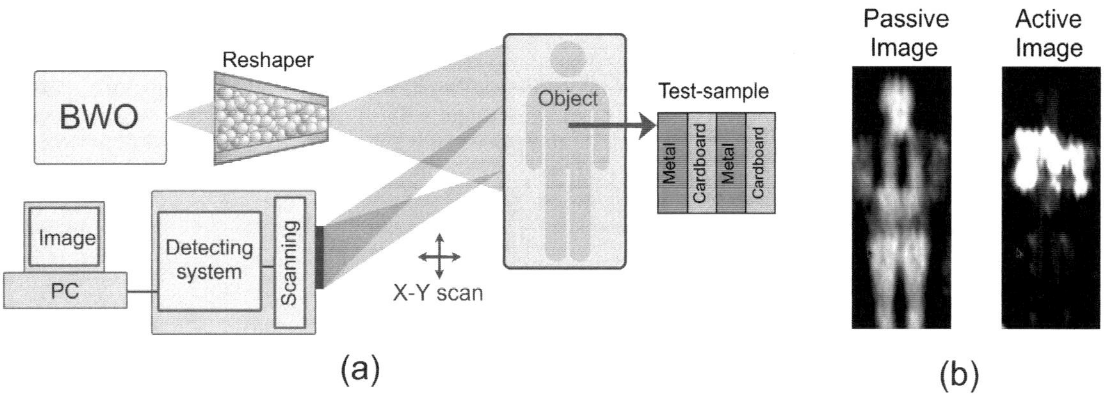

Figure 1. (a) Schematic representation of experimental setup used to study THz imaging in the hybrid mode, BWO stands for the backward-wave oscillator, and PC stands for the personal computer; (b) Examples of passive and active THz images of the test object, placed on the human body beneath the T-shirt.

Active and passive modes have the same detection system based on the heterodyne principle. Schottky and Gunn diodes are used as the photo-mixer and local oscillator, respectively. The THz image is provided via the raster scanning of the object plane by mechanical mirror-based scanning. To make active images we used the the backward-wave oscillator (BWO) as a source of quasi-monochromatic continuous-wave THz radiation. The operational frequency of the BWO is tunable in the range of 0.2 to 0.3 THz and is set near 0.25 THz to lock it to the local oscillator. The output power of the BWO THz beam for this conditions is $\sim$ 50 mW. The detecting system includes lens with focal length 50.0 mm. The beam shape of BWO is rather nonhomogeneous, thus, the studied object is illuminated non-uniformly. To correct it and to use maximal output beam we applied the reshaper (or THz homogenizer). It consists of hollow metal-coated prism with small plastic particles inside. These particles cause scattering which leads to source beam homogeneity.

The object was remote from the system at a distance of 4 meters. It constituted of test-samples on a human body behind different clothes materials: T-shirt, jacket, and coat. Test-samples consisted of cardboard base with metal stripes of several width. Thus the spatial frequency of the test-samples was different: $f = 0.035$, 0.052, 0.714, and 0.105 cm^{-1}. To enable the active and passive THz imaging techniques to be compared, images of the test objects were detected in the same position within a short time period.

3. Active and Passive Image Quality. Hybrid Imaging Synthesis

In the figure 1(b) active and passive images are shown. We calculated the images intensity cross-sections, which were used to understand image quality of two imaging modalities. The example of intensity cross-sections are presented in figure 2(a).

We received such cross-sections for all test-samples and all clothes materials in order to study image contrast of different objects and scattering conditions. For the obtained curves we applied the harmonic approximation of the intensity curves by the use of the Gaussian least squares method:

$$I = A + B\sin(2\pi f x + C), \tag{1}$$

where x is the horizontal coordinate, the frequency-dependent coefficients A, B, C are the mean intensity, the amplitude of intensity oscillation, and the harmonic phase shift, respectively.

By finding values of coefficients A and B for each case, we can calculate the image contrast for exact spatial frequency as $K = B/A$ and receive MTF in each imaging condition as the function $K(f)$ of spatial frequency. Four test-samples are enough to estimate the MTF character in active and passive modes. The obtained results are shown in the figure 2(b–d). The same character and tendency are obvious for these curves. There are intersection points which divide the whole frequency range into two parts. In the left one the active MTF curve lies above the passive curve and active THz imaging provides higher contrast for visualizing the low-spatial frequencies of

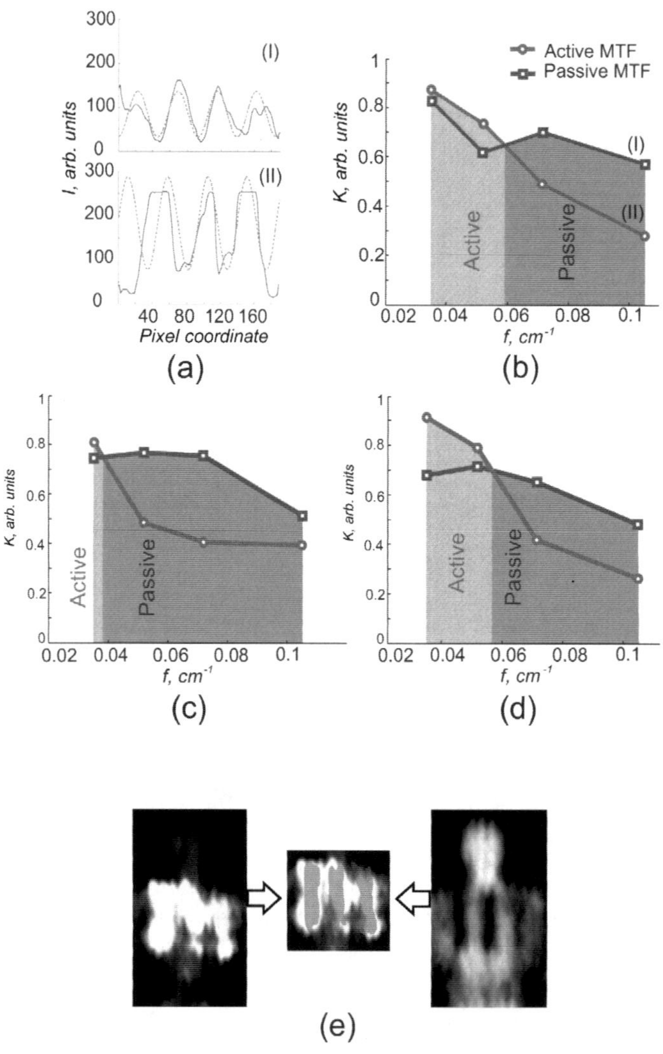

Figure 2. (a) Image intensity cross section (solid curves) and the appropriate harmonic approximation (dashed curves) for images from the figure 1(b), respectively. (b–d) Active and passive MTFs, $K(f)$, for the test samples placed under T-shirt (b), jacket (c), and coat (d). Red and blue regions indicate the spatial frequencies for which the active or passive mode is more effective, respectively. (e) Result of image synthesis (hybrid THz image) of the test sample covered with T-shirt.

the object, in the the right one the passive MTF is higher than the active values and it allows us to detect small details of the object more effectively. Physically, the differences between the active- and passive-mode MTFs appear mainly due to the variation in THz-wave scattering for different cover materials. Matter, such as clothing fabric with its partially-ordered structure, causes THz radiation to scatter in a complex manner [21–25], which leads to decreased resolution of the THz images.

Thus there are no definitely preferred imaging modality in the whole spacial frequency range. But the combination of them in a hybrid system can cause increasing of the imaging quality. In this case it is possible to obtain the highest envelope of MTF.

To make it clear, we propose one of the possible imaging synthesis algorithm based on the threshold filtration procedure. The hybrid image is resented in the false-color (figure 2(e)), but other methods can be used in order to extract more information about the object. In our case of metal-striped test-samples, the hybrid image helps us to detect metal objects under clothes by the used of the high contrast of the active imaging and to resolve the object details by the used of the passive imaging.

4. Conclusions

In the present paper we have compared active and passive THz imaging modalities, and experimentally demonstrated the advantage of their combination on a hybrid imaging system. The obtained MTF curves for different imaging condition illustrate the spatial frequency regions of effectively using of active or passive modes. The demonstrated synthesised false-color THz image justify the potential of hybrid imaging in various THz applications, such as biomedical systems, security applications, the medical and pharmaceutical sciences, chemical physics, astrophysics, and materials sciences.

Acknowledgements

This work was supported by the Russian Scientific Foundation, Project # 14-19-01083.

References

[1] Pickwell E, Wallace V P 2006 *J. Phys. D: Appl. Phys.* **39** R301
[2] Zaytsev K I, Kudrin K G, Karasik V E, Reshetov I V, and Yurchenko S O 2015 *Appl. Phys. Lett.* **106** 053702
[3] Zaytsev K I, Kudrin K G, Reshetov I V, Gavdush A A, Chernomyrdin N V, Karasik V E, and Yurchenko S O 2015 *J. Phys.: Conf. Ser* **584** 012023
[4] Stoik C D, Bohn M J, and Blackshire J L 2008 *Opt. Exp.* **16** 17039
[5] Yakovlev E V, Zaytsev K I, Fokina I N, Karasik V E, and Yurchenko S O 2014 *J. Phys.: Conf. Ser* **486** 012008
[6] Watanabe Y, Kawase K, Ikari T, Ito H, Ishikawa Y, and Minamide H 2003 *App. Phys. Lett.* **83** 800
[7] Murrill S R, Jacobs E L, Moyer S K, Halford C E, Griffin S T, De Lucia F C, Petkie D T, and Franck C C 2008 *Appl. Opt.* **47** 1286
[8] Zhang L, Shen J, Zhong H, Deng C, Zhang C, and Zhao Y 2008 *Appl. Phys. Lett.* **92** 091117
[9] Kemp M C, Glauser A, and Baker C 2006 *Proc. SPIE* **6212** 62120T
[10] Karasik B S, McGrath W R, LeDuc H G, and Gershenson M E 1999 *Superconduct. Sci. Tech.* **12** 745
[11] Mittleman D M, Jacobsen R H, and Nuss M C 1996 *IEEE J. Sel. Top. Quant. Electr.* **2** 679
[12] Zaytsev K I, Karasik V E, Fokina I N, and Alekhnovich V I 2013 *Opt. Engineering* **52** 068203
[13] Karpowicz N, Zhong H, Zhang C, Lin K-I, Hwang J-S, Xu J, and Zhang X-C 2005 *Appl. Phys. Lett.* **86** 054105
[14] Petkie D T, Casto C, De Lucia F C, Murrill S R, Redman B, Espinola R L, Franck C C, Jacobs E L, Griffin S T, Halford C E, Reynolds J, O'Brien S, and Tofsted D, 2008 *J. Opt. Soc. Am. B* **25** 1523
[15] Lee A, Qin Q, Kumar S, Williams B S, Hu Q, and Reno J L 2006 *Appl. Phys. Lett.* **89** 141125
[16] Perov A N, Zaytsev K I, Fokina I N, Karasik V E, Yakovlev E V, and Yurchenko S O 2014 *J. Phys.: Conf. Ser.* **486** 012027
[17] Appleby R and Wallace H B 2007 *IEEE Trans. Antennas and Propagation* **55** 2944
[18] Scheller M 2011 *Opt. Expr.* **19** 10647

[19] Grossman E, Dietlein C, Ala-Laurinaho J, Leivo M, Gronberg L, Gronholm M, Lappalainen P, Rautiainen A, Tamminen A, and Luukanen A 2010 *Appl. Opt.* **49** E106
[20] Jacobs E L and Furxhi O 2010 *Appl. Opt.* **49** E94
[21] Dolganova I N, Neganova A S, Karasik V E, Zaytsev K I, and Yurchenko S O 2015 *J. Phys.: Conf. Ser.* **584** 012019
[22] Fokina I N, Karasik V E, Orlov V M, and Budak V P 2014 *J. Quant. Spectroscop. Radiat. Transfer* **149** 108
[23] Yakovlev E V, Dolganova I N, Zaytsev K I, Kuliev T I, Kruchkov N P, and Yurchenko S O 2014 *J. Phys.: Conf. Ser.* **584** 012025
[24] Fokina I N, Zaytsev K I, Karasik V E, and Tsapenko K P 2013 *Proc. SPIE* **8846** 88460A
[25] Dolganova I N, Yurchenko S O, and Karasik V E 2015 *Int. J. High Speed Electronics and Systems* **24** 1520002

In vivo terahertz pulsed spectroscopy of dysplastic and non-dysplastic skin nevi

Kirill I. Zaytsev[a,b,c], **Nikita V. Chernomyrdin**[a,b,c], **Konstantin G. Kudrin**[a,b,c], **Arseniy A. Gavdush**[a], **Pavel A. Nosov**[a], **Stanislav O. Yurchenko**[a,c], and **Igor V. Reshetov**[a,b,c]

[a]Bauman Moscow State Technical University, 2nd Baumanskaya str. 5, Moscow, Russia, 105005

[b]Institute of Improvement of Professional Skill of the Federal Medico-Biological Agency of Russia, Volokolamskoe shosse 91, Moscow, Russia, 125371

[c]I. M. Sechenov First Moscow State Medical University, Bolshaya Pirogovskaya st. 2, Moscow, Russia, 119991

E-mail: kirzay@gmail.com

Abstract. The results of the *in vivo* terahertz (THz) pulsed spectroscopy (TPS) of pigmentary skin nevi are reported. Observed THz dielectric permittivity of healthy skin and dysplastic and non-dysplastic skin nevi exhibits significant contrast in THz frequency range. Dysplastic skin nevus is a precursor of melanoma, which is reportedly the most dangerous cancer of the skin. Therefore, the THz dielectric spectroscopy is potentially an effective tool for non-invasive early diagnosis of melanomas of the skin.

1. Introduction

Terahertz (THz) pulsed spectroscopy (TPS) is an effective instrument for characterizing the THz dielectric permittivity, or the THz material parameters of media [1–6], as well as for reconstructing the internal structure of dielectric objects [7–13]. In particular, TPS techniques have been recently applied for medical diagnosis of oral [14], skin [15–17], and colon [18] cancers, and for intraoperative diagnosis of breast cancer [19]. However, numerous medical applications of TPS have not been considered yet, for instance, diagnosis of dysplastic skin nevi. This problem is dramatically important because the dysplastic skin nevus is considered to be a melanoma precursor [20], and the latter is reportedly the most dangerous cancer of the skin [21].

In this paper we show the results of studying *in vivo* the THz dielectric characteristics of healthy skin, dysplastic and non-dysplastic skin nevi from four patients. We demonstrate that THz dielectric characteristics of dysplastic and non-dysplastic nevi significantly differ. Thus, the ability to use the TPS for early non-invasive diagnosis of dysplastic skin nevi and, as a consequence, of melanomas of the skin, is demonstrated.

2. Materials and Methods

In the present work we use the TPS setup operating in reflection–mode [9]. Generation and detection of THz pulses are performed in LT-GaAs photoconductive antenna and electrooptical detector based on ZnTe–crystal, respectively. The setup allows us to measure the THz reflectivity of the sample, and to reconstruct its THz dielectric permittivity (or THz material parameters) in a wide frequency range of 0.1 to 2.5 THz.

During the experimental study we use a reference 1–mm–thick SiO_2–window to fix the tissue. Three THz waveforms are detected for each tissue sample: the waveform $E_r(t)$ reflected from an empty SiO_2–window; the waveform $E_m(t)$ reflected from a SiO_2–window with a gold mirror placed behind it; and the waveform $E_s(t)$ reflected from a SiO_2–window with the sample of interest placed behind it. Let us define the Fourier spectrums of these waveforms as $\widetilde{E}_r(\nu)$, $\widetilde{E}_m(\nu)$, and $\widetilde{E}_s(\nu)$, where ν stands for an electromagnetic wave frequency.

We reconstruct the THz dielectric permittivity of the sample $\widetilde{\varepsilon}(\nu)$, such as

$$\widetilde{\varepsilon}(\nu) = \varepsilon'(\nu) - i\varepsilon''(\nu), \tag{1}$$

where $\varepsilon'(\nu)$ and $\varepsilon''(\nu)$ are the real and imaginary parts, respectively. The solution of this ill-posed inverse problem should minimize an error functional [1, 6]:

$$\widetilde{\varepsilon} = argmin_{\widetilde{\varepsilon}}[\Phi], \qquad \Phi = \left| \left|\widetilde{H}_{exp}\right| - \left|\widetilde{H}_{th}\right| \right|^2 + \left| \phi\left[\widetilde{H}_{exp}\right] - \phi\left[\widetilde{H}_{th}\right] \right|^2, \tag{2}$$

where $\widetilde{H}_{exp}(\nu)$ and $\widetilde{H}_{th}(\nu, \widetilde{\varepsilon})$ are experimental and theoretical transfer functions, and $|...|$ and $\phi[...]$ are operators to extract the modulus and the phase.

We define the experimental transfer function as

$$\widetilde{H}_{exp} = \frac{\widetilde{E}_s - \widetilde{E}_r}{\widetilde{E}_m - \widetilde{E}_r}. \tag{3}$$

The theoretical transfer function assumes multiple THz wave reflections in the SiO_2–window:

$$\widetilde{H}_{th} = \frac{\widetilde{R}_{12} - \widetilde{R}_{10} + \sum_{j=1}^{N}\left(\widetilde{R}_{12}^{j+1} - \widetilde{R}_{10}^{j+1}\right)\widetilde{R}_{10}^{j}\widetilde{P}_1^{2j}}{\widetilde{R}_{13} - \widetilde{R}_{10} + \sum_{j=1}^{N}\left(\widetilde{R}_{13}^{j+1} - \widetilde{R}_{10}^{j+1}\right)\widetilde{R}_{10}^{j}\widetilde{P}_1^{2j}}, \tag{4}$$

where $\widetilde{R}_{mk}(\nu, \widetilde{\varepsilon}_m, \widetilde{\varepsilon}_k)$ describes Fresnel reflection at the interface of the mth and k^{th} media:

$$\widetilde{R}_{mk} = \frac{\sqrt{\widetilde{\varepsilon}_k} - \sqrt{\widetilde{\varepsilon}_m}}{\sqrt{\widetilde{\varepsilon}_m} + \sqrt{\widetilde{\varepsilon}_k}}. \tag{5}$$

The indexes $m = 0, 1, 2$, and 3 correspond to the air, SiO_2, tissue, and gold mirror media, respectively. $\widetilde{P}_1 = \widetilde{P}_1(\nu, \widetilde{\varepsilon}_1, l)$ is based on the Bouguer–Lambert–Beer law describing the THz–wave propagation along a distance l in the SiO_2–window:

$$\widetilde{P}_1 = exp\left(-i\frac{2\pi\nu}{c}\sqrt{\widetilde{\varepsilon}_1}l\right), \tag{6}$$

where $c \simeq 3 \times 10^8$ m/s is the speed of light in vacuum.

In Eqs. (4)–(6), the dielectric permittivity of air and SiO_2 ($\widetilde{\varepsilon}_0$ and $\widetilde{\varepsilon}_1$); the effective dielectric permittivity of mirror ($\widetilde{\varepsilon}_3$); the number of satellite pulses (N); and the SiO_2–window thickness (l) are known *a priori*, and the THz permittivity of the sample ($\widetilde{\varepsilon} = \widetilde{\varepsilon}_2$), is estimated via Eq. (2). In contrast to the well-known methods used to determine the THz dielectric permittivity via the reflection-mode measurements, for example, see [4], in the described approach we assume multiple THz-wave reflections in the reference window, thereby allowing us to increase the reconstruction accuracy [6].

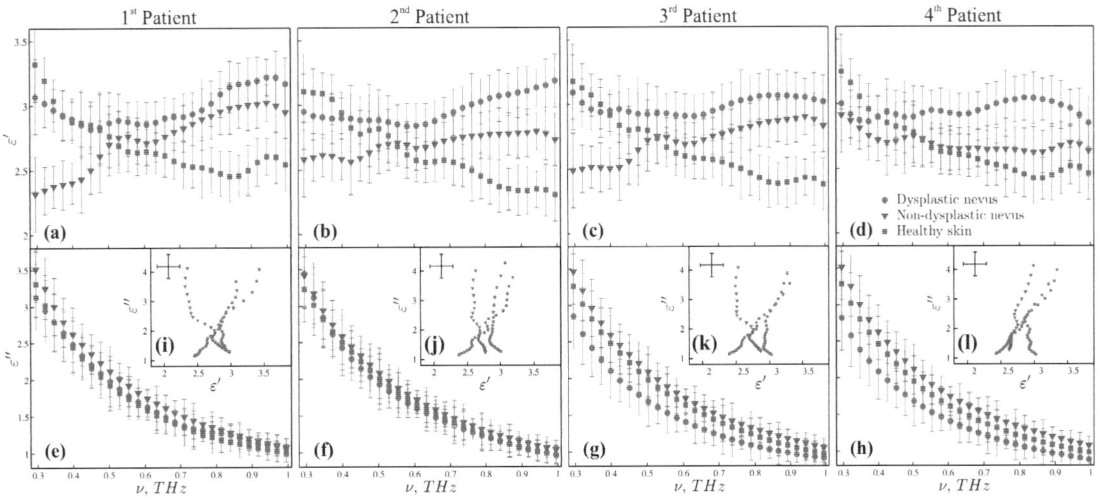

Figure 1. THz dielectric characteristics of healthy skin and dysplastic and non-dysplastic skin nevi *in vivo* from four patients: (a)–(d) and (e)–(h) show the real $\varepsilon'(\nu)$ and the imaginary $\varepsilon''(\nu)$ parts of the complex dielectric permittivity, respectively; and the insets from (i) to (l) show the Cole-Cole diagrams $\varepsilon''/\varepsilon'(\nu)$ [22, 23].

3. Results

Figure 1 shows the THz dielectric permittivity of healthy skin, dysplastic and non-dysplastic skin nevi from four patients: panels from (a) to (d) and from (e) to (h) show the real $\varepsilon'(\nu)$ and the imaginary $\varepsilon''(\nu)$ parts of the complex dielectric permittivity, respectively; and the insets from (i) to (l) show the Cole-Cole diagrams $\varepsilon''/\varepsilon'(\nu)$ [22, 23]. The contrast in the experimental data is observed at low (from 0.3 to 0.45 THz) and at high (from 0.8 to 0.95 THz) frequencies. This significant contrast demonstrates the ability to differentiate the dysplastic and non-dysplastic nevi of the skin using the THz dielectric spectroscopy. Because the THz permittivity of biological tissue is described by the Debye [22, 23], double-Debye [24–26] models of dielectric dispersion, or the dispersion models characterized with continuous density function $g(\tau)$ of relaxation times τ [13]

$$\widetilde{\varepsilon} = \varepsilon_\infty + \int_0^\infty \frac{g(\tau)\, d\tau}{1 + 2\pi i \nu \tau}, \tag{7}$$

where ε_∞ is the dielectric permittivity for $\nu \to \infty$, the differences in the experimental curves (Fig. 1) are associated with differences in the picosecond dynamics in media [27].

In Refs. [1, 28] the principal component analysis was implemented to highlight the ability for differential diagnosis of dysplastic and non-dysplastic skin nevi based on THz dielectric dispersion. The principal components were introduced on the basis of the gradients (average slopes) of $\varepsilon'(\nu)$ or $\varepsilon''/\varepsilon'(\nu)$–curves, and the simple thresholding procedure was applied to differentiate the nevi from all four patient.

The results of the present study demonstrate the potential of TPS use for non-invasive early diagnosis of the dysplastic skin nevi and the melanomas of the skin.

Acknowledgments

This work was supported by the Russian Foundation for Basic Research, Projects #14-02-00781, 14-08-31102, 14-15-00758, 14-08-31124, 14-02-00256.

References

[1] Zaytsev K I, Kudrin K G, Karasik V E, Reshetov I V and Yurchenko S O 2015 *Applied Physics Letters* **106** 053702
[2] Duvillaret L, Garet F and Coutaz J L 1999 *Applied Optics* **38** 409–415
[3] Dorney T D, Baraniuk R G and Mittleman D M 2001 *Journal of the Optical Society of America A* **18** 1562–1571
[4] Huang S, Ashworth P C, Kan K W, Chen Y, Wallace V P, Zhang Y T and Pickwell-MacPherson E 2009 *Optics Express* **17** 3848–3854
[5] Zaytsev K I, Gavdush A A, Karasik V E, Alekhnovich V I, Nosov P A, Lazarev V A, Reshetov I V and Yurchenko S O 2014 *Journal of Applied Physics* **115** 193105
[6] Zaytsev K, Gavdush A, Chernomyrdin N and Yurchenko S 2015 *IEEE Transactions on Terahertz Science and Technology* **5** 1–11
[7] Mittleman D M, Hunsche S, Boivin L and Nuss M C 1997 *Optics Letters* **22** 904–906
[8] Chen Y, Huang S and Pickwell-MacPherson E 2010 *Optics Express* **18** 1177–1190
[9] Zaytsev K I, Karasik V E, Fokina I N and Alekhnovich V I 2013 *Optical Engineering* **52** 068203
[10] Zaytsev K I, Gavdush A A, Lebedev S P, Karasik V E and Yurchenko S 2015 *Optics and Spectroscopy* **118** 552–562
[11] Zaytsev K I, Gavdush A A, Chernomyrdin N V, Kruchkov N P, Kudrin K G, Nosov P A, Karasik V E and Yurchenko S O 2014 *Journal of Physics: Conference Series* **536**
[12] Yakovlev E V, Zaytsev K I, Fokina I N, Karasik V E and Yurchenko S O 2014 *Journal of Physics: Conference Series* **486**
[13] Yakovlev E V, Zaytsev K I, Dolganova I N and Yurchenko S O 2015 *IEEE Transactions on Terahertz Science and Technology* **5** 1–7
[14] Sim Y C, Park J Y, Ahn K M, Park C and Son J H 2013 *Biomedical Optics Express* **4** 1413–1421
[15] Woodward R M, Wallace V P, Pye R J, Cole B E, Arnone D D, Lineld E H and Pepper M 2003 *Journal of Investigative Dermatology* **120** 72–78
[16] Wallace V P, Fitzgerald A J, Shankar S, Flanagan N, Pye R, Cluff J and Arnone D D 2004 *British Journal of Dermatology* **151** 424–432
[17] Nakajima S, Hoshina H, Yamashita M, Otani C and Miyoshi N 2007 *Applied Physics Letters* **90** 041102
[18] Reid C B, Fitzgerald A, Reese G, Goldin R, Tekkis P, O'Kelly P S, Pickwell-MacPherson E, Gibson A P and Wallace V P 2011 *Physics in Medicine and Biology* **56** 4333–4353
[19] Ashworth P C, Pickwell-MacPherson E, Provenzano E, Pinder S E, Purushotham A D, Pepper M and Wallace V P 2009 *Optics Express* **17** 12444–12454
[20] Arumi-Uria M, McNutt N S and Finnerty B 2003 *Modern Pathology* **16** 764–771
[21] Barnhill R L, Fine J A, Roush G C and Berwick M 1996 *Cancer* **78** 427–432
[22] Cole K S and Cole R H 1941 *The Journal of Chemical Physics* **9** 341–351
[23] Cole K S and Cole R H 1942 *The Journal of Chemical Physics* **10** 98–105
[24] Pickwell E, Cole B E, Fitzgerald A J, Pepper M and Wallace V P 2004 *Physics in Medicine and Biology* **49** 1595
[25] Pickwell E, Cole B E, Fitzgerald A J, Wallace V P and Pepper M 2004 *Applied Physics Letters* **84** 2190–2192
[26] Yurchenko S O and Zaytsev K I 2014 *Journal of Applied Physics* **116** 113508
[27] Acbas G, Niessen K A, Snell E H and Markelz A G 2014 *Nature Communications* **5** 3076
[28] Zaytsev K I, Chernomyrdin N V, Kudrin K G, Reshetov I V and Yurchenko S O 2015 *Optics and Spectroscopy* **119** 430–437

Using of optical splitters in quantum random number generators, based on fluctuations of vacuum

A E Ivanova[1], **S A Chivilikhin**[1] **and A V Gleim**[1,2]
[1]ITMO University, Kronverkskiy pr. 49, St.Petersburg 197101, Russian Federation,
[2]Kazan National Research Technical University, K.Marx st. 10, Kazan 420111, Russian Federation

E-mail: newiva@mail.ru

Abstract. In this paper mathematical description of using of fiber Y-splitter and optical splitter with two input and two output ports in quantum random number generation systems were obtained.

1. Introduction

Quantum random number generators are based on non-deterministic physical processes. They allows to obtain sequences of random numbers, that can be used in applications requiring high degree of randomness, for example, in cryptography [1].

Existing approaches to quantum random number generation include the use of separation of radiation [2], entangled photon states [3], quantum noise of a laser [4], processes of photon emission and detection [5]. An alternative approach is quantum random number generators based on quantum fluctuations of vacuum [6, 7].

In quantum random number generation schemes based on quantum fluctuations(Fig. 1) beam splitters with two inputs and two outputs are normally used. The purpose of this research was a comparison of quantum descriptions of such beam splitter and a fiber splitter with one input and two outputs. If these quantum descriptions will be equal, it will allow us to use the Y-splitter for the implementation of a quantum random number generation system based on quantum fluctuations of vacuum.

2. Beam splitter

Beam splitter is a key element for quantum random number generation schemes based on vacuum fluctuations [6, 7]. In these schemes a strong coherent emission from the laser comes to the one input of beam splitter and quantum fluctuations of vacuum come to another input. The beam splitter prepares the mixture of these signals, and then separates it to two signals, arriving at balanced detector. One of output signals after detection is subtracted from the other, and thus, the final signal is only quantum vacuum noise, which is completely random and can be used to random number generation.

Mathematical description of a beam splitter(Fig. 2), when strong laser signal, described by the Poisson distribution, arrives to one of its inputs and vacuum state arrives to other, has been received in the operator form. In this description mean photon number of laser signal α, the

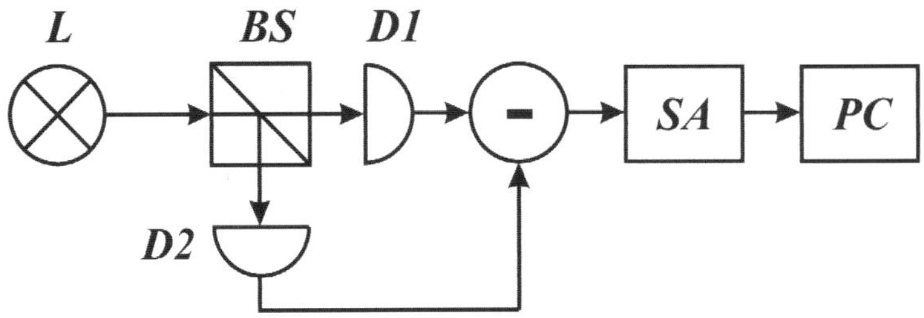

Figure 1. Quantum random number generation scheme, based on vacuum fluctuations: L - laser, BS - beam splitter, D1, D2 - detectors, SA - spectrum analyzer, PC - computer.

angle of beam splitter θ and quantum efficiencies of detectors γ_1 and γ_2 were taken into account. If signals a_1 and a_2 comes to beam splitter inputs, as shown on fig. 2, then signals at outputs,

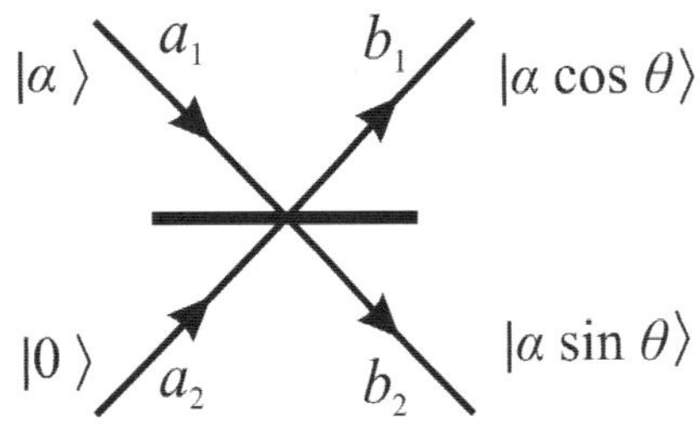

Figure 2. Scheme of a beam splitter with angle θ, where to the 1st splitter input a coherent state is send, and to other input - a vacuum state.

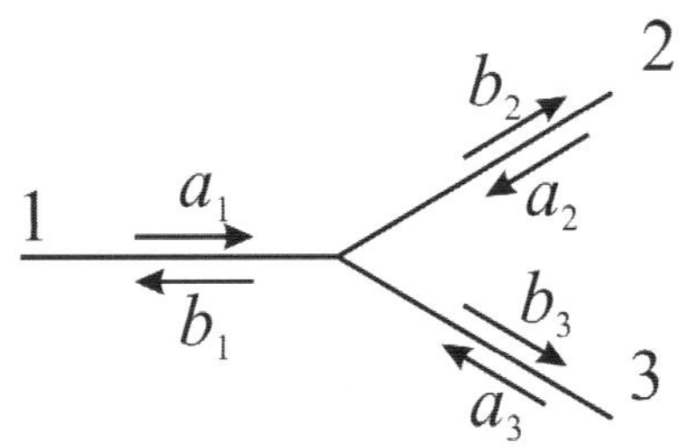

Figure 3. Optical Y-splitter. a_1, a_2, a_3 - input signals of 1st, 2nd and 3rd ports, respectively, b_1, b_2, b_3 - output signals from the splitter

b_1 and b_2 can be described by this formula

$$\begin{cases} b_1 = a_1 \cos\theta - a_2 \sin\theta, \\ b_2 = a_1 \sin\theta + a_2 \cos\theta. \end{cases} \tag{1}$$

Laser radiation at first input is characterized by Poisson distribution with parameter α (describing mean photon number), which in operator form is described as follows

$$|\alpha\rangle = e^{\alpha a_1^+ - \alpha^* a_1}|0\rangle, \tag{2}$$

where a_1^+ and a_1 - photon creation and annihilation operators at first input of beam splitter, $|\alpha\rangle$ - coherent state, $|0\rangle$ - vacuum state.

When to first splitter input a coherent state is send, and to second splitter input a vacuum state is send, then beam splitter input signal is expressed as a tensor product

$$|\alpha\rangle|0\rangle = e^{\alpha a_1^+ - \alpha^* a_1}|0\rangle_1|0\rangle_2. \tag{3}$$

In this case, if angle of beam splitter is θ, then one of beam splitter outputs is characterized by Poisson distribution with parameter$|a\cos\theta\rangle$, and second is characterized by Poisson distribution with parameter$|a\sin\theta\rangle$.

In case of symmetric beam splitter we obtain expression, describing signals at both outputs

$$b_1^+ = b_2^+ = \frac{1}{\sqrt{2}} a_1^+. \tag{4}$$

Differential current after detecting can be defined as follows

$$\Delta i = i_2 - i_1 = \gamma_2 b_2^+ b_2 - \gamma_1 b_1^+ b_1, \tag{5}$$

where i_1, i_2 are photocurrents at first and second detectors, γ_1, γ_2 are quantum efficiencies of detectors.

For a symmetric beam splitter and detectors with equal quantum efficiencies, mean value of differential current is determined to be zero, and amplitude of difference current deviation is directly proportional to intensity of incident radiation.

In case of using an asymmetric beam splitter and detectors with different quantum efficiencies, mean value of differential current $\langle \Delta i \rangle$ and amplitude of differential current deviationδi are characterized by the following equations

$$\langle \Delta i \rangle = \alpha^2(\gamma_2 \sin^2\theta - \gamma_1 \cos^2\theta), \delta i = \alpha\sqrt{\gamma_2^2 \sin^2\theta + \gamma_1^2 \cos^2\theta}, \tag{6}$$

where γ_1, γ_2 are quantum efficiencies of the detectors, θ is beam splitter angle.

3. Y-splitter

Using Y-splitter as a basic element for the homodyne detection allows getting lower level of determined nature noise and reducing the size of device without compromising the degree of randomness of generated sequences or the rate of generation. We consider Y-splitter as a system with three inputs and three outputs (fig. 3), because input and output signals can pass through one channel.

Relationship between input and output signals in Y-splitter, that allows showing the correlation between each pair of signals, can be described by following expression:

$$\begin{pmatrix} b_1 \\ b_2 \\ b_3 \end{pmatrix} = \begin{pmatrix} -\sqrt{1-2\lambda^2} & \beta & \beta \\ \lambda & -\gamma & \sqrt{1-\beta^2-\gamma^2} \\ \lambda & \sqrt{1-\beta^2-\gamma^2} & -\gamma \end{pmatrix} \begin{pmatrix} a_1 \\ a_2 \\ a_3 \end{pmatrix}. \tag{7}$$

where λ is proportionality factor, connecting input signal at 1st port and output signals at 2nd and 3rd ports; β is proportionality factor, connecting output signal at 1st port and input signals at 2nd or 3rd ports; γ is proportionality factor , connecting input and output signals of 2nd or 3rd ports.

These coefficients are selected in accordance with requirements of unitary property of matrix. Next expressions are also followed from unitarity conditions

$$\begin{cases} -\sqrt{1-2\lambda^2}\beta - \lambda\gamma + \lambda\sqrt{1-\beta^2-\gamma^2} = 0, \\ -\sqrt{1-2\lambda^2}\beta + \lambda\sqrt{1-\beta^2-\gamma^2} - \lambda\gamma = 0, \\ \beta^2 - 2\gamma\sqrt{1-\beta^2-\gamma^2} = 0. \end{cases} \tag{8}$$

After selection of matrix proportionality coefficients it is possible to simplify matrix form, using fact, that parameters λ and β can be expressed from this system through the γ:

$$\alpha = \beta = \sqrt{2\gamma(1-\gamma)}. \tag{9}$$

We consider special case when signal from 1st input port is distributed only between ports 2 and 3.In this case, $\sqrt{1-2\lambda^2}=0$, and $\lambda=\frac{1}{\sqrt{2}}$ and by using expressions was obtained above, we can receive values $\beta=\frac{1}{\sqrt{2}}$ and $\gamma=\frac{1}{2}$.

If signal a_1 is send to 1st port of Y-splitter, then signals from outputs 2 and 3 can be described by using matrix with this proportionality factors

$$\begin{pmatrix} b_1 \\ b_2 \\ b_3 \end{pmatrix} = \begin{pmatrix} 0 & \frac{1}{\sqrt{2}} & \frac{1}{\sqrt{2}} \\ \frac{1}{\sqrt{2}} & -\frac{1}{2} & \frac{1}{2} \\ \frac{1}{\sqrt{2}} & \frac{1}{2} & -\frac{1}{2} \end{pmatrix} \begin{pmatrix} a_1 \\ 0 \\ 0 \end{pmatrix} = \begin{pmatrix} 0 \\ \frac{1}{\sqrt{2}}a_1 \\ \frac{1}{\sqrt{2}}a_1 \end{pmatrix}. \tag{10}$$

Then we can consider matrix elements, describing interconnection between signal at 1st input port of Y-splitter and signals, emanating from 2nd and 3rd ports, thus

$$b_2^+ = b_3^+ = \frac{1}{\sqrt{2}}a_1^+. \tag{11}$$

This expression coincides with result, that we can obtain at output ports of symmetric beam splitter, when coherent state a_1 is send to the first splitter input, and a vacuum state - to the other. Thus, as description for beam splitter and Y-splitter are equal, we can use results for beam splitter, obtained earlier to evaluate work of quantum random generation systems, based on vacuum fluctuations using the Y-splitter.

4. Conclusions

In this research quantum mathematic descriptions of using Y-splitter and beam splitter in quantum random generation schemes based on fluctuation of vacuum were obtained. For beam splitter we described relationship between input radiation and differential current for quantum random number generation scheme using homodyne detection. We also derived expressions allowing estimation of scheme parameters imperfection impact on measurement results.

A comparison of results showed that for two types of optical splitters mathematical quantum description of resulting output signals is identical. That allows using fiber splitter with one input and two output ports at quantum random number generator based on fluctuations of the vacuum, instead of optical beam splitter.

Acknowledgments

This work was financially supported by Government of Russian Federation, Grant 074-U01.

References

[1] Scarani V, Bechmann-Pasquinucci H et al 2009 *Rev. Mod. Phys.* **81** 1301
[2] Jennewein T, Achleitner U et al 2000 *Rev. Sci. Instrum* **71** 1675
[3] Kwon O, Cho Y-W and Kim Y-H 2009 *Appl. Opt.* B **48** 1774
[4] Qi B, Chi Y-M, Lo H-K and Qian L 2010 *Optics Letters* **35** 312
[5] Dynes J F, Yuan L et al 2008 *Appl. Phys. Lett.* **93** 031109
[6] Shen Y, Tian L and Zou H 2010 *Phys. Rev.* **81** 063814
[7] Gabriel C, Wittmann C et al 2010 *Nature Phot* **4** 711

Journal of Physics: Conference Series **735** (2016) 012078 doi:10.1088/1742-6596/735/1/012078

Conversion circularly polarized beam shifting optical vortices with a fractional topological charges in a uniaxial crystal

A O Pogrebnaya, S I Halilov and A F Rubass
V.I. Vernadsky Crimean Federal University, Vernadsky Prospekt, 4, Simferopol, 295007, Russian Federation.

E-mail: kurioza.9@gmail.com

Abstract. In this work we have studied the distribution of a circularly polarized beam carrying the optical vortex with fractional topological charge equal to ½ in a uniaxial crystal. We have found that by increasing the angle of inclination of the beam relative to the optical axis of the crystal to $\alpha = 1.75$ °, mixed dislocation movement observed wave front interference pattern to beam periphery. Experimental research has shown that when the angle $\alpha = 2$ ° in the central region of the beam, we are seeing the emergence of "fork", optical vortex with a topological charge of the order of 1. The results show depolarization of the beam and the transition to the spin angular momentum of the orbital angular momentum. The intensity of the RCP and LCP component in the beam carrying the optical vortex with fractional topological charge oscillate. The total intensity of the beam as the sum of two orthogonally polarized components does not change.

1. Introduction

As know, at the moment, we investigate a new type of singular beams carrying optical vortices with fractional topological charges. Earlier work [1] had shown that there are four types of optical vortices with fractional topological charges: even, odd characters with opposite topological charges. The sum or difference of even and odd optical vortex with fractional topological charge, the standard form of the vortex (or vortex) beam with a topological charge of the whole order. All optical vortices with a fractional topological charges in a beam annihilated. This means that the wavelength of the optical vortex state with fractional topological charge is structurally unstable in the propagation of the beam in free space. It has been shown that the structurally stable optical vortex with fractional topological charge must be, first and foremost, a vector field with circularly polarized components having a half-order of topological charges. In addition, the medium must have a birefringent properties, which essentially forms the tensor field with fractional topological index.

The aim is to analyze the structural features of a circularly polarized beam carrying the optical vortex with fractional topological charge in a uniaxial crystal.

2. Experimental study of the conversion of a circularly polarized beam shifting optical vortices with a fractional topological charges in a uniaxial crystal

Methods of experimental studies of the structure of the singular beam is to obtain a picture of the intensity distribution of the beam and the interference pattern.

The study was conducted in the experimental plant shown in Figure 1

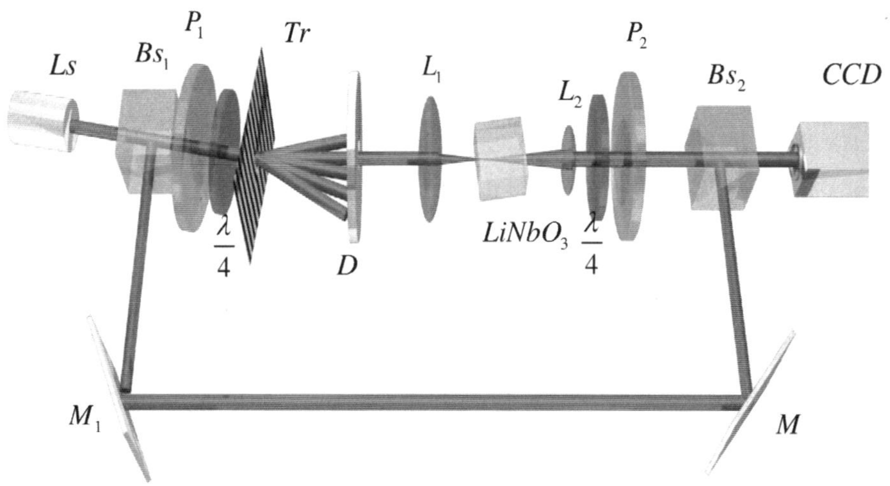

Figure 1. Experimental setup: (Ls) laser, (Bs) beam splitter, (P) a polarizer, (λ / 4) wave plate, (Tr) phase transparency, (D) aperture, (L) lens, (LiNbO3) uniaxial crystal, (M) mirror, (CCD) camera.

The experimental setup is a MachZander interferometer in which the right-circularly polarized Gaussian beam passing through a phase transparency, converted to optical vortex with fractional topological charge. The light then passes through a uniaxial crystal is placed on the progress, allowing the crystal to rotate in the horizontal plane with a pitch 0,03 °. The data recorded by a CCD camera are transmitted to a personal computer.

In the experimental study, to get a picture of the intensity distribution of interference and the right-circularly polarized beam, carrying the optical vortex with fractional topological charge equal to ½, in a uniaxial crystal, when the angle between the optical axis of the beam and the optical axis of the crystal in the range $\Delta\alpha = 0\ 01 \div 2$ °, (Figure 2).

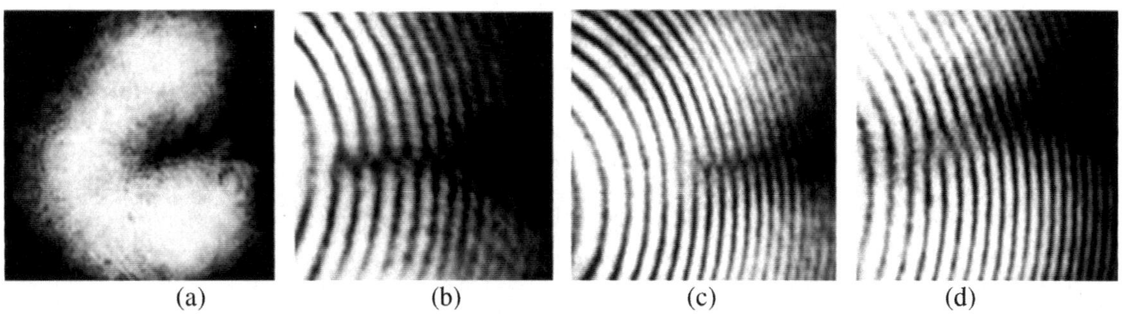

Figure 2. The intensity of the (a) and interference patterns (b, c, d) a circularly polarized beam carrying the optical vortex with fractional topological charge equal to ½ in a uniaxial crystal, when the angle between the optical axis of the beam and the optical axis of the crystal (b) α = 0,75 °, (c) α =1.75°, (d) α = 2 °, with the beam waist radius w= 100 mkm and a length of the crystal z=2 cm.

Figure 2 shows that an increase in the angle of the beam relative to the optical axis of the crystal to α = 1,75 °, there is a mixed movement of wave front dislocations [2,3] in the interference pattern to the periphery of the beam. Upon reaching the angle α = 2 ° in the central region of the beam, we are seeing the emergence of "fork", i.e. optical vortex with a topological charge of the order of 1. The results show depolarization of the beam and the transition to the spin angular momentum of the orbital angular momentum.

According to the results of the experiment were the curves of the intensity $I_{\pm}$ on the angle α for the value of waist radius w= 100 mkm. This graph is shown in Figure 3.

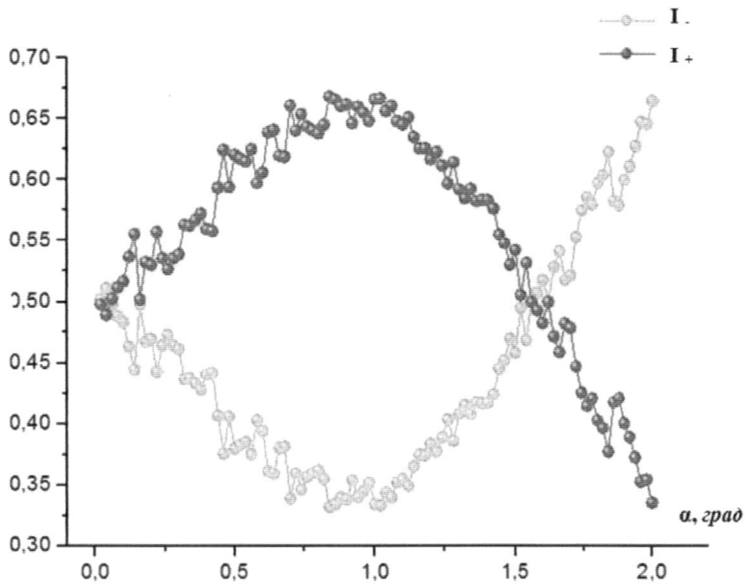

Figure 3. The intensity of the angle between the beam axis and the axis of the crystal α. I_- - the intensity of the left circularly polarized (LCP) components, I_+- the intensity of the right circularly polarized (RCP) component.

Figure 3 shows that the phase difference that occurs between the RCP and LCP orthogonal circularly polarized components of the beam, resulting in energy transfer. Since this transfer changes the value of the spin angular momentum, it is based on the spin-orbit coupling. The intensity of the RCP and LCP component in the beam carrying the optical vortex with fractional topological charge oscillate. The total intensity of the beam as the sum of two orthogonally polarized components does not change.

References

[1] Alexander V Volyar 2013 *Do optical quarks exist in free space: scalar treatment? Ukrainian Journal of Physical Optics*; **Vol. 14**, p.31

[2] A Kovalyova; A Markovskyy; T Fadeyeva; A Rubass 2013 *Generation of fractional optical vortices at the edge of the phase wedge, SPIE*

[3] T A Fadeyeva, A F Rubass, R V Aleksandrov, A V Volyar 2014 *Does the optical angular momentum change smoothly in fractional-charged vortex beams? JOSA B 31 (4)*, p.798-805

Journal of Physics: Conference Series **735** (2016) 012079 doi:10.1088/1742-6596/735/1/012079

Study of signal demultiplexing in beam Laguerre-Gaussian after passing low mode fibers

A.O. Pogrebnaya, S.I. Halilov and A.F. Rubass
V.I. Vernadsky Crimean Federal University, Vernadsky Prospekt, 4, Simferopol, 295007, Russian Federation.

E-mail: kurioza.9@gmail.com

Abstract. In the experimental study demonstrated the possibility of demultiplexing complex signal in the optical low mode fiber, feed beams of different wavelengths, with different polarization states with different orbital angular moments. We get a picture of the distribution of polarization possible to identify each of the beams with different wavelengths on the type of mode. Thus, analyzing the superposition of the beams on the mode composition, we can distinguish between the input signal to carry out his final demultiplexing.

1. Introduction

Currently, the main tasks of fiber optics include finding ways to increase the capacity of communication lines, increasing the data rate through them. Known methods of increasing the capacity of information are processes multiplexing and demultiplexing signals in the optical fiber. Earlier these methods were used standard Gaussian beams. The object of our study are the Laguerre-Gaussian beams, which due to the removal of degeneracy of the signal in the fiber caused by the spin-orbit coupling of the angular momenta of the beam, advantageously allows to increase the number of channels of the signal in the fiber. Such parameters Laguerre-Gaussian as the wavelength of the radiation, the type of polarization and orbital angular momentum of the beam allows us to more fully explore the process of adding the signals in an optical fiber.

The purpose of experimentally demonstrate the possibility of demultiplexing of the complex signal supplied beams of different wavelengths, the mode composition of the state of the field.

2. The process of signal demultiplexing in beam Laguerre-Gaussian after passing little mode fibers

Principle experimental study is as follows: after multiplexing signals emitted from two light sources with different wavelengths and polarization states fields different topological charges carried by the optical vortex beams may complete isolation of each input signal at the output of the fiber. Unlike the beams output from the optical fiber by parameters such as the wavelength is performed using a prism or diffraction grating. The difference between the beams of the orbital angular momentum (in the topological charge of optical vortices) is carried out interference methods. Identification of the beams on the state of the field, it is possible the measurement of the polarization states using a polarizing filter. The main problem is that in the fiber due to the spin-orbit interaction of the angular momentum of the beam, the state of the topological charges of optical vortices and the polarization state of the beams are mixed. To solve this problem we use Stokes polarimetry, allowing to determine the overall state of polarization of the beam, from which you can select a basic mode.

The study was conducted in the experimental setup shown in Figure 1

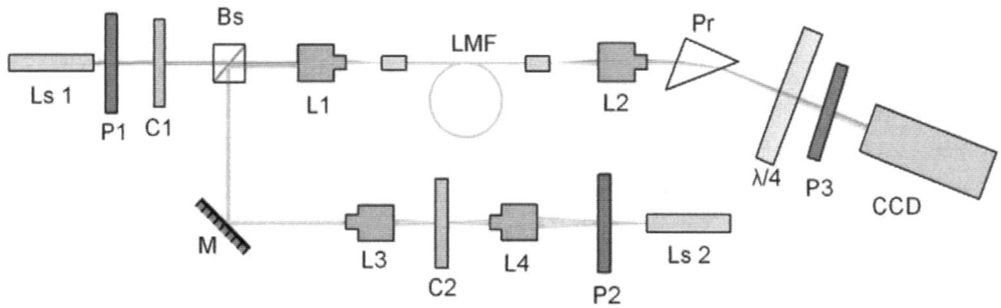

Figure 1. Experimental setup: (Ls) laser, (C) dielectric wedge, (Bs) beam splitter, (P) a polarizer, (λ/4) wave plate, (L) lens, (M) mirror, (LMF) low mode fibers, (Pr) prism, (CCD) camera.

The beams emanating from the light sources (Ls1) of wavelength λ1=0.6328 mkm and (Ls2) of wavelength λ2=0.532 mkm, passing through the dielectric wedges (C1) and (C2) are diffracted, giving rise to optical vortices, as shown in work [1] (Figure 2). Next multiplexing beams carrying optical vortices with a beam splitter (Bs). The resulting beam lens (L1) projecting coaxially on the input end of a low mode fibers (LMF). After passing the fiber beam falling on the prism (Pr), which is the demultiplexer is divided into two beams of different wavelengths (λ1 and λ2). Next, the measurement of the polarization states of these beams by Stokes polarimetry, on the basis of the Stokes parameters introduced in reference [2]. CCD camera records the results to a personal computer.

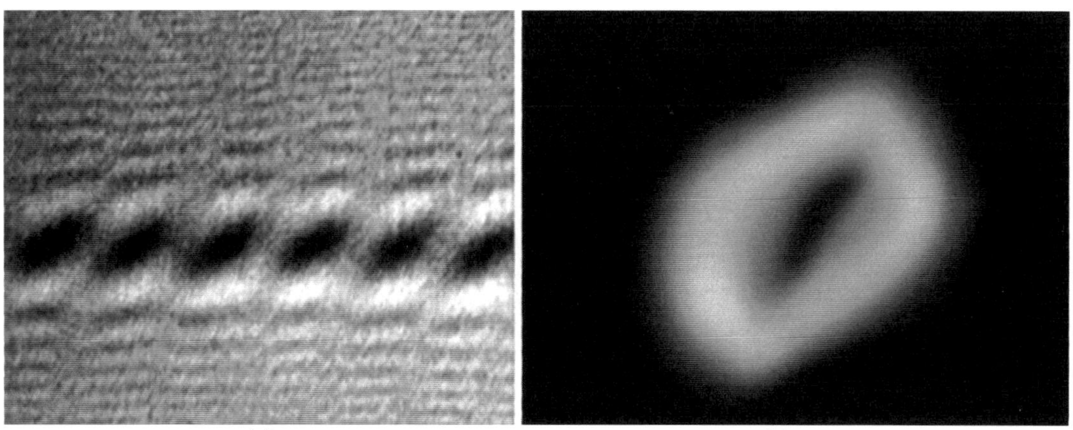

(a) (b)

Figure 2. The intensity of the beams passed dielectric wedge. (a) a chain of optical vortices, (b) an optical vortex with the topological charge l = 1

The resulting picture of the distribution of polarization of the light beam with a wavelength λ1= 0.6328 mkm and a beam of light with a wavelength λ2 = 0.532 mkm, carrying optical vortices and past the optical fiber shown in Figure 3.

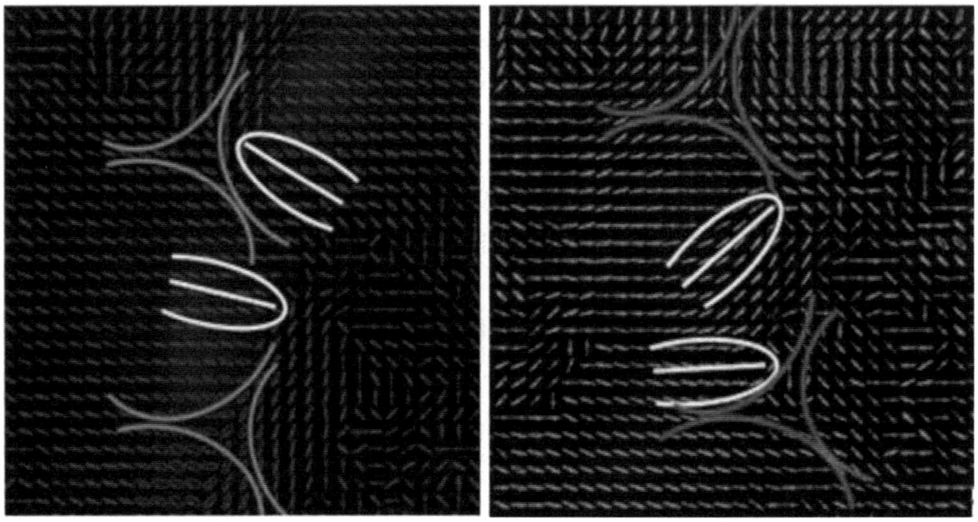

(a) (b)

Figure 3. The distribution pattern of polarization. (a) for a light beam with a wavelength λ1 = 0.6328 mkm, (b) for a light beam with a wavelength λ2 = 0.532 mkm

Analysis of the paintings presented in Figure 3 the distribution of polarization allows us to identify each of the beams with wavelengths λ1 = 0.6328 mkm and λ2 = 0.532 mkm on the type of mode. Thus, analyzing the superposition of the beams on the mode composition, we can distinguish between the input signal to carry out his final demultiplexing.

References

[1] V G Shvedov, Y V Izdebskaya, A N Alekseev, AV Volyar 2002 Formation of optical vortices in the process of diffraction of light by dielectric wedge. Technical Physics Letters Volume 28, Issue 6, pages 87-94

[2] M Born, E Wolf, Principles of Optics 1973 - M .: Science, p. 720

Fabrication and characterization of biotissue-mimicking phantoms in the THz frequency range

E Liakhov[1,2], O Smolyanskaya[1], A Popov[1,3], E Odlyanitskiy[1], N Balbekin[1] and M Khodzitsky[1]

[1] ITMO University, 199034, Kadetskaya 3, Saint-Petersburg, Russia
[2] Friedrich-Schiller-Universität Jena, Fürstengraben 1, 07743 Jena, Germany
[3] Optoelectronics and Measurement Techniques Laboratory, University of Oulu, P.O. Box 4500, FI-90014, Oulu, Finland

E-mail: o_smolyanskaya@mail.ru

Abstract. The study revealed the most promising candidates for phantoms mimicking different biological tissues in the terahertz frequency range. Closest to biological tissues in terms of the refractive index appeared to be gelatin-based gels; in terms of the absorption coefficient they were agar-based gels. Gelatin is more stable in time, but requires special storage conditions to limit water evaporation. The dense structure of the agar-based phantom allows its use without mold and risk of damage. However, agar is a nutrient medium for bacteria and its parameters degrade even when the phantom form and water content are retained. Use of liquid suspensions of lecithin and milk powder are found to be extremely limited.

1. Introduction

A number of applications demand for stable objects with preset properties. New biomedical devices require tests to reveal their performance and, later on, periodic calibration to maintain the properties over time at the proper level. Familiarization of young researchers from biomedical engineering field with functioning principles of biological tissues and organs finds obstacles in many technical universities and laboratories due to lack of equipment and facilities necessary for working with biological objects such as animal or human cultured cells [1], blood [2] etc [3]. In particular, specific requirements are set for educational labs, where students are trained. In case of optics and photonics, test objects termed *phantoms* should possess optical properties of biological tissues and be stable in time.

A phantom is a biotissue-mimicking material molded with a specific shape having a range of real-tissue properties, such as elasticity, refractive index, scattering and absorption coefficients, stability over time etc. For most types of modern medical diagnostics a variety of phantoms are commercially available [4-6]. However, only the first steps are performed in this direction at the moment for terahertz (THz) frequency range, mainly in a form of a catalog of parameters of various biological and organic gels. This work shows characteristics of various biological tissues at the THz spectral range and suggests the most promising candidates for mimicking them.

2. Materials and methods

2.1 Terahertz time-domain spectrometer

Study on phantoms and biological tissues was performed by time-domain spectroscopy using a system developed in ITMO University [7-9]. Broadband pulsed terahertz radiation was generated by a semiconductor undoped crystal of indium arsenide upon its irradiation with femtosecond pulses of KYW:Yb laser (wavelength of 1040 nm, duration of 120 fs, pulse frequency of 70 MHz, power of 1.2 W). Characteristics of the emitted THz radiation are the following: the frequency ranges from 0.1 to 1.8 THz, average power of 0.3 μW and pulse duration of 2.7 ps [9,10]. Most of the power was spread within the 0.1 to 0.6 THz frequency range. In the frequency range from 1.0 to 1.8 THz, the signal did not exceed 1% of the maximum value. The sample was fixed perpendicularly to the optical axis in the focal plane using a two-axis motorized translation stage. Electro-optical detection was carried out using a quarter-wave plate, a Wollaston prism, a balanced photodetector and a lock-in amplifier. Filtered and amplified signals were transmitted to a computer via a digital voltmeter. Spectral resolution during measurements was about 7 GHz in the transmission mode and 15 GHz in the reflection mode. System control was performed by an in-house written code in LabVIEW software.

Biological samples and phantoms were placed at the bottom of a plastic one-well plate. The samples were covered with a silicon window (thickness is 1.04 mm, refractive index of 3.5). Measurements were performed in the reflection mode. Each point of the scanned pulse was averaged 100 times.

2.2 Preparation of biological tissue

Tissue samples were taken from different animals, *post mortem*. A blood vessel was cut out from a fresh sheep heart ($N_{exp} = 5$). The sample was dried and straightened. An adipose tissue was cut out of a cow kidney ($N_{exp} = 10$). Peripheral blood was drawn from tails of white laboratory mice. Blood was not centrifuged ($N_{exp} = 4$). Pig eye cornea was investigated ($N_{exp} = 3$). A human skin sample was taken from a front wall of a breast of a fifty-two-year-old woman.

2.3 Preparation of phantoms

Gelatin is a partially hydrolyzed collagen. It is a resultant product of processing or denature of animal connective tissue. Collagen comprises about ~20% of all proteins of the human body, and therefore gelatin was chosen as the best study object. In order to create a phantom, 10 g gelatin was first soaked in cold water (15 °C) to form a gelled granular paste at concentrations of 1:25, 10:125 and 100:625. After that, the temperature was elevated to 60 °C, with constant stirring, until formation of a pale-yellow liquid poured into a mold and cooled for 12 hours. The mold was aligned to ensure parallelism of upper and lower phantom surfaces. Its melting point is about 35 °C, so the sample was stored in the refrigerator between the experiments.

Agar is a mixture of polysaccharides of agarose and agaropectin extracted from red and brown algae growing on the costs of the Black Sea, the White Sea and the Pacific Ocean, which forms a dense jelly substance in water solutions. Agar is insoluble in cold water. 5 g powder was soaked in the ratio of 5:167, 5:50 and 5:30 for one hour. The resulting mixture was heated to a boiling point, since agar is fully dissolved in water at the temperature of 95-100 °C. The hot solution is transparent and partially viscous. Upon cooling to 35-40 °C it becomes a clear strong heat-reversible gel. After that the solution was cast into a mold, with an adjusted level. The viscous mixture thickened within a few seconds, but for full curing the mold was placed into a cooling chamber at least for one hour. When pouring, the thickening liquid was stirred to avoid appearance of bubbles. The melting point of such phantom is above 75 °C, so the phantom is not necessarily be stored in the refrigerator.

For the preparation of liquid phantoms, two dry powders were used: lecithin (phospholipid) and powdered milk. 50 mg powder lecithin or milk powder was stirred in water in 1:10 ratio. Such phantom is very durable because of quickly water evaporation from the surface.

3. Results and Discussion

Optical characteristics of five different biological tissues within a 0.3-1.0 THz frequency range were retrieved. Phantoms were created with the parameters closest to biotissues (Figs. 1-7). The closest in terms of the refractive index were gelatin gels (4% for skin and cornea, 8% for blood and 16% for a blood vessel). The closest in terms of absorption coefficient were agar gels (10% for skin and a blood vessel) and gelatin gels (16% for blood).

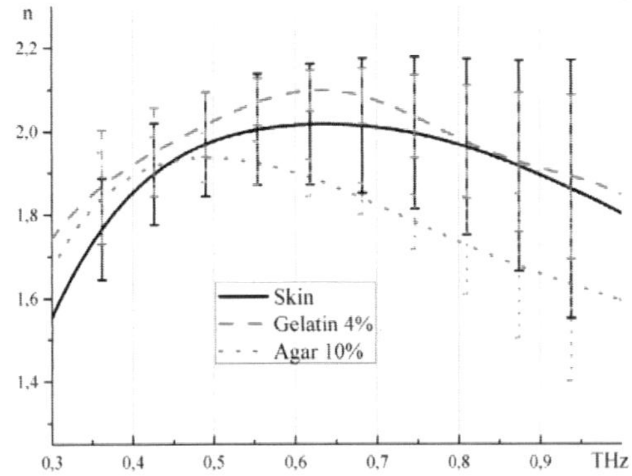

Figure 1. Dispersion of the refractive index of skin, gelatin 4% and agar 10%.

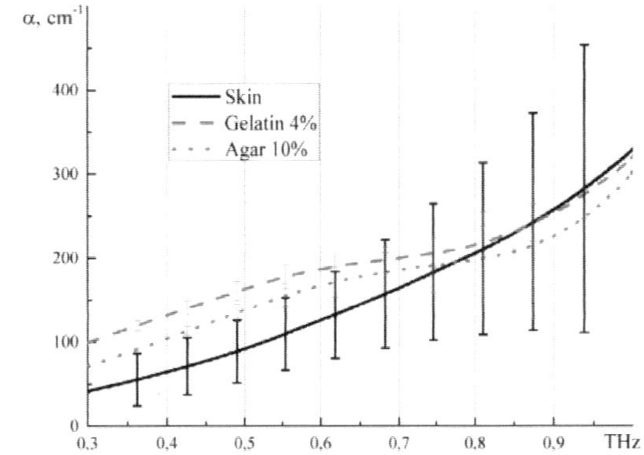

Figure 2. Dispersion of the absorption coefficient of skin, gelatin 4% and agar 10%.

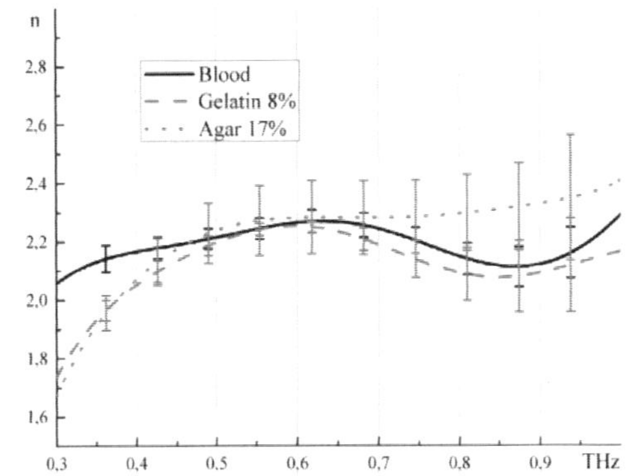

Figure 3. Dispersion of the refractive index of blood, gelatin 8% and agar 17%.

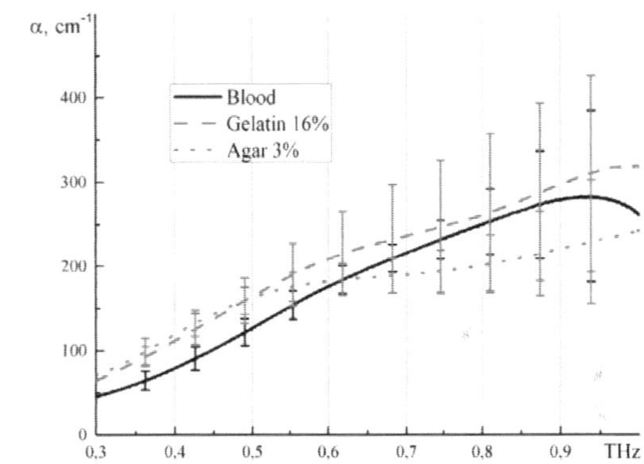

Figure 4. Dispersion of the absorption coefficient of blood, gelatin 16% and agar 3%.

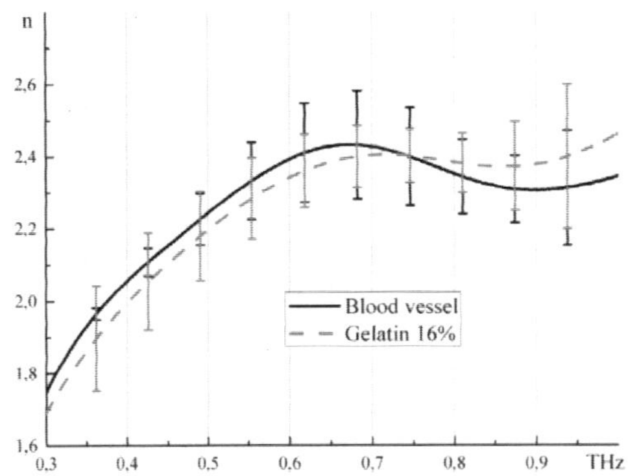

Figure 5. Dispersion of the refractive index of a blood vessel and gelatin 16%.

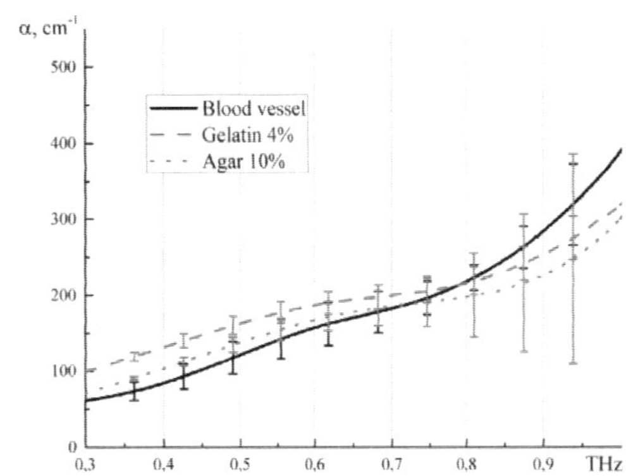

Figure 6. Dispersion of the absorption coefficient of a blood vessel, gelatin 4% , agar 10%.

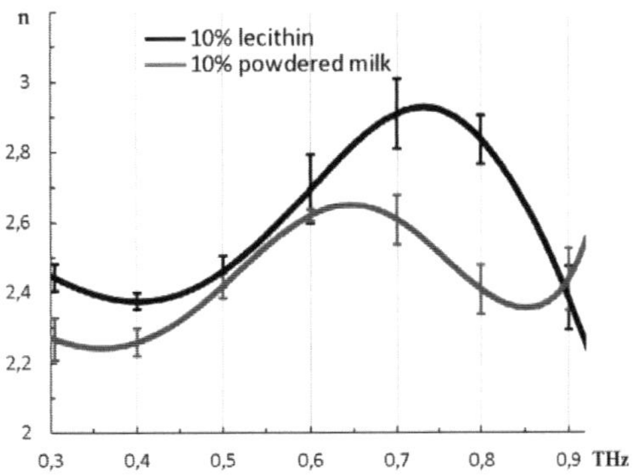

Figure 7. Dispersion of the refractive index of lecithin 10% and powdered milk 10%.

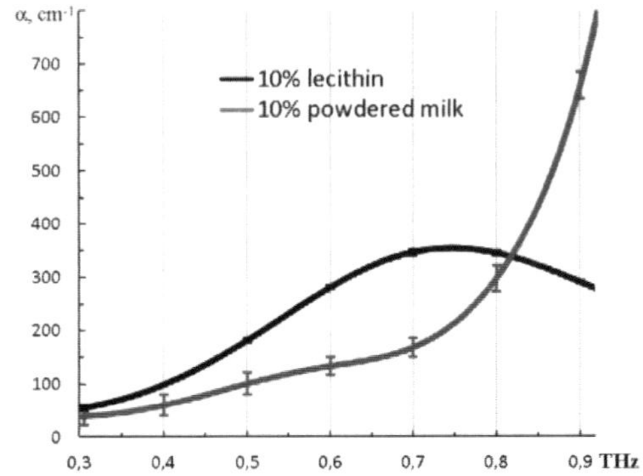

Figure 8. Dispersion of the absorption coefficient of lecithin 10% and powdered milk 10%.

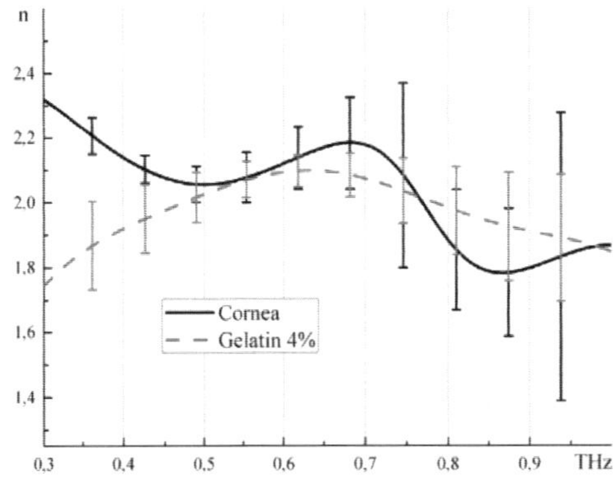

Figure 9. Dispersion of the refractive index of cornea and gelatin 4%.

4. Conclusion

The optimal phantoms for five types of biological tissues in the THz frequency range were identified during these researches. The gelatin gels were the closest in terms of the refractive index, agar gels were the closest in terms of the absorption coefficient. The gelatin gel is the optimal composite with density similar to many tissues; the cheapest and easiest to prepare. Its jelly-like structure provides a good optical contact with the substrate but it is very fragile and can be damaged during the measurement process. Gelatin is more stable in time, but it requires special handling to limit water evaporation. The dense structure of the agar phantom allows using it without mold and risk of damage. However, agar is a nutrient medium for bacteria, and its parameters degrade rapidly even with retained shape and water content. Use of aqueous suspensions of milk powder lecithin is found to be extremely limited.

Acknowledgments

This work was supported by the Government of Russian Federation (Grant 074 – U01) and Academy of Finland (Grants № 260321, 290596).

References

[1] Duka M. V., Dvoretskaya L. N., Babelkin N. S., Khodzitskii M. K., Chivilikhin S. A., Smolyanskaya O. A. 2014 *Quantum Electronics* **44(8)** 707

[2] Gusev S. I., Borovkova M. A., Strepitov M. A., Khodzitsky M. K. 2015 *European Conferences*

on Biomedical Optics **9537** 95372A

[3] Geyko I. A., Smolyanskaya O. A., Sulatsky M. I., Parakhuda S. E., Sedykh E. A., Odlyanitskiy E. L., Zabolotniy A. G. 2015 *European Conferences on Biomedical Optics* **9542** 95420E

[4] Wróbel M S, Popov A P, Bykov A V, Kinnunen M, Jędrzejewska-Szczerska M, Tuchin V V 2015 *J. Innov. Opt. Health Sci.* **8 (3)** 1541005.

[5] Wróbel M S, Popov A P, Bykov A V, Kinnunen M, Jędrzejewska-Szczerska M, Tuchin V V 2015 *J. Biomed. Opt.* **20 (4)** 045004.

[6] Jędrzejewska-Szczerska M, Wróbel M S, Galla S, Piechowski L, Popov A P, Bykov A V, Tuchin V V, Cenian 2015 *J. Biomed. Optics* **20 (8)** 085003

[7] Strepitov E A, Prozheev I V, Balbekin N S, Sulatsky M I, Khodzitsky M K, Smolyanskaya O A, Trulioff A S, Serebryakova M K 2014 *Progress in Electromagnetics Research Symposium* 1707-1710

[8] Strepitov E A, Liakhov E P, Balbekin N S, Khodzitsky M K, Smolyanskaya O A, Trulyov AS, Serebryakova M K 2015 *Proceedings of SPIE* **9542** 95420M

[9] Balbekin N S, Grachev Y V, Smirnov S V, Bespalov V G 2015 *Journal of Physics: Conference Series* **584 (1)** 012010

[10] Balbekin N S, Novoselov E V, Pavlov P V, Bespalov V G, Petrov N V 2015 *Proc. SPIE* **9448** 94482D-8.

THz spectroscopy of whole blood, plasma and cells in mice of SHR line with various pathology

A Panchenko[1,2,3], **M Tyndyk**[1,2], **O Smolyanskaya**[1], **M Sulatskiy**[1], **O Kravtsenyuk**[1], **N Balbekin**[1], M **Khodzitsky**[1]
[1] ITMO University, 197101, Kadetskaya 3, Saint-Petersburg, Russia
[2] N N Petrov Research institute of oncology of Russian Ministry of Health, 197758 Leningradskaya str., 68, Pesochny, Saint-Petersburg, Russia
[3] Federal Almazov North-West Medical Research Centre of Russian Ministry of Health, 197341, Akkuratova str., 2, Saint-Petersburg, Russia

E-mail: o_smolyanskaya@mail.ru

Abstract. This paper is devoted to studying of optical properties of whole blood and blood plasma in SHR mice grafted Ehrlich's carcinoma and mice with chronic inflammation at the terahertz frequency range. Additionally physiological saline solution suspension of ascites Ehrlich's carcinoma cells was explored.

1. Introduction

Over the last few years an opportunity of using pulsed terahertz (THz) sources for medical biomedicine practice including low-intensity systems has become extensively investigated [1-5]. Pulsed terahertz radiation has a great potential for medical diagnostics, it allows analyzing a broad spectral range of the investigated object within a short scan period of time. Special properties of terahertz radiation are exhibited in its broad frequency range and low intensity.

2. Materials and methods

2.1 Animals

Outbred Swiss-derived (9 male and 36 female) SHR mice (NN Petrov Research Institute of Oncology) three month of age (25-30 g) were used. Animals were kept in the ventilated vivarium in the cages of T2 type during 12 hours of light and 12 hours of darkness at 22 ± 2 °C, they received laboratory chow (Laboratorkorm, Moscow, Russia) and tap water ad libitum.

2.2. Tumor model

Ascitic Ehrlich's carcinoma of mice (NN Petrov Research Institute of Oncology) was used. Mice were subcutaneously inoculated with suspension of tumor cells (10^6 cells per 0.2 ml of 0.9 % NaCl solution, Dalchimpharm, Russia) for induction of solid tumors. Tumor growth was estimated, the day of tumor grafting was accounted as "0": the tumor volume was calculated by ellipsoid formula:

$$V = \frac{ab^2}{2},$$

where a is a length, and b is a width of tumor node.

2.3 Inflammation model

The model of a chronic inflammation in mice was used [6], 12-O-tetradecanoylphorbol 13-acetate, TPA (Sigma, USA) was applied (20 μl 0.01% weight/volume in acetone) on each ear (10 μl per each side of an ear) at day 0, 2, 4, 6 and 8. Control animals were treated with the same volume of acetone (analytical grade, Ecos, Russia). The ear swelling was estimated.

2.4 Blood samples

Blood samples were obtained from a tail cut into K_2EDTA MiniCollect tubes (Greiner Bio-One). Blood counts were performed (Mindray 2800Vet, China) prior to Terahertz spectroscopy. Blood plasma samples were prepared from blood obtained during the decapitation procedure into K_2EDTA MiniCollect tubes and centrifuged (5 minutes at 1500 rpm) after visual control of clotting.

2.5. Experimental procedures, distribution of animals in groups

Experimental procedure consists of three set:

1. whole blood measurements in intact 9 male (group 1) and 4 female (group 2) mice;
2. whole blood measurements in female mice with subcutaneous Ehrlich's carcinoma and with inflammation: 4 mice injected physiological solution at day 0 (group 3); 4 mice with Ehrlich's carcinoma (group 4); 4 mice with Ehrlich's carcinoma and treated by cyclophosphamide intraperitoneally 48 hours after tumor grafting (200 mg/kg, Baxter Oncology, Germany) (group 5); 5 mice with inflammation (group 6); 5 mice with acetone treatment (group 7, control to group 6);
3. blood plasma and tumor cells suspension measurements in 4 control mice (group 8) and 6 mice with ascites Ehrlich's carcinoma (group 9).

2.6 Terahertz time-domain spectrometer

The study was performed by time-domain spectroscopy using a system developed in ITMO University [7]. Characteristics of the emitted THz radiation are the following: the frequency ranges from 0.1 to 1.8 THz, average power of 0.3 μW and pulse duration of 2.7 ps. Most of the power was spread within the 0.1 to 0.6 THz frequency range. In the frequency range from 1.0 to 1.8 THz the signal did not exceed 1% of the maximum value. The sample was fixed perpendicularly to the optical axis in the focal plane using a two-axis motorized translation stage. Spectral resolution during measurements was about 15 GHz in the reflection mode. System control was performed by an in-house written code in LabVIEW environment.

2.7. Statistical analysis of results

Experimental results were statistically processed using the Student's test (MS Excel 2010, Origin Pro 2015 programs). Distinction was proposed as reliable at $p < 0.05$.

3. Results and discussion

The refractive and absorption indices (Fig. 1) of whole blood for males and females coincide in THz frequency of 0.3 ÷ 0.9 THz. The differences in blood count were not found (data not shown).

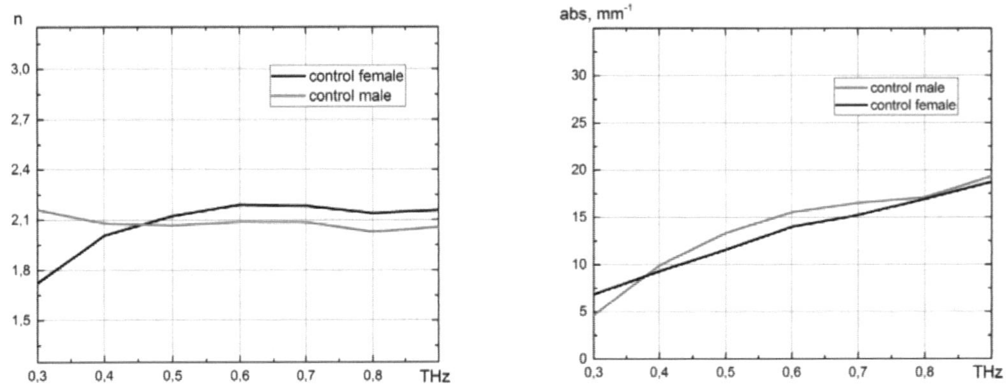

(a) **(b)**

Figure 1. Refractive (a) and absorption (b) indices of whole blood in the THz region on SHR mice

The second set results are presented on Fig. 2 and Fig. 3. The typical tumor growth was observed in mice with Ehrlich's carcinoma (group 4, mean tumor volume at the 14-th day was 1.56 $\pm$ 0.32 cm^3) and this growth was inhibited after cyclophosphamide injection (group 5, tumor volume at the 14-th day was 0.19 $\pm$ 0.14, $p < 0.01$). Refractive and absorption indices of whole blood are shown on Figure 2. It is obvious that optical properties of whole blood in (0.3 ÷ 0.9) THz frequency range for control mice (group 3), mice grafted by Ehrlich's carcinoma (group 4) and mice with Ehrlich's carcinoma injected by cyclophosphamide (group 5) practically coincide. The difference was found only in absorption coefficient of whole blood in mice of group 5 on 14-th day. The comparison of group 5 with day "0" and with group 3 isn't shown any differences in blood count. Moreover the recovery of white blood cells decreased by cyclophosphamide injection on the 7-th day was observed on 14-th day. So, leucopenia does not influence on whole blood refractive and absorption indices at the THz frequency range. The difference found in group 5 whole blood absorption coefficient on 14-th day cannot be explained by changes in hematological parameters.

Whole blood optical properties (refractive and absorption indices) at terahertz frequency range were also assessed at 10-th day of inflammation in mice. The obvious oedema and inflammation reaction were observed after TPA application (group 6) and they increased during the experiment (mean ear thickness at day 0 was 0.30 $\pm$ 0.02 mm for right ear and 0.30 $\pm$ 0.00 mm for left ear; at day 10 these values were 0.62 $\pm$ 0.19 mm, $p < 0.05$ and 0.56 $\pm$ 0.13 mm, $p<0.05$ accordingly). Blood count at day 10 revealed granulocytosis: 4.10 $\pm$ 0.57 (control) and 6.80 $\pm$ 1.77 (inflammation), $p < 0.05$. Differences in optical characteristics (absorption and refractive indices) of whole blood in mice with inflammation and corresponding control group as well as with Ehrlich's carcinoma bearing mice were not found.

Absence of influence of white blood changes on optical properties can be caused by huge amount of erythrocytes whose absorption and refractive indices are proportional to their concentration [8].

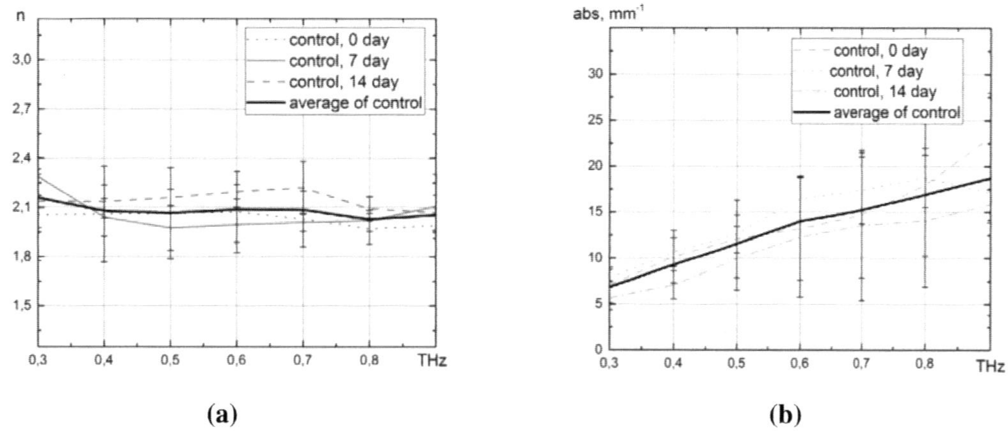

Figure 2. Refractive **(a)** and absorption **(b)** indices of whole blood in the terahertz region on SHR mice (days 0, 7 and 14 after grafting). Control group is group 3.

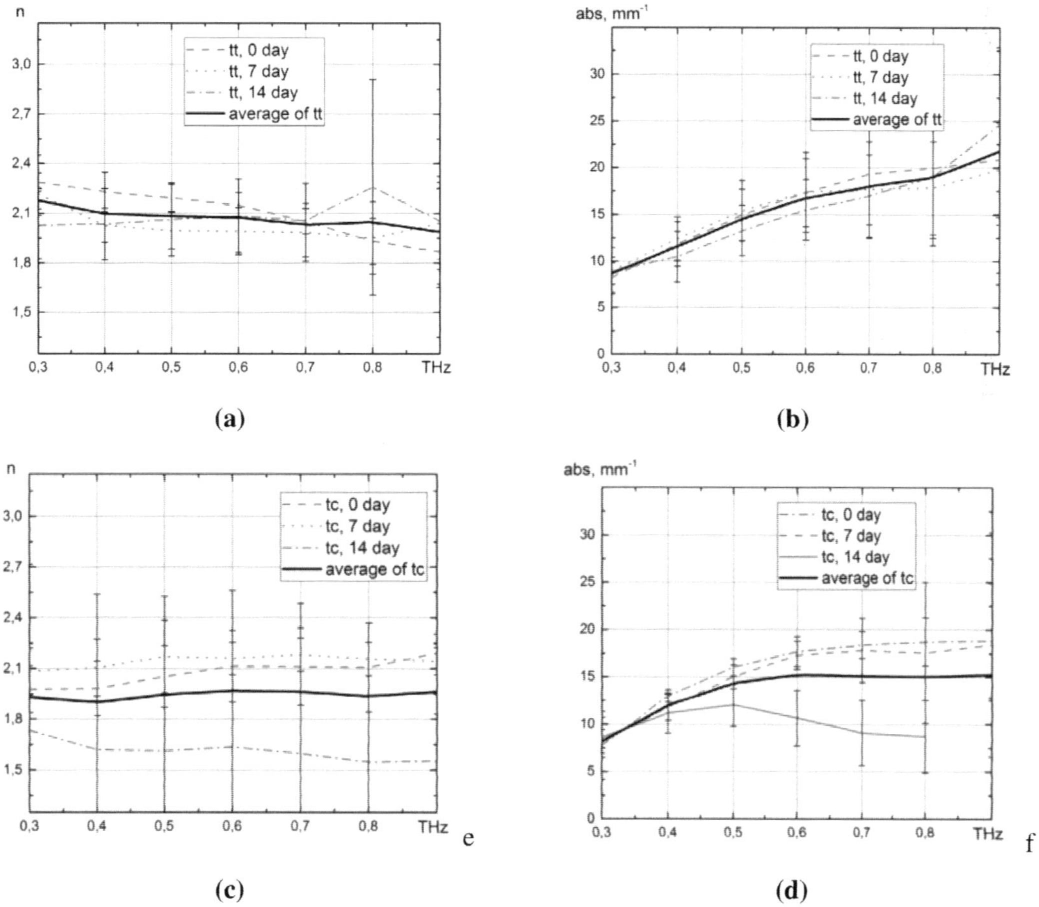

Figure 3 Refractive (left) and absorption (right) indices of whole blood in the terahertz region on SHR mice (days 0, 7 and 14 after grafting). **(c, d)** Control group 4:tt – mice grafted by Ehrlich's carcinoma. **(e, f)** Control group 4: tc - mice with Ehrlich's carcinoma injected by cyclophosphamide.

The third set of experiments was done to evaluate blood plasma and suspension of tumor cells of Ehrlich's carcinoma in physiological saline solution (Fig. 4). The refractive and absorption indices of tumor cells (Ehrlich's carcinoma) suspension in THz frequency range did not depend on their concentration in solution. The difference between blood plasma and tumor cells suspension is not significant in term of refractive indices. But the difference between plasma and tumor cells suspension absorption indices was found for all investigated frequency ranges (0.3 ÷ 0.9 THz). Differences in optical properties of blood plasma in group 8 and group 9 were found at the frequencies 0.7 and 0.8 THz. The values of absorption and scattering indices of samples were grown with water concentration. This observation is consistent with the results of the study Jeong et al. [8].

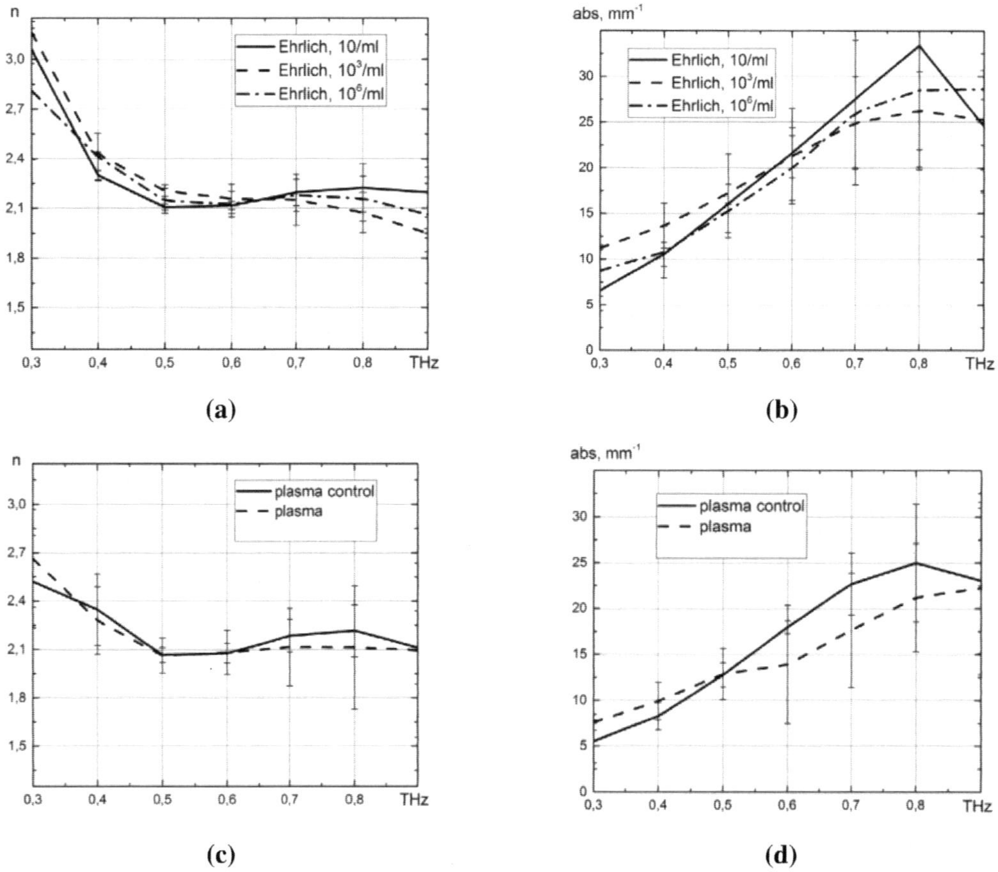

Figure 3. Refractive (left) and absorption (right) indices of object in the THz frequency range on suspen-sion of tumor cells (a,b) and blood plasma (c,d), group 8 and 9.

4. Conclusion

The whole blood optical properties (refractive and absorption indices) do not depend on the growth of Ehrlich's tumor and inflammation process. Changes in white blood cells characteristics do not affect the optical properties of whole blood at THz spectral range. The optical properties of biological systems at THz spectral range are highly dependent on the water concentration in the system.

Acknowledgments

This work was supported by the Government of Russian Federation (Grant 074 – U01).

References

[1] Strepitov E A, Prozheev I V, Balbekin N S, Sulatsky M I, Khodzitsky M K, Smolyanskaya O A, Trulioff A S, Serebryakova M K 2014 *Progress in Electromagnetics Research Symposium* 1707-1710.

[2] Strepitov E A, Liakhov E P, Balbekin N S, Khodzitsky M K, Smolyanskaya O A, Trulyov AS, Serebryakova M K 2015 *Proceedings of SPIE* **9542** 95420M.

[3] Duka M. V., Dvoretskaya L. N., Babelkin N. S., Khodzitskii M. K., Chivilikhin S. A., Smolyanskaya O. A. 2014 *Quantum Electronics* **44(8)** 707

[4] Geyko I. A., Smolyanskaya O. A., Sulatsky M. I., Parakhuda S. E., Sedykh E. A., Odlyanitskiy E. L., Zabolotniy A. G. 2015 *European Conferences on Biomedical Optics* **9542** 95420E

[5] Gusev S. I., Borovkova M. A., Strepitov M. A., Khodzitsky M. K. 2015 *European Conferences on Biomedical Optics* **9537** 95372A

[6] Stanley P L, Steiner S, Havens M, Tramposch K M 1991 *Skin Pharmaco.* **4(4)** 262-71.

[7] Balbekin N S, Grachev Y V, Smirnov S V, Bespalov V G 2015 *Journal of Physics: Conference Series* **584 (1)** 012010

[8] Jeong K, Huh Y M, Kim S H, Park Y, Son J H, Oh S J, Suh J S 2013 *J. Biomedical Optics* **18(10)** 107008.

The conversion of phase structure of singular beams spreading in uniaxial crystal

B Sokolenko[1], D Poletaev[2], A Rubass [1], A Volyar [1]
[1] General Physics department, V I Vernadsky Crimean Federal University, Simferopol, 295007, Russia
[2] Radiophysics and Electronics department, V I Vernadsky Crimean Federal University, Simferopol, 295007, Russia

E-mail: simplex@crimea.edu

Abstract. The transformation of the intensity and phase of paraxial optical beams passed uniaxial crystal strictly orthogonal to the optical axis is analysed. Imbedded optical vortex in such case imputes structural disturbance to the phase and intensity distribution after anisotropic media. Considering Left and Right circular polarized components of light, we theoretically and numerically shown the dynamics of phase shaping within the rotating birefringent crystal due to anisotropic diffraction effect. Off-axial vortex experienced tangential shift at the beam component and stimulates appearance of topological pairs in vicinity of beam axis.

1. Introduction

An investigation of optical vortex beams propagation and their transformation in anisotropic media plays a special role in singular optics of anisotropic threads [1 – 4]; it allows to form the beam shape and phase by the anisotropy of crystals [3 – 5] provoking appearance of polarization singularities. Changing of crystal properties and geometry makes possible to control the number and relative position of singular beams in front of observation plane, which can be widely used for modulation devices, optical tweezers and microscopy [6]. As it was shown in related articles, named to the orthogonal beam propagation in anisotropic media [7], the beams experienced elliptical deformation of cross-section, the value of ellipticity is proportional to the relation between the refractive indices for the ordinary and extraordinary beams. Nevertheless, the question about the features of the spatial phase structure was not fully disclosed. In particular, the forming of singular beam conoscopic patterns deserves special consideration due to the rich variety of effects caused by the interference between ordinary and extraordinary beams.
Physical mechanisms of the formation of complex vector fields in crystals were previously studied as propagation of uniformly polarized paraxial beams along the crystal optical axis [4]. In our case, the rotation of vortex around perpendicular to the optical axis allows to specify required configuration of phase singularities in the field after crystal. Thus, the question about beam shaping and phase control of optical vortices is a topical problem of modern singular optics.

2. Paraxial wave equations for singular beam spreading through the uniaxial crystal

Let us consider the uniaxial crystal with optical axis directed strictly perpendicular to the beam direction. The crystal can be represented as uniaxial homogeneous and unlimited media, where main crystallographic axes are tied to the laboratory axes linked with beam direction.

The dielectric permittivity tensor can be written as [1]:

$$\hat{\varepsilon} = \begin{pmatrix} \varepsilon_1 & 0 & 0 \\ 0 & \varepsilon_2 & 0 \\ 0 & 0 & \varepsilon_1 \end{pmatrix}, \tag{1}$$

where ε_1 is a relative dielectric permittivity coefficient for the ordinary E_x beam and ε_2 is for the extraordinary E_y beam respectively. The beam components are orthogonally polarized fundamental Gaussian modes with linear polarization and, in this case, do not interfere with each other. In paraxial approximation the beam components can be expressed as:

$$E_x = \tilde{E}_x(\mathrm{x,y,z})\mathrm{e}^{(-\mathrm{i}k_1 z)}, \tag{2}$$

$$E_y = \tilde{E}_y(\mathrm{x,y,z})\mathrm{e}^{(-\mathrm{i}k_2 z)}. \tag{3}$$

Then, for each transverse polarized component we can write wave equation in a form:

$$\partial_x^2 \tilde{E}_x + \partial_y^2 \tilde{E}_x - 2ik_1 \partial_z \tilde{E}_x = 0, \tag{4}$$

$$\partial_x^2 \tilde{E}_y + \frac{\varepsilon_2}{\varepsilon_1}\partial_y^2 \tilde{E}_y - 2ik_2 \partial_z \tilde{E}_y = 0, \tag{5}$$

where $k_1 = k_z^{(x)} = k_0 n_1$ and $k_2 = k_z^{(y)} = k_0 n_2$. Let crystal rotate and introduce new variables: ψ – is the precession angle and φ – is an angle of internal vortex rotation around beam axis. We restrict ourselves to the case when initial beam has circular cross-section ($w_o = w_e$) at the plane $z=0$. The solutions of wave equations (4) and (5) can be written as electric field components:

$$\tilde{E}_x = \left(\frac{X - i\xi Y}{w_o \sigma_o} - a e^{-i\xi(\psi-\varphi)} \right) \times \exp\left[-\left(X^2 + Y^2 \right) / w_o^2 \sigma_o \right] / \sigma_o, \tag{6}$$

$$\tilde{E}_y = \frac{i}{\sqrt{\sigma_x \sigma_y}} \left(\frac{X}{w_e \sigma_x} - i\xi \frac{Y}{w_e \sigma_y} - a e^{-i\xi(\psi-\varphi)} \right) \times \exp\left[-\frac{X^2}{w_e^2 \sigma_x} - \frac{Y^2}{w_e^2 \sigma_y} \right], \tag{7}$$

where we introduced: $z_o = k_1 w_o^2 / 2$, $z_x = k_2 w_e^2 / 2$, $z_y = k_2 w_e^2 n_1^2 / 2n_2^2$, $\sigma_o = 1 - iz/z_o$, $\sigma_x = 1 - iz/z_x$, $\sigma_y = 1 - iz/z_y$, $X = x\cos\psi - y\sin\psi$, $\xi = \pm 1$. Obviously, the shapes of ordinary and extraordinary beams due to the crystal's anisotropy and unequal Rayleigh length will be different and, as result, we get displacement of optical vortex in the beam cross section of each circularly polarized component: $E_+ = E_x - iE_y$, $E_- = E_x + iE_y$. The phase portrait shown in figure 1 (a,b) indicates the presence of a vortex in the beam central region where direction of the helix refers to the sign of topological charge.

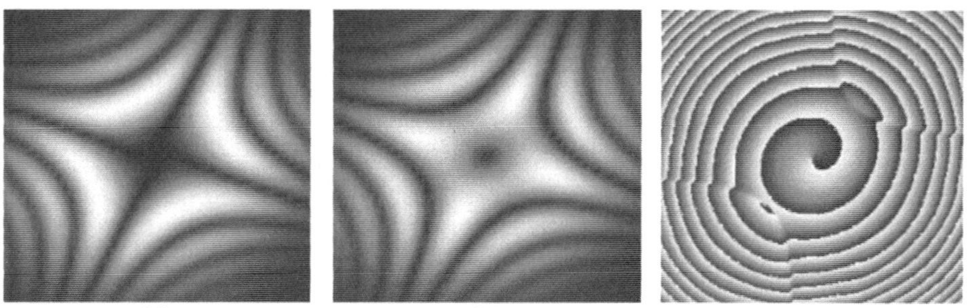

(a) (b) (c)

Figure 1. The intensity distribution of left E_- (a) and right E_+ (b) circularly polarized components and phase pattern (c) of the beam passed a crystal strictly orthogonal to the optical axis. Beam parameters are: $a = 0$, $\omega_0 = 20\,\mu m$, $z = 20\,mm$.

A phase shift on the beam periphery indicates presence of constant phase difference between ordinary and extraordinary beams (see figure 1 (c)).

As it shown in figure 2 (a,b), when initial vortex shifts from the beam center on a distance equal to a half of the waist radius, vortex displacement in the ordinary vortex beam after uniaxial crystal corresponds to one in the original beam, while the position of vortex in extraordinary elliptically deformed beam is mealy shifted towards periphery.

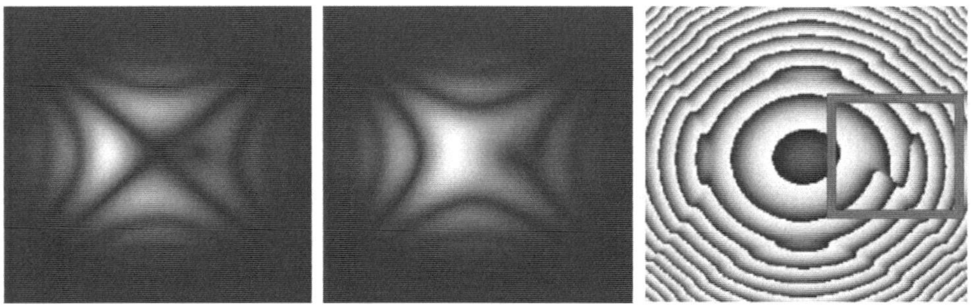

(a) (b) (c)

Figure 2. The intensity distribution of left E_- (a) and right E_+ (b) circularly polarized components and phase pattern (c) of the beam with off-axis singularity. The rectangle highlights a pair of phase dislocations. Beam parameters are: $a = 0.5$, $\omega_0 = 10\,\mu m$, $z = 20\,mm$, $\xi = -1$, $n_o = 1.54$, $n_e = 1.55$.

In other words, elliptical deformation (caused by anisotropy of medium) increases displacement of the singularity in the extraordinary beam.

3. The dynamics of phase singularities in circularly polarized components after the crystal

The fine structure of conoscopic pattern contributes an additional distortion to the phase where difference between isoclines has a constant value equal to π. Accordingly the presence of off-axis vortex at the beam introduces additional phase shift to the conoscopic structure and causes the formation of topological dipole as shown in figure 3 (a).

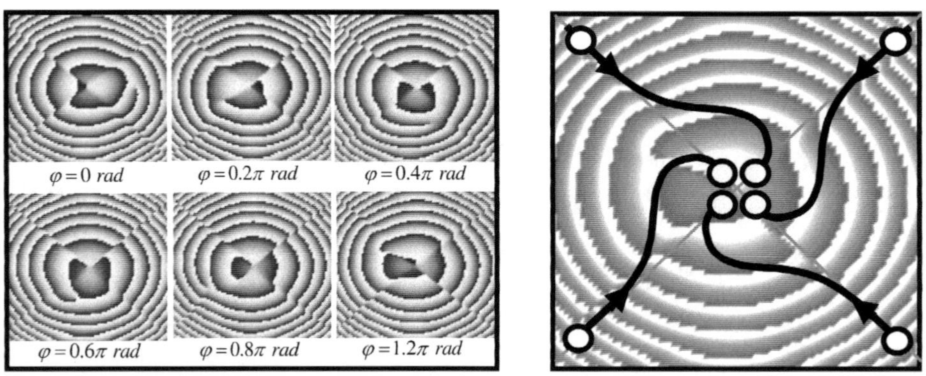

(a) (b)

Figure 3. Phase evolution of left circular polarized component in the vicinity of vortex core: (a) phase patterns of off-axial optical vortex in case of different angles φ and the trajectory of phase singularities (b): white circles correspond to areas where vortices are born and semi black circles indicates areas of topological reaction.

Spatial orientation of conoscopic pattern and the relative displacement of optical vortex leads to a number of topological reactions and local phase unfolding. By the controlling of initial vortex position with parameter a and azimuthal angle φ, we can increase the slight offset of vortex from the beam axis which is accompanied with misalignment of phase profile between singularity and conoscopic pattern. In this case, the sequential changes in position of optical vortex results on a dynamics of birth and annihilation events for topological dipoles as shown in figure 3 (b). The black arrows indicate a direction of singular points moving from the beam periphery to its center.

The initial optical vortex asymptotically approaches to the beam axis, where it annihilates with vortex of opposite topological charge; meanwhile a new topological pair is born on the beam's periphery. First vortex is located near the beam axis, while second comes from infinity and replace the original vortex. Upon further vortex rotation, the process repeats. It is worth noting to see the vortex motion appears along the lines marking the families of isoclines where phase difference is discontinuous, and have gap equal to π. As could be expected, due to the initial phase singularity introduced into the beam, new vortices can appear only in the vicinity of phase gap. This process is illustrated in figure 4.

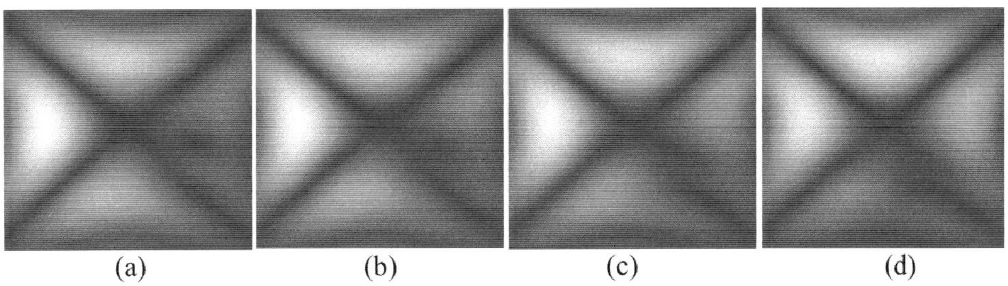

Figure 4. The intensity distribution of left circularly polarized component E_- in case of different rotational angles φ: $\varphi = 0.2\pi\ rad$ (a); $\varphi = 0.26\pi\ rad$ (b); $\varphi = 0.3\pi\ rad$ (c); $\varphi = 0.36\pi\ rad$ (d). Other parameters of the beam are: $a = 0.5$, $\omega_0 = 10\,\mu m$, $z = 20\ mm$, $\xi = -1$, $n_o = 1.54$, $n_e = 1.55$.

All the physical parameters was used in a calculation correspond to the SiO_2 crystal and weakly focused Gaussian beam. Thus, such conditions support a paraxial regime and approximately the same beam waists along the crystal. As we can see from the figure 4 (a-d), the initial vortex formed a new topological pair which moves to the beam center and annihilate. The identical process of topological reactions occur at the orthogonal circularly polarized component E_+, where local trajectories of topological dipoles depend on the displacement of the initial vortex towards the beam axis.

Topological reaction in component E_+ occurs, mainly, due to the interaction between an original vortex and isoclines. Such process results on a splitting of the intensity line with distortion of the phase portrait as illustrated in figure 5.

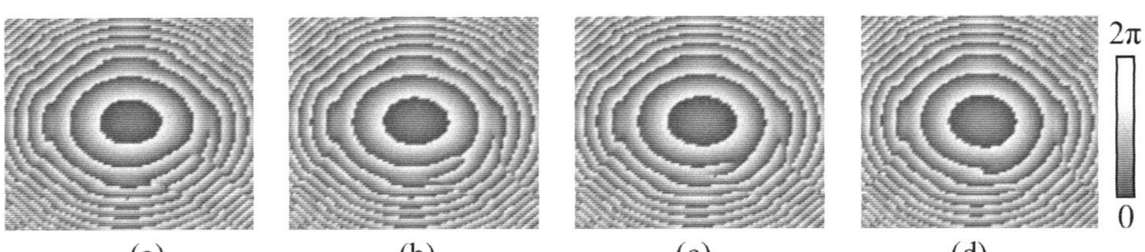

Figure 4. The phase portrait of E_+ component in case of different rotational angles φ: $\varphi = 0.1\pi\ rad$ (a); $\varphi = 0.16\pi\ rad$ (b); $\varphi = 0.22\pi\ rad$ (c); $\varphi = 0.52\pi\ rad$ (d) and: $a = 0.5$, $\omega_0 = 10\,\mu m$, $z = 20\ mm$, $\xi = -1$, $n_o = 1.54$, $n_e = 1.55$.

Here, we can observe a typical "forks", which correspond to the phase singularities and topological dipoles. When angle φ changes, it results on a splitting of intensity lines localized close to an off-axis vortex path. Thus, the phase singularity brings an excitation of phase in the structure of the conoscopic pattern, causing reaction of the medium and appearance of topological pairs

4. Conclusion

Classical conoscopic pattern formed by the interference of the ordinary and extraordinary beams has a high sensitivity to phase distortions where optical vortices, edge dislocations, and other inhomogeneities are born.

It is shown the displacement of the optical beam axis relatively to the vortex causes the formation of phase singularities that interact with the original vortex affecting its trajectory. Such effects raise, mainly, due to the spatial divergence of singularities in the ordinary and extraordinary beams.

It is demonstrated that in circularly polarized beam components both ordinary and extraordinary beams interfere and provide complex reorganization of phase structure. Even if initial vortex shifts from the beam center on $0.5w$ it stimulates appearance of topological reactions between vortices in ordinary and extraordinary beams. The birth of topological pairs, their annihilation and spatial reorientation indicates the medium response, and hence, this effects can be used for sensors producing, analyzers and phase modulators.

References

[1] Fadeyeva T A, Rubass A F, Sokolenko B V, Volyar A V 2009 *J. Opt. A: Pure Appl. Opt.* **11** 53
[2] Bekshaev A Ya, Soskin M S and Vasnetsov M V 2008 *Paraxial light beams with angular momentum* (New York: Nova Science Publishers)
[3] Sokolenko B V, Rubass A F, Volyar A V 2013, *Semiconductor Physics, Quantum Electronics and Optoelectronics* **16** 344-348
[4] Volyar A V, Shvedov V G, Fadeyeva T A 2009 *Optics Express* **14** 3724
[5] Sokolenko B V, Rubass A F, Lapaeva S N, Glumova M V, Volyar A V 2013 *Proceedings of SPIE* **9066** 1
[6] Popiolek-Masajada A, Sokolenko B, Augustyniak I, Masajada J, Khoroshun A, Bacia M 2014 *Optics and Lasers in Engineering* **55** 105–112
[7] Ciattoni A, Palma C 2008 *J. Opt. Soc. Am. A* **20** 2163-71

Journal of Physics: Conference Series **735** (2016) 012083 doi:10.1088/1742-6596/735/1/012083

QUANTUM ADDER OF CLASSICAL NUMBERS

Cherkas A.V., Chivilikhin S.A.

ITMO University, Saint-Petersburg, Russia

E-mail: alina-cherkas@mail.ru

Abstract. In this article we show the precise algorithm of functioning of quantum adder on the example of addition of two 2-bit numbers. It consists of the quantum Fourier transformer and conditional rotation gates that let us use the minimum number of qubits to get the addition realization of the sum. Despite the fact that the algorithm uses the minimum amount of operations this accelerates the process of the adder.

1. Introduction

Recently quantum technologies have developed headily. There are almost no fundamental problems left.

The benefit of quantum technologies nowadays is pretty obvious. Those things that some years ago were just theoretical now are experimentally approved and efficient. Although in modern science there is always a place for new ideas. One of those ideas is a quantum adder.

A quantum adder is a physical machine that can accept input states which represent a coherent superposition of many different possible inputs and subsequently evolve them into a corresponding superposition of outputs. Computation, i.e. a sequence of unitary transformations, affects simultaneously each element of the superposition, generating a massive parallel data processing albeit within one piece of quantum hardware [1]. This way quantum computers can efficiently solve some problems which are believed to be intractable on any classical computer [2,3].

First quantum addition algorithms mirrored their classical counterparts [4,5,6,] with necessary extensions for a reversible computation. Further quantum addition algorithms were based on the use of the complementary qubit transfer [7,8], but it still followed the classical model. The ideal addition algorithm for a quantum computer may not be similar to its classical counterpart, so that's why the quantum adder on the base of quantum Fourier transformation (QFT) was invented (Figure.1).

The horizontal lines are wires that carry the qubits. In the upper lines is encoded number A, presented in binary type. The number B is encoded in the lowest lines. Computing the Quantum Fourier Transformer may be done according to the following wire diagram (Figure.2). Each gate can be depicted as a matrix form. (Figure.3).

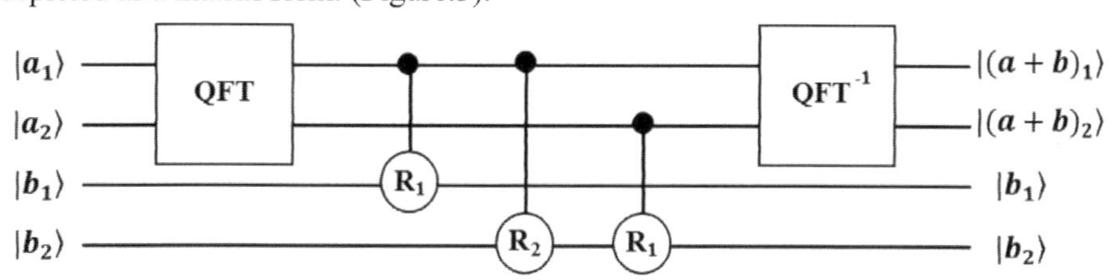

Figure 1. The quantum addition scheme based on the quantum Fourier transformer.

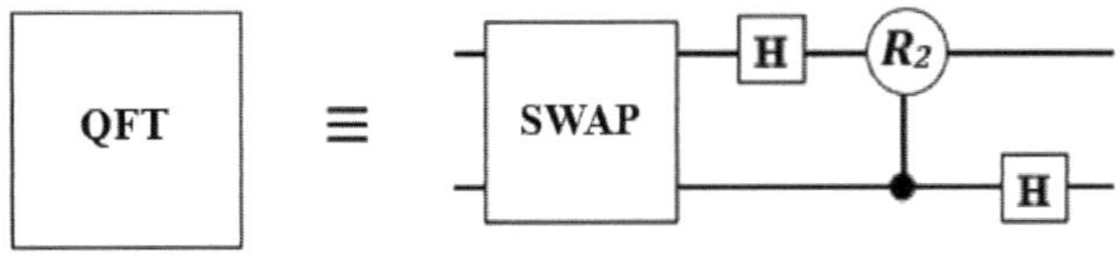

$$\equiv \frac{1}{2}\begin{pmatrix} 1 & 1 & 1 & 1 \\ 1 & i & -1 & -i \\ 1 & -1 & 1 & -1 \\ 1 & -i & -1 & i \end{pmatrix}$$

Figure 2. a) The Quantum Fourier Transformer scheme and it's mathematical form.

a) $\mathbf{SWAP} = \begin{pmatrix} 1 & 0 & 0 & 0 \\ 0 & 0 & 1 & 0 \\ 0 & 1 & 0 & 0 \\ 0 & 0 & 0 & 1 \end{pmatrix}$ b) $\mathbf{H} = \frac{1}{\sqrt{2}}\begin{pmatrix} 1 & 1 \\ 1 & -1 \end{pmatrix}$ c) $\mathbf{R_k} = \begin{pmatrix} 1 & 0 & 0 & 0 \\ 0 & 1 & 0 & 0 \\ 0 & 0 & 1 & 0 \\ 0 & 0 & 0 & e^{2i\pi\left(\frac{1}{2^k}\right)} \end{pmatrix}$

Figure 3. a) Swap gate. b) Hadamard transform. c) Conditional rotation.

2. Example «1+2»

For better understanding let's look upon the algorithm of addiction of two classical numbers 1 and 2. Thus, the number A is a 1, and the number B is a 2. Let's see what will happen when those numbers enter the quantum adder. Decimal number "1" in binary system is shown as "01". We have this number encoded by two qubits. In order to have quantum-mechanical characters for our qubits, we tensorially multiply them and get the next vector.

After this, our numbers enter QFT where they can be depicted as a matrix form. To know what will happen to our number afterwards, we multiply the matrix and the vector and get the following state. Then we take one qubit of "A" number and one qubit of "B" number. The result of their tensorial multiplication we apply to the conditional rotation gate.

We point out that all R1 can be done simultaneously as they include different qubits. This also applies to R2, etc. As a result, the number of operations is still the same, but the time necessary to the calculation declines in N times and can be taken as O (log2n).

But how can we distinguish the answer? In order to do it we use the reversal quantum Fourier transformation. Having multiplied the matrix to the obtained vector we will receive the final result.

You can see below the precise description of the following calculations.

2.1. Tensor product

$$\begin{pmatrix} 1 \\ 0 \end{pmatrix} \otimes \begin{pmatrix} 0 \\ 1 \end{pmatrix} = \begin{pmatrix} 0 \\ 1 \\ 0 \\ 0 \end{pmatrix}$$

2.2. Application of the QFT

$$\frac{1}{2}\begin{pmatrix} 1 & 1 & 1 & 1 \\ 1 & \mathrm{i} & -1 & -\mathrm{i} \\ 1 & -1 & 1 & -1 \\ 1 & -\mathrm{i} & -1 & \mathrm{i} \end{pmatrix}\begin{pmatrix} 0 \\ 1 \\ 0 \\ 0 \end{pmatrix} = \frac{1}{2}\begin{pmatrix} 1 \\ \mathrm{i} \\ -1 \\ -\mathrm{i} \end{pmatrix} = \frac{1}{\sqrt{2}}\begin{pmatrix} 1 \\ \mathrm{i} \end{pmatrix} + \frac{1}{\sqrt{2}}\begin{pmatrix} 1 \\ -1 \end{pmatrix}$$

2.3. Tensor product

$$\begin{pmatrix} 0 \\ 1 \end{pmatrix} \otimes \frac{1}{\sqrt{2}}\begin{pmatrix} 1 \\ \mathrm{i} \end{pmatrix} = \frac{1}{\sqrt{2}}\begin{pmatrix} 0 \\ 0 \\ 1 \\ \mathrm{i} \end{pmatrix}$$

2.4. Application of the controlled phase modification operator (k=1)

$$\frac{1}{\sqrt{2}}\begin{pmatrix} 1 & 0 & 0 & 0 \\ 0 & 1 & 0 & 0 \\ 0 & 0 & 1 & 0 \\ 0 & 0 & 0 & -1 \end{pmatrix}\begin{pmatrix} 0 \\ 0 \\ 1 \\ \mathrm{i} \end{pmatrix} = \frac{1}{\sqrt{2}}\begin{pmatrix} 0 \\ 0 \\ 1 \\ -\mathrm{i} \end{pmatrix} = \frac{1}{\sqrt{2}}\begin{pmatrix} 1 \\ -\mathrm{i} \end{pmatrix} \otimes \begin{pmatrix} 0 \\ 1 \end{pmatrix}$$

2.5. Tensor product

$$\begin{pmatrix}1\\0\end{pmatrix}\otimes\frac{1}{\sqrt{2}}\begin{pmatrix}1\\-i\end{pmatrix}=\frac{1}{\sqrt{2}}\begin{pmatrix}1\\-i\\0\\0\end{pmatrix}$$

2.6. Application of the controlled phase modification operator (k=2)

$$\frac{1}{\sqrt{2}}\begin{pmatrix}1&0&0&0\\0&1&0&0\\0&0&1&0\\0&0&0&i\end{pmatrix}\begin{pmatrix}1\\-i\\0\\0\end{pmatrix}=\frac{1}{\sqrt{2}}\begin{pmatrix}1\\-i\\0\\0\end{pmatrix}=\frac{1}{\sqrt{2}}\begin{pmatrix}1\\-i\end{pmatrix}\otimes\begin{pmatrix}1\\0\end{pmatrix}$$

2.7. Tensor product

$$\frac{1}{\sqrt{2}}\begin{pmatrix}1\\-1\end{pmatrix}\otimes\begin{pmatrix}1\\0\end{pmatrix}=\begin{pmatrix}1\\-1\\0\\0\end{pmatrix}$$

2.8. Application of the controlled phase modification operator (k=1)

$$\frac{1}{\sqrt{2}}\begin{pmatrix}1&0&0&0\\0&1&0&0\\0&0&1&0\\0&0&0&-1\end{pmatrix}\begin{pmatrix}1\\-1\\0\\0\end{pmatrix}=\begin{pmatrix}1\\-1\\0\\0\end{pmatrix}=\frac{1}{\sqrt{2}}\begin{pmatrix}1\\-1\end{pmatrix}\otimes\begin{pmatrix}1\\0\end{pmatrix}$$

2.9. Tensor product

$$\frac{1}{\sqrt{2}}\begin{pmatrix}1\\-i\end{pmatrix}\otimes\frac{1}{\sqrt{2}}\begin{pmatrix}1\\-1\end{pmatrix}=\frac{1}{2}\begin{pmatrix}1\\-i\\-1\\i\end{pmatrix}$$

2.10. Application of the QFT^{-1}

$$\frac{1}{4}\begin{pmatrix}1&1&1&1\\1&i&-1&-i\\1&-1&1&-1\\1&-i&-1&i\end{pmatrix}\begin{pmatrix}1\\-i\\-1\\i\end{pmatrix}=\frac{1}{4}\begin{pmatrix}0\\0\\0\\4\end{pmatrix}=\begin{pmatrix}0\\0\\0\\1\end{pmatrix}=\begin{pmatrix}0\\1\end{pmatrix}\otimes\begin{pmatrix}0\\1\end{pmatrix}=|1\rangle\otimes|1\rangle=11_2=3_{10}$$

3. Conclusions

In this research we demonstrated the simplest scheme of computations on the quantum computer. The presented algorithm helps to understand in the best way how the quantum computations of classical numbers are made. This article has the educational character and aims to convince that the quantum informatics is a very appealing subject and it's not that hard to learn if you try to do those calculations by yourself with the help of the quantum adder. On the example of that research, we've shown you how quantum technologies can be applied for calculations of greater numbers and for solving more complicated tasks. On the base of that algorithm, we plan to elaborate the quantum multiplier, which will be especially useful in quantum cryptography.

4. Acknowledgments

This work was financially supported by the Government of Russia Federation, Grant 074-U01.

References

[1] Deutsch D 1985 Proc R Soc Lond A 400 97
[2] Deutsch D, Jozsa R 1992 Proc R Soc Lond A 439 553

[3] Shor P 1994 IEEE Computer Society Press 124
[4] Vedral V, Barenco A, Ekert A 1995 Phys Rev 54 147
[5] Beckman D, Chari A, Devabhaktuni S, Preskill J 1996 Phys Rev A 54 1034
[6] Miquel C, Paz J, Perazzo R 1996 Phys Rev A 54 2605
[7] Gossett P 1998 "Quantum carry-save arithmetic" arXiv: quant-ph/9808061.
[8] Zalka C 1998 "Fast versions of Shor's quantum factoring algorithm" arXiv: quant-ph/9806084

The study of plant tissue by optical coherent microscopy method

V V Chirskaya[1], N B Margaryants[1], E V Zhukova[1]
[1]ITMO University, 49 Kronverksky pr., St. Petersburg, Russia, 197101

E-mail: fosp@grv.ifmo.ru

Abstract. The article presents the results of application of the optical coherent microscopy technique using a high-resolution automatic Linnik interference microscope to study the structure of plant tissues exemplified by surface periderm layers of a tuberous nightshade (*solánum tuberosum*) bulb. The results of 3D visualization of the structure of the sample under examination are provided. Scanning depth was 32 μm, with axial and lateral resolution of the device 1 μm.

1. Introduction

Research into surface and internal structure of various tissues are in demand in many areas of science and technique. To study biological objects, it would be of prime importance to save integrity of a sample under examination. The technique of optical coherent microscopy (OCM) offers the possibility to visualize not only an object surface, but also its internal structure, without disturbance of the same.

2. Method and appliance

The OCM applied possesses high axial and lateral resolutions at the depth of optical radiation penetration, which often surpass characteristics of similar optical and other methods of examination. It works according to the principle of measurement of the extent to which optical radiation reflects from the object examined at different depths, based on light interferometry. In the course of measurements, an object is lit with optical radiation, which, while penetrating the medium reflects from layers of the sample varying in depth to different extent. Determination of reflectivity for each point of certain layers of the medium examined permits to get information on internal structure of an object of examination, and to restore its three-dimensional structure. In this paper, the samples examined were illuminated with visible range beam, with a halogen lamp used as a source of radiation. A two-beam interferometer is a key element of OCM system. In this research, a high resolution automatic Linnik interference microscope was used running in two modes, combining advantages of a traditional two-beam interferometer and of a high-magnification optical microscope (up to 500^{x}). The area of the observation surface was 200x160 μm. Therefore, it allows for examination of both an interference pattern and surface of the object examined [1]. Lateral and axial resolution of the device was 1 μm.

3. Object of study

This research considered near-surface layers of cover tissue of a bulb of tuberous nightshade (*solánum tuberosum*), i.e. periderm. Periderm of the object investigated represents a multilayer structure built of tissue of different types and structures, at different depths from surface. Considering the functional purpose of periderm, OCM technique may reveal and measure its structural elements. In the course of work, different sections of several samples were examined, namely, those with no visible defects, those with mechanical damage, and areas with swellings and other structural abnormities. The tomograms revealed differences between structure of defective tissues and healthy periderm layers [2].

4. Research and analysis of the visualization results

At the initial stage of research, plant samples were examined under microscope by transmitted light, using a bright-field method to detect the areas of interest. The samples selected for further research were scanned with variable parameters to choose scanning criteria allowing for the best visualization under current conditions. The rated values provided scanning depth 32 μm. Applying the OCM technique and software processing of video image data array resulted in construction of tomograms of the areas researched, which can be interpreted as images of cross sections of the samples scanned. Tomograms of surface periderm layer are discussed in [2]. Visualization of internal structure allows not only for a conclusion of an abnormity of tissues without disturbing thereof, but also for further observation over progress of a disease as exemplified by the same specific area.
Further software processing of experimental data was followed by three-dimensional pictures of the areas scanned, one of which is represented in Fig, 1.

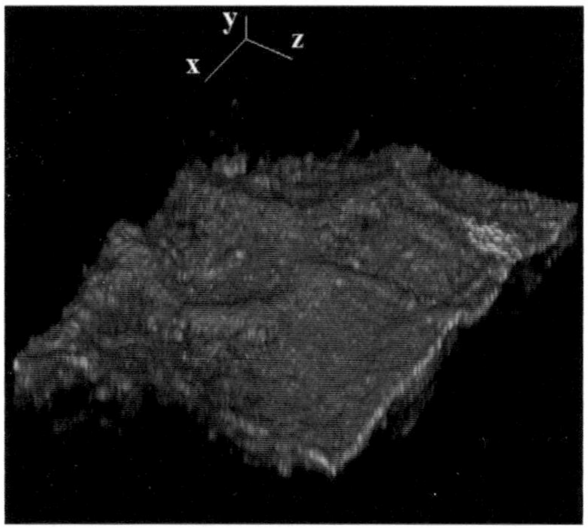

Figure 1. Reconstructed three-dimensional model in the scanned area

The 3D-patterns constructed permitted not only to view surface of the samples from different perspectives, but also to get tomograms of the objects in any section of interest. Therefore, 3D images made it possible to detect and identify certain structural elements of periderm of the samples visualized. Besides, three-dimensional visualization results offer the possibility to detect and measure structural elements, both observable in pictures of surface, and inaccessible for traditional microscopy. Thus, the initial stage of the research involved a healthy area of periderm. The still picture of surface obtained through a high resolution automatic Linnik interference microscope showed no defects or structural elements. However, upon viewing the three-dimensional pattern of the area concerned from the same perspective, a so called lenticel in the sample periderm was observed. This is a structural element participating in a metabolic process between the bulb and environment.

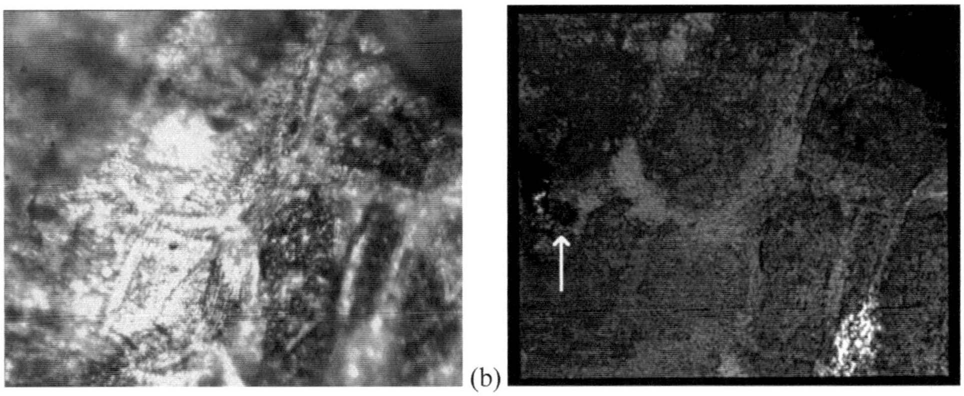

(a) (b)

Figure 2. Image of the surface observed in the microspore mode (a), image of a horizontal profile of a three-dimensional digital model (b)

Detection of the lenticel determined more detailed study of the area concerned, construction of tomogram in specific cross-sections, and further measurement of geometric parameters of the structural element detected. Based on the tomograms of the lenticel shown in Fig. 2 in different sections, the diameter of the lenticel was measured (4 μm).

These measurements let us conclude that the sample examined is healthy, since according to biological research [3], diameter of a lenticel of a healthy bulb of tuberous nightshade ranges from 3 to 8 μm depending on the grade, which complies with the measurements. Control over diameter of lenticels in the bulb provides diagnostics of viral and other diseases of tuberous nightshade. Thus, using a noninvasive optical coherent microscopy technique, we can watch changing in a state of lenticels of the same bulb, without interference with the process of disease. Therefore, the said technique is applicable not only for study of structure of plants, but also for research into different abnormities thereof.

5. Conclusions

The combination of two high-resolution digital optical methods for diagnostics of diseases and assessment sizes of periderm surface defect may be useful for investigations of other plants' periderm. The study of periderm by the high-resolving full-field optical coherence microscopy method showed that in order to observe the microstructure of layers located deeper from surface, it is necessary to use a more powerful source with emission in the near IR spectrum area and greater scanning depth. Also, in order to enhance the conditions of plant texture visualization, it is necessary to carry out additional studies of spectral characteristics of individual tissue textures of the object. We believe that the results of the study can be helpful to a wide range of specialists.

6. References

[1] Alarousu E, Gurov I, Kalinina N, Karpets A, Margariants N, Myllylä R, Prykäri T and Vorobeva E 2008 *Proc. SPIE of Advanced Laser Technologies* (Finland, Levi, 03 September 2007) 7022120

[2] Chirskaya V V, Margaryants N B, Zhukova E V 2014 The study of plant tissue by digital microscopy methods *Journal of Physics: Conference Series* **536** **1** 012014

[3] Khassanov V.T., Muranets A.P., Beisembina B., Fida M.A. 2014 Studying cyto-anatomical changes in tissues of virus infected Solanum tuberosum and Nicotiana tabacum plants *Zhivye I biokosnye sistemy* **9**

Achieving high visibility in subcarrier wave quantum key distribution system

V V Chistyakov, S V Smirnov, Yu V Nazarov, S M Kynev, A V Gleim

St.Petersburg National Research University of Information Technologies, Mechanics and Optics, Kronverkskiy pr. 49, St.Petersburg 197101, Russian Federation

E-mail: v_chistyakov@corp.ifmo.ru

Abstract. We study influence of quantum signal polarization distortions in the optical fiber on the interference pattern visibility in a subcarrier wave quantum key distribution system. An optical scheme of the polarization compensation unit is suggested, and dynamics of the QBER depending on the unit architecture is explored.

1. Introduction

Quantum cryptography allows distributing secure keys in a way that any eavesdropping in the channel is inevitably detected [1]. An important practical problem in the development of quantum cryptography systems is to enable their integration to the existing telecommunications lines. A subcarrier wave quantum key distribution system (SCW QKD) is one the most perspective solutions of that problem [2-5]. In this system a quantum signal is not generated directly by the source, but it is put to the sideband as a result of the phase-frequency modulation. Its advantages include high phase stability, broad multiplexing capabilities, and possibility to function in standard telecom optical cables.

It is well known that conventional optical fibers have birefringence, which randomly fluctuates in space and time. Moreover, the electro-optical phase modulators used, in practical implementations of fiber-optics phase-coding quantum key distribution systems, are sensitive to the polarization of light. Therefore, the distortions of the quantum signal leads to a decrease in visibility of the interference pattern and a subsequent raise of quantum bit error rate (QBER), which is used as a criterion for the detection of eavesdropping in QKD systems.

In order to compensate the polarization distortion introduced by the fiber in the SCW QKD system, an improved double-pass scheme was earlier proposed [6, 7]. The main principle was in connection of two electro-optical modulators in the receiving module with a Faraday mirror. After passing through the modulators, the signal was reflected from the Faraday mirror, where its polarization was rotated, and the signal was send back. In this design both polarizing modes were modulated, and the interference pattern visibility increased to 95%. However, a significant disadvantage of this setup was high insertion losses induced by two passes through modulators (about 6.5 dB, equal to additional 38 km of Corning ULL optical fiber [8]).

The aim of this work is the development of the optimal scheme of the polarization distortion compensation unit (PDCU) for SCW QKD, and studying the impact of polarization distortions on QBER and the visibility.

2. Theoretical analysis

In practice, the limiting factors for the security of quantum key distribution systems are the loss ratio in the channel, and QBER. It is known that an eavesdropper (Eve) is able to mask her actions by errors that inevitably arise in the process of key distribution. QBER value depends on the design of the optical circuit and the equipment used. In particular, the most important are the parameters of the receiving device, such as loss of system components. Threshold performance for the quantum key distribution system is an error rate of 6.4% for the four states with "strong reference" protocol [9]. The error rate of the cryptographic key in a four-state phase system is determined by the following formula:

$$QBER = \frac{1-V}{2} + \frac{p}{4\mu\eta\cdot 10^{-\frac{\alpha L+\beta}{10}}}, \quad (1)$$

where V is interference pattern visibility, β are the losses in the receiving module, p is the dark count probability per bit, μ is the mean photon number per pulse, α is the optical fiber attenuation coefficient at the central wavelength of the photon source, L is the optical fiber length and η is the detection efficiency.

There are two terms in (1): the first one depends on the quality of optical phase matching in the Alice (sender) and Bob (receiver) modules and is responsible for incorrect photon registration due to imperfect interference or depolarization. The second one characterizes the dark count as a fraction of all the registered counts. The latter increases with distance, since the number of photons reaching the detector decreases as transmission loss grows, whilst the dark count rate remains unchanged. The factor of four is introduced because the photons have a probability of being in the correct basis and a probability of returning from sidebands to the carrier. Thus, it is possible to form the lower threshold of the quantum bit error, in the absence of external factors [1]:

$$QBER_{min} = \frac{1-V_{mean}}{2}, \quad (2)$$

where V_{mean} is the mean value of visibility obtained from the experiment.

To assess the impact of distortions, a polarization controller (PC) was connected into the transmission channel. In order to measure the angles of induced polarization rotation, a standard protractor with a range of angles from 0 to 180 degrees was attached to each of the PC plates used for manipulation. In the experiment, we used in the step of 30 degrees. Thus, each of the plates had 7 possible positions. The combined system of two PC plates gave 49 options. In each position of the PC the value of constructive and destructive interference of signals in the quantum channel was measured, thus determined the visibility of the interference pattern [1]:

$$V = \frac{I_{max}-I_{min}}{I_{max}+I_{min}}, \quad (3)$$

where I_{max} – constructive interference signal's intensity, I_{min} – destructive interference signal's intensity.

3. Experimental results

For the experimental investigation of the mechanism of compensation of the polarization signal distortion in fiber-optic communication lines in SCW QKD we assembled experimental models of PDCU in three versions:

- SCW QKD system without any PDCU;
- SCW QKD system with double-pass PDCU described in the introduction;
- SCW QKD system with single-pass PDCU proposed in this paper.

Let us discuss the proposed scheme. The main principle of this subsystem is to divide signal coming from the communication line into two orthogonal polarization components for subsequent independent modulation of each polarization mode. This is achieved by the use of fiber polarization beam splitter with a division ratio of 50/50. As shown in the diagram (Fig. 1) after the beam splitter the polarization modes of the original signal follow two different arms. All fiber connections inside the system are polarization-preserving. Each arm contains a phase modulator, which provides a sinusoidal high frequency modulation signal, and optical circulator in conjunction with faraday mirror. We assume that in the first arm of a preferred direction in a lithium niobate crystal coincides with the polarization mode of the signal coming from the polarization beam splitter.

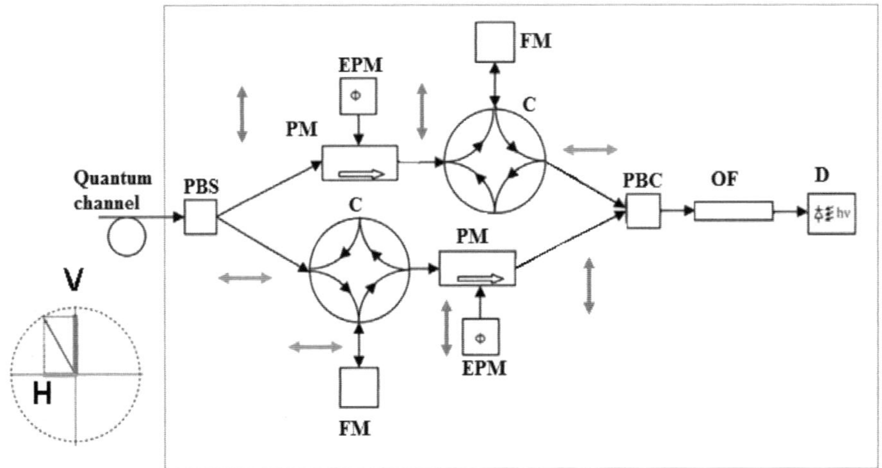

Figure 1. Principal scheme of receiving part of the subcarrier wave quantum cryptography system with single-pass polarization distortion compensation unit (PBS - polarizing beam splitter, PM – phase modulator, EPM – electrical phase modulator, PBC – polarizing beam combiner, C – circulator, OF – optical filter, D – detector. On the insets one can see signal polarization modes at different ports).

The second arm contains an important difference. The beam splitter divides the signal into two polarization modes. The mode distributed in the first arm coincides with the preferred direction in the phase modulator. After modulation it should be rotated by FM, so it becomes orthogonal to incoming mode. Therefore, the mode in the second arm needs to be rotated by 90° in order to coincide with the crystal axis and introduce no difference in modulation efficiency of the components. His rotation is achieved by a Faraday mirror and a circulator, as shown in Fig. 1. Thanks to this addition, both polarization modes undergo equal phase shifts, if we take into account delays between PMs in different arms and fix them by EPM. After the modulation, the pulse components from the shoulders are connected by a polarization maintaining optical coupler. And because of FMs they don't interfere with each other. Then the spectral filter separates sideband frequencies and the reference carrier, according to the SCW protocol. The result of sidebands interference from Alice and Bob is measured on a single photon detector.

The received experimental data in the form of six surfaces is presented on two graphs (Fig. 2, 3) corresponding to the values of the visibility and QBER. The average value of visibility of the interference pattern without compensation (purple surface) is 88%, which corresponds to QBER 6%. Thus, the QBER value is close to the theoretical limit, which significantly decreases the secure key rate. For the double-pass PDCU (blue surface) the average value of the visibility of the interference pattern is equal to 95.5%, which corresponds to QBER - 2.21%. Finally, the single-pass PDCU proposed in this work the average value of the visibility of the interference pattern is 98.8%, with QBER as low as 0.58%.

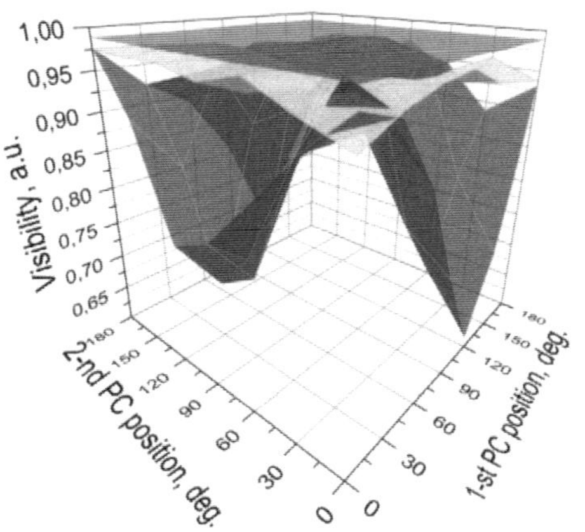

Figure 2. Dependence of the interference pattern visibility on polarization controller's position for SCW-QKD without any PDCU (violet), with double-pass PDCU (blue), with single-pass PDCU (red).

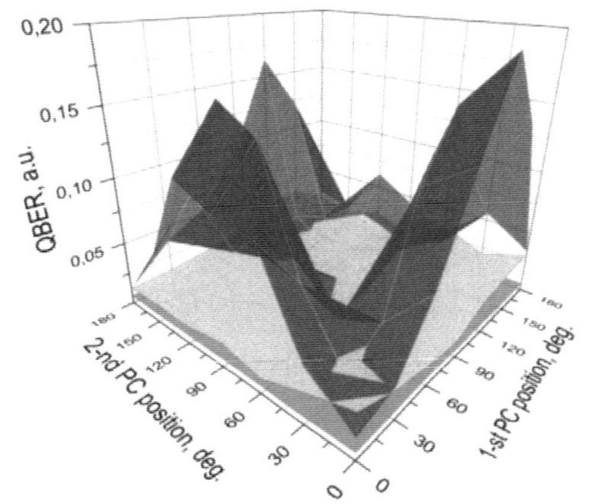

Figure 3. Dependence of QBER value on polarization controller's position for SCW-QKD without any PDCU (violet), with double-pass PDCU (blue), with single-pass PDCU (red).

4. Conclusion

The main motivation of this work lies in the necessity of integrating QKD devices into existing telecommunication environment. We developed a SCW QKD system module that makes it fully insensitive to polarization fluctuations and robust to environmental changes. The feasibility of single-pass polarization distortion compensation unit was proved experimentally using as criteria the interference pattern visibility that describes the quality of optical phase matching in the Alice and Bob modules; and QBER that is responsible for errors in the quantum key distribution process and robustness to the eavesdropping. The proposed module also decreases the insertion losses in the receiving party. This will allow increasing the distance of SCW QKD in standard telecommunication optical fibers.

Acknowledgements
This work was financially supported by Government of Russian Federation, Grant 074-U01.

References

[1] Scarani V, Bechmann-Pasquinucci H, Cerf N J *et al.* 2009 *Rev. Mod. Phys.* **81** 1301–1350
[2] Mazurenko Yu T, Mérolla J-M and Goedgebuer J-P 1999 *Optics and spectroscopy* **86** 181
[3] Guerreau O L, Mérolla J-M, Soujaeff A *et al.* 2003 *IEEE J. Selected Topics in Quantum Electronics* **9** 1533
[4] Bloch M, McLaughlin S W and Mérolla J-M 2007 *Opt. Lett.* **32** 301
[5] Rupasov A V, Gleim A V, Egorov V I and Mazurenko Yu T *Scientific and Technical Journal of Information Technologies, Mechanics and Optics* 2011 **02** 95
[6] Egorov V I, Gleim A V and Rupasov A V *Uchenye Zapiski Kazanskogo Universiteta. Seriya Fiziko-Matematicheskie Nauki* 2013 **155** 59–66
[7] Kynev S M, Ivanova A E, Egorov V I, Gleim A V, Rupasov A V and Chivilikhin S A 2012 *Nanosystems:Physics, Chemistry, Mathematics* 2012 **3** 92–98
[8] Corning SMF-28 ULL Product information http://www.corning.com/opticalfiber/products
[9] Guerreau O L, Malassenet F J, McLaughlin S W and Mérolla J-M *2005 IEEE Photon. Technol. Lett.* **17** 1755 - 1757

Terahertz generator on basis on basis of magnetic system with high localized magnetic field values

S E Azbite, A Kh Denisultanov, M K Khodsitsky

Department of Photonics and Optical Information Technologies, ITMO University, 49 Kronverksky pr., St. Petersburg, Russian Federation

E-mail: azbite@mail.ru

Abstract. In this paper terahertz generator design was proposed. In this THz generator magnetic system on permanent magnets was used to create which create localized magnetic field. Analytical calculation was made for this magnetic system and simulated was done for it. One can obtain point magnetic field with flux magnetic field value over 2.5 T due to the magnetic system. Beyond that analysis of influence of semiconductor crystals properties was done. The semiconductor crystals are used in the magnetic system for terahertz radiation generation.

1. Introduction

At the present time scientific researches in terahertz (THz) frequency range attract more interest. This is due to use of terahertz radiation in different fields of science and to applications in physics, biomedicine, defense and security [1-2] and metamaterials science [3-6]. The special interest is devoted to the terahertz time-domain spectroscopy (THz TDS) since amplitude and a phase may be simultaneously obtained. Scientific application for static THz TDS is spectroscopy of solid-state materials, water and aqueous solutions, semiconductor nanostructures [7]. Terahertz radiation is obtained by variety of ways, which include ultrabroadband THz pulse generation by a femtosecond laser. This method can be realized by photoconductive antennas, optical rectification or optical-induced breakdown [8].

2. Experimental facilities

The theoretical calculation of terahertz generator parameters was implemented for experimental facilities that is shown in figure 1 [9]. The femtoseconds-laser produces the laser beam that is splitted by two ones in the ratio of 10% to 90%. Probe beam transfers 10% energy and pump beam transfers 90% one. Optical delay line controls the pump beam optical path. This beam is modulated by chopper at the frequency of 667 Hz. The generator of THz radiation is based on the crystal in the magnetic system. After passing through IR radiation filter THz beam incidents on the sample. Then, THz radiation induces birefringence in electro-optical crystal. Wollaston prism splits the probe IR beam on two ones with orthogonal polarization, which are detected by balanced photodiodes. THz spectrometer has following parameters: the pump pulse power of 1W, the pump pulse duration of 200 fs, the frequency range is 0.1-1.2 THz, the average THz radiation power is 30 μW [10]. In this setup the Dember effect is used. The Dember effect is one of THz generation methods by illumination of ultrashort laser pulses [11]. It is effect of electric field origination in a semiconductor by illumination it owing to difference in

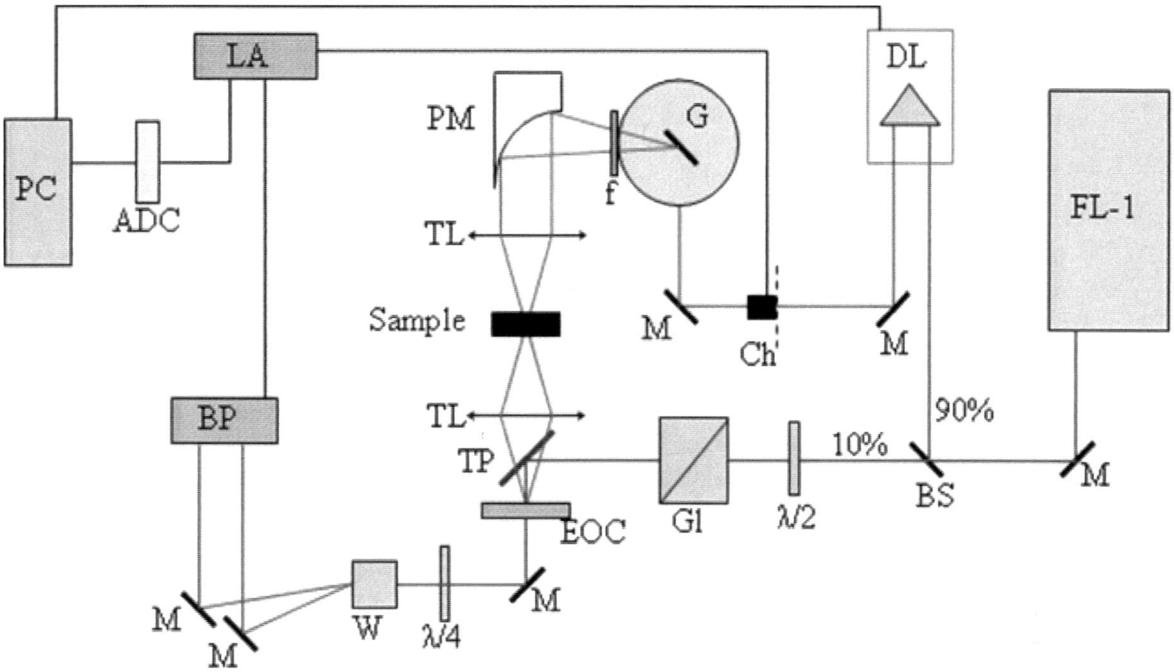

Figure 1. Scheme of the time-domain THz spectrometer in transmission mode. FL-1 is laser of femtosecond pulses, M are mirrors, BS is beam splitter, DL is optical delay line, G is generator of THz radiation based on the crystal InAs, PM are parabolic mirrors, TL are lenses for THz radiation, f is filter of infrared (IR) radiation, TP is translucent silicon plate, Ch is chopper, EOC is electro optical crystal of CdTe, W is Wollaston prism, BP is balanced photodiodes, LA is lock-in amplifier, ADC is analog-to-digital converter, PC is personal computer.

electrons and holes diffusion coefficient. With a strong absorption of laser pulses a strongly nonhomogeneous distribution of electron-hole pairs emerges in a surface layer of semiconductor. Charge carriers diffuse into a material. Inasmuch as electron mobility is more then hole mobility, holes diffuse slowly and in the surface layer charge separation and electromagnetic field occur. The field can be transformed in a free propagating emission. Oscillations of carrier density occur perpendicular to the semiconductor surface and emission effectiveness is not large. For more effective generation one applies matching elements like lenses, prisms or special methods for a turn of a directional radiation pattern of an elementary dipole at a tangent to semiconductor surface. For example, a strong magnetic field can be used as in the described experimental setup.

3. Magnetic systems

The magnetic system produces a magnetic field which is perpendicular to magnet surface. It does not change the magnitude of the carrier acceleration. The B field converts some of the current initially flowing along the normal z to surface to the component parallel to the surface x, when we consider a 3-dimensional space. The p-polarized THz field is described

$$E_p(t) \sim \theta(t)e^{-\frac{t}{\tau}}(\cos\omega_c t + \gamma\sin\omega_c t)\sin\theta, \tag{1}$$

where θ is angle, under that laser pulse impinges, t is time, τ is the mean scattering time, ω_c is cyclotron frequency, $\gamma = (\epsilon - \sin 2\theta)^{1/2}/\sin\theta$, is a parameter, which relates to the radiation efficiency. For typical photoconductive materials with $\epsilon \gg 1$ parameter $\gamma \gg 1$, the B field enhances THz emission [12]. The enhancement can be explored by defining a power-enhancement factor

$$\eta_p(\gamma, g = \mu_0 B) = \frac{\gamma^2}{2} + \frac{3}{2} + (\frac{1}{2} - \frac{\gamma^2}{2} + \gamma g)/(1+g^2) - (2+\gamma g)/(1+\frac{g^2}{4}), \quad (2)$$

which depends on the radiation factor $\gamma, g = \mu_0 B = \omega_c t$, where B is field strength and $\mu_0 = e\tau/m^*$ is carrier mobility. A strong enhancement is predicted for the actual THz dielectric constant, for example $\epsilon = 14.6$ for an InAs, but in the same conditionals little effect is expected for a material with $\epsilon = 1$. It is necessary to have a magnetic system with high magnetic field at the room temperature. In our paper the magnetic system was simulated NdFeB [13]. The system consists of two pairs of magnets (type 1 and type 2). The shape of type 1 magnet is represented a cone, and the shape of type 2 magnets is presented a hemisphere with a cut in the form of a cone. Both magnets are divided into eight equal sectors by radial planes passing through the axis of cones. Magnetization vector in each part is parallel to dihedral angle bisector plane and it is directed at the angle to cone axis. The magnetic field strength distribution is described by the formula:

$$H(\alpha, \phi_1, \phi_2) = [2\pi\sigma_{av}\sin 2\alpha + 2n \cdot 2M_s(\sigma_1 \sin\alpha + \sigma_2(1-\sin\alpha))] \cdot \ln(R/r), \quad (3)$$

where α is an angle at the cone vertex, φ_1 and φ_2 is the angles between the vector Ms of the first and the second magnets types, correspondingly, and the cone axis. For the segments number $n = 8$ the parameters $\alpha = 54°, \varphi_1 = -107°, \varphi_2 = 14°$ allow to obtain the most effective system design. The outward of the magnetic system and the numerical magnetic field distribution are shown in figure 2 a) and b) respectively. [14] As seen from figure 2, b) we can obtain the magnetic

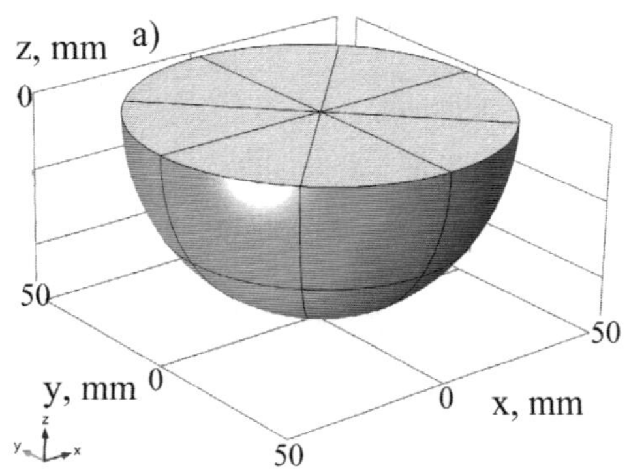

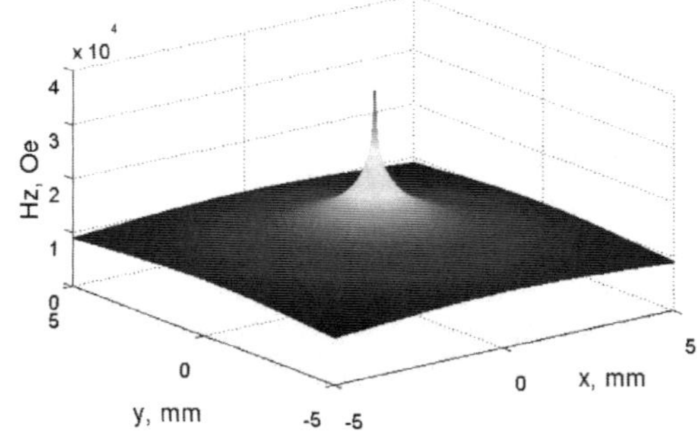

Figure 2. Suggested magnetic system: a) outward and b) analytical magnetic field strength distributions.

field B above 2.5 T by this system design at the room temperature. The magnetic system is a part of a THz radiation generator.

4. Semiconductor crystals in THz generator

The THz radiation generator is shown on figure 3. The pumping beam is generated by the femtosecond laser Yb3+:KYW and goes through a lens that focuses this beam on the semiconductor InAs crystal than THz radiation and reflected IR radiation go through IR-filter and THz collimating lens. InAs crystal is placed into the area of the maximum external

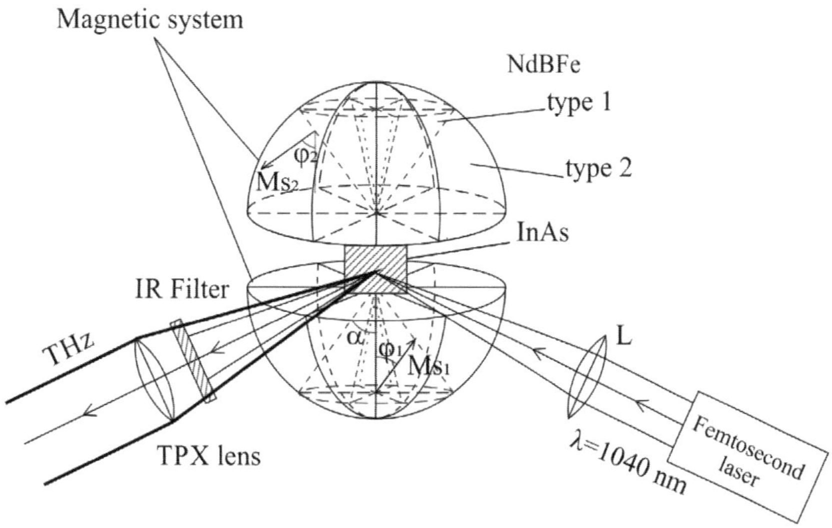

Figure 3. Scheme of the THz radiation generator.

magnetic field that is caused by the magnetic system [15]. The parameters of the THz generator are following: the wavelength of the pump radiation, $\lambda_p = 1040nm$, the THz radiation wavelength generated by the InAs-crystal, $\lambda_g = 0,3..3mm$, the focus point, f, the diameter of the pump beam, $D_p = 2mm$, the diameter of the spot at the focal point on the InAs-crystal surface $d = 1,27f\lambda_p/D_p$ (in our setup $d = 1,5mm$), the diffraction-limited beam divergence, $\theta_d = \frac{1}{\pi}\lambda_g/(d/2)$, the diameter of the generated THz beam at the distance L: $D_g = 2L\theta_d$. As seen from the formula (2), the power-enhancement factor depends not only on B-field but on semiconductor THz dielectric constant, ϵ, and inferred carrier mobility, μ_0. Usually for THz generation one uses semiconductor crystal the following GaSb, InP, GaAs, InAs, InSb. The properties are presented in table 1. The power-enhancement factor dependence is shown on

Table 1. Properties of semiconductors

Property	GaSb	InP	GaAs	InAs	InSb
THz dielectric constant ϵ	15.7	12.4	13.2	14.6	17.7
Inferred carrier mobility, $\mu_0, cm^2/Vs$	2200	5600	9300	8000	19000

figure 4 for the semiconductors. Carrier mobility $\mu 0$ differs for each semiconductor, for this reason on the absciss $g = \mu_0 B$ is, points on plots mean value of the power-enhancement factor for magnetic field B 2.5 T. As seen from figure 4, the most effective semiconductor for THz generation is InSb.

5. Conclusions

Therefore, in this paper the development of the magnetic system with point type magnetic field for terahertz time-domain spectroscopy was developed and simulated. The magnetic field approaches to the values higher 2.5 T. It was shown that this system improve efficiency THz

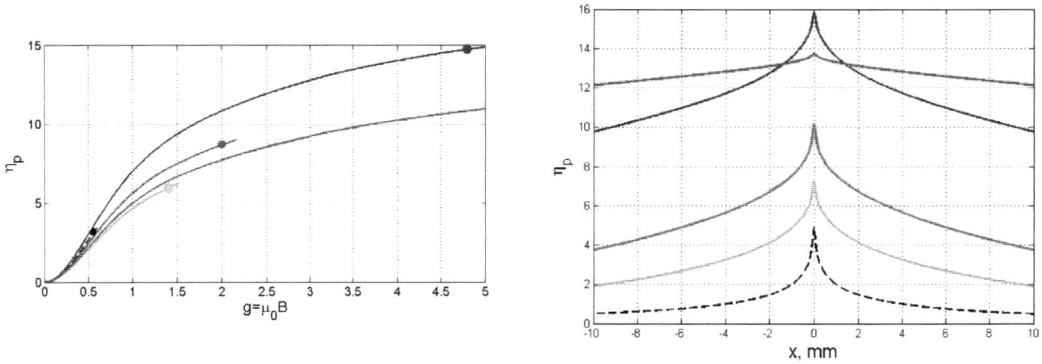

Figure 4. Dependence of the power-enhancement factor a) on magnetic field B and carrier mobility μ_0 and b) on coordinate x at the surface of the magnet; blue line is InSb, black line is GaSb, purple line is InAs, red line is GaAs, green line is InP.

radiation generation. The power-enhancement factor was calculated for typical semiconductors which are often used for the generation. The most efficiency material is InSb.

Acknowledgments
This work was financially supported by Government of Russian Federation, Grant 074-U01.

References
[1] Plusquellic D, Siegrist K, Heilweil E and Esenturk O 2007 *Chemphyschem* **8** 2412–31.
[2] Jepsen P, Cooke D, and Koch M 2011 *SLaser Photonics Rev*, **5**, 124–166.
[3] Gurvitz E A, Vozianova A V, Khodzitsky M K 2014 *J. of Physics:Conference Series* **541(1)** 012067.
[4] Denisultanov A K, Khodzitsky M K 2013 *SPIE NanoScience + Engineering*, **8806** 88062I.
[5] Terekhov Y E, Khodzitsky M K, Grachev Y V, Sedykh E A, Belokopytov G V, Zhang X C 2013 *SPIE NanoScience+ Engineering* **8806** 88062Q.
[6] Girich A, Khodzitsky M, Nedukh S, Tarapov S 2011 *Terahertz and Mid Infrared Radiation* 159–164.
[7] Bespalov V et al 2008 *J. of Opt. Tech.* **75** 636–642.
[8] Mittleman D et al 1999 *App. Phys.*, **68** 1085–1094.
[9] Denisultanov A K et al 2014 *Prog. in Elect. Res. Sym., PIERS* 2022–2025.
[10] Bespalov V et al 2002 *Opt. and Spect.* **93** 148–152.
[11] Zhang X-Ch and Jingzhou Xu 2010 *Introduction to THz wave photonics* (New York: Springer)
[12] Shan J et al 2001 *Optics letters* **26** 849–851.
[13] Azbite S E, Denisultanov A K, Khodzitsky M K 2015 Magnetic system design for investigation of magneto-optical properties of objects in strong magnetic field in terahertz frequency range *J. Phys.: Conf. Series* **643** 012065 (*Preprint*)
[14] Samofalov V, Belozorov D, Ravlik A, 2006 *The Phys. of Metals and Metallogr.* **102** 494–505.
[15] Azbite S E, Denisultanov A K and Khodzitsky M K 2015 Development of magnetic system with high-anisotropy localized magnetic field for terahertz time-domain spectroscopy. *SPIE Optical Engineering+ Applications* 95850Z–95850Z. International Society for Optics and Photonics.

Modeling and experimental demonstration of terahertz frequency tunable metamaterial absorber

D A Gomon[1], E A Sedykh[1], V V Gill[1], K I Zaitsev[2], M K Khodzitsky[1]
[1] ITMO University, Saint-Petersburg, Russia
[2] Bauman Moscow State Technical University, Moscow, Russia

E-mail: GomonDA89@ya.ru

Abstract. The aim of this research is design of tunable narrowband THz metamaterial absorber, which achieves a resonant absorptivity of 100%. This structure is a promising candidate for selective reflective mirror for THz narrowband absorption.

1. Introduction

At a recent time scientists make huge efforts in research and creation of electromagnetic (EM) materials and new alternative methods of devices components design based on metamaterials [1]. Metamaterials (MM) are artificial composite structures, which EM properties originate from oscillating electrons in unit cell composed of conductive definite form of components, made of metal e.g. copper or gold. The effective electromagnetic properties of metamaterials at any frequency can be engineered to take on arbitrary values, including those not appearing in nature, such as negative permittivity, negative permeability, negative refractive index. It allows obtaining unusual physical effects such as negative refraction [2], backward wave, superlens, superprism effects, etc [3-5]. Sub wavelength metamaterial unit cell duplicates to forming material, which allows receiving resonance response of metamaterial electric and magnetic properties. Application metamaterials for THz frequency range is perspective, because THz radiation could be applied for different science researches and practical applications, such as security systems, detectors, microobjects imaging, different medicine applications, communication systems, etc [6-13].

Design, fabrication and characterization of metamaterial absorbers (MA) were shown in reports [14-18]. Our work is devoted to development of ideal (absorptivity up to unity) metamaterial absorber for terahertz frequency range.

2. Simulation

The structure based on ERR was chosen for creation of tunable absorber for terahertz frequency range, analogous to structures of reports provided earlier. For simulation the parameters of structure were chosen in such a way as to achieve the absorptivity of A=0.99 at resonant peaks in the frequency range from 0.25 to 1.75 THz. Unit cell of proposed absorber is illustrated in Fig. 1; it consists of 4 layers:

1) silicon substrate;
2) lower metallic copper film, to operate as reflective layer;
3) dielectric layer of polymide;
4) copper double ERR.

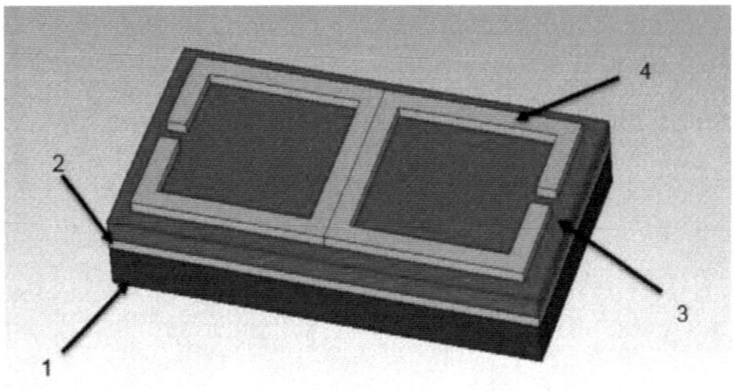

Figure 1. Model of MA unit cell, perspective view.

Table 1. Parameters of the MA unit cell..

layer	Material	Layer thickness h, μm
1	Silicon	320
2	Copper	0.5
3	Photoresist SU-8	12
4	Copper	0.5

Design of the MA unit cell with double ERR is illustrated in Fig. 2. It consists of two ERR connected to each other by one side. They could be defining by a set of parameters:

- g – a ERR gap;
- w – a ERR lines width;
- a, b – a ERR length and width;
- x, y– a MA unit cell length and width;
- p – a distance between ERR edge and unit cell border.

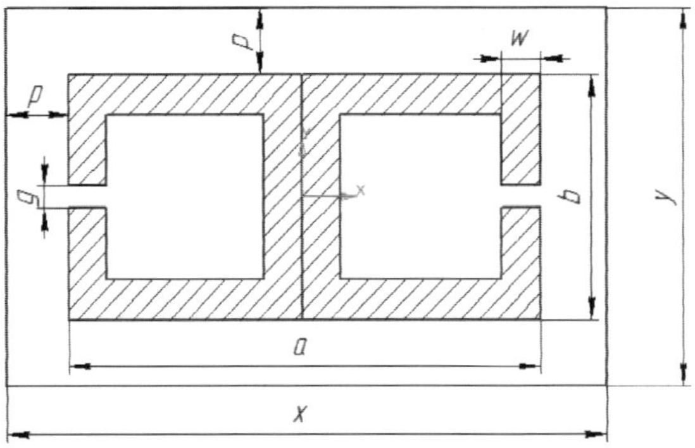

Figure 2 – Design of MA unit cell, top view.

The influence of polymide layer thickness on structure spectrum is shown in Figure 3. Maximal absorptivity is achieved at the thickness of 18 μm; the bigger difference between optimal and real layer thickness, the lower absorptivity. This is an implicit dependence between layer thickness and peak frequency.

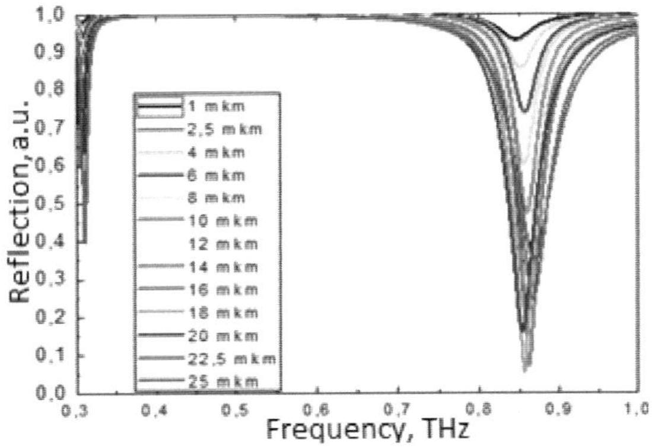

Figure 3. Influence of polymide layer thickness on structure spectrum (in case of TE-polarization).

Table 2 presents 5 variants of MA with different geometric parameters.

Table 2. Geometric parameters of simulated MA

	Geometric parameters				
№ MA sample	a, μm	b, μm	g, μm	p, μm	w, μm
1	180	90	10	10	12
2	220	110	10	10	18
3	260	130	10	10	25
4	300	150	10	10	32
5	340	170	10	10	40

All variants of MA were optimized during the simulation. They demonstrate high absorption of THz radiation with absorptivity approximately 1 (Fig. 4). The enlargement of geometrical parameters redshifts peak frequency.

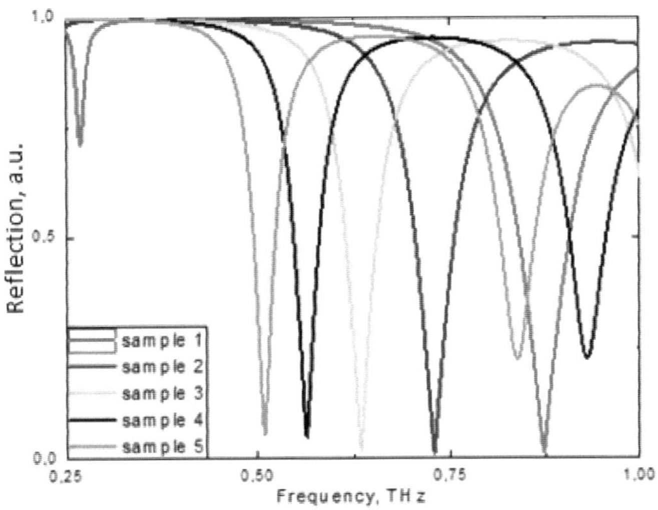

Figure 4. Spectra of 5 samples with geometric parameters from Table 2 (in case of TE-polarization).

3. Sample fabrication

Process of experimental samples manufacturing consists in the following:

1. substrate preparing: firstly thin chrome layer (20 nm) and after that copper layer (500 nm) was sputtered on silicon substrate, using ion-vacuum spattering method;
2. liquid polyimide solution SU-8 was sputtered using centrifuge;
3. photoresist layer was applied by centrifuge and image was formed by photolithography;
4. another copper layer was sputtered (0,5 mm) as electronic resonant rings;
5. photoresist deleting.

Process of MA sample fabrication schematically is sketched in Fig. 5. Microimage of fabricated sample presented in Fig. 6.

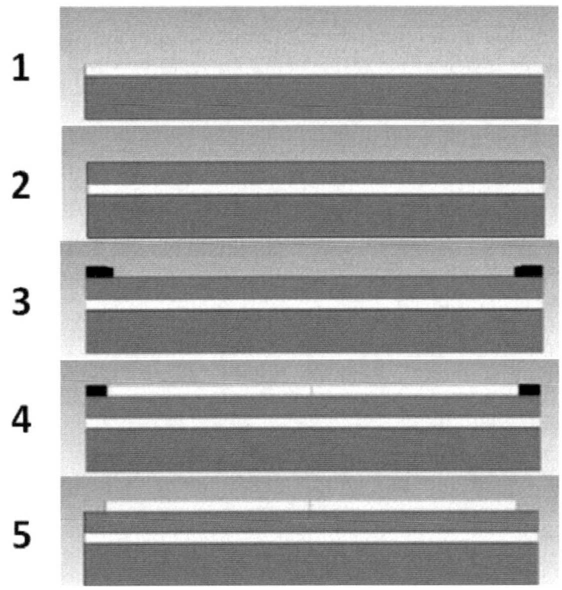

Figure 5. 5 steps of MA sample fabrication (violet - silicon, yellow - copper, red - polyimide, black - photoresist).

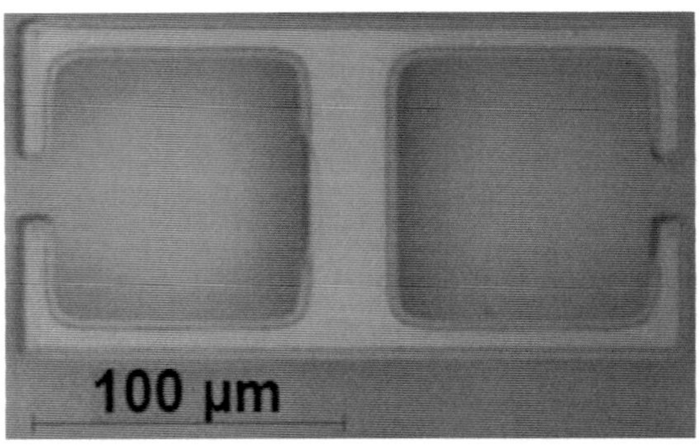

Figure 6. Microimage of fabricated MA sample with set of parameters: a=208μm, b=98μm, p=13μm, g=23μm, w=6μm.

4. Experiment

In the present work compact TPS (terahertz pulsed spectroscopy) setup was used (analogous setup was used [19]). The setup operates in reflection mode and allows measuring the THz spectra between 0.25 and 1.75 THz. Fig. 7 represents the experimental setup design and illustrates the process of TPS waveform detection.

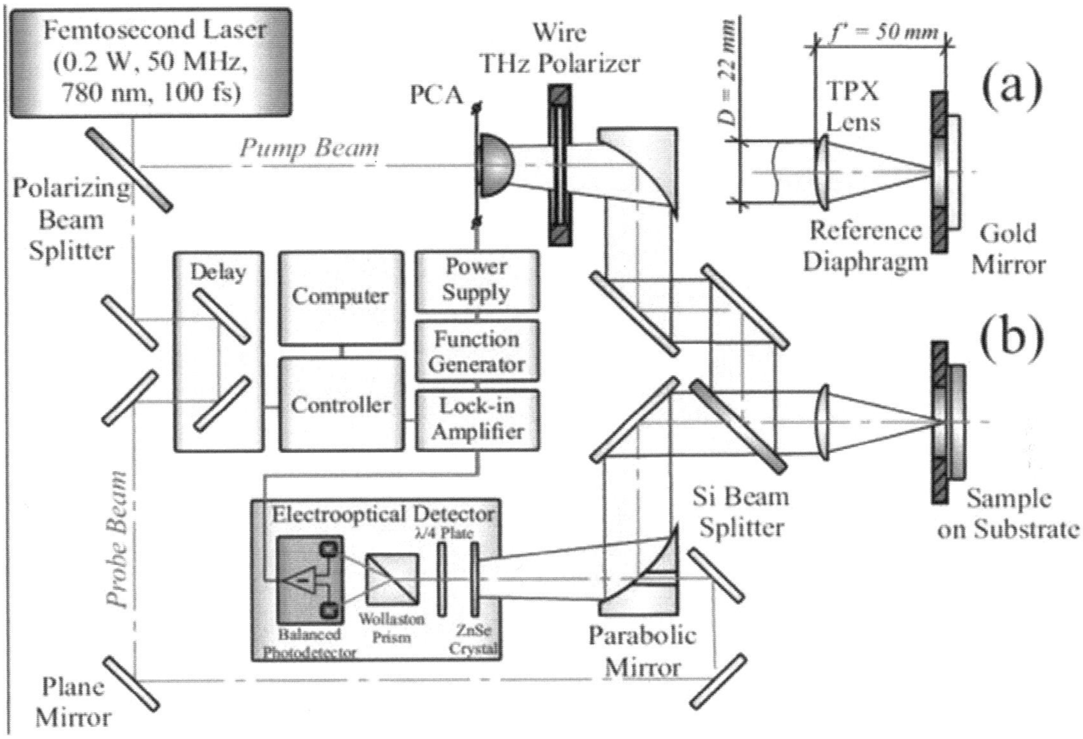

Figure 7. Design of reflective THz spectrometer.

The spectrum of the fabricated MA sample (with set of parameters: a=208μm, b=98μm, p=13μm, g=23μm, w=6μm) was investigated using reflective time-domain THz spectrometer [19] (Fig. 8).

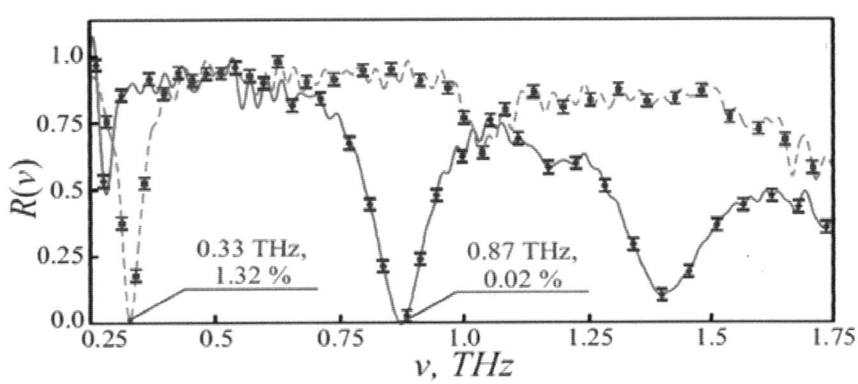

Figure 8. Spectrum of fabricated MA sample for TE-polarization (blue curve) and TM-polarization (red curve).

So absorption 99,98% was demonstrated at the frequency 0.87 THz for TE-polarization (blue curve) and 99,67% at the frequency 0,33 THz for TM polarization (red curve).

5. Conclusions

The results of geometrical parameters and polyimide layer thickness influence on absorption spectrum of MA based on ERR were shown by numerical method FDTD:

- theoretical research of polyimide layer thickness influence on absorption spectrum shows that maximal MA absorptivity is up to 100% at layer thickness of 18 μm;
- enlargement of MA geometrical parameters redshifts the absorption resonant frequency, it's caused by the increasing of ERR inductance.

Experimental results show absorption of 99,98% at the frequency of 0,87 THz for TE-polarization and 98,68% at the frequency of 0,33 THz for TM-polarization.

Acknowledgements
This work was supported by Government of Russian Federation (Grant 074-U01).

References

[1] Gurvitz E. A., Vozianova A. V., Khodzitsky M. K. 2014 *Journal of Physics: Conference Series* **541(1)** 012067
[2] Khodzitsky M. K., Kharchenko A. A., Strashevskyi A. V., Tarapov S. I. 2009 *Telecommunications and Radio Engineering* **68(7)** 561-566
[3] Withayachumnankul W, Abbott D 2009 *IEEE Photonics J.* **1** 99
[4] Khodzitsky M. K., Kharchenko A. A., Strashevskyi A. V., Tarapov S. I. 2009 *Telecommunications and Radio Engineering* **68(7)** 561-566
[5] Gurvitz E. A., Vozianova A. V., Khodzitsky M. K. 2014 *Journal of Physics: Conference Series* **541(1)** 012067
[6] Mitrofanov O., Brener I, Harel R., Wynn J. D., Pfeiffer L. N., West K. W., Feder C. I. 2000 *Applied Physics Letters* **77(22)** 3496
[7] Lu Q. Y., Slivken S., Bandyopadhyay N., Bai Y. and Razeghi M. 2014 *Appl. Phys. Lett.* **105** 201102
[8] Nagel M., Haring Bolivar P., Brucherseifer M. and Kurz H. 2002 *Applied Physics Letters* **80** 154
[9] Wood C., Cunningham J., Upadhya P. C., Linfield E. H., Hunter I. C., Davies A. G., Missous M. 2006 *Applied Physics Letters* **88** 142103
[10] Denisultanov A. K., Khodzitsky M. K. 2013 *SPIE NanoScience + Engineering* **8806** 880621
[11] Terekhov Y. E., Khodzitsky M. K., Grachev Y. V., Sedykh E. A., Belokopytov G. V., Zhang, X. C. 2013 *SPIE NanoScience+ Engineering* **8806** 88062Q
[12] Girich A., Khodzitsky M., Nedukh S., Tarapov S. 2011 *Terahertz and Mid Infrared Radiation* 159-164
[13] Hatem O., Cunningham J., Linfield E. H, Wood C. D., Davies A. G., Cannard P. J., Robertson M. J., Moodie D. G. 2011 *Applied Physics Letters* **98** 121107
[14] Tao H, Landy N I, Bingham C M, Zhang X, Averitt R D & Padilla W J 2008 *Optics express* ***16*(10)** 7181
[15] Hu Tao et al 2010 *J. Phys. D: Appl. Phys.* **43(22)**. 225102
[16] Hu F, Wang L, Quan B, Xu X, Li Z, Wu Z, & Pan X 2013 *Journal of Physics D: Applied Physics* **46(19)** 195103
[17] Huang, Li, et al 2012 *Optics letters* **37(2)** 154
[18] Wang B X, Zhai X, Wang G Z, Huang W Q, & Wang L L 2015 *Optical Materials Express* **5(2)** 227
[19] Zaitsev, K. I., et al. 2015 *Optics and Spectroscopy* **119(3)** 404-410

Influence of creatinine and triglycerides concentrations on blood optical properties of diabetics in THz frequency range

Gusev S I [1], Balbekin N S [1], Sedykh E A [1], Kononova Yu A [2], Litvinenko E V [2], Goryachuk A A[1], Begaeva V A[1], Babenko A Yu [2], Grineva E N [2],Khodzitsky M K [1]

[1] Department of Photonics and Optoinformatics, ITMO University, Kronverkskiy pr. 49, St. Petersburg, 197101, Russia

[2] Federal Almazov North-West Medical Research Centre Parkhomenko st. 15, St. Petersburg, 194156, Russia

E-mail: mail@gusev-spb.ru

Abstract: This paper is devoted to research influence of creatinine and triglycerides on THz optical properties of blood for diabetes care purposes.

1. Introduction

The amount of diabetics is growing every day. The number of patients suffering from diabetes had amounted to more than 230 million people by the year 2010, which is estimated as 3.5% of the whole world adult population [1]. According to expert forecasts, this number is projected to double by the year 2025, which is going to be 7% of whole Earth population. It was calculated that every 10 seconds someone in the world dies due to diabetes and its complications that consists 3 million people per year. Late diagnosis of diabetes and poor control are the main causes of diabetes complications, which determine the morbidity and mortality of diabetics. The escalation of hypoglycemia, even subclinical, associated with an increase in cardiovascular mortality by 2-2.5 times, and cardiovascular events 4 times [2]. At the same time, episodes of hypoglycemia were present in 39% of study patients with CGMS control. Morbidity and mortality from macrovascular diseases also highly associated with different characteristics of blood glucose levels: hyperglycemia, hypoglycemia, glycemic variability. Thus, increasing of the glycated hemoglobin level from 6% to 9% provides the microvascular complications risk increasing in 4 times [3].

At the moment, the blood glucose level measurements are performed by glucometers [4,5]. This method requires making a finger puncture for every measurement. About five punctures per day should be done for proper glucose monitoring, which is about 1800 punctures per year. Besides, each measurement by glucometer requires a distinct test strip. Expenses for 1800 test strips could be estimated as about 450 euros per year. It is also necessary to take into account that each puncture has a risk of blood poisoning. Using non-invasive techniques for glucose level control could reduce the amount of possible risky manipulations by 1800 per year. Moreover, it is worth mentioning that only eight of ten fingers are suitable for puncturing, and the constant skin damage which cannot be avoided is quite annoying for the patients.

Last decade the most popular topics in terahertz science is a development of THz filters [6, 7], which may be used in biomedicine diagnostics due to change of filter resonant frequency caused by the surrounding. At pathology of blood it may be revealed the changing of relaxation time of biomolecules namely the varying of dispersion of refractive index of whole blood in terahertz frequency range. However it needs to take in account the influence of THz radiation on biological objects [8,10]. The previous results of this scientific work were devoted to construction calibration graph for detecting glucose levels in blood using refractive index in order to control diabetes [9]. The main aim of this investigation is to determine the influence of creatinine and triglycerides on THz optical properties of blood for diabetes care purpose.

2. The investigated biological objects overview

The samples with different concentration of creatinine and triglycerides were chosen for investigation because of simple detection these components levels in diagnostic laboratories using common tests. Low female level of creatinine concentration is considered to vary from 53.0 to 70.7 μmol/l (equiv. 954 to 1272.6 μg/dl), normal level from 70.8 to 106 μmol/l (equiv. 1274.4 to 1908 mg/dl) of the blood plasma (this type of glucose concentration measuring system is generally used in medical practice). Also low level of triglycerides concentration is considered to be lower than 1.77 mmol/l (equiv. 31.86 mg/dl), normal female level higher than 1.77 mmol/l (equiv. 31.86 mg/dl) of the blood plasma.

Compared samples had almost same compositions of blood plasma, but they had significant differences in creatinine or triglycerides levels. In addition, the results were observed at different glucose levels in blood plasma. The concentration of components was determined by professional medical laboratory of Federal Almazov North-West Medical Research Centre [11].

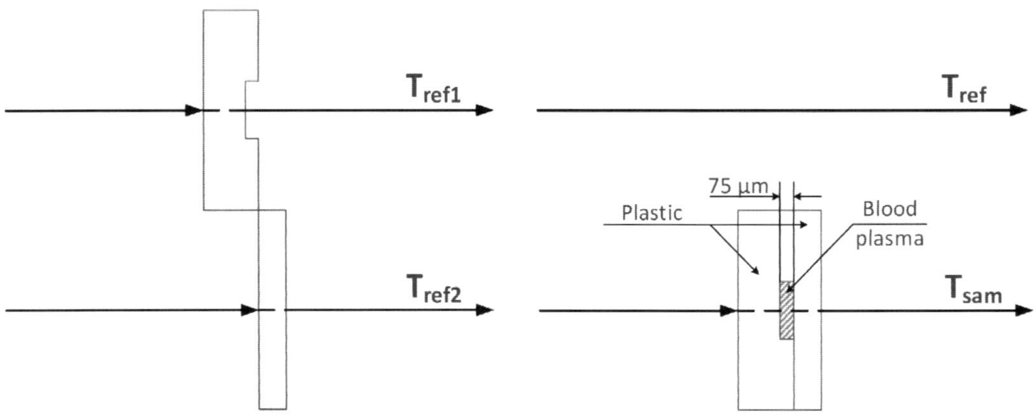

Figure 1. The schematic sample preparing for measurement process.

The sample consists on “Goryaev camera” type container (see the Figures 1, 2) with drop of blood plasma. This type of container provides us stable thickness of sample and protects drying of biological fluid.

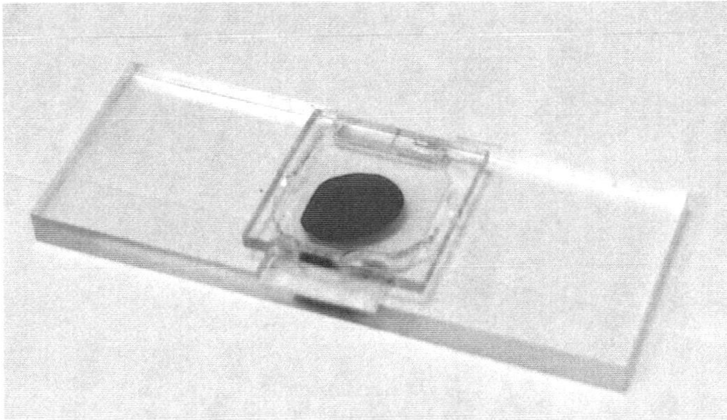

Figure 2. "Goryaev camera" type container with blood.

The blood plasma samples were compared because blood contains solid inclusions that distort the signal, while the plasma has uniform consistency.

3. Experimental setup

The optical properties of blood plasma samples were studied in the frequency range of 0.1-1 THz using the time-domain spectrometer in transmission mode. The photo and scheme of the setup were shown in Fig. 3 and Fig. 4 correspondingly.

Figure 3. Pulsed THz spectrometer at the THz Biomedicine Laboratory, operating in the transmission mode.

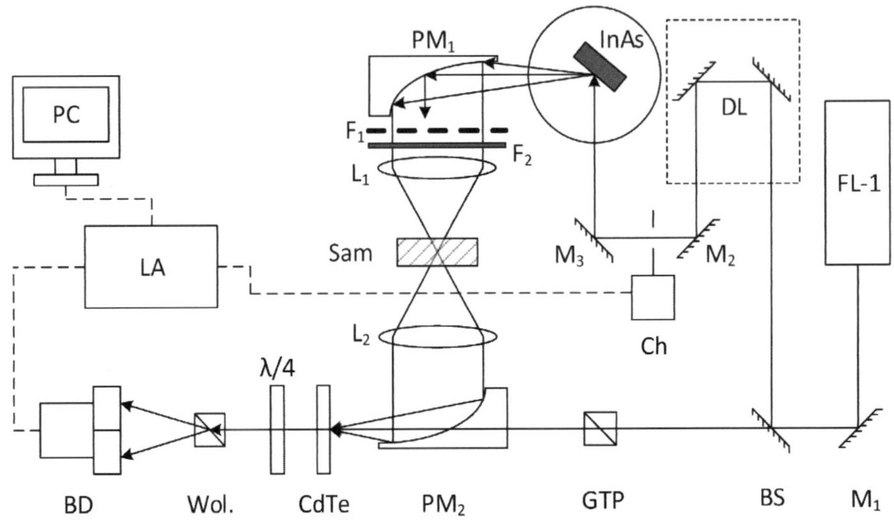

Figure 4. Schematic diagram of the set up (FL-1 - femtosecond laser based on potassium-yttrium tungstate crystal activated with ytterbium (Yb: KYW), generating femtosecond pulses; $F_{1,2}$ – a set of teflon filters for IR wavelength range cutting off, BS – beamsplitter, DL – optical delay line, $M_{1,2,3}$ – mirrors, Sam – investigated sample, Wol. – Wollaston prism, CdTe – electro optical cadmium-telluric crystal, BD – balanced detector, LA – lock-in amplifier, PC – personal computer, GTP - Glan-Taylor prism, $PM_{1,2}$ – parabolic mirrors, Ch - chopper.

Broadband pulsed THz radiation was generated using a photoconductive antenna (undoped InAs crystal) by irradiating it with femtosecond pulses of an ytterbium laser (wavelength - 1040 nm, pulse duration - 120 fs, pulse repetition frequency - 75MHz, power - 1 W). THz radiation had the following output characteristics: spectral range from 0.05 to 2 THz, average power - 30 uW, pulse duration - 2.7 ps. The main power was concentrated at the frequency range from 0.12 to 1.1 THz. THz radiation was generated by a gallium arsenide crystal in the magnetic system, then passed through a teflon filter (which cuts wavelengths shorter than 50 um). After that, the radiation passed through the sample fixed in a focal plane perpendicularly to the beam. With the simultaneous incidence of the femtosecond probe beam and THz beam passed through the sample on the electro optical CdTe crystal, THz pulse induced birefringence of the probe beam in the crystal due to the electrooptical effect. The birefringence magnitude is directly proportional to the intensity of terahertz wave electric field in the time point $E(t)$. These data are required to calculating $E(\omega)$ using Fourier transform.

4.Results

During the experiments the signals passed through an object and air were recorded (reference signal through air, object signal through bottom of camera, object signal through top of camera, object signal through camera filled blood plasma). Each of them is the dependence of the amplitude on time.

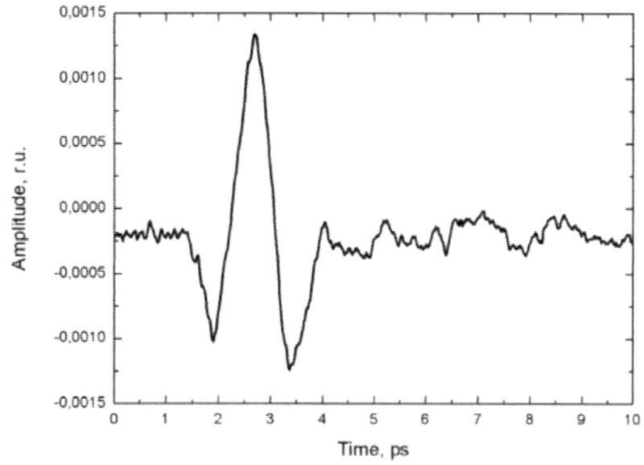

Figure 5. Example of typical THz temporal waveform.

Further, the THz waveforms were processed using Spectrina M5 program (see the Figure 6) which performs the Fourier transform and the extraction of optical properties [12, 13-16].

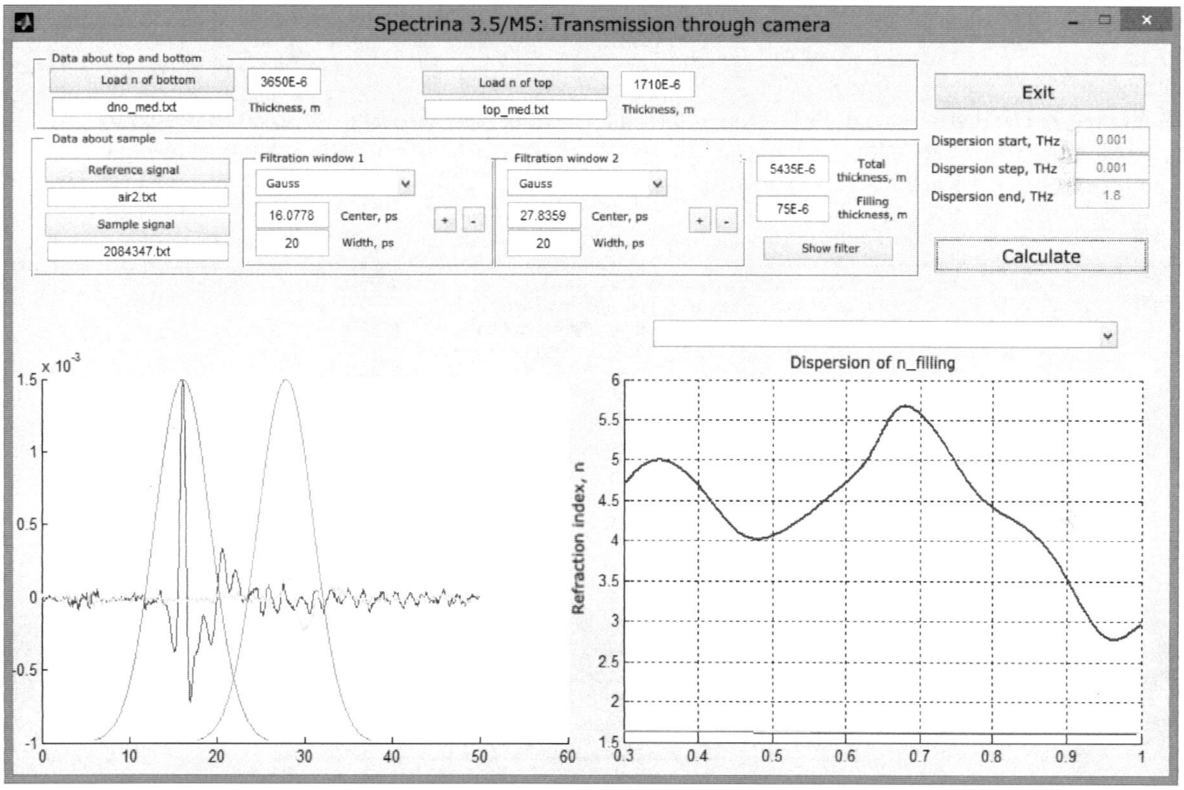

Figure 6. Screenshot of the Spectrina program interface.

As a result, such properties of the blood plasma were obtained: dispersion of the complex refractive index, the absorption coefficient, the penetration depth, the complex permittivity.

After consideration of results, the comparison was made between the refractive indices as a function of concentration categories of creatinine and triglycerides in the sample.

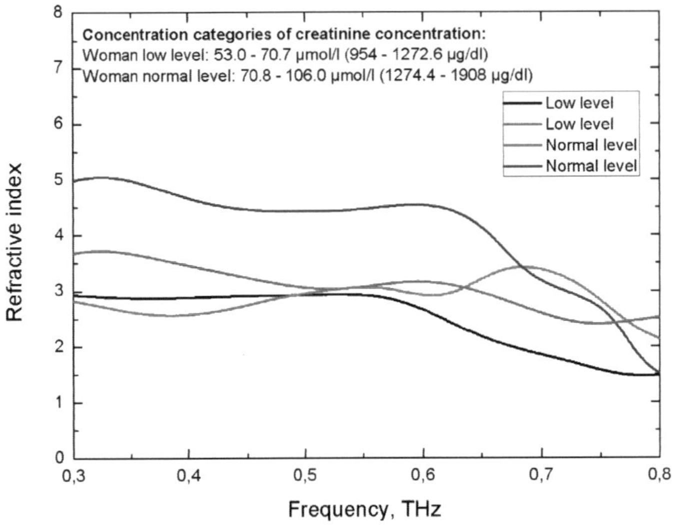

Figure 7. The dependence of dispersion of the real part of the blood plasma refractive index on the creatinine concentration level at woman. The concentration of glucose in these samples is located between 4.5 – 5.5 mmol/l (81 – 99 mg/dl).

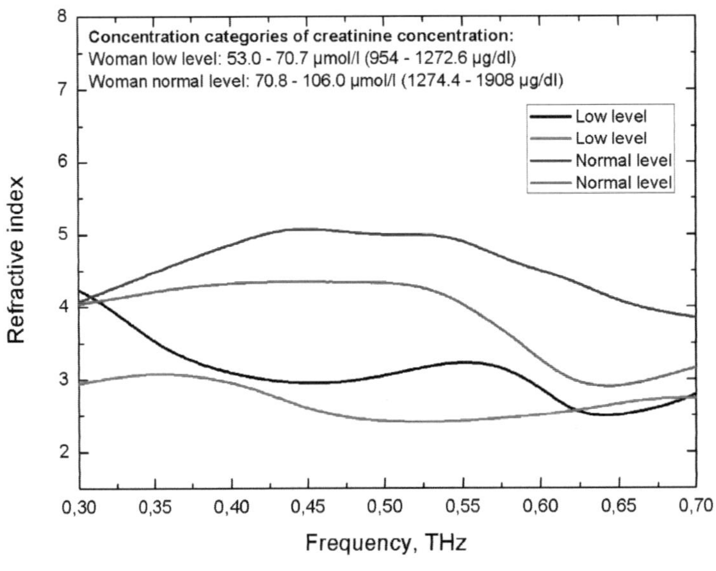

Figure 8. The dependence of dispersion of the real part of the blood plasma refractive index on the creatinine concentration level at woman. The concentration of glucose in these samples is located between 8.5 – 9.5 mmol/l (153 – 171 mg/dl).

Journal of Physics: Conference Series **735** (2016) 012088 doi:10.1088/1742-6596/735/1/012088

As you can see, the increasing of creatinine concentration provides increasing the blood refraction index. This influence is evident at both glucose concentration groups.

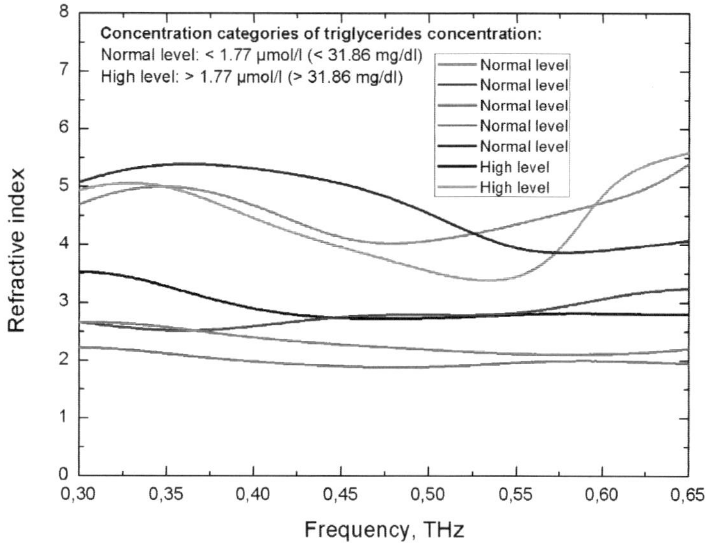

Figure 9. The dependence of dispersion of the real part of the blood plasma refractive index on the triglycerides concentration level. The concentration of glucose in these samples is located between 4.5 – 5.5 mmol/l (81 – 99 mg/dl).

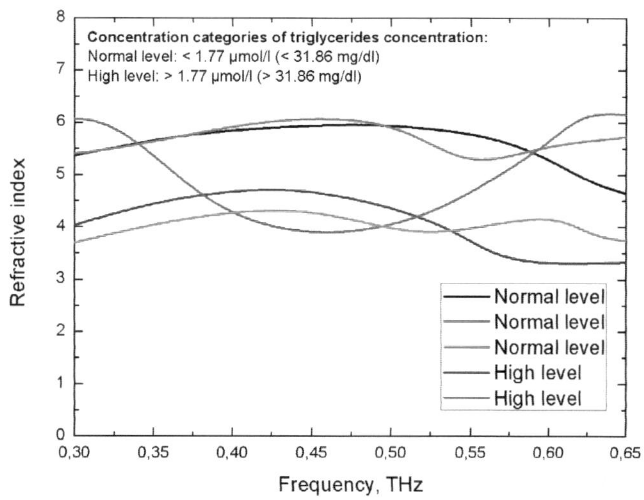

Figure 10. The dependence of dispersion of the real part of the blood plasma refractive index on the triglycerides concentration level. The concentration of glucose in these samples is located between 6.5 – 7.5 mmol/l (153 – 135 mg/dl).

Journal of Physics: Conference Series **735** (2016) 012088 doi:10.1088/1742-6596/735/1/012088

These graphs show us that the increasing of triglycerides concentration have not specific dependence on the blood refraction index.

5. Conclusion

The dependence of the refractive index of blood on the creatinine concentration was shown. The dependence of the refractive index on the triglycerides levels was not observed. There results confirm that each patient required individual calibration of perspective glucometric device. These data may be useful for development of non-invasive terahertz diagnostic methods of diabetes.

Acknowledgment

This work was financially supported by Government of Russian Federation, Grant 074-U01.

References

[1] Braun K. et al. 2012 *Diabetes research and clinical practice* **98.1** 164-168

[2] Stratton I. et al. "Association of glycaemia with macrovascular and microvascular complications of type 2 diabetes (UKPDS 35): prospective observational study," Bmj, 321.7258, 405-412 (2000).

[3] Zoungas, Sophia, et al. "Severe hypoglycemia and risks of vascular events and death," New England Journal of Medicine 363.15, 1410-1418 (2010).

[4] Cohen M. et al. "A comparison of blood glucose meters in Australia," Diabetes research and clinical practice Vol. 71(2), 113-118 (2006).

[5] Engel L., Delaney C., Cohen M. "Blood glucose meters: an independent head-to-head comparison," Practical diabetes international Vol. 15(1), 15-18 (1998).

[6] Terekhov Y. E., Khodzitsky M. K., Grachev Y. V., Sedykh E. A., Belokopytov G. V., Zhang, X. C. 2013 *SPIE NanoScience+ Engineering* **8806** 88062Q

[7] Girich A., Khodzitsky M., Nedukh S., Tarapov S. 2011 *Terahertz and Mid Infrared Radiation* 159-164

[8] Geyko I. A., Smolyanskaya O. A., Sulatsky M. I., Parakhuda S. E., Sedykh E. A., Odlyanitskiy E. L., Zabolotniy A. G. 2015 *European Conferences on Biomedical Optics* **9542** 95420E

[9] Gusev S. I., Borovkova M. A., Strepitov M. A., Khodzitsky M. K. 2015 *European Conferences on Biomedical Optics* **9537** 95372A

[10] Duka M. V., Dvoretskaya L. N., Babelkin N. S., Khodzitskii M. K., Chivilikhin S. A., Smolyanskaya O. A. 2014 *Quantum Electronics* **44(8)** 707

[11] Federal Almazov North-West Medical Research Centre webpage," Federal Almazov North-West Medical Research Centre, 25 September 2015, <http://www.almazovcentre.ru/?lang=en> (25 September 2015).

[12] Zhang X. C. and Xu J., [Introduction to THz wave photonics], Springer, New York, 49-52 (2010).

[13] Gurvitz, E. A., et al. "Development of 3D Anisotropic Artificial Dielectric Metamaterial for THz Frequency Range," PIERS Proc., 2022-2025 (2014).

[14] Denisultanov, A. K., et al. "Optical properties of graphene on quartz and polyethylene substrates in terahertz frequency range," PIERS Proc., 2022-2025 (2014).

[15] Onushchenko, P. A., et al. "Observation of terahertz radiation absorption in CdSe quantum dots," Proc. SPIE 9502, 950216-950216-7 (2015).

[16] Strepitov, E. A., et al. "Investigation of the optical properties of normal fibroblasts and fibroblasts cultured with cancer cells in terahertz frequency range," European Conferences on Biomedical Optics, 95420M-95420M-6 (2015).

Simulation of polarizer based on chiral medium for terahertz frequency range

S Yu Korolenko, A N Grebenchukov, M S Masyukov, A V Vozianova, M K Khodzitsky
Photonics and Optical Information Technologies Dept., ITMO University, Russian Federation

Email: grebenchukov_a@mail.ru

Abstract. The work is devoted to development of the polarizer for terahertz frequency range. Chiral medium was simulated using the software package CST Microwave Studio by the method of Finite-Differences in Frequency Domain. The influence of geometry of chiral unit cell on the polarization of incident plane wave was investigated.

1. Introduction

Nowadays there is a great interest to explorations of materials, which have chirality effect. An object or a structure is chiral if it is distinguishable from its mirror image, it can't be superposed onto it, at no translation or rotation. Conversely, a mirror image of an achiral object, such as a sphere, cannot be distinguished from the object. It is important to note that it is not just a mathematical model. It exists in real world. DNA molecules, some amino acids, liquid crystals, spiral metallic conductors and other molecules have this feature.

Chiral elements have several interesting properties such as mutual bianisotropy [1], optical activity [2], cross-polarization [3], which makes them promising for various applications. The most important demonstrations of bianisotropy are birefringence and gyrotropy. Each of these effects is caused by the presence of certain parts (недиагональные компоненты) in permittivity and permeability tensors. For birefringence it is symmetrical real part, for gyrotropy it is asymmetric imaginary part. Optical rotation (also known as optical activity or rotary polarization) is the turning of the plane of linearly polarized light relative to the direction of propagation as the light travels through certain materials. Cross-polarization is a phenomenon in which there is a change of reflected wave polarization. It lies in the fact that orthogonally polarized component to incident radiation is appeared in reflected wave.

Nowadays, terahertz (THz) frequency range and THz technology are extremely attractive research fields [4-9]. Because of fast development of biomedicine, necessity of design of THz compact controllable components for time-domain THz spectroscopy/ imaging / tomography, radiation generators has appeared. As well as there is need to develop methods of timely diagnostics and therapy in THz frequency range. One of the most important components for solving these problems are polarizers. Such polarizers can be realized by using chiral media. Single chiral elements such as intersection crosses [2], U-shaped resonators [10], and spiral [11] were considered by others researchers. In addition, there are other ways of using chiral materials in THz frequency range, for example, space-time cloaking devices based on chiral structure cause interest.

2. Description of model

In this paper, the simulation of terahertz polarizer based on optical activity of chiral material in the form of rosettes was considered. Chiral elements of similar form were investigated at microwaves [12]. In THz frequency range, the research of these structures has not been done yet.

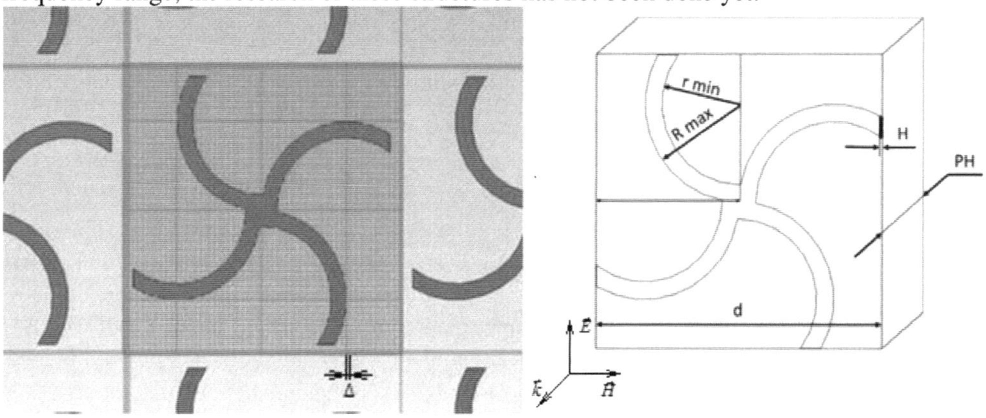

Figure 1. Scheme of chiral structure.

The Figure 1 shows the single chiral element. The side of the foursquare, in which the single chiral element placed (d) was 450 μm. Rosette consists of 4 elements. For basic structure that element was a part of the circle with the following dimensions: the external radius (R max) of 150 μm, the internal radius (r min) of 123.3 μm, the thickness (H) of 0.1 μm. The thickness of substrate (PH) was 150 μm. The material of chiral element was perfect electric conductor (PEC). As a substrate the dielectric with permittivity of ε=3.67 was used.

"Unit cell" was chosen as boundary conditions in performed modeling (see Fig.1). The size of mesh cell was $\Delta = d/N = 450/90 = 5$ μm. The simulation of two-dimensional (planar) array of chiral elements is more interesting than a single element due to the fact that it is more applicable in practice. The "Unit cell" boundary condition allows to simulate periodic structure of identical single elements.

Primarily, the basic structure of chiral element has been numerically simulated for preliminary estimate of the chiral effect. THz radiation normally incidents on to the object under study, i.e. the emission ports are located so that the wave vector $\vec{k}$ is perpendicular to the plane of chiral element. For the simulation the plane-wave was used for the frequency range of 0.1 - 1 THz. Spectral characteristics of reflection and transmission coefficients can be obtained by using this simulation.

3. Simulation results

As a result of such simulations for single basic chiral structure there were some features in the spectrum – the typical frequencies, where given material shows itself like a chiral material. Therefore, the electric field distribution was simulated for the structure. The results show, that the electric field vector $\vec{E}$ changes the polarization from linear to elliptical at these frequencies. This effect confirms that the structure really shows the chiral properties at the following frequencies: 0.778, 0.865, 0.980 THz. At consideration of the magnetic field vector $\vec{H}$ polarization changes were not revealed.

In order to find the form, which most strongly affects on incident wave, the geometry of structural unit of chiral element was modified.

The first modification as "circle" is a scaled version of the original structure. Here the radius of underlying circle was reduced in 2 times and it has a form of half-circle (see Fig 2b). The second modification as "ellipse" in which the element of rosette has a form of half-ellipse (see Fig 2c). The third modification as "circle-ellipse" is a mixture of the first and the second modifications (see Fig 2d).

The same steps that were used to obtain the spectral characteristics of basic chiral element were performed for various modifications of the basic structure.

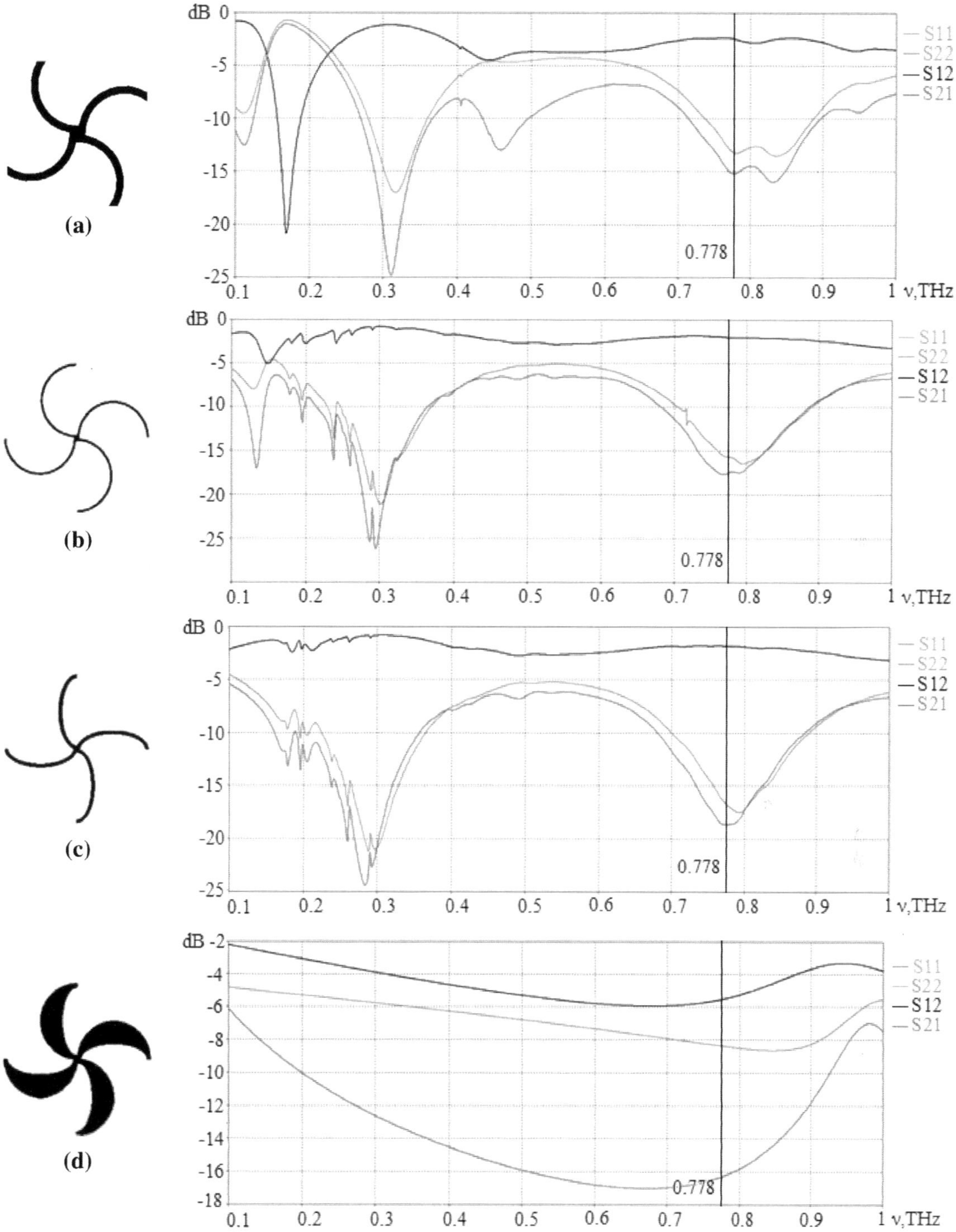

Figure 2. Spectra of reflection and transmission coefficients for 4 different chiral structures. (a) Basic structure. (b) "Enlarged circle" structure. (c) "Ellipse" structure. (d) "Circle-ellipse" structure.

On Figure 2, S11 and S22 lines are graphs of reflection coefficient. S11 line means that the radiation emerged from the first port and registered by the first port. S22 line means that the radiation emerged from the second port and registered in the second port. S12 and S21 lines were graphs of transmission coefficient. S12 line means than the radiation emerged from the first port and registered in the second port. S21 line means that the radiation emerged from the second port and registered in the first port.

By comparing the obtained results it can be noted that the best chiral properties were shown by the second modification of the structure “ellipse”. Because for this modification there is the least reflection of 18.5 dB at 0.778 THz among all structures. The rotation of electric field vector after passing through the structure is shown for all modifications (see Fig 3).

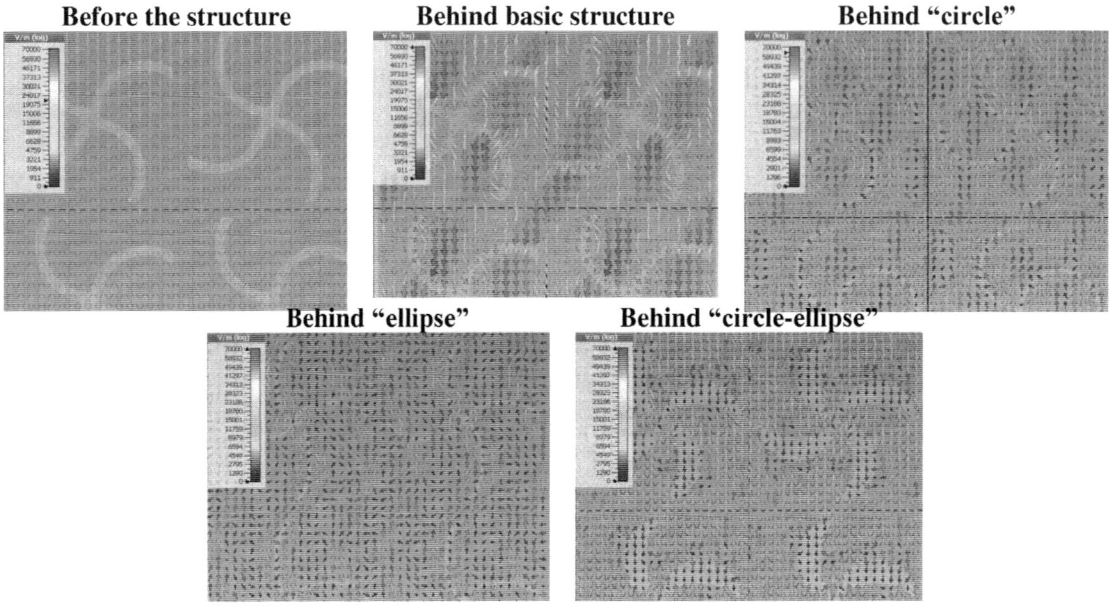

Figure 3. Rotation of the electric field vector $\vec{E}$.

The rotation of the electric field vector $\vec{E}$ at 0.778 THz is illustrated in Figure 3 before and behind the structure for four types of chiral media. After passage of the chiral structure the field vector $\vec{E}$ becomes elliptically polarized.

4. Conclusion

Thus, the chiral structure based on rosettes for the THz frequency range was designed and the change of polarization of the electric field $\vec{E}$ from linear to elliptical was observed. The various modifications of the chiral elements such as “circle”, “ellipse” and “circle-ellipse” have been investigated to identify the most effective one. The results of simulation showed that the chirality effect is most strongly manifested for “ellipse” modification. These results may be used for development of a polarizer for time-domain THz spectroscopy/ imaging / tomography.

Acknowledgments

This work was supported by Government of Russian Federation (Grant 074-U01).

References

[1] Ivanov O V 2010 *Propagation of electromagnetic waves in anisotropic and bianisotropic layered structures* (Ulyanovsk: UlSTU) 262 p.

[2] Ji Dong, Ji Zhou, T Koschny, C Soukoulis *Bi-layer cross chiral structure with strong optical activity and negative refractive index.*

[3] Ludwig A 1973 *The definition of cross polarization California Institute of Technology* (Pasadena, CA, USA) pp 116-119

[4] Xiang Xiong, Wei-Hua Sun 2010 *Construction of a chiral metamaterial with a U-shaped resonator assembly.*

[5] Khodzitsky M. K., Kharchenko A. A., Strashevskyi A. V., Tarapov S. I. 2009

Telecommunications and Radio Engineering **68(7)** 561-566

[6] Gurvitz E. A., Vozianova A. V., Khodzitsky M. K. 2014 *Journal of Physics: Conference Series* **541(1)** 012067

[7] Denisultanov A. K., Khodzitsky M. K. 2013 *SPIE NanoScience + Engineering* **8806** 88062I

[8] Terekhov Y. E., Khodzitsky M. K., Grachev Y. V., Sedykh E. A., Belokopytov G. V., Zhang, X. C. 2013 *SPIE NanoScience+ Engineering* **8806** 88062Q

[9] Girich A., Khodzitsky M., Nedukh S., Tarapov S. 2011 *Terahertz and Mid Infrared Radiation* 159-164

[10] Gurvitz E. A., Vozianova A. V., Khodzitskiy M. K. 2014 *SPIE NanoScience + Engineering* **9160** 91602C

[11] S. Zhang, J. Zhou, Y.-S. Park, J. Rho 2012 *Photoinduced handedness switching in terahertz chiral metamolecules.*

[12] S.Y. Polevoy, S.L. Prosvirnin, S.I. Tarapov 2013 *Resonance properties of a planar metamaterial formed by an array of rosettes on a non-magnetized ferrodielectric substrate.*

The numerical simulation and investigation of plasmonic properties of clusters consisted of two nanoparticles

A V Vozianova and A V Chernyadiev

Dept. of Photonics and Optical Information Technologies, ITMO University, 49 Kronverskiy pr. Saint-Petersburg, Russia, 197101

E-mail: vozianova@gmail.com

Abstract. Resonant properties of subwavelength spherical particles in strongly anisotropic metamaterials were investigated in this paper by means of spherical harmonics expansion. Numerical simulation of such problem using finite element method was demonstrated in COMSOL Multiphysics software.

1. Introduction

In the recent years scientists are occupied with studying of metamaterials (MM). There are artificially created materials that can realize some interesting effects at any frequency [1-3]. Different devices components can be created from MM [4,5]. There are some approaches for studying different MM [6]. Now the nanoplasmonics is topical because of one can see localization and amplification of the optical field (luminous edge effect) due to the substance in a nanoscale space. On the other side metal particles have eigen modes on frequencies ranged from ultraviolet to infrared wavelength range. Both of these peculiarities determine an attractive and complicated physics one observes, that serves as a basis for many applications [7-9]. Clusters containing several nanoparticles (so-called determinate structures or nanoparticle arrays/gratings) are of special interest to scientists by the side of the single (isolated) plasmonic nanoparticles [10] which properties are particularly interesting and have a number of significant applications. This interest is caused by that clusters unlike single nanoparticles have at least two various parameters: the size of nanoparticles and the distance between them. If we vary any of these parameters plasmonic properties of nanoparticles will essentially change, that enables us to exercise significant influence over them and spectrum of plasmon-polariton oscillations, to ensure coupling between clusters containing plasmonic nanoparticles with atoms, molecules and other nanoscale devices. Such versatility affords an opportunity to create the most diverse nanostructures [11-14]. Furthermore, there are strong local fields in the clearance between nanoparticles that can be several orders greater than incident fields and be required for a plenty of applications as well. The most fundamental significance has clusters containing two subwavelength particles, which this paper is devoted to, namely clusters containing two subwavelength spherical one size particles. To be more exact, it is supposed to consider an anisotropy of both subwavelength spherical particles and the medium apart from the already studied isotropic nanoparticles in the optically isotropic medium. Investigation of the nature of nontrivial eigen resonances and their impact on the local field correction for emission processes in THz frequency range will be studied here.

2. Quasistatic Model for two couple spherical particles inside anisotropic uniaxial medium

Plasmon modes of cluster located in the anisotropic medium and containing two subwavelength particles can be found by means of «ε-method» as well as in case of cluster located in the dielectric (isotropic) background.

It demands a solution of eigen modes of the system in the quasistatic limit by adopting Laplace equation for electric potential with continuity of it and the normal component of the displacement at the boundary:

$$E = -\nabla\varphi,\ div(\hat{\varepsilon} grad\vec{\varphi}) = 0,$$
$$[\varphi]_S = 0,\ \left[\varepsilon\frac{\partial\varphi}{\partial n}\right]_S = 0, \tag{1}$$

where $\hat{\varepsilon}$ represented by eq. 2 ($\hat{\varepsilon}_{out}$ is $\hat{\varepsilon}$ outside the particle, $\hat{\varepsilon}_{in}$ is $\hat{\varepsilon}$ inside the particle).

$$\hat{\varepsilon} = \begin{cases} \hat{\varepsilon}_{out} = diag(\varepsilon_1, \varepsilon_2, \varepsilon_3) \\ \hat{\varepsilon}_{in} \end{cases}. \tag{2}$$

The equation $div(\hat{\varepsilon} grad\vec{\varphi}) = 0$ in (1) for the eigen modes determination problem for two subwavelength particles located in the anisotropic medium disintegrates into system of two equations, for the potential inside and outside the particles (spheres) respectively:

$$\begin{cases} \dfrac{\partial^2\varphi}{\partial x^2} + \dfrac{\partial^2\varphi}{\partial y^2} + \dfrac{\partial^2\varphi}{\partial z^2} = 0, \quad \text{inside the particles.} \\ \varepsilon_1\dfrac{\partial^2\varphi}{\partial x^2} + \varepsilon_2\dfrac{\partial^2\varphi}{\partial y^2} + \varepsilon_3\dfrac{\partial^2\varphi}{\partial z^2} = 0, \quad \text{outside the particles.} \end{cases} \tag{3}$$

It will be more convenient to solve this problem in bispherical coordinates which are obtained by rotating a two-dimensional bipolar coordinate system about the axis connecting its two poles. Coordinates in such system ($-\infty < \eta < \infty,\ \ 0 < \xi \le \pi,\ \ 0 \le \phi < 2\pi$) and the Cartesian coordinates are related as follows:

$$x = a\frac{\sin\xi\cos\phi}{\cosh\eta - \cos\xi},\ y = a\frac{\sin\xi\sin\phi}{\cosh\eta - \cos\xi},\ z = a\frac{\sinh\eta}{\cosh\eta - \cos\xi}. \tag{4}$$

Lamé coefficients of such coordinate system are:

$$h_\xi = h_\eta = \frac{a}{\cosh\eta - \cos\xi}, h_\phi = \frac{a\sin\xi}{\cosh\eta - \cos\xi}. \tag{5}$$

The first equation of the system (3) in bispherical coordinates converts into the following equation:

$$\Delta\varphi = \frac{1}{h_\eta^3}\left[\frac{\partial}{\partial\eta}\left(h_\eta\frac{\partial\varphi}{\partial\eta}\right) + \frac{1}{\sin\xi}\frac{\partial}{\partial\xi}\left(h_\eta\sin\xi\frac{\partial\varphi}{\partial\xi}\right) + \frac{h_\eta}{\sin^2\xi}\frac{\partial^2\varphi}{\partial\phi^2}\right] = 0. \tag{6}$$

The solution of this equation is well-known [15, 16]. Variables in the Laplace's equation (6) can be separated only after the substitution $\varphi = \sqrt{\cosh\eta - \cos\xi}\,F$, where $F = M(\eta)H(\xi)\Phi(\phi)$. Then, the equation (6) transforms into equation (7):

$$\frac{\partial^2 F}{\partial\eta^2} + \frac{1}{\sin\xi}\frac{\partial}{\partial\xi}\left(\sin\xi\frac{\partial F}{\partial\xi}\right) + \frac{1}{\sin^2\xi}\frac{\partial^2 F}{\partial\phi^2} - \frac{1}{4}F = 0. \tag{7}$$

The anisotropy $\hat{\varepsilon}_{out}$ represented by the tensor $diag(\varepsilon_1, \varepsilon_2, \varepsilon_3)$ complicates the solution of the second equation of the system (3). The above-mentioned tensor in bispherical coordinates converts into the following:

$$D\begin{pmatrix} \varepsilon\sin^2\xi\sinh^2\eta+\varepsilon_3 L^2 & -\varepsilon\sin\xi\sinh\eta\, L+\varepsilon_3\sinh\eta\sin\xi\, L & 0 \\ -\varepsilon\sin\xi\sinh\eta\, L+\varepsilon_3\sinh\eta\sin\xi\, L & \varepsilon L^2+\varepsilon_3\sinh^2\eta\sin^2\xi & 0 \\ 0 & 0 & \varepsilon(\cosh\eta-\cos\xi)^2 \end{pmatrix},$$

where factor $L=\cos\xi\cosh\eta-1$, $D=\frac{1}{(\cosh\eta-\cos\xi)^2}$.

The presumption $\varepsilon_1=\varepsilon_2=\varepsilon$ was made to simplify the problem. The distribution of normalized electric field was simulated at the terahertz frequency range by the means of the Radio Frequency module of the COMSOL Multiphysics software, see Fig.1.

3. Numerial simulation results

Figure 1 (a) shows simulation of the distribution of normalized electric field component of two subwavelength particles isotropic background at the frequency of 1.0009 THz with the following parameters: the radius of the particles R=0.5 mm, the distance between the particles R_{12}=1.17 mm, the permittivity of the particles $\varepsilon_{in}=1$, the permittivity of the dielectric (isotropic) background $\varepsilon_{out}=-0.5$. In Figure 1 (b) it is demonstrated the simulation of the distribution of normalized electric field component of two subwavelength particles anisotropic background at the frequency of 1.0031 THz with the radius of the particles R=0.5 mm, the distance between the particles R_{12}=1.17 mm, the permittivity of the particles $\varepsilon_{in}=1$, the permittivity of the anisotropic background $\hat{\varepsilon}_{out}=diag(0.5,0.5,-1.5)$.

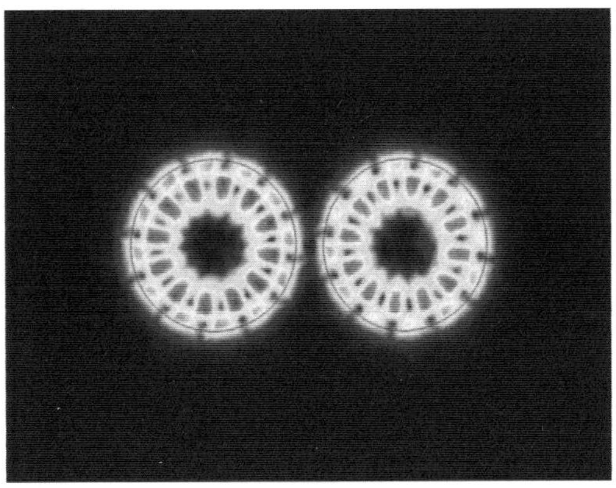

(a)

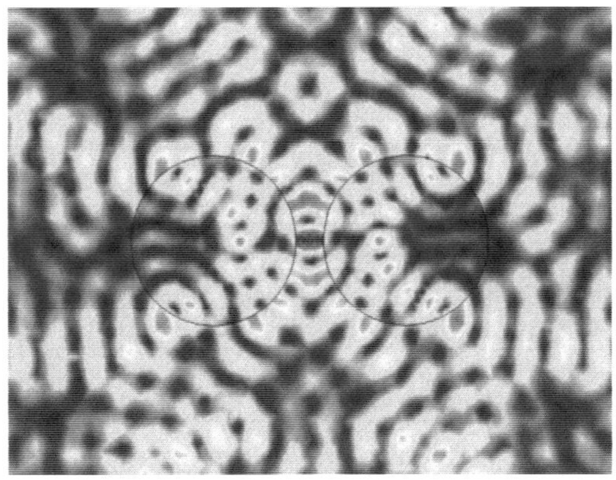

(b)

Figure 1. Distribution of normalized electric field component in the cluster containing two subwavelength particles: (a) dielectric (isotropic) background, (b) anisotropic background.

4. Conclusions

In this paper resonant properties of two subwavelength particles in anisotropic metamaterials by means of spherical harmonics expansion were investigated. These harmonics are nontrivially coupled by strong anisotropy and corresponding boundary conditions. The permittivity tensor in bispherical coordinates for anisotropic background was found. The electric field inside and outside particles was considered by means of spherical harmonics expansion. The field distribution simulation for the cluster containing two subwavelength particles and located in both anisotropic and isotropic background was performed by the means of the COMSOL Multiphysics software.

Acknowledgement

This work was financially supported by Government of Russian Federation, Grant 074-U01.

References

[1] Khodzitsky M. K., Tarapov S. I., Belozorov D. P., Pogorily A. M., Tovstolytkin A. I., Belous A.

G., Solopan S. A. 2010 *Applied Physics Letters* **97(13)** 131912

[2] Khodzitsky M. K., Kharchenko A. A., Strashevskyi A. V., Tarapov S. I. 2009 *Telecommunications and Radio Engineering* **68(7)** 561-566

[3] Terekhov Y. E., Khodzitsky M. K., Grachev Y. V., Sedykh E. A., Belokopytov G. V., Zhang, X. C. 2013 *SPIE NanoScience+ Engineering* **8806** 88062Q

[4] Denisultanov A. K., Khodzitsky M. K. 2013 *SPIE NanoScience + Engineering* **8806** 88062I

[5] Girich A., Khodzitsky M., Nedukh S., Tarapov S. 2011 *Terahertz and Mid Infrared Radiation* 159-164

[6] Gurvitz E. A., Vozianova A. V., Khodzitsky M. K. 2014 *Journal of Physics: Conference Series* **541(1)** 012067

[7] Prober J M, Trainor G L, Dam R J, Hobbs F W, Robertson C W, Zagursky R J, Cocuzza A J, Jensen M A and Baumeister K 1987 *Science*. **238** 336-343

[8] Copland J A, Eghtedari M, Popov V L, Kotov N, Mamedova N, Motamedi M and Oraevsky A A 2004 *Mol. Imaging Biol.* **6** 341-349

[9] Anger P, Bharadwaj P, and Novotny L Enchancement and Quenching of Single-Molecule Fluorescence 2006 *Phys. Rev.* Lett. **96** 113002

[10] Vozianova A V, Ginzburg P B and Poddubny A N 2014 *PIERS Proc., Cambridge: The Electromagnetics Academy.*. 2044-2046

[11] Hirsh L, Jackson J, Lee A, Halas N J and West J A Whole Blood 2003 *Analytical Chemistry*. **4** 2377-2381

[12] Greulich K O and Thiel E 2001 *Single Mol.* **2** 1

[13] Smolyaninov I I, Hung, Y-J and Davis C C 2007 *Science*. **315** 1699-1701

[14] Chang E et al 2005 *Biochemical and Biophysical Research Communications*. **334** 1317-1321

[15] Morse P M and Feshbach H 1953 **1–2**. (McGraw-Hill New York)

[16] Klimov V V Nanoplasmonics 2008 *UFN*. **178:8** 875–880